COMPUTATIONAL CHEMISTRY

*Introduction to the Theory and Applications of
Molecular and Quantum Mechanics*

Errol Lewars
*Chemistry Department
Trent University
Peterborough, Ontario
Canada*

KLUWER ACADEMIC PUBLISHERS
Boston / Dordrecht / London

Distributors for North, Central and South America:
Kluwer Academic Publishers
101 Philip Drive
Assinippi Park
Norwell, Massachusetts 02061 USA
Telephone (781) 871-6600
Fax (781) 681-9045
E-Mail <kluwer@wkap.com>

Distributors for all other countries:
Kluwer Academic Publishers Group
Post Office Box 322
3300 AH Dordrecht, THE NETHERLANDS
Telephone: 31 786 576 000
Fax: 31 786 576 254
E-Mail services@wkap.nl

 Electronic Services <http://www.wkap.nl>

Library of Congress Cataloging in-Publication Data

Computational Chemistry: Introduction to the Theory and Applications of
Molecular and Quantum Mechanics by Errol Lewars
ISBN 1-4020-7285-6

To Anne and John

CONTENTS

PREFACE

Every attempt to employ mathematical methods in the study of chemical questions must be considered profoundly irrational and contrary to the spirit of chemistry. If mathematical analysis should ever hold a prominent place in chemistry – an aberration which is happily almost impossible – it would occasion a rapid and widespread degeneration of that science.
Augustus Compte, French philosopher, 1798–1857; in *Philosophie Positive*,
1830.

A dissenting view:

The more progress the physical sciences make, the more they tend to enter the domain of mathematics, which is a kind of center to which they all converge. We may even judge the degree of perfection to which a science has arrived by the facility to which it may be submitted to calculation.
Adolphe Quetelet, French astronomer, mathematician, statistician, and
sociologist, 1796–1874, writing in 1828.

The purpose of this book is to teach the basics of the core concepts and methods of computational chemistry. Some fundamental concepts are the idea of a potential energy surface, the mechanical picture of a molecule as used in molecular mechanics, and the Schrödinger equation and its elegant taming with matrix methods to give energy levels and molecular orbitals. All the needed matrix algebra is explained before it is used. The fundamental methods of computational chemistry are molecular mechanics, *ab initio*, semiempirical, and density functional methods. Molecular dynamics and Monte Carlo methods are only mentioned; while these are important, they utilize fundamental concepts and methods treated here. I wrote the book because there seemed to be no text quite right for an introductory course in computational chemistry suitable for a fairly general chemical audience; I hope it will be useful to anyone who wants to learn enough about the subject to start reading the literature and to start doing computational chemistry. There are excellent books on the field, but evidently none that seeks to familiarize the general student of chemistry with computational chemistry in the same sense that standard textbooks of those subjects make organic or physical chemistry accessible. To that end the mathematics has been held on a leash (no attempt is made to show that molecular orbitals are vectors in Hilbert space, or that a finite-dimensional inner-product space must have an orthonormal basis), and the only sections that the nonspecialist may justifiably view with some trepidation are the (outlined) derivation of the Hartree–Fock and, to a lesser extent, the Kohn–Sham equations. These sections should be read, if only to get the flavor of the procedures, but should not stop anyone from getting on with the rest of the book.

Computational chemistry has become a tool used in much the same spirit as IR or NMR spectroscopy, and to use it sensibly it is no more necessary to be able to write your own programs than the fruitful use of IR or NMR spectroscopy requires you to be able to able to build your own spectrometer. I have tried to give enough theory to provide a reasonably good idea of how the programs work. In this regard, the concept of constructing and diagonalizing a Fock matrix is introduced early, and there is little talk of secular determinants (except for historical reasons in connection with the simple Hückel method). Many results of actual computations, most of them specifically for this book, are given. Almost all the assertions in these pages are accompanied by literature references, which should make the text useful to researchers who need to track down methods or results, and students (i.e. anyone who is still learning anything) who wish to delve deeper. The material should be suitable for senior undergraduates, graduate students, and novice researchers in computational chemistry. A knowledge of the shapes of molecules, covalent and ionic bonds, spectroscopy, and some familiarity with thermodynamics at about the level provided by second-year undergraduate courses is assumed. Some readers may wish to review basic concepts from physical and organic chemistry.

The reader, then, should be able to acquire the basic theory and a fair idea of the kinds of results to be obtained from the common computational chemistry techniques. You will learn how one can calculate the geometry of a molecule, its IR and UV spectra and its thermodynamic and kinetic stability, and other information needed to make a plausible guess at its chemistry.

Computational chemistry is accessible. Hardware has become far cheaper than it was even a few years ago, and powerful programs previously available only for expensive workstations have been adapted to run on relatively inexpensive personal computers. The actual use of a program is best explained by its manuals and by books written for a specific program, and the actual *directions* for setting up the various computations are not given here. Information on various programs is provided in chapter 8. Read the book, get some programs and go out and do computational chemistry.

It is a real pleasure acknowledge the help of many people: Professor Imre Csizmadia of the University of Toronto, who gave unstintingly of his time and experience, the students in my computational and other courses, the generous and knowledgeable people who subscribe to CCL, the computational chemistry list, an exceedingly helpful forum for anyone seriously interested in the subject; and my editor at Kluwer, Dr Emma Roberts, who was always most helpful and encouraging.

E. Lewars
Department of Chemistry
Trent University
Peterborough, Ontario
Canada

Chapter 1

An Outline of What Computational Chemistry is All About

1.1 WHAT YOU CAN DO WITH COMPUTATIONAL CHEMISTRY

Computational chemistry (also called molecular modelling; the two terms mean about the same thing) is a set of techniques for investigating chemical problems on a computer. Questions commonly investigated computationally are:

Molecular geometry: The shapes of molecules – bond lengths, angles, and dihedrals.

Energies of molecules and transition states: This tells us which isomer is favored at equilibrium, and (from transition state and reactant energies) how fast a reaction should go.

Chemical reactivity: For example, knowing where the electrons are concentrated (nucleophilic sites) and where they want to go (electrophilic sites) enables us to predict where various kinds of reagents will attack a molecule.

IR, UV, and NMR spectra: These can be calculated, and if the molecule is unknown, someone trying to make it knows what to look for.

The interaction of a substrate with an enzyme: Seeing how a molecule fits into the active site of an enzyme is one approach to designing better drugs.

The physical properties of substances: These depend on the properties of individual molecules and on how the molecules interact in the bulk material. For example, the strength and melting point of a polymer (e.g. a plastic) depend on how well the molecules fit together and on how strong the forces between them are. People who investigate things like this work in the field of materials science.

1.2   THE TOOLS OF COMPUTATIONAL CHEMISTRY

In studying these questions computational chemists have a selection of methods at their disposal. The main tools available belong to five broad classes as described below.

Molecular mechanics (MM) is based on a model of a molecule as a collection of balls (atoms) held together by springs (bonds). If we know the normal spring lengths and the angles between them, and how much energy it takes to stretch and bend the springs, we can calculate the energy of a given collection of balls and springs, i.e. of a given molecule; changing the geometry until the lowest energy is found enables us to do a *geometry optimization*, i.e. to calculate a geometry for the molecule.

Molecular mechanics is fast: a fairly large molecule like a steroid (e.g. cholesterol, $C_{27}H_{46}O$) can be optimized in seconds on a powerful desktop computer (a workstation); on a personal computer the job might also take only a few seconds.

Ab initio calculations (*ab initio* is from the Latin: "from first principles") are based on the Schrödinger equation. This is a one of the fundamental equations of modern physics and describes, among other things, how the electrons in a molecule behave. The *ab initio* method solves the Schrödinger equation for a molecule and gives us the molecule's energy and *wavefunction*. The wavefunction is a mathematical function that can be used to calculate the electron distribution (and, in theory at least, anything else about the molecule). From the electron distribution we can tell things like how polar the molecule is, and which parts of it are likely to be attacked by nucleophiles or electrophiles.

The Schrödinger equation cannot be solved exactly for any molecule with more than one (!) electron. Thus approximations are used; the less serious these are, the "higher" the level of the *ab initio* calculation is said to be. Regardless of its level, an *ab initio* calculation is based only on basic physical theory (quantum mechanics) and is in this sense "from first principles". *Ab initio* calculations are relatively slow: the geometry and IR spectra (= the vibrational frequencies) of propane can be calculated at a reasonably high level in minutes on a Pentium-type machine, but a fairly large molecule, like a steroid, could take perhaps weeks. The latest personal computers (like a Pentium or a PowerMac), with a GB of RAM and several GB of disk space, are serious computational tools and now compete with UNIX workstations even for the demanding tasks associated with high-level *ab initio* calculations. Such calculations on a well-outfitted personal computer (ca. \$4000) are perhaps a few times slower than on an average UNIX workstation (ca. \$15 000). The distinction between workstations and high-end PCs has blurred.

Semiempirical (SE) calculations are, like *ab initio*, based on the Schrödinger equation. However, more approximations are made in solving it, and the very complicated integrals that must be calculated in the *ab initio* method are not actually evaluated in SE calculations: instead, the program draws on a kind of library of integrals that was compiled by finding the best fit of some *calculated* entity like geometry or energy (heat of formation) to the *experimental* values. This plugging of experimental values into a mathematical procedure to get the best calculated values is called *parameterization* (or *parametrization*). It is the mixing of theory and experiment that makes the method "semiempirical": it is based on the Schrödinger equation, but parameterized with experimental values (*empirical* means experimental). Of course one hopes that

SE calculations will give good answers for molecules for which the program has *not* been parameterized (otherwise why not just look up the experimental results?) and this is often the case (MM, too, is parameterized).

Semiempirical calculations are slower than MM but much faster than *ab initio* calculations. SE calculations take roughly 100 times as long as MM calculations, and *ab initio* calculations take roughly 100–1000 times as long as SE. A SE geometry optimization on a steroid might take minutes on a Pentium-type machine.

Density functional calculations (often called density functional theory (DFT) calculations) are, like *ab initio* and SE calculations, based on the Schrödinger equation. However, unlike the other two methods, DFT does not calculate a wavefunction, but rather derives the electron distribution (electron *density* function) directly. A *functional* is a mathematical entity related to a function.

Density functional calculations are usually faster than *ab initio*, but slower than SE. DFT is relatively new (serious DFT computational chemistry goes back to the 1980's, while computational chemistry with the *ab initio* and SE approaches was being done in the 1960s).

Molecular dynamics calculations apply the laws of motion to molecules. Thus one can simulate the motion of an enzyme as it changes shape on binding to a substrate, or the motion of a swarm of water molecules around a molecule of solute.

1.3   PUTTING IT ALL TOGETHER

Very large molecules can be studied only with MM, because other methods (*quantum mechanical* methods, based on the Schrödinger equation: SE, *ab initio* and DFT) would take too long. Novel molecules, with unusual structures, are best investigated with *ab initio* or possibly DFT calculations, since the parameterization inherent in MM or SE methods makes them unreliable for molecules that are very different from those used in the parameterization. DFT is relatively new and its limitations are still unclear.

Calculations on the structure of large molecules like proteins or DNA are done with MM. The motions of these large biomolecules can be studied with molecular dynamics. Key *portions* of a large molecule, like the active site of an enzyme, can be studied with SE or even *ab initio* methods. Moderately large molecules, like steroids, can be studied with SE calculations, or if one is willing to invest the time, with *ab initio* calculations. Of course MM can be used with these too, but note that this technique does not give information on electron distribution, so chemical questions connected with nucleophilic or electrophilic behaviour, say, cannot be addressed by MM alone.

The energies of molecules can be calculated by MM, SE, *ab initio* or DFT. The method chosen depends very much on the particular problem. Reactivity, which depends largely on electron distribution, must usually be studied with a quantum-mechanical method (SE, *ab initio* or DFT). Spectra are most reliably calculated by *ab initio* methods, but useful results can be obtained with SE methods, and some MM programs will calculate fairly good IR spectra (balls attached to springs vibrate!).

Docking a molecule into the active site of an enzyme to see how it fits is an extremely important application of computational chemistry. One manipulates the substrate with a mouse or a kind of joystick and tries to fit it (dock it) into the active site (automated docking is also possible); with some computer systems a feedback device enables you to *feel* the forces acting on the molecule being docked. This work is usually done with MM, because of the large molecules involved, although selected portions of the biomolecules could be studied by one of the quantum mechanical methods. The results of such docking experiments serve as a guide to designing better drugs, molecules that will interact better with the desired enzymes but be ignored by other enzymes.

Computational chemistry is valuable in studying the properties of materials, i.e. in materials science. Semiconductors, superconductors, plastics, ceramics – all these have been investigated with the aid of computational chemistry. Such studies tend to involve a knowledge of solid-state physics and to be somewhat specialized.

Computational chemistry is fairly cheap, it is fast compared to experiment, and it is environmentally safe. It does not replace experiment, which remains the final arbiter of truth about Nature. Furthermore, to *make* something – new drugs, new materials – one has to go into the lab. However, computation has become so reliable in some respects that, more and more, scientists in general are employing it before embarking on an experimental project, and the day may come when to obtain a grant for some kinds of experimental work you will have to show to what extent you have computationally explored the feasibility of the proposal.

1.4   THE PHILOSOPHY OF COMPUTATIONAL CHEMISTRY

Computational chemistry is the culmination (to date) of the view that chemistry is best understood as the manifestation of the behavior of atoms and molecules, and that these are real entities rather than merely convenient intellectual models [1]. It is a detailed physical and mathematical affirmation of a trend that hitherto found its boldest expression in the structural formulas of organic chemistry [2], and it is the unequivocal negation of the till recently trendy assertion [3] that science is a kind of game played with "paradigms" [4].

In computational chemistry we take the view that we are simulating the behaviour of real physical entities, albeit with the aid of intellectual models; and that as our models improve they reflect more accurately the behavior of atoms and molecules in the real world.

1.5   SUMMARY OF CHAPTER 1

Computational chemistry allows one to calculate molecular geometries, reactivities, spectra, and other properties. It employs:

- Molecular mechanics – based on a ball-and-springs model of molecules;
- *Ab initio* methods – based on approximate solutions of the Schrödinger equation without appeal to fitting to experiment;
- Semiempirical methods – based on approximate solutions of the Schrödinger equation with appeal to fitting to experiment (i.e. using parameterization);

- DFT methods – based on approximate solutions of the Schrödinger equation, bypassing the wavefunction that is a central feature of *ab initio* and semiempirical methods;
- Molecular dynamics methods study molecules in motion.

Ab initio and the faster DFT enable novel molecules of theoretical interest to be studied, provided they are not too big. Semiempirical methods, which are much faster than *ab initio* or even DFT, can be applied to fairly large molecules (e.g. cholesterol, $C_{27}H_{46}O$), while MM will calculate geometries and energies of very large molecules such as proteins and nucleic acids; however, MM does not give information on electronic properties. Computational chemistry is widely used in the pharmaceutical industry to explore the interactions of potential drugs with biomolecules, for example by docking a candidate drug into the active site of an enzyme. It is also used to investigate the properties of solids (e.g. plastics) in materials science.

REFERENCES

[1] The physical chemist Wilhelm Ostwald (Nobel Prize 1909) was a disciple of the philosopher Ernst Mach. Like Mach, Ostwald attacked the notion of the reality of atoms and molecules ("Nobel Laureates in Chemistry, 1901–1992," L. K. James, Ed., American Chemical Society and the Chemical Heritage Foundation, Washington, DC, 1993) and it was only the work of Jean Perrin, published in 1913, that finally convinced him, perhaps the last eminent holdout against the atomic theory, that these entities really existed (Perrin showed that the number of tiny particles suspended in water dropped off with height exactly as predicted in 1905 by Einstein, who had derived an equation assuming the existence of atoms). Ostwald's philosophical outlook stands in contrast to that of another outstanding physical chemist, Johannes van der Waals, who staunchly defended the atomic/molecular theory and was outraged by the Machian positivism of people like Ostwald. See "Van der Waals and Molecular Science," A. Ya. Kipnis, B. F. Yavelov and J. S. Powlinson, Oxford University Press, New York, 1996.

For the opposition to and acceptance of atoms in physics see: D. Lindley, "Boltzmann's Atom. The Great Debate that Launched a Revolution in Physics," Free Press, New York, 2001; C. Cercignani, "Ludwig Boltzmann: The Man who Trusted Atoms," Oxford University Press, New York, 1998.

Of course, to anyone who knew anything about organic chemistry, the existence of atoms was in little doubt by 1910, since that science had by that time achieved significant success in the field of synthesis, and a rational synthesis is predicated on assembling atoms in a definite way.

[2] For accounts of the history of the development of structural formulas see M. J. Nye, "From Chemical Philosophy to Theoretical Chemistry," University of California Press, 1993; C. A. Russell, "Edward Frankland: Chemistry, Controversy and Conspiracy in Victorian England," Cambridge University Press, Cambridge, 1996.

[3] (a) An assertion of the some adherents of the "postmodernist" school of social studies; see P. Gross and N. Levitt, "The Academic Left and its Quarrels with Science," John Hopkins University Press, 1994. (b) For an account of the exposure of the intellectual vacuity of some members of this school by physicist Alan Sokal's hoax see M. Gardner, "Skeptical Inquirer," 1996, *20*(6), 14.

[4] (a) A trendy word popularized by the late Thomas Kuhn in his book "The Structure of Scientific Revolutions," University of Chicago Press, 1970. For a trenchant comment on Kuhn, see Ref. [3b]. (b) For a kinder perspective on Kuhn, see S. Weinberg, "Facing Up," Harvard University Press, 2001, chapter 17.

EASIER QUESTIONS

1. What does the term *computational chemistry* mean?
2. What kinds of questions can computational chemistry answer?
3. Name the main tools available to the computational chemist. Outline (a few sentences for each) the characteristics of each.
4. Generally speaking, which is the fastest computational chemistry method (tool), and which is the slowest?
5. Why is computational chemistry useful in industry?
6. Basically, what does the Schrödinger equation describe, from the chemist's viewpoint?
7. What is the limit to the kind of molecule for which we can get an exact solution to the Schrödinger equation?
8. What is parameterization?
9. What advantages does computational chemistry have over "wet chemistry"?
10. Why cannot computational chemistry replace "wet chemistry"?

HARDER QUESTIONS

Discuss the following and justify your conclusions.
1. Was there computational chemistry before electronic computers were available?
2. Can "conventional" physical chemistry, such as the study of kinetics, thermodynamics, spectroscopy and electrochemistry, be regarded as a kind of computational chemistry?
3. The properties of a molecule that are most frequently calculated are geometry, energy (compared to that of other isomers), and spectra. Why is it more of a challenge to calculate "simple" properties like melting point and density?
   Hint: Is there a difference between a molecule X and the substance X?
4. Is it surprising that the geometry and energy (compared to that of other isomers) of a molecule can often be accurately calculated by a ball-and-springs model (MM)?
5. What kinds of properties might you expect MM to be unable to calculate?
6. Should calculations from first principles (*ab initio*) necessarily be preferred to those which make some use of experimental data (semiempirical)?
7. Both experiments and calculations can give wrong answers. Why then should experiment have the last word?
8. Consider the docking of a potential drug molecule X into the active site of an enzyme: a factor influencing how well X will "hold" is clearly the shape of X; can you think of another factor?
   Hint: Molecules consist of nuclei and electrons.

9. In recent years the technique of *combinatorial chemistry* has been used to quickly synthesize a variety of related compounds which are then tested for pharmacological activity (S. Borman, Chemical & Engineering News: 2001, 27 August, p. 49; 2000, 15 May, p. 53; 1999, 8 March, p. 33). What are the advantages and disadvantages of this method of finding drug candidates, compared with the "rational design" method of studying, with the aid of computational chemistry, how a molecule interacts with an enzyme?

10. Think up some unusual molecule which might be investigated computationally. What is it that makes your molecule unusual?

Chapter 2

The Concept of the Potential Energy Surface

Everything should be made as simple as possible, but not simpler.
Albert Einstein

2.1 PERSPECTIVE

We begin a more detailed look at computational chemistry with the potential energy surface (PES) because this is central to the subject. Many important concepts that might appear to be mathematically challenging can be grasped intuitively with the insight provided by the idea of the PES [1].

Consider a diatomic molecule AB. In some ways a molecule behaves like balls (atoms) held together by springs (chemical bonds); in fact, this simple picture is the basis of the important method molecular mechanics, discussed in chapter 3. If we take a macroscopic balls-and-spring model of our diatomic molecule in its normal geometry (the equilibrium geometry), grasp the "atoms" and distort the model by stretching or compressing the "bonds," we increase the potential energy of the molecular model (Fig. 2.1). The stretched or compressed spring possesses energy, by definition, since we moved a force through a distance to distort it. Since the model is motionless while we hold it at the new geometry, this energy is not kinetic and so is by default *potential* ("depending on position"). The graph of potential energy against bond length is an example of a PES (we will soon see an example of an actual *surface* rather than the line of Fig. 2.1).

Real molecules behave similarly, but they differ from our macroscopic model in two relevant ways:

1. They vibrate incessantly (as we would expect from Heisenberg's uncertainty principle: a stationary molecule would have an exactly defined momentum and position) about the equilibrium bond length, so that they always possess kinetic energy (T) and/or potential energy (V): as the bond length passes through the equilibrium length, $V = 0$, while at the limit of the vibrational amplitude, $T = 0$; at all other positions both T and V are nonzero. The fact that a molecule is never actually stationary with zero kinetic energy (it always has *zero-point energy* (ZPE); section 2.5) is usually shown on potential energy/bond length diagrams by drawing a series of

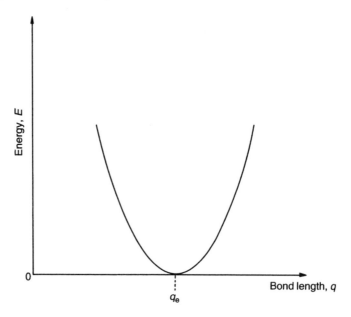

Figure 2.1. The PES for a diatomic molecule. The potential energy increases if the bond length q is stretched or compressed away from its equilibrium value q_e. The potential energy at q_e (zero distortion of the bond length) has been chosen here as the zero of energy.

lines above the bottom of the curve (Fig. 2.2) to indicate the possible amounts of vibrational energy the molecule can have (the *vibrational levels* it can occupy). A molecule never sits at the bottom of the curve, but rather occupies one of the vibrational levels, and in a collection of molecules the levels are populated according to their spacing and the temperature [2]. We will usually ignore the vibrational levels and consider molecules to rest on the actual potential energy curves or (below) surfaces, and

2. Near the equilibrium bond length q_e the potential energy/bond length curve for a macroscopic balls-and-spring model or a real molecule is described fairly well by a quadratic equation, that of the simple harmonic oscillator ($E = (1/2)k(q - q_e)^2$, where k is the force constant of the spring). However, the potential energy deviates from the quadratic (q^2) curve as we move away from q_e (Fig. 2.2). The deviations from molecular reality represented by this *anharmonicity* are not important to our discussion.

Figure 2.1 represents a *one-dimensional* (1D) PES (a line is a 1D "surface") in the 2D graph of E vs. q. A diatomic molecule AB has only one geometric parameter for us to vary, the bond length q_{AB}. Suppose we have a molecule with more than one geometric parameter, e.g. water: the geometry is defined by two bond lengths and a bond angle. If we reasonably content ourselves with allowing the two bond lengths to be the same, i.e. if we limit ourselves to C_{2v} symmetry (two planes of symmetry and a two-fold symmetry axis; see section 2.6) then the PES for this triatomic molecule is a graph of E vs. two geometric parameters, q_1 = the O–H bond length, and q_2 = the H–O–H

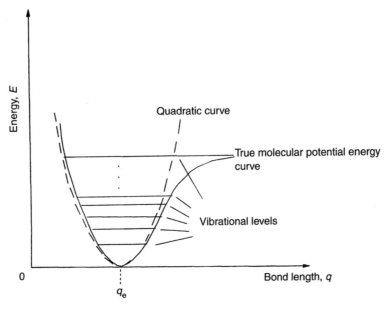

Figure 2.2. Actual molecules do not sit still at the bottom of the potential energy curve, but instead occupy vibrational levels. Also, only near q_e, the equilibrium bond length, does the quadratic curve approximate the true potential energy curve.

bond angle (Fig. 2.3). Figure 2.3 represents a 2D PES (a normal surface is a 2D object) in the 3D-graph; we could make an actual 3D model of this drawing of a 3D graph of E vs. q_1 and q_2.

We can go beyond water and consider a triatomic molecule of lower symmetry, such as HOF. This has three geometric parameters, the H–O and O–F lengths and the H–O–F angle. To construct a Cartesian PES graph for HOF analogous to that for H_2O would require us to plot E vs. $q_1 =$ H–O, $q_2 =$ O–F, and $q_3 =$ angle H–O–F. We would need four mutually perpendicular axes (for E, q_1, q_2, q_3, Fig. 2.4), and since such a 4D graph cannot be constructed in our 3D space we cannot accurately draw it. The HOF PES is a 3D "surface" of more than two dimensions in 4D space: it is a hypersurface, and PESs are sometimes called potential energy hypersurfaces. Despite the problem of drawing a hypersurface, we can define the *equation $E = f(q_1, q_2, q_3)$* as the PES for HOF, where f is the function that describes how E varies with the q's, and treat the hypersurface mathematically. For example, in the AB diatomic molecule PES (a line) of Fig. 2.1 the minimum potential energy geometry is the point at which $dE/dq = 0$. On the H_2O PES (Fig. 2.3) the minimum energy geometry is defined by the point P_m, corresponding to the equilibrium values of q_1 and q_2; at this point $dE/dq_1 = dE/dq_2 = 0$. Although hypersurfaces cannot be faithfully rendered pictorially, it is very useful to a computational chemist to develop an intuitive understanding of them. This can be gained with the aid of diagrams like Figs 2.1 and 2.3, where we content ourselves with a line or a 2D surface, in effect using a slice of a multidimensional diagram. This can be understood by analogy: Fig. 2.5 shows how 2D slices can be made of the 3D diagram for water.

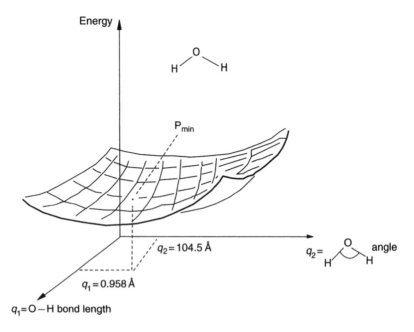

Figure 2.3. The H_2O PES. The point P_{min} corresponds to the minimum-energy geometry for the three atoms, i.e. to the equilibrium geometry of the water molecule.

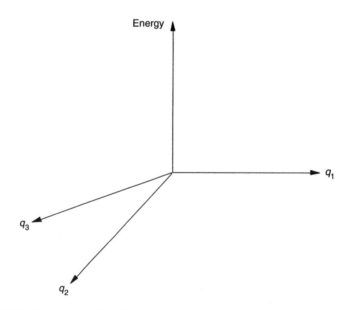

Figure 2.4. To plot energy against three geometric parameters in a Cartesian coordinate system we would need four *mutually perpendicular* axes. Such a coordinate system cannot be actually constructed in our 3D space. However, we can work with such coordinate systems, and the PESs in them, mathematically.

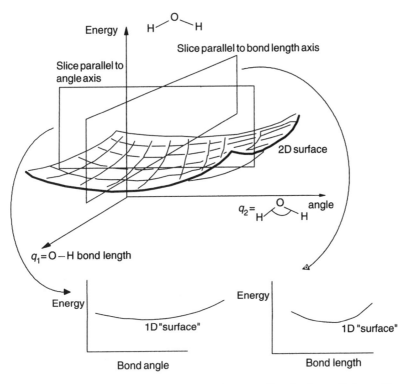

Figure 2.5. Slices through a 2D PES give 1D surfaces. A slice that is parallel to neither axis would give a plot of geometry vs. a composite of bond angle and bond length, a kind of average geometry.

The slice could be made holding one or the other of the two geometric parameters constant, or it could involve both of them, giving a diagram in which the geometry axis is a composite of more than one geometric parameter. Analogously, we can take a 3D slice of the hypersurface for HOF (Fig. 2.6) or even a more complex molecule and use an E vs. q_1, q_2 diagram to represent the PES; we could even use a simple 2D diagram, with q representing one, two or all of the geometric parameters. We shall see that these 2D and particularly 3D graphs preserve qualitative and even quantitative features of the mathematically rigorous but unvisualizable $E = f(q_1, q_2, \ldots, q_n)$ n-dimensional hypersurface.

2.2 STATIONARY POINTS

Potential energy surfaces are important because they aid us in visualizing and understanding the relationship between potential energy and molecular geometry, and in understanding how computational chemistry programs locate and characterize structures of interest. Among the main tasks of computational chemistry are to determine the structure and energy of molecules and of the transition states involved in chemical

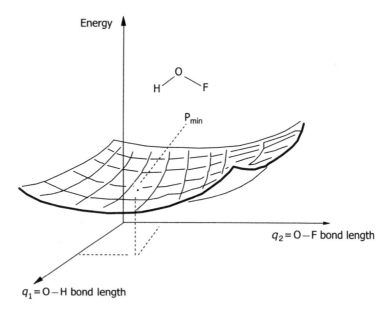

Figure 2.6. A PES for HOF. Here the HOF angle is not shown. This picture could represent one of two possibilities: the angle might be the same (some constant, reasonable value) for every calculated point on the surface; this would be an *unrelaxed* or *rigid* PES. Alternatively, for each calculated point the geometry might be that for the best angle corresponding to the other two parameters, i.e. the geometry for each calculated point might be fully optimized (section 2.4); this would be a *relaxed* PES.

reactions: our "structures of interest" are molecules and the transition states linking them. Consider the reaction

Ozone   Transition state   Isoozone

Reaction 1

A priori, it seems reasonable that ozone might have an isomer (call it isoozone) and that the two could interconvert by a transition state as shown in reaction (1). We can depict this process on a PES. The potential energy E must be plotted against only two geometric parameters, the bond length (we may reasonably assume that the two O–O bonds of ozone are equivalent, and that these bond lengths remain equal throughout the reaction) and the O–O–O bond angle. Figure 2.7 shows the PES for reaction (1), as calculated by the AM1 semiempirical method (chapter 6; the AM1 method is unsuitable for *quantitative* treatment of this problem, but the PES shown makes the point), and shows how a 2D slice from this 3D diagram gives the energy/reaction coordinate type of diagram commonly used by chemists. The slice goes along the lowest-energy path connecting ozone, isoozone and the transition state, i.e. along the *reaction coordinate*,

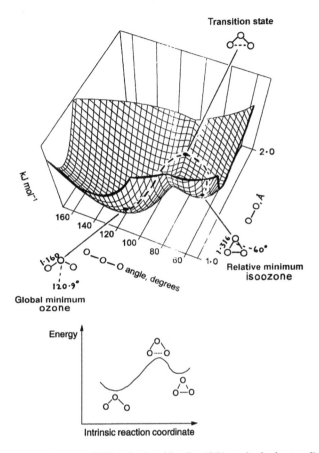

Figure 2.7. The ozone/isoozone PES (calculated by the AM1 method; chapter 6), a 2D surface in a 3D diagram. The dashed line on the surface is the reaction coordinate (IRC). A slice through the reaction coordinate gives a 1D "surface" in a 2D diagram. The diagram is not meant to be quantitatively accurate.

and the horizontal axis (the reaction coordinate) of the 2D diagram is a composite of O–O bond length and O–O–O angle. In most discussions this horizontal axis is left quantitatively undefined; qualitatively, the reaction coordinate represents the progress of the reaction. The three species of interest, ozone, isoozone, and the transition state linking these two, are called *stationary points*. A stationary point on a PES is a point at which the surface is flat, i.e. parallel to the horizontal line corresponding to the one geometric parameter (or to the plane corresponding to two geometric parameters, or to the hyperplane corresponding to more than two geometric parameters). A marble placed on a stationary point will remain balanced, i.e. stationary (in principle; for a transition state the balancing would have to be exquisite indeed). At any other point on a potential surface the marble will roll toward a region of lower potential energy.

Mathematically, a stationary point is one at which the first derivative of the potential energy with respect to each geometric parameter is zero:

$$\frac{\partial E}{\partial q_1} = \frac{\partial E}{\partial q_2} = \cdots = 0 \qquad (2.1)$$

Partial derivatives, $\partial E/\partial q$, are written here rather than dE/dq, to emphasize that each derivative is with respect to just one of the variables q of which E is a function. Stationary points that correspond to actual molecules with a finite lifetime (in contrast to transition states, which exist only for an instant), like ozone or isoozone, are *minima*, or *energy minima*: each occupies the lowest-energy point in its region of the PES, and any small change in the geometry increases the energy, as indicated in Fig. 2.7. Ozone is a *global minimum*, since it is the lowest-energy minimum on the whole PES, while isoozone is a *relative minimum*, a minimum compared only to *nearby* points on the surface. The lowest-energy pathway linking the two minima, the reaction coordinate or *intrinsic reaction coordinate* (IRC; dashed line in Fig. 2.7) is the path that would be followed by a molecule in going from one minimum to another should it acquire just enough energy to overcome the activation barrier, pass through the transition state, and reach the other minimum. Not all reacting molecules follow the IRC exactly: a molecule with sufficient energy can stray outside the IRC to some extent [3].

Inspection of Fig. 2.7 shows that the transition state linking the two minima represents a maximum along the direction of the IRC, but along all other directions it is a minimum. This is a characteristic of a saddle-shaped surface, and the transition state is called a *saddle point* (Fig. 2.8). The saddle point lies at the "center" of the saddle-shaped region and is, like a minimum, a stationary point, since the PES at that point is parallel to the plane defined by the geometry parameter axes: we can see that a marble placed (precisely) there will balance. Mathematically, minima and saddle points differ in that although both are stationary points (they have zero first derivatives; Eq. (2.1)), a minimum is a minimum in all directions, but a saddle point is a maximum along the

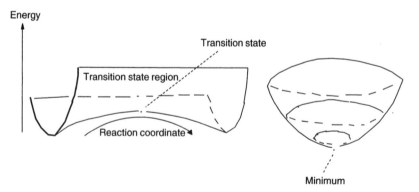

Figure 2.8. A transition state or saddle point and a minimum. At both the transition state and the minimum $\partial E/\partial q = 0$ for all geometric coordinates q (along all directions). At the transition state $\partial E^2/\partial q^2 < 0$ for $q =$ the reaction coordinate and > 0 for all other q (along all other directions). At a minimum $\partial E^2/\partial q^2 > 0$ for all q (along all directions).

reaction coordinate and a minimum in all other directions (examine Fig. 2.8). Recalling that minima and maxima can be distinguished by their second derivatives, we can write:

For a minimum

$$\frac{\partial^2 E}{\partial q^2} > 0 \qquad (2.2)$$

for all q.

For a transition state

$$\frac{\partial^2 E}{\partial q^2} > 0 \qquad (2.3)$$

for all q, *except along the reaction coordinate*, and

$$\frac{\partial^2 E}{\partial q^2} < 0 \qquad (2.4)$$

along the reaction coordinate.

The distinction is sometimes made between a *transition state* and a *transition structure* [4]. Strictly speaking, a transition *state* is a thermodynamic concept, the species an ensemble of which are in a kind of equilibrium with the reactants in Eyring's[1] transition-state theory [5]. Since equilibrium constants are determined by free energy differences, the transition state, within the strict use of the term, is a free energy maximum along the reaction coordinate (in so far as a single species can be considered representative of the ensemble). This species is also often (but not always [5]) also called an activated complex. A transition *structure*, in strict usage, is the saddle point (Fig. 2.8) on a theoretically calculated (e.g. Fig. 2.7) PES. Normally such a surface is drawn through a set of points each of which represents the enthalpy of a molecular species at a certain geometry; recall that free energy differs from enthalpy by temperature times entropy. The transition structure is thus a saddle point on an enthalpy surface. However, the energy of each of the calculated points does not normally include the vibrational energy, and even at 0 K a molecule has such energy (ZPE: Fig. 2.2, and section 2.5). The usual calculated PES is thus a hypothetical, physically unrealistic surface in that it neglects vibrational energy, but it should qualitatively, and even semiquantitatively, resemble the vibrationally-corrected one since in considering *relative* enthalpies ZPEs at least roughly cancel. In accurate work ZPEs are calculated for stationary points and added to the "frozen-nuclei" energy of the species at the bottom of the reaction coordinate curve in an attempt to give improved relative energies which represent enthalpy differences at 0 K (and thus, at this temperature where entropy is zero, free energy differences also; Fig. 2.19). It is also possible to calculate enthalpy and entropy differences, and thus free energy differences, at, say, room temperature (section 5.5.2). Many chemists do not routinely distinguish between two terms, and in this book the commoner term, transition state, is used. Unless indicated otherwise, it will mean a calculated geometry, the saddle point on a hypothetical vibrational-energy-free PES.

---

[1] Henry Eyring, American chemist. Born Colonia Juarárez, Mexico, 1901. Ph.D. University of California, Berkeley, 1927. Professor Princeton, University of Utah. Known for his work on the theory of reaction rates and on potential energy surfaces. Died Salt Lake City, Utah, 1981.

The geometric parameter corresponding to the reaction coordinate is usually a composite of several parameters (bond lengths, angles and dihedrals), although for some reactions one or two may predominate. In Fig. 2.7, the reaction coordinate is a composite of the O–O bond length and the O–O–O bond angle.

A saddle point, the point on a PES where the second derivative of energy with respect to one and only geometric coordinate (possibly a composite coordinate) is negative, corresponds to a transition state. Some PES's have points where the second derivative of energy with respect to more than one coordinate is negative; these are *higher-order saddle points* or *hilltops*: e.g. a *second*-order saddle point is a point on the PES which is a maximum along *two* paths connecting stationary points. The propane PES, Fig. 2.9, provides examples of a minimum, a transition state and a hilltop – a *second*-order saddle point in this case. Figure 2.10 shows the three stationary points in more detail. The "doubly-eclipsed" conformation (A), in which there is eclipsing as viewed along the C1–C2 and the C3–C2 bonds (the dihedral angles are 0° viewed along these bonds) is a second-order saddle point because single bonds do nor like to eclipse single bonds and rotation about the C1–C2 and the C3–C2 bonds will remove this eclipsing: there are *two* possible directions along the PES which lead, without a barrier, to lower-energy regions, i.e. changing the H–C1/C2–C3 dihedral and changing the H–C3/C2–C1 dihedral. Changing *one* of these leads to a "singly-eclipsed" conformation (B) with only one offending eclipsing CH_3–CH_2 arrangement, and this

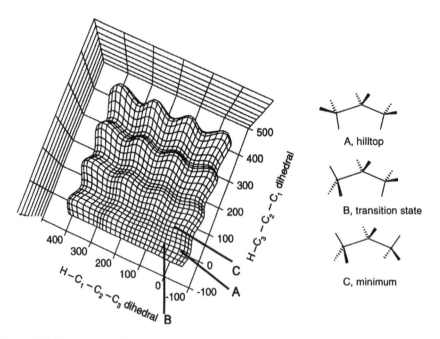

Figure 2.9. The propane PES as the two HCCC dihedrals are varied (calculated by the AM1 method, chapter 6). Bond lengths and angles were not optimized as the dihedrals were varied, so this is not a relaxed PES; however, changes in bond lengths and angles from one propane conformation to another are small, and the relaxed PES should be very similar to this one.

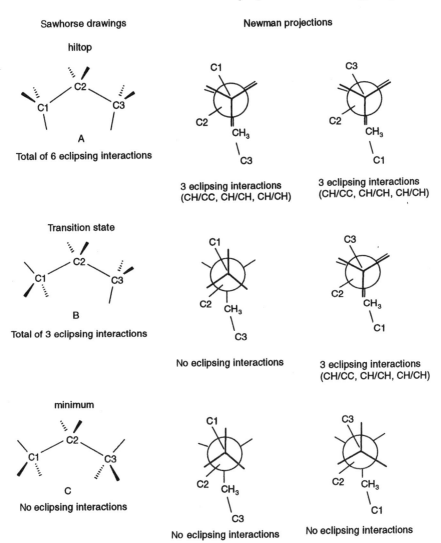

Figure 2.10. The stationary points on the propane PES. Hydrogens at the end of CH bonds are omitted for clarity.

is a first-order saddle point, since there is now only *one* direction along the PES which leads to relief of the eclipsing interactions (rotation around C3–C2). This route gives a conformation C which has no eclipsing interactions and is therefore a minimum. There are no lower-energy structures on the C_3H_8 PES and so C is the global minimum.

The geometry of propane depends on more than just two dihedral angles, of course; there are several bond lengths and bond angles and the potential energy will vary with changes in all of them. Figure 2.9 was calculated by varying only the dihedral angles associated with the C1–C2 and C2–C3 bonds, keeping the other geometrical parameters the same as they are in the all-staggered conformation. If at every point on the dihedral/dihedral grid all the other parameters (bond lengths and angles) had been optimized (adjusted to give the lowest possible energy, for that particular calculational method; section 2.4), the result would have been a *relaxed* PES. In Fig. 2.9 this was not done, but because bond lengths and angles change only slightly with changes in dihedral angles the PES would not be altered much, while the time required for the calculation (for the *PES scan*) would have been much greater. Figure 2.9 is a nonrelaxed or rigid PES, albeit not very different, in this case, from a relaxed one.

Chemistry is essentially the study of the stationary points on a PES: in studying more or less stable molecules we focus on minima, and in investigating chemical reactions we study the passage of a molecule from a minimum through a transition state to another minimum. There are four known forces in nature: the gravitational force, the strong and the weak nuclear forces, and the electromagnetic force. Celestial mechanics studies the motion of stars and planets under the influence of the gravitational force and nuclear physics studies the behavior of subatomic particles subject to the nuclear forces. Chemistry is concerned with aggregates of nuclei and electrons (with molecules) held together by the electromagnetic force, and with the shuffling of nuclei, followed by their obedient retinue of electrons, around a PES under the influence of this force (with chemical reactions).

The concept of the chemical PES apparently originated with R. Marcelin [6]: in a dissertation-long paper (111 pages) he laid the groundwork for transition-state theory 20 years before the much better-known work of Eyring [5,7]. The importance of Marcelin's work is acknowledged by Rudolph Marcus in his Nobel Prize (1992) speech, where he refers to " . . . Marcelin's classic 1915 theory which came within one small step of the transition state theory of 1935." The paper was published the year after the death of the author, who seems to have died in World War I, as indicated by the footnote "Tué à l'ennemi en sept 1914". The first PES was calculated in 1931 by Eyring and Polanyi,[2] using a mixture of experiment and theory [8].

2.3   THE BORN–OPPENHEIMER APPROXIMATION

A PES is a plot of the energy of a collection of nuclei and electrons against the geometric coordinates of the nuclei – essentially a plot of molecular energy vs. molecular geometry (or it may be regarded as the mathematical equation that gives the energy as a function of the nuclear coordinates). The nature (minimum, saddle point or neither) of each point was discussed in terms of the response of the energy (first and second derivatives) to changes in nuclear coordinates. But if a molecule is a collection of nuclei and electrons why plot energy vs. *nuclear* coordinates – why not against *electron* coordinates? In other words, why are nuclear coordinates the parameters that define molecular geometry? The answer to this question lies in the Born–Oppenheimer approximation.

---

[2]Michael Polanyi, Hungarian–British chemist, economist, and philosopher. Born Budapest, 1891. Doctor of medicine 1913, Ph.D. University of Budapest, 1917. Researcher Kaiser-Wilhelm Institute, Berlin, 1920–1933. Professor of chemistry, Manchester, 1933–1948; of social studies, Manchester, 1948–1958. Professor Oxford, 1958–1976. Best known for book "Personal Knowledge," 1958. Died Northampton, England, 1976.

Born[3] and Oppenheimer[4] showed in 1927 [9] that to a very good approximation the nuclei in a molecule are stationary with respect to the electrons. This is a qualitative expression of the principle; mathematically, the approximation states that the Schrödinger equation (chapter 4) for a molecule may be separated into an electronic and a nuclear equation. One consequence of this is that all (!) we have to do to calculate the energy of a molecule is to solve the *electronic* Schrödinger equation and then add the electronic energy to the internuclear repulsion (this latter quantity is trivial to calculate) to get the total internal energy (see section 4.4.1). A deeper consequence of the Born–Oppenheimer approximation is that a molecule has a shape.

The nuclei see the electrons as a smeared-out cloud of negative charge which binds them in fixed relative positions (because of the mutual attraction between electrons and nuclei in the internuclear region) and which defines the (somewhat fuzzy) surface [10] of the molecule (see Fig. 2.11). Because of the rapid motion of the electrons compared to the nuclei the "permanent" geometric parameters of the molecule are the *nuclear* coordinates. The energy (and the other properties) of a molecule is a *function* of the electron coordinates ($E = \Psi(x, y, z$ of each electron); section 5.2), but depends only *parametrically* on the nuclear coordinates, i.e. for each geometry 1, 2, ... there is a particular energy: $E_1 = \Psi_1(x, y, z, \ldots)$, $E_2 = \Psi_2(x, y, z, \ldots)$; cf. x^n, which is a function of x but depends only parametrically on n. Actually, the nuclei are not

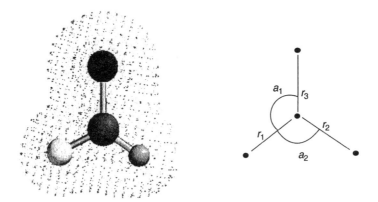

Figure 2.11. The nuclei in a molecule see a time-averaged electron cloud. The nuclei vibrate about equilibrium points which define the molecular geometry; this geometry can be expressed simply as the nuclear cartesian coordinates, or alternatively as bond lengths and angles (r and a here) and dihedrals, i.e. as internal coordinates. As far as size goes, the experimentally determined van der Waals surface encloses about 98 percent of the electron density of a molecule.

---

[3]Max Born, German–British physicist. Born in Breslau (now Wroclaw, Poland), 1882, died in Göttingen, 1970. Professor Berlin, Cambridge, Edinburgh. Nobel prize, 1954. One of the founders of quantum mechanics, originator of the probability interpretation of the (square of the) wavefunction (chapter 4).

[4]J. Robert Oppenheimer, American physicist. Born in New York, 1904, died in Princeton, 1967. Professor California Institute of Technology. Fermi award for nuclear research, 1963. Important contributions to nuclear physics. Director of the Manhattan Project 1943–1945. Victimized as a security risk by senator Joseph McCarthy's Un-American Activities Committee in 1954. Central figure of the eponymous PBS TV series (Oppenheimer: Sam Waterston).

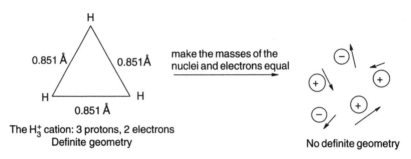

Figure 2.12. A molecule has a definite shape because unlike the electrons, the nuclei are (relatively) stationary (since they are much more massive). If the masses of the nuclei and the electrons could be made equal, the distinction in lethargy would be lost, and the molecular geometry would dissolve.

stationary, but execute vibrations of small amplitude about equilibrium positions; it is these equilibrium positions that we mean by the "fixed" nuclear positions. It is only because it is meaningful to speak of (almost) fixed nuclear coordinates that the concepts of molecular geometry or shape [11] and of the PES are valid. The nuclei are much more sluggish than the electrons because they are much more massive (a hydrogen nucleus is about 2000 times more massive than an electron).

Consider the molecule H_3^+, made up of three protons and two electrons. *Ab initio* calculations assign it the geometry shown in Fig. 2.12. The equilibrium positions of the nuclei (the protons) lie at the corners of an equilateral triangle and H_3^+ has a definite shape. But suppose the protons were replaced by positrons, which have the same mass as electrons. The distinction between nuclei and electrons, which in molecules rests on mass and not on some kind of charge chauvinism, would vanish. We would have a quivering cloud of flitting particles to which a shape could not be assigned on a macroscopic time scale.

A calculated PES, which we might call a Born–Oppenheimer surface, is normally the set of points representing the geometries, and the corresponding energies, of a collection of atomic nuclei; the electrons are taken into account in the calculations as needed to assign charge and multiplicity (multiplicity is connected with the number of unpaired electrons). Each point corresponds to a set of stationary nuclei, and in this sense the surface is somewhat unrealistic (see section 2.5).

2.4   GEOMETRY OPTIMIZATION

The characterization (the "location" or "locating") of a stationary point on a PES, i.e. demonstrating that the point in question exists and calculating its geometry and energy, is a *geometry optimization*. The stationary point of interest might be a minimum, a transition state, or, occasionally, a higher-order saddle point. Locating a minimum is often called an energy minimization or simply a minimization, and locating a transition state is often referred to specifically as a transition state optimization. Geometry optimizations are done by starting with an input structure that is believed to resemble (the closer the better) the desired stationary point and submitting this plausible structure to a computer algorithm that systematically changes the geometry until it has found a stationary point. The curvature of the PES at the stationary point, i.e. the second derivatives of energy with respect to the geometric parameters (section 2.2) may then be determined (section 2.5) to characterize the structure as a minimum or as some kind of saddle point.

Let us consider a problem that arose in connection with an experimental study. Propanone (acetone) was subjected to ionization followed by neutralization of the radical cation, and the products were frozen in an inert matrix and studied by IR spectroscopy [12]. The spectrum of the mixture suggested the presence of the enol isomer of propanone, 1-propen-2-ol:

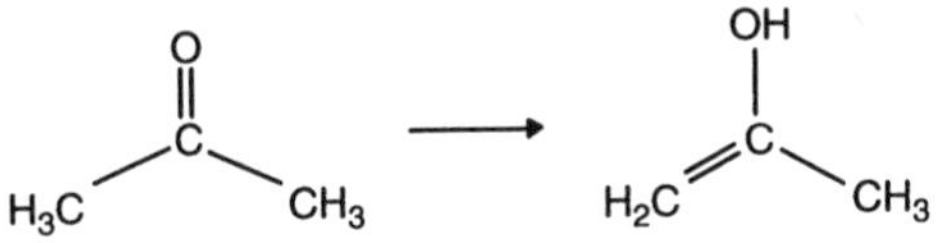

Reaction 2

To confirm (or refute) this the IR spectrum of the enol might be calculated (see section 2.5 and the discussions of the calculation of IR spectra in subsequent chapters). But which conformer should one choose for the calculation? Rotation about the C–O and C–C bonds creates six plausible stationary points (Fig. 2.13), and a PES scan (Fig. 2.14)

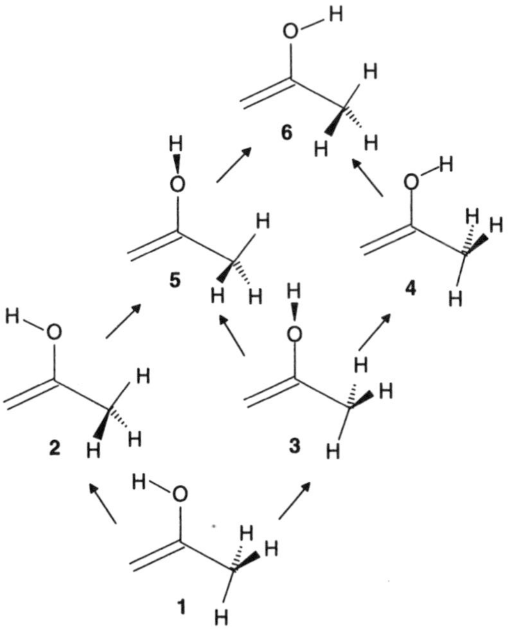

Figure 2.13. The plausible stationary points on the propenol PES. A PES scan (Fig. 2.14) indicated that **1** is the global minimum and **4** is a relative minimum, while **2** and **3** are transition states and **5** and **6** are hilltops. AM1 calculations gave relative energies for **1, 2, 3** and **4** of 0, 0.6, 14 and 6.5 kJ mol^{-1}, respectively (**5** and **6** were not optimized). The arrows represent one-step (rotation about one bond) conversion of one species into another.

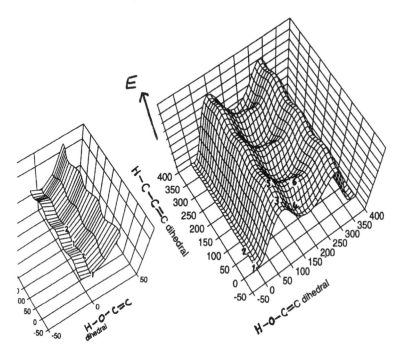

Figure 2.14. The 1-propen-2-ol PES (calculated by the AM1 method). See Fig. 2.13.

indicated that there are indeed six such species. Examination of this PES shows that the global minimum is structure **1** and that there is a relative minimum corresponding to structure **4**. Geometry optimization starting from an input structure resembling **1** gave a minimum corresponding to **1**, while optimization starting from a structure resembling **4** gave another, higher-energy minimum, resembling **4**. Transition-state optimizations starting from appropriate structures yielded the transition states **2** and **3**. These stationary points were all characterized as minima or transition states by second-derivative calculations (section 2.5) (the species **5** and **6** were not located). The calculated IR spectrum of **1** (using the ab initio HF/6-31G* method – chapter 5) was in excellent agreement with the observed spectrum of the putative propenol.

This illustrates a general principle: the optimized structure one obtains is that closest in geometry on the PES to the input structure (Fig. 2.15). To be sure we have found a *global* minimum we must (except for very simple or very rigid molecules) *search* a PES (there are algorithms that will do this and locate the various minima). Of course we may not be interested in the global minimum; e.g. if we wish to study the cyclic isomer of ozone (section 2.2) we will use as input an equilateral triangle structure, probably with bond lengths about those of an O–O single bond.

In the propenol example, the PES scan suggested that to obtain the global minimum we should start with an input structure resembling **1**, but the exact values of the various bond lengths and angles were unknown (the exact values of even the dihedrals was not known with certainty, although general chemical knowledge made H–O–C–C = H–C–C = C = 0° seem plausible). The actual creation of input structures is

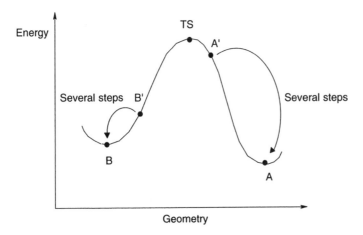

Figure 2.15. Geometry optimization to a minimum gives the minimum closest to the input structure. The input structure A′ is moved toward the minimum A, and B′ toward B. To locate a transition state a special algorithm is usually used: this moves the initial structure A′ toward the transition state TS. Optimization to each of the stationary points would probably actually require several steps (see Fig. 2.16).

usually done nowadays with an interactive mouse-driven program, in much the same spirit that one constructs plastic models or draws structures on paper. An older alternative is to specify the geometry by defining the various bond lengths, angles and dihedrals, i.e. by using a so-called Z-matrix (internal coordinates).

To move along the PES from the input structure to the nearest minimum is obviously trivial on the 1D PES of a diatomic molecule: one simply changes the bond length till that corresponding to the lowest energy is found. On any other surface, efficient geometry optimization requires a sophisticated algorithm. One would like to know in which direction to move, and how far in that direction (Fig. 2.16). It is not possible, in general, to go from the input structure to the proximate minimum in just one step, but modern geometry optimization algorithms commonly reach the minimum in about 10 steps, given a reasonable input geometry. The most widely-used algorithms for geometry optimization [13] use the first and second derivatives of the energy with respect to the geometric parameters. To get a feel for how this works, consider the simple case of a 1D PES, as for a diatomic molecule (Fig. 2.17). The input structure is at the point $P_i(E_i, q_i)$ and the proximate minimum, corresponding to the optimized structure being sought, is at the point $P_o(E_o, q_o)$. Before the optimization has been carried out the values of E_o and q_o are of course unknown. If we assume that near a minimum the potential energy is a quadratic function of q, which is a fairly good approximation, then

$$E - E_o = k(q - q_o)^2 \tag{2.5}$$

At the input point

$$(dE/dq)_i = 2k(q_i - q_o) \tag{2.6}$$

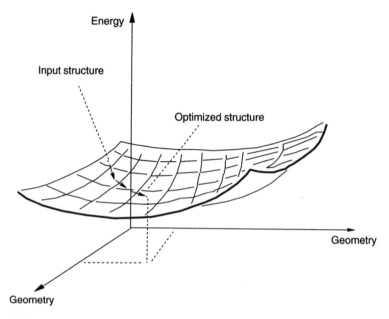

Figure 2.16. An efficient optimization algorithm knows approximately in which direction to move and how far to step, in an attempt to reach the optimized structure in relatively few (commonly about 5–10) steps.

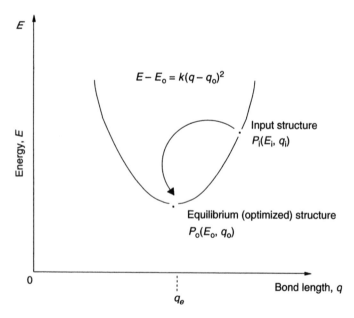

Figure 2.17. The potential energy of a diatomic molecule near the equilibrium geometry is approximately a quadratic function of the bond length. Given an input structure (i.e. given the bond length q_i), a simple algorithm would enable the bond length of the optimized structure to be found in one step, if the function were strictly quadratic.

At all points

$$d^2 E/dq^2 = 2k \ (= \text{force constant}) \tag{2.7}$$

From (2.6) and (2.7),

$$(dE/dq)_i = (d^2 E/dq^2)(q_i - q_0) \tag{2.8}$$

and

$$q_0 = q_i - (dE/dq)_i/(d^2 E/dq^2) \tag{2.9}$$

Equation (2.9) shows that if we know $(dE/dq)_i$, the slope or gradient of the PES at the point of the initial structure, $(d^2 E/dq^2)$, the curvature of the PES (which for a quadratic curve $E(q)$ is independent of q), and q_i, the initial geometry, we can calculate q_0, the optimized geometry. The second derivative of potential energy with respect to geometric displacement is the force constant for motion along that geometric coordinate; as we will see later, this is an important concept in connection with calculating vibrational spectra.

For multidimensional PES's, i.e. for almost all real cases, far more sophisticated algorithms are used, and several steps are needed since the curvature is not exactly quadratic. The first step results in a new point on the PES that is (probably) closer to the minimum than was the initial structure. This new point then serves as the initial point for a second step toward the minimum, etc. Nevertheless, most modern geometry optimization methods do depend on calculating the first and second derivatives of the energy at the point on the PES corresponding to the input structure. Since the PES is not strictly quadratic, the second derivatives vary from point to point and are updated as the optimization proceeds.

In the illustration of an optimization algorithm using a diatomic molecule, Eq. (2.9) referred to the calculation of first and second derivatives with respect to bond length, which latter is an *internal coordinate* (inside the molecule). Optimizations are actually commonly done using Cartesian coordinates x, y, z. Consider the optimization of a triatomic molecule like water or ozone in a Cartesian coordinate system. Each of the three atoms has an x, y and z coordinate, giving 9 geometric parameters, $q_1, q_2, \ldots, q_9$; the PES would be a 9-dimensional hypersurface on a 10D graph. We need the first and second derivatives of E with respect to each of the 9 q's, and these derivatives are manipulated as matrices. Matrices are discussed in section 4.3.3; here we need only know that a matrix is a rectangular array of numbers that can be manipulated mathematically, and that they provide a convenient way of handling sets of linear equations. The first-derivative matrix, the gradient matrix, for the input structure can be written as a column matrix

$$\mathbf{g_i} = \begin{bmatrix} (\partial E/\partial q_1)_i \\ (\partial E/\partial q_2)_i \\ \vdots \\ (\partial E/\partial q_9)_i \end{bmatrix} \tag{2.10}$$

and the second-derivative matrix, the force constant matrix, is

$$
\mathbf{H} = \begin{bmatrix} \partial^2 E/\partial q_1 q_1 & \partial^2 E/\partial q_1 q_2 & \cdots & \partial^2 E/\partial q_1 q_9 \\ \partial^2 E/\partial q_2 q_1 & \partial^2 E/\partial q_2 q_2 & \cdots & \partial^2 E/\partial q_2 q_9 \\ \vdots & \vdots & \cdots & \vdots \\ \partial^2 E/\partial q_9 q_1 & \partial^2 E/\partial q_9 q_2 & \cdots & \partial^2 E/\partial q_9 q_9 \end{bmatrix} \tag{2.11}
$$

The force constant matrix is called the *Hessian*.[5] The Hessian is particularly important, not only for geometry optimization, but also for the characterization of stationary points as minima, transition states or hilltops, and for the calculation of IR spectra (section 2.5). In the Hessian $\partial^2 E/\partial q_1 q_2 = \partial^2 E/\partial q_2 q_1$, as is true for all well-behaved functions, but this systematic notation is preferable: the first subscript refers to the row and the second to the column. The geometry coordinate matrices for the initial and optimized structures are

$$
\mathbf{q_i} = \begin{bmatrix} q_{i1} \\ q_{i2} \\ \vdots \\ q_{i9} \end{bmatrix} \tag{2.12}
$$

and

$$
\mathbf{q_o} = \begin{bmatrix} q_{o1} \\ q_{o2} \\ \vdots \\ q_{o9} \end{bmatrix} \tag{2.13}
$$

The matrix equation for the general case can be shown to be:

$$
\mathbf{q_o} = \mathbf{q_i} - \mathbf{H}^{-1}\mathbf{g_i} \tag{2.14}
$$

which is somewhat similar to Eq. (2.9) for the optimization of a diatomic molecule. For n atoms we have $3n$ Cartesians; $\mathbf{q_o}$, $\mathbf{q_i}$ and $\mathbf{g_i}$ are $3n \times 1$ column matrices and $\mathbf{H}$ is a $3n \times 3n$ square matrix; multiplication by the inverse of $\mathbf{H}$ rather than division by $\mathbf{H}$ is used because matrix division is not defined.

Equation (2.14) shows that for an efficient geometry optimization we need an initial structure (for $\mathbf{q_i}$), initial gradients (for $\mathbf{g_i}$) and second derivatives (for $\mathbf{H}$). With an initial "guess" for the geometry (e.g. from a model-building program followed by molecular mechanics) as input, gradients can be readily calculated analytically (from the derivatives of certain integrals). An approximate initial Hessian is often calculated from molecular mechanics (chapter 3). Since the PES is not really exactly quadratic, the first step does not take us all the way to the optimized geometry, corresponding to the matrix $\mathbf{q_o}$. Rather, we arrive at $\mathbf{q_1}$, the first calculated geometry; using this geometry a new gradient matrix and a new Hessian are calculated (the gradients are calculated analytically and the second derivatives are updated using the changes in the gradients – see below). Using $\mathbf{q_1}$ and the new gradient and Hessian matrices a new approximate

---

[5]Ludwig Otto Hesse, 1811–1874, German mathematician.

geometry matrix $\mathbf{q}_2$ is calculated. The process is continued until the geometry and/or the gradients (or with some programs possibly the energy) have ceased to change appreciably.

As the optimization proceeds the Hessian is updated by approximating each second derivative as a ratio of finite increments:

$$\frac{\partial^2 E}{\partial q_i \partial q_j} \approx \frac{\Delta(\partial E/\partial q_j)}{\Delta q_i} \tag{2.15}$$

i.e. as the change in the gradient divided by the change in geometry, on going from the previous structure to the latest one. Analytic calculation of second derivatives is relatively time-consuming and is not routinely done for each point along the optimization sequence, in contrast to analytic calculation of gradients. A fast lower-level optimization, for a minimum or a transition state, usually provides a good Hessian and geometry for input to a higher-level optimization [14]. Finding a transition state (i.e. optimizing an input structure to a transition state structure) is a more challenging computational problem than finding a minimum, as the characteristics of the PES at the former are more complicated than at a minimum: at the transition state the surface is a maximum in one direction and a minimum in all others, rather than simply a minimum in all directions. Nevertheless, modifications of the minimum-search algorithm enable transitions states to be located, albeit often with less ease than minima.

2.5   STATIONARY POINTS AND NORMAL-MODE VIBRATIONS: ZPE

Once a stationary point has been found by geometry optimization, it is usually desirable to check whether it is a minimum, a transition state, or a hilltop. This is done by calculating the vibrational frequencies. Such a calculation involves finding the *normal-mode* frequencies; these are the simplest vibrations of the molecule, which, in combination, can be considered to result in the actual, complex vibrations that a real molecule undergoes. In a normal-mode vibration all the atoms move in phase with the same frequency: they all reach their maximum and minimum displacements and their equilibrium positions at the same moment. The other vibrations of the molecule are combinations of these simple vibrations. Essentially, a normal-modes calculation is a calculation of the infrared spectrum, although the experimental spectrum is likely to contain extra bands resulting from interactions among normal-mode vibrations.

A nonlinear molecule with n atoms has $3n - 6$ normal modes: the motion of each atom can be described by 3 vectors, along the x, y, and z axes of a Cartesian coordinate system; after removing the 3 vectors describing the translational motion of the molecule as a whole (the translation of its center of mass) and the 3 vectors describing the rotation of the molecule (around the 3 principal axes needed to describe rotation for a 3D object of general geometry), we are left with $3n - 6$ independent vibrational motions. Arranging these in appropriate combinations gives $3n - 6$ normal modes. A linear molecule has $3n - 5$ normal modes, since we need subtract only three translational and *two* rotational vectors, as rotation about the molecular axis does not produce a recognizable

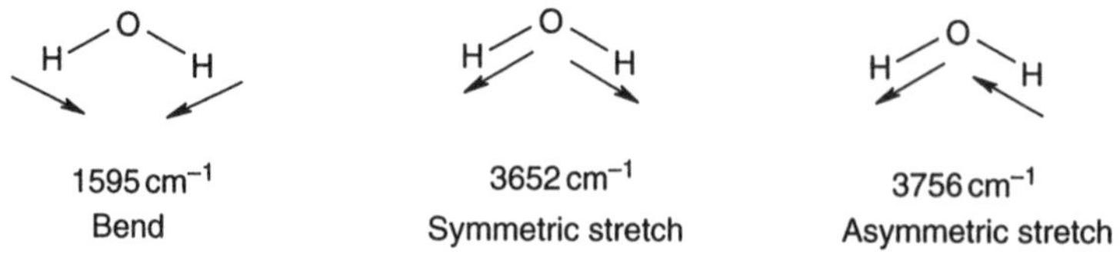

| 1595 cm^{-1} | 3652 cm^{-1} | 3756 cm^{-1} |
| Bend | Symmetric stretch | Asymmetric stretch |

Figure 2.18. The normal-mode vibrations of water. The arrows indicate the directions in which the atoms move; on reaching the maximum amplitude these directions are reversed.

change in the nuclear array. So water has $3n - 6 = 3(3) - 6 = 3$ normal modes, and HCN has $3n - 5 = 3(3) - 5 = 4$ normal modes. For water (Fig. 2.18) mode 1 is a bending mode (the H–O–H angle decreases and increases), mode 2 is a symmetric stretching mode (both O–H bonds stretch and contract simultaneously) and mode 3 is an asymmetric stretching mode (as the O–H$_1$ bond stretches the O–H$_2$ bond contracts, and vice versa). At any moment an actual molecule of water will be undergoing a complicated stretching/bending motion, but this motion can be considered to be a combination of the three simple normal-mode motions.

Consider a diatomic molecule A–B; the normal-mode frequency (there is only one for a diatomic, of course) is given by [15]:

$$\tilde{\nu} = \frac{1}{2\pi c} \left(\frac{k}{\mu} \right)^{1/2} \tag{2.16}$$

where $\tilde{\nu}$ = vibrational "frequency," actually wavenumber, in cm^{-1}; from deference to convention we use cm^{-1} although the cm is not an SI unit, and so the other units will also be non-SI; $\tilde{\nu}$ signifies the number of wavelengths that will fit into one cm. The symbol ν is the Greek letter nu, which resembles an angular vee; $\tilde{\nu}$ could be read "nu tilde"; $\bar{\nu}$, "nu bar," has been used less frequently. c = velocity of light; k = force constant for the vibration; μ = reduced mass of the molecule $= (m_A m_B)/(m_A + m_B)$; m_A and m_B are the masses of A and B.

The force constant k of a vibrational mode is a measure of the "stiffness" of the molecule toward that vibrational mode – the harder it is to stretch or bend the molecule in the manner of that mode, the bigger is that force constant (for a diatomic molecule k simply corresponds to the stiffness of the one bond). The fact that the frequency of a vibrational mode is related to the force constant for the mode suggests that it might be possible to calculate the normal-mode frequencies of a molecule, i.e. the directions and frequencies of the atomic motions, from its force constant matrix (its Hessian). This is indeed possible: *matrix diagonalization* of the Hessian gives the directional characteristics (which way the atoms are moving), and the force constants themselves, for the vibrations. Matrix diagonalization (section 4.3.3) is a process in which a square matrix **A** is decomposed into three square matrices, **P**, **D**, and **P**$^{-1}$: $\mathbf{A} = \mathbf{PDP}^{-1}$. **D** is a diagonal matrix: as with **k** in Eq. (2.17) all its off-diagonal elements are zero. **P** is a premultiplying matrix and **P**$^{-1}$ is the inverse of **P**. When matrix algebra is applied to physical problems, the diagonal row elements of **D** are the magnitudes of some physical quantity, and each column of **P** is a set of coordinates which give a direction associated with that physical quantity. These ideas are made more concrete in the discussion accompanying Eq. (2.17), which shows the diagonalization of the Hessian matrix for a triatomic molecule, e.g. H_2O.

$$\mathbf{H} = \begin{bmatrix} \partial^2 E/\partial q_1 q_1 & \partial^2 E/\partial q_1 q_2 & \cdots & \partial^2 E/\partial q_1 q_9 \\ \partial^2 E/\partial q_2 q_1 & \partial^2 E/\partial q_2 q_2 & \cdots & \partial^2 E/\partial q_2 q_9 \\ \vdots & \vdots & \cdots & \vdots \\ \partial^2 E/\partial q_9 q_1 & \partial^2 E/\partial q_9 q_2 & \cdots & \partial^2 E/\partial q_9 q_9 \end{bmatrix}$$

$$= \underbrace{\begin{bmatrix} q_{11} & q_{12} & \cdots & q_{19} \\ q_{21} & q_{22} & \cdots & q_{29} \\ \vdots & & & \\ q_{91} & q_{92} & \cdots & q_{99} \end{bmatrix}}_{\mathbf{P}} \underbrace{\begin{bmatrix} k_1 & 0 & \cdots & 0 \\ 0 & k_2 & \cdots & 0 \\ \vdots & & & \\ 0 & 0 & \cdots & k_9 \end{bmatrix}}_{\mathbf{k}} \mathbf{P}^{-1} \qquad (2.17)$$

Equation (2.17) is of the form $\mathbf{A} = \mathbf{PDP}^{-1}$. The 9×9 Hessian for a triatomic molecule (three Cartesian coordinates for each atom) is decomposed by diagonalization into a $\mathbf{P}$ matrix whose columns are "direction vectors" for the vibrations whose force constants are given by the $\mathbf{k}$ matrix. Actually, columns 1, 2 and 3 of $\mathbf{P}$ and the corresponding k_1, k_2 and k_3 of $\mathbf{k}$ refer to *translational* motion of the molecule (motion of the whole molecule from one place to another in space); these three "force constants" are nearly zero. Columns 4, 5 and 6 of $\mathbf{P}$ and the corresponding k_4, k_5 and k_6 of $\mathbf{k}$ refer to *rotational* motion about the three principal axes of rotation, and are also nearly zero. Columns 7, 8 and 9 of $\mathbf{P}$ and the corresponding k_7, k_8 and k_9 of $\mathbf{k}$ are the direction vectors and force constants, respectively, for the normal-mode vibrations: k_7, k_8 and k_9 refer to vibrational modes 1, 2 and 3, while the 7th, 8th, and 9th columns of $\mathbf{P}$ are composed of the x, y and z components of vectors for motion of the three atoms in mode 1 (column 7), mode 2 (column 8), and mode 3 (column 9). "Mass-weighting" the force constants, i.e. taking into account the effect of the masses of the atoms (cf. Eq. (2.16) for the simple case of a diatomic molecule), gives the vibrational frequencies. The $\mathbf{P}$ matrix is the *eigenvector* matrix and the $\mathbf{k}$ matrix is the *eigenvalue* matrix from diagonalization of the Hessian $\mathbf{H}$. "Eigen" is a German prefix meaning "appropriate, suitable, actual" and is used in this context to denote mathematically appropriate entities for the solution of a matrix equation. Thus the directions of the normal-mode frequencies are the eigenvectors, and their magnitudes are the mass-weighted eigenvalues, of the Hessian.

Vibrational frequencies are calculated to obtain IR spectra, to characterize stationary points, and to obtain zero point energies (below). The calculation of meaningful frequencies is valid only at a stationary point and only using the same method that was used to optimize to that stationary point (e.g. an *ab initio* method with a particular correlation level and basis set – see chapter 5). This is because (1) the use of second derivatives as force constants presupposes that the PES is quadratically curved along each geometric coordinate q (Fig. 2.2) but it is only near a stationary point that this is true, and (2) use of a method other than that used to obtain the stationary point presupposes that the PES's of the two methods are parallel (that they have the same curvature) at the stationary point. Of course, "provisional" force constants at nonstationary points are used in the optimization process, as the Hessian is updated from step to step. Calculated IR frequencies are usually somewhat too high, but (at least for ab initio and density functional calculations) can be brought into reasonable agreement with experiment by multiplying them by an empirically determined factor, commonly about 0.9 [16] (see the discussion of frequencies in chapters 5–7).

A minimum on the PES has all the normal-mode force constants (all the eigenvalues of the Hessian) positive: for each vibrational mode there is a restoring force, like that of a spring. As the atoms execute the motion, the force pulls and slows them till they move in the opposite direction; each vibration is periodic, over and over. The species corresponding to the minimum sits in a well and vibrates forever (or until it acquires enough energy to react). For a transition state, however, one of the vibrations, that along the reaction coordinate, is different: motion of the atoms corresponding to *this* mode takes the transition state toward the product or toward the reactant, without a restoring force. This one "vibration" is not a periodic motion but rather takes the species through the transition state geometry on a one-way journey. Now, the force constant is the first derivative of the gradient or slope (the derivative of the first derivative); examination of Fig. 2.8 shows that along the reaction coordinate the surface slopes downward, so the force constant for this mode is *negative*. A transition state (a first-order saddle point) has one and only one negative normal-mode force constant (one negative eigenvalue of the Hessian). Since a frequency calculation involves taking the square root of a force constant (Eq. (2.16)), and the square root of a negative number is an imaginary number, a transition state has one imaginary frequency, corresponding to the reaction coordinate. In general an nth-order saddle point (an nth-order hilltop) has n negative normal-mode force constants and so n imaginary frequencies, corresponding to motion from one stationary point of some kind to another.

A stationary point could of course be characterized just from the number of negative force constants, but the mass-weighting requires much less time than calculating the force constants, and the frequencies themselves are often wanted anyway, e.g. for comparison with experiment. In practice one usually checks the nature of a stationary point by calculating the frequencies and seeing how many imaginary frequencies are present; a minimum has none, a transition state one, and a hilltop more than one. If one is seeking a particular transition state the criteria to be satisfied are:

1. It should look right. The structure of a transition state should lie somewhere between that of the reactants and the products; e.g. the transition state for the unimolecular isomerization of HCN to HNC shows an H bonded to both C and N by an unusually long bond, and the CN bond length is in-between that of HCN and HNC.

2. It must have one and only one imaginary frequency (some programs indicate this as a negative frequency, e.g. $-1900\,\mathrm{cm}^{-1}$ instead of the correct $1900i$ ($i = \sqrt{(-1)}$).

3. The imaginary frequency must correspond to the reaction coordinate. This is usually clear from animation of the frequency (the motion, stretching, bending, twisting, corresponding to a frequency may be visualized with a variety of programs). For example, the transition state for the unimolecular isomerization of HCN to HNC shows an imaginary frequency which when animated clearly shows the H migrating between the C and the N. Should it not be clear from animation which two species the

transition state connects, one may resort to an IRC calculation [17]. This procedure follows the transition state downhill along the IRC (section 2.2), generating a series of structures along the path to the reactant or product. Usually it is clear where the transition state is going without following it all the way to a stationary point.

4. The energy of the transition state must be higher than that of the two species it connects.

Besides indicating the IR spectrum and providing a check on the nature of stationary points, the calculation of vibrational frequencies also provides the ZPE (most programs will calculate this automatically as part of a frequency job). The ZPE is the energy a molecule has even at absolute zero (Fig. 2.2), as a consequence of the fact that even at this temperature it still vibrates [2]. The ZPE of a species is usually not small compared to activation energies or reaction energies, but ZPEs tend to cancel out when these are calculated (by subtraction), since for a given reaction the ZPE of the reactant, transition state and product tend to be roughly the same. However, for accurate work the ZPE should be added to the "total" (electronic + nuclear repulsion) energies of species and the ZPE-corrected energies should then be compared (Fig. 2.19). Like the frequencies, the ZPE is usually corrected by multiplying it by an empirical factor; this is sometimes the same as the frequency correction factor, but slightly different factors have been recommended [16].

The Hessian that results from a geometry optimization was built up in steps from one geometry to the next, approximating second derivatives from the changes in gradients (Eq. (2.15)). This Hessian is not accurate enough for the calculation of frequencies and ZPE's. The calculation of an accurate Hessian for a stationary point can be done analytically or numerically. Accurate numerical evaluation approximates the second derivative as in Eq. (2.15), but instead of $\Delta(\partial V/\partial q)$ and Δq being taken from optimization iteration steps, they are obtained by changing the position of each atom of the optimized structure slightly (Δq = about 0.01 Å) and calculating analytically the change in the gradient at each geometry; subtraction gives $\Delta(\partial V/\partial q)$. This can be done for a change in one direction only for each atom (method of forward differences) or more accurately by going in two directions around the equilibrium position and averaging the gradient change (method of central differences). Analytical calculation of *ab initio* frequencies is much faster than numerical evaluation, but demands on computer hard drive space may make numerical calculation the only recourse at high *ab initio* levels (chapter 5).

2.6 SYMMETRY

Symmetry is important in theoretical chemistry (and even more so in theoretical physics), but our interest in it here is bounded by modest considerations: we want to see why symmetry is relevant to setting up a calculation and interpreting the results, and to make sense of terms like C_{2v}, C_s, etc., which are used in various places in this book. Excellent expositions of symmetry are given by, e.g. Atkins [18] and Levine [19].

The symmetry of a molecule is most easily described by using one of the standard designations like C_{2v}, C_s. These are called *point groups* (Schoenflies point groups) because when symmetry operations (below) are carried out on a molecule (on any object)

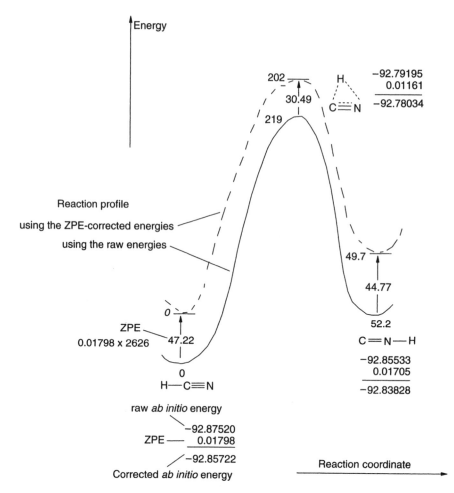

Figure 2.19. Correcting relative energies for ZPE. These are *ab initio* HF/6-31G* (chapter 5) results for the HCN → HNC reaction. The corrections are most simply made by adding the ZPE to the raw energy (in energy units called hartrees or atomic units), to get the corrected energies. Using corrected or uncorrected energies, relative energies are obtained by setting the energy of one species (usually that of lowest energy) equal to zero. Finally, energy differences in hartrees were multiplied by 2626 to get kJ mol^{-1}. The ZPEs are also shown here in kJ mol^{-1}, just to emphasize that they are not small compared to reaction energies or activation energies, but tend to cancel; for accurate work ZPE-corrected energies should be used.

with symmetry, at least one point is left unchanged. The classification is according to the presence of symmetry elements and corresponding symmetry operations. The main symmetry elements are mirror planes (symmetry planes), symmetry axes, and an inversion center; other symmetry elements are the entire object, and an improper rotation axis. The operation corresponding to a mirror plane is reflection in that plane, the operation corresponding to a symmetry axis is rotation about that axis, and the operation corresponding to an inversion center is moving each point in the molecule along a straight line to that center then moving it further, along the line, an equal distance beyond the center. The "entire object" element corresponds to doing nothing (a null operation); in common parlance an object with only this symmetry element would be said to have no symmetry. The improper rotation axis corresponds to rotation followed by a reflection through a plane perpendicular to that rotation axis. We are concerned mainly with the first three symmetry elements. The main point groups are exemplified in Fig. 2.20.

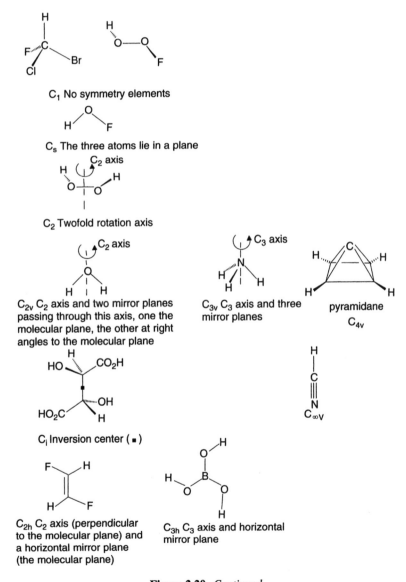

Figure 2.20. *Continued*

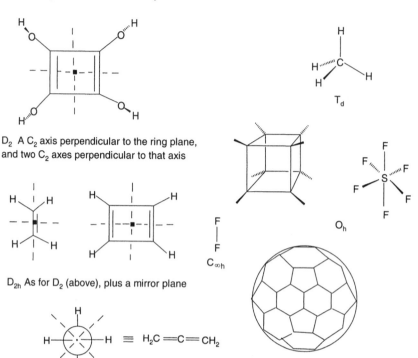

D$_2$ A C$_2$ axis perpendicular to the ring plane, and two C$_2$ axes perpendicular to that axis

D$_{2h}$ As for D$_2$ (above), plus a mirror plane

H_2C=C=CH_2

D$_{2d}$ As for D$_2$ (above), plus two dihedral mirror planes: the two planes that contain the HCH groups bisect the two axes that are shown as dashed lines (the third axis passes through CCC).

buckminsterfullerene, C$_{60}$
The bonds are of two kinds, with lengths ca. 1.39 and 1.46 Å

Figure 2.20. Examples of molecules with various symmetry elements (belonging to various point groups).

C$_1$ A molecule with no symmetry elements at all is said to belong to the group C$_1$ (to have "C$_1$ symmetry"). The only symmetry operation such a molecule permits is the null operation – this is the only operation that leaves it unmoved. An example is CHBrClF, with a so-called asymmetric atom; in fact, *most* molecules have no symmetry – just think of steroids, alkaloids, proteins, most drugs. Note that a molecule does not need an "asymmetric atom" to have C$_1$ symmetry: HOOF in the conformation shown is C$_1$ (has no symmetry).

C$_s$ A molecule with only a mirror plane belongs to the group C$_s$. Example: HOF. Reflection in this plane leaves the molecule apparently unmoved.

C$_2$ A molecule with only a C$_2$ axis belongs to the group C$_2$. Example: H$_2$O$_2$ in the conformation shown. Rotation about this axis through 360° gives the same orientation twice. Similarly C$_3$, C$_4$, etc. are possible.

C$_{2v}$ A molecule with two mirror planes whose intersection forms a C$_2$ axis belongs to the C$_{2v}$ group. Example: H$_2$O. Similarly NH$_3$ is C$_{3v}$, pyramidane is C$_{4v}$, and HCN is C$_{\infty v}$.

C_i   A molecule with only an inversion center (center of symmetry) belongs to the group C_i. Example: *meso*-tartaric acid in the conformation shown. Moving any point in the molecule along a straight line to this center, then continuing on an equal distance leaves the molecule apparently unchanged.

C_{2h}   A molecule with a C_2 axis and a mirror plane horizontal to this axis is C_{2h} (a C_{2h} object will also perforce have an inversion center). Example: (*E*)-1,2-difluoroethene. Similarly $B(OH)3$ is C_{3h}.

D_2   A molecule with a C_2 axis and two more C_2 axes, perpendicular to that axis, has D_2 symmetry. Example: the tetrahydroxycyclobutadiene shown. Similarly, a molecule with a C_3 axis (the *principal axis*) and three other perpendicular C_2 axes is D_3.

D_{2h}   A molecule with a C_2 axis and two perpendicular C_2 axes (as for D_2 above), plus a mirror plane is D_{2h}. Examples: ethene, cyclobutadiene. Similarly, a C_3 axis (the principal axis), three perpendicular C_2 axes and a mirror plane horizontal to the principal axis confer D_{3h} symmetry, as in the cyclopropenyl cation. Similarly, benzene is D_{6h}, and F_2 is $D_{\infty h}$.

D_{2d}   A molecule is D_{2d} if it has a C_2 axis and two perpendicular C_2 axes (as for D_2 above), plus two "dihedral" mirror planes; these are mirror planes that bisect two C_2 axes (in general, that bisect the C_2 axes perpendicular to the principal axis). Example: allene (propadiene). Staggered ethane is D_{3d} (it has D_3 symmetry elements plus three dihedral mirror planes. D_{nd} symmetry can be hard to spot.

Molecules belonging to the *cubic point groups* can, in some sense, be fitted symmetrically inside a cube. The commonest of these are T_d, O_h and I; they will be simply exemplified:

T_d   This is tetrahedral symmetry. Example: CH_4.

O_h   This might be considered "cubic symmetry". Example: cubane, SF_6.

I   Also called icosahedral symmetry. Example: buckminsterfullerene.

Less-common groups are S_4, and the cubic groups T, T_h (dodecahedrane is T_h) and O (see [18,19]). Atkins [18] and Levine [19] give flow charts which make it relatively simple to assign a molecule to its point group, and Atkins provides pictures of objects of various symmetries which often make it possible to assign a point group without having to examine the molecule for its symmetry elements.

We saw above that most molecules have no symmetry. So why is a knowledge of symmetry important in chemistry? Symmetry considerations are essential in the theory of molecular electronic (UV) spectroscopy and sometimes in analyzing in detail molecular wavefunctions (chapter 4), but for us the reasons are more pragmatic. A calculation run on a molecule whose input structure has the exact symmetry that the molecule should have will tend to be faster and will yield a "better" (see below) geometry than one run on an approximate structure, however close this may be to the exact one. Input molecular structures for a calculation are usually created with an interactive graphical program and a computer mouse: atoms are assembled into molecules much as with a model kit, or the molecule might be drawn on the computer screen. If the molecule has symmetry (if it is not is not C_1) this can be imposed by optimizing the geometry with molecular mechanics (chapter 3). Now consider water: we would of course normally input the H_2O molecule with its exact equilibrium C_{2v} symmetry, but we could also alter the input structure slightly making the symmetry C_s (three atoms must lie in a plane).

The C_{2v} structure has two degrees of freedom: a bond length (the two bonds are the same length) and a bond angle. The C_s structure has three degrees of freedom: two bond lengths and a bond angle. The optimization algorithm has more variables to cope with in the case of the lower-symmetry structure. A moderately high-level geometry optimization and frequencies job on C_{2v} $(CH_3)_2O$, dimethyl ether, took 5.7 min, but on the C_s ether 6.8 min (actually, small molecules like water, and low-level calculations, show a levelling effect, taking only seconds and requiring about the same time regardless of symmetry).

What do we mean by a better geometry? Although a successful geometry optimization will give essentially the same geometry from a slightly distorted input structure as from one with the perfect symmetry of the molecule in question, corresponding bond lengths and angles (e.g. the four C–H bonds and the two HCH angles of ethene) will not be *exactly* the same. This can confuse an analysis of the geometry, and carries over into the calculation of other properties like, say, charges on atoms – corresponding atoms should have exactly the same charges. Thus both esthetic and practical considerations encourage us to aim for the exact symmetry that the molecule should possess.

2.7   SUMMARY OF CHAPTER 2

The PES is a central concept in computational chemistry. A PES is the relationship – mathematical or graphical – between the energy of a molecule (or a collection of molecules) and its geometry.

Stationary points on a PES are points where $\partial E/\partial q = 0$ for all q, where q is a geometric parameter. The stationary points of chemical interest are minima ($\partial^2 E/\partial q_i q_j > 0$ for all q) and transition states or first-order saddle points; $\partial^2 E/\partial q_i q_j < 0$ for one q, along the reaction coordinate (IRC), and > 0 for all other q. Chemistry is the study of PES stationary points and the pathways connecting them.

The Born–Oppenheimer approximation says that in a molecule the nuclei are essentially stationary compared to the electrons. This is one of the cornerstones of computational chemistry because it makes the concept of molecular shape (geometry) meaningful, makes possible the concept of a PES, and simplifies the application of the Schrödinger equation to molecules by allowing us to focus on the electronic energy and add in the nuclear repulsion energy later.

Geometry optimization is the process of starting with an input structure "guess" and finding a stationary point on the PES. The stationary point found will normally be the one closest to the input structure, not necessarily the global minimum. A transition state optimization usually requires a special algorithm, since it is more demanding than that required to find a minimum. Modern optimization algorithms use analytic first derivatives and (usually numerical) second derivatives.

It is usually wise to check that a stationary point is the desired species (a minimum or a transition state) by calculating its vibrational spectrum (its normal-mode vibrations). The algorithm for this works by calculating an accurate Hessian (force constant matrix) and diagonalizing it to give a matrix with the "direction vectors" of the normal modes, and a diagonal matrix with the force constants of these modes. A procedure of "mass-weighting" the force constants gives the normal-mode vibrational frequencies. For a minimum all the vibrations are real, while a transition state has one imaginary vibration, corresponding to motion along the reaction coordinate. The criteria for a transition state are appearance, the presence of one imaginary frequency corresponding to the reaction coordinate, and an energy above that of the reactant and the product. Besides serving to characterize the stationary point, calculation of the vibrational frequencies enables one to predict an IR spectrum and provides the ZPE. The ZPE is needed for accurate comparisons of the energies of isomeric species. The accurate Hessian required for calculation of frequencies and ZPE's can be obtained either numerically or analytically (faster, but much more demanding of hard drive space).

REFERENCES

[1] (a) S. S. Shaik, H. B. Schlegel, and S. Wolfe, "Theoretical Aspects of Physical Organic Chemistry: the S_N2 Mechanism," Wiley, New York, 1992. See particularly Introduction and chapters 1 and 2. (b) R. A. Marcus, Science, 1992, 256, 1523. (c) For a very abstract and mathematical but interesting treatment, see P. G. Mezey, "Potential Energy Hypersurfaces," Elsevier, New York, 1987. (d) J. I. Steinfeld, J. S. Francisco, and W. L. Hase, "Chemical Kinetics and Dynamics," 2nd edn., Prentice Hall, Upper Saddle River, New Jersey, 1999.

[2] I. N. Levine, "Quantum Chemistry," 5th edn., Prentice Hall, Upper Saddle River, NJ, 2000, section 4.3.

[3] Reference [1a], pp. 50–51.

[4] K. N. Houk, Y. Li, and J. D. Evanseck, Angew. Chem. Int. Ed. Engl., 1992, *31*, 682.

[5] P. Atkins, "Physical Chemistry," 6th edn, Freeman, New York, 1998, pp. 830–844.

[6] R. Marcelin, Annales de Physique, 1915, 3, 152. Potential energy surface: p. 158.

[7] H. Eyring, J. Chem Phys., 1935, *3*, 107.

[8] H. Eyring and M. Polanyi, Z. Physik Chem., 1931, *B*, *12*, 279.

[9] M. Born and J. R. Oppenheimer, Ann. Physik., 1927, *84*, 457.

[10] A standard molecular surface, corresponding to the size as determined experimentally (e.g. by X-ray diffraction) encloses about 98 per cent of the electron density. See e.g. R. F. W. Bader, M. T. Carroll, M. T. Cheeseman, and C. Chang, J. Am. Chem. Soc., 1987, *109*, 7968.

[11] For some rarefied but interesting ideas about molecular shape see P. G Mezey, "Shape in Chemistry," VCH, New York, 1993.

[12] X. K. Zhang, J. M. Parnis, E. G. Lewars, and R. E. March, Can. J. Chem., 1997, *75*, 276.

[13] See e.g. (a) A. R. Leach, "Molecular Modelling. Principles and Applications," Longman, Essex, UK, 1996, chapter 4. (b) F. Jensen, "Introduction to Computational Chemistry," Wiley, New York, 1999, chapter 14.

[14] W. J. Hehre, "Practical Strategies for Electronic Structure Calculations," Wavefunction Inc., Irvine, CA, 1995, p. 9.

[15] I. N. Levine, "Quantum Chemistry," 5th edn, Prentice Hall, Upper Saddle River, NJ, 2000, p. 65.

[16] A. P. Scott and L. Radom, J. Phys. Chem., 1996, *100*, 16502.

[17] J. B. Foresman and Æ. Frisch, "Exploring Chemistry with Electronic Structure Methods," 2nd edn., Gaussian Inc., Pittsburgh, PA, 1996, pp. 173–211.

[18] P. Atkins, "Physical Chemistry," 6th edn, Freeman, New York, 1998, chapter 15.

[19] I. N. Levine, "Quantum Chemistry," 5th edn, Prentice Hall, Upper Saddle River, NJ, 2000, chapter 12.

EASIER QUESTIONS

1. What is a PES (give the two viewpoints)?
2. Explain the difference between a relaxed PES and a rigid PES.
3. What is a stationary point? What kinds of stationary points are of interest to chemists, and how do they differ?
4. What is a reaction coordinate?
5. Show with a sketch why it is not correct to say that a transition state is a maximum on a PES.
6. What is the Born-Oppenheimer approximation, and why is it important?
7. Explain, for a reaction A → B, how the potential energy change on a PES is related to the enthalpy change of the reaction. What would be the problem with calculating a free energy/geometry surface?
   Hint: Vibrational frequencies are normally calculated only for stationary points.
8. What is geometry optimization? Why is this process for transition states (often called transition state optimization) more challenging than for minima?
9. What is a Hessian? What uses does it have in computational chemistry?
10. Why is it usually good practice to calculate vibrational frequencies where practical, although this often takes considerably longer than geometry optimization?

HARDER QUESTIONS

1. The Born–Oppenheimer principle is often said to be a prerequisite for the concept of a PES. Yet the idea of a PES (Marcelin, 1915) predates the Born–Oppenheimer principle (1927). Discuss.
2. How high would you have to lift a mole of water for its gravitational potential energy to be equivalent to the energy needed to dissociate it completely into hydroxyl radicals and hydrogen atoms? The strength of the O–H bond is about $400 \, kJ \, mol^{-1}$; the gravitational acceleration g at the Earth's surface (and out to hundreds of km) is about $10 \, m \, s^{-2}$. What does this indicate about the role of gravity in chemistry?
3. If gravity plays no role in chemistry, why are vibrational frequencies different for, say, C–H and C–D bonds?
4. We assumed that the two bond lengths of water are equal. *Must* an acyclic molecule AB_2 have equal A–B bond lengths? What about a cyclic molecule AB_2?
5. Why are chemists but rarely interested in finding and characterizing second-order and higher saddle points (hilltops)?
6. What kind(s) of stationary points do you think a second-order saddle point connects?
7. If a species has one calculated frequency very close to $0 \, cm^{-1}$ what does that tell you about the (calculated) PES in that region?

8. The ZPE of many molecules is greater than the energy needed to break a bond; e.g. the ZPE of hexane is about $530 \, kJ \, mol^{-1}$, while the strength of a C–C or a C–H bond is only about $400 \, kJ \, mol^{-1}$. Why then do such molecules not spontaneously decompose?

9. Only certain parts of a PES are chemically interesting: some regions are flat and featureless, while yet other parts rise steeply and are thus energetically inaccessible. Explain.

10. Consider two PESs for the HCN $\rightleftharpoons$ reaction: *A*, a plot of energy vs. the H–C bond length, and *B*, a plot of energy vs. the HCN angle. Recalling that HNC is the higher-energy species (Fig. 2.19), sketch qualitatively the diagrams *A* and *B*.

Chapter 3

Molecular Mechanics

We don't give a damn where the electrons are.
Words to the author, from the president of a
well-known chemical company, emphasizing
his firm's position on basic research.

3.1 PERSPECTIVE

Molecular mechanics (MM) [1] is based on a mathematical model of a molecule as a collection of balls (corresponding to the atoms) held together by springs (corresponding to the bonds) (Fig. 3.1). Within the framework of this model, the energy of the molecule changes with geometry because the springs resist being stretched or bent away from some "natural" length or angle, and the balls resist being pushed too closely together. The *mathematical* model is thus conceptually very close to the intuitive feel for molecular energetics that one obtains when manipulating molecular models of plastic or metal: the model resists distortions (it may break!) from the "natural" geometry that corresponds to the bond lengths and angles imposed by the manufacturer, and in the case of space-filling models, atoms cannot be forced too closely together. The MM model clearly ignores electrons.

The principle behind MM is to express the energy of a molecule as a function of its resistance toward bond stretching, bond bending, and atom crowding, and to use this energy equation to find the bond lengths, angles, and dihedrals corresponding to the minimum-energy geometry – or more precisely, to the various possible potential energy surface minima (chapter 2). In other words, MM uses a conceptually *mechanical* model of a molecule to find its minimum-energy geometry (for flexible molecules, the geometries of the various conformers). The form of the mathematical expression for the energy, and the parameters in it, constitute a *forcefield*, and MM methods are sometimes called forcefield methods. The term arises because the negative of the first derivative of the potential energy of a particle with respect to displacement along some direction is the force on the particle; a "forcefield" $E(x, y, z$ coordinates of atoms) can be differentiated to give the force on each atom.

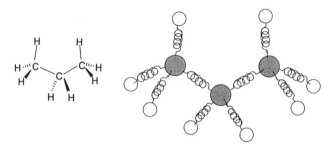

Figure 3.1. Molecular mechanics (the forcefield method) considers a molecule to be a collection of balls (the atoms) held together by springs (the bonds).

The method makes no reference to electrons, and so cannot (except by some kind of empirical algorithm) throw light on *electronic* properties like charge distributions or nucleophilic and electrophilic behaviour. Note that MM implicitly uses the Born–Oppenheimer approximation, for only if the nuclei experience what amounts to a static attractive force, whether from electrons or springs, does a molecule have a distinct geometry (section 2.3).

An important point, which students sometimes have a problem with, is that the concept of a *bond* is central to MM, but not essential – although often useful – in *electronic* structure calculations. In MM a molecule is defined by the atoms and the bonds, which latter are regarded almost literally as springs holding the atoms together. Usually, bonds are placed where the rules for drawing structural formulas require them, and to do a MM calculation you must specify each bond as single, double, etc., since this tells the program how strong a bond to use (sections 3.2.1 and 3.2.2). In an electronic structure calculation – *ab initio* (chapter 5), semiempirical (SE) (chapter 6), and density functional theory (chapter 7) – a molecule is defined by the relative positions of its atomic nuclei, the charge, and the "multiplicity" (which follows easily from the number of unpaired electrons). An oxygen nucleus and two protons with the right x, y, z coordinates, no charge, and multiplicity one (no unpaired electrons) is a water molecule. There is no need to mention bonds here, although the chemist might wish to somehow extract this useful concept from this picture of nuclei and electrons. This can be done by calculating the electron density and associating a bond with, for example, a path along which electron density is concentrated, but there is no unique definition of a bond in electronic structure theory. It is worth noting, too, that in some graphical interfaces used in computational chemistry bonds are specified by the user, while in others they are shown by the program depending on the separation of pairs of atoms. The novice may find it disconcerting to see a specified bond still displayed even when a change in geometry has moved a pair of atoms far apart, or to see a bond vanish when a pair has moved beyond the distance recognized by some fudge factor.

Historically [2], MM seems to have begun as an attempt to obtain quantitative information about chemical reactions at a time when the possibility of doing quantitative quantum mechanical (chapter 4) calculations on anything much bigger than the hydrogen molecule seemed remote. Specifically, the principles of MM, as a potentially general method for studying the variation of the energy of molecular systems with their geometry, were formulated in 1946 by Westheimer[1] and Meyer [3a], and by Hill [3b]. In this same year Dostrovsky, Hughes[2] and Ingold[3] independently applied MM concepts of to the quantitative analysis of the S_N2 reaction, but they do not seem to have recognized the potentially wide applicability of this approach [3c]. In 1947 Westheimer [3d] published detailed calculations in which MM was used to estimate the activation energy for the racemization of biphenyls.

Major contributors to the development of MM have been Schleyer[4] [2b,c] and Allinger[5] [1c,d]; one of Allinger's publications on MM [1d] is, according to the Citation Index, one of the most frequently cited chemistry papers. The Allinger group has, since the 1960s, been responsible for the development of the "MM-series" of programs, commencing with MM1 and continuing with the currently widely-used MM2 and MM3, and the recent MM4 [4]. MM programs [5] like Sybyl and UFF will handle molecules involving much of the periodic table, albeit with some loss of accuracy that one might expect for trading breadth for depth, and MM is the most widely-used method for computing the geometries and energies of large biological molecules like proteins and nucleic acids (although recently SE (chapter 6) and even *ab initio* (chapter 5) methods have begun to be applied to these large molecules).

3.2   THE BASIC PRINCIPLES OF MM

3.2.1   Developing a forcefield

The potential energy of a molecule can be written

$$E = \sum_{\text{bonds}} E_{\text{stretch}} + \sum_{\text{angles}} E_{\text{bend}} + \sum_{\text{dihedrals}} E_{\text{torsion}} + \sum_{\text{pairs}} E_{\text{nonbond}} \quad (3.1)$$

where E_{stretch} etc. are energy contributions from bond stretching, angle bending, torsional motion (rotation) around single bonds, and interactions between atoms or groups which are nonbonded (not directly bonded together). The sums are over all the bonds, all the angles defined by three atoms A−B−C, all the dihedral angles defined by four atoms A−B−C−D, and all pairs of significant nonbonded interactions. The mathematical form of these terms and the parameters in them constitute a particular forcefield. We can make this clear by being more specific; let us consider each of these four terms.

The bond stretching term. The increase in the energy of a spring (remember that we are modelling the molecule as a collection of balls held together by springs) when it is

---

[1] Frank H. Westheimer, born Baltimore, Maryland, 1912. Ph.D. Harvard 1935. Professor University of Chicago, Harvard.

[2] Edward D. Hughes, born Wales, 1906. Ph.D. University of Wales, D.Sc. University of London. Professor, London. Died 1963.

[3] Christopher K. Ingold, born London 1893. D.Sc. London 1921. Professor Leeds, London. Knighted 1958. Died London 1970.

[4] Paul von R. Schleyer, born Cleveland, Ohio, 1930. Ph.D. Harvard 1957. Professor Princeton; institute codirector and professor University of Erlangen-Nürnberg, 1976–1998. Professor University of Georgia.

[5] Norman L. Allinger, born Rochester New York, 1930. Ph.D. University of California at Los Angeles, 1954. Professor Wayne State University, University of Georgia.

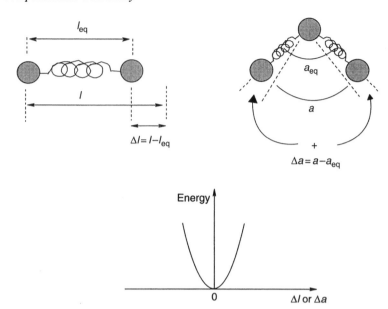

Figure 3.2. Changes in bond lengths or in bond angles result in changes in the energy of a molecule. Such changes are handled by the $E_{stretch}$ and E_{bend} terms in the MM forcefield. The energy is approximately a quadratic function of the change in bond length or angle.

stretched (Fig. 3.2) is approximately proportional to the square of the extension:

$$\Delta E_{stretch} = k_{stretch}(l - l_{eq})^2$$

where $k_{stretch}$ is the proportionality constant (actually one-half the *force constant* of the spring or bond [6]; but note the warning about identifying MM force constants with the traditional force constant from, say, spectroscopy – see section 3.3); the bigger $k_{stretch}$, the stiffer the bond/spring – the more it resists being stretched; l is the length of the bond when stretched; and l_{eq} is the equilibrium length of the bond, its "natural" length.

If we take the energy corresponding to the equilibrium length l_{eq} as the zero of energy, we can replace $\Delta E_{stretch}$ by $E_{stretch}$:

$$E_{stretch} = k_{stretch}(l - l_{eq})^2 \tag{3.2}$$

The angle bending term. The increase in energy of system ball–spring–ball–spring–ball, corresponding to the triatomic unit A–B–C (the increase in "angle energy") is approximately proportional to the square of the increase in the angle (Fig. 3.2); analogously to Eq. (3.2):

$$E_{bend} = k_{bend}(a - a_{eq})^2 \tag{3.3}$$

where k_{bend} is a proportionality constant (one-half the angle bending force constant [6]; note the warning about identifying MM force constants with the traditional force constant from, say, spectroscopy – see section 3.3); a is the size of the angle when distorted; and a_{eq} is the equilibrium size of the angle, its "natural" value.

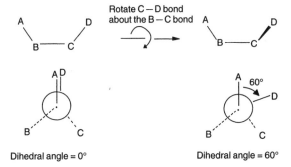

Figure 3.3. Dihedral angles (torsional angles) affect molecular geometries and energies. The energy is a periodic (cosine or combination of cosine functions) function of the dihedral angle; see, for example, Figs. 3.4 and 3.5.

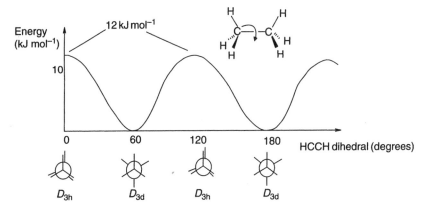

Figure 3.4. Variation of the energy of ethane with dihedral angle. The curve can be represented as a cosine function.

The torsional term. Consider four atoms sequentially bonded: A–B–C–D (Fig. 3.3). The dihedral angle or torsional angle of the system is the angle between the A–B bond and the C–D bond as viewed along the B–C bond. Conventionally this angle is considered positive if regarded as arising from clockwise rotation (starting with A–B covering or eclipsing C–D) of the back bond (C–D) with respect to the front bond (A–B). Thus in Fig. 3.3 the dihedral angle A–B–C–D is 60° (it could also be considered as being −300°). Since the geometry repeats itself every 360°, the energy varies with the dihedral angle in a sine or cosine pattern, as shown in Fig. 3.4 for the simple case of ethane. For systems A–B–C–D of lower symmetry, like butane (Fig. 3.5), the torsional potential energy curve is more complicated, but a *combination* of sine or cosine functions will reproduce the curve:

$$E_{torsion} = k_0 + \sum_{r=1}^{n} k_r[1 + \cos(r\theta)] \tag{3.4}$$

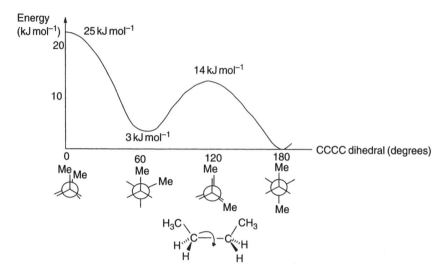

Figure 3.5. Variation of the energy of butane with dihedral angle. The curve can be represented by a sum of cosine functions.

The nonbonded interactions term. This represents the change in potential energy with distance apart of atoms A and B that are not directly bonded (as in A–B) and are not bonded to a common atom (as in A–X–B); these atoms, separated by at least two atoms (A–X–Y–B) or even in different molecules, are said to be nonbonded (with respect to each other). Note that the A–B case is accounted for by the bond stretching term $E_{stretch}$, and the A–X–B term by the angle bending term E_{bend}, but the nonbonded term $E_{nonbond}$ is, for the A–X–Y–B case, superimposed upon the torsional term $E_{torsion}$: we can think of $E_{torsion}$ as representing some factor inherent to resistance to rotation about a (usually single) bond X–Y (MM does not attempt to explain the theoretical, electronic basis of this or any other effect), while for certain atoms attached to X and Y there may also be nonbonded interactions.

The potential energy curve for two nonpolar nonbonded atoms has the general form shown in Fig. 3.6. A simple way to approximate this is by the so-called Lennard–Jones 12-6 potential [7]:

$$E_{nonbond} = k_{nb} \left[\left(\frac{\sigma}{r} \right)^{12} - \left(\frac{\sigma}{r} \right)^{6} \right] \tag{3.5}$$

where r is the distance between the centers of the nonbonded atoms or groups.

The function reproduces the small attractive dip in the curve (represented by the negative term) as the atoms or groups approach one another, then the very steep rise in potential energy (represented by the raising the positive, repulsive term raised to a large power) as they are pushed together closer than their van der Waals radii. Setting $dE/dr = 0$, we find that for the energy minimum in the curve the corresponding value of r is $r_{min} = 2^{1/6}\sigma$, i.e.

$$\sigma = 2^{-1/6} r_{min} \tag{3.6}$$

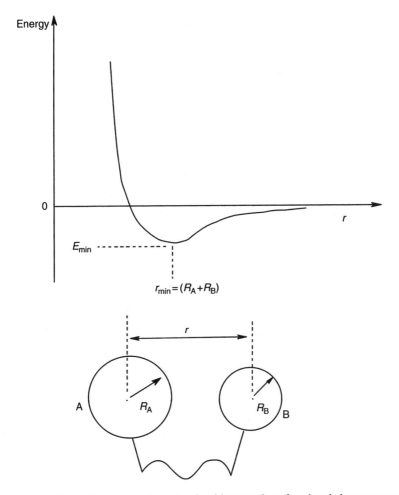

Figure 3.6. Variation of the energy of a molecule with separation of nonbonded atoms or groups. Atoms/groups A and B may be in the same molecule (as indicated here) or the interaction may be intermolecular. The minimum energy occurs at van der Waals contact. For small nonpolar atoms or groups the minimum energy point represents a drop of a few kJ mol^{-1} ($E_{min} = -1.2$ kJ mol^{-1} for CH_4/CH_4), but short distances can make nonbonded interactions destabilize a molecule by many kJ mol^{-1}.

If we assume that this minimum corresponds to van der Waals contact of the nonbonded groups, then $r_{min} = (R_A + R_B)$, the sum of the van der Waals radii of the groups A and B. So

$$2^{1/6}\sigma = (R_A + R_B)$$

and so

$$\sigma = 2^{-1/6}(R_A + R_B) = 0.89(R_A + R_B) \tag{3.7}$$

Thus σ can be calculated from r_{min} or estimated from the van der Waals radii.

Setting $E = 0$, we find that for this point on the curve $r = \sigma$, i.e.

$$\sigma = r(E = 0) \tag{3.8}$$

If we set $r = r_{min} = 2^{1/6}\sigma$ (from Eq. (3.6)) in Eq. (3.5), we find

$$E(r = r_{min}) = (-1/4)k_{nb}$$

i.e.

$$k_{nb} = -4E(r = r_{min}) \tag{3.9}$$

So k_{nb} can be calculated from the depth of the energy minimum.

In deciding to use equations of the form (3.2)–(3.5) we have decided on a particular MM forcefield. There are many alternative forcefields. For example, we might have chosen to approximate $E_{stretch}$ by the sum of a quadratic and a cubic term:

$$E_{stretch} = k_{stretch}(l - l_{eq})^2 + k(l - l_{eq})^3$$

This gives a somewhat more accurate representation of the variation of energy with length. Again, we might have represented the nonbonded interaction energy by a more complicated expression than the simple 12-6 potential of Eq. (3.5) (which is by no means the best form for nonbonded repulsions). Such changes would represent changes in the forcefield.

3.2.2   Parameterizing a forcefield

We can now consider putting actual numbers, $k_{stretch}$, l_{eq}, k_{bend}, etc., into Eqs (3.2)–(3.5), to give expressions that we can actually use. The process of finding these numbers is called *parameterizing* (or parametrizing) the forcefield. The set of molecules used for parameterization, perhaps 100 for a good forcefield, is called the *training set*. In the purely illustrative example below we use just ethane, methane and butane.

Parameterizing the bond stretching term. A forcefield can be parameterized by reference to experiment (empirical parameterization) or by getting the numbers from high-level *ab initio* or density functional calculations, or by a combination of both approaches. For the bond stretching term of Eq. (3.2) we need $k_{stretch}$ and l_{eq}. Experimentally, $k_{stretch}$ could be obtained from IR spectra, as the stretching frequency of a bond depends on the force constant (and the masses of the atoms involved) [8], and l_{eq} could be derived from X-ray diffraction, electron diffraction, or microwave spectroscopy [9].

Let us find $k_{stretch}$ for the C/C bond of ethane by *ab initio* (chapter 5) calculations. Normally high-level *ab initio* calculations would be used to parameterize a forcefield, but for illustrative purposes we can use the low-level but fast STO-3G method [10]. Eq. (3.2) shows that a plot of $E_{stretch}$ against $(l - l_{eq})^2$ should be linear with a slope of $k_{stretch}$. Table 3.1 and Fig. 3.7 show the variation of the energy of ethane with stretching of the C/C bond, as calculated by the *ab initio* STO-3G method. The equilibrium bond length has been taken as the STO-3G length:

$$l_{eq}(C-C) = 1.538 \, \text{Å} \tag{3.10}$$

Table 3.1. Change in energy as the C–C bond in CH_3–CH_3 is stretched away from its equilibrium length

C–C length, l	$l - l_{eq}$	$(l - l_{eq})^2$	$E_{stretch}$ (kJ mol^{-1})
1.538	0	0	0
1.550	0.012	0.00014	0.29
1.560	0.022	0.00048	0.89
1.570	0.032	0.00102	1.86
1.580	0.042	0.00176	3.15
1.590	0.052	0.00270	4.75
1.600	0.062	0.00384	6.67

The calculations are *ab initio* (STO-3G; chapter 5). Bond lengths are in Å.

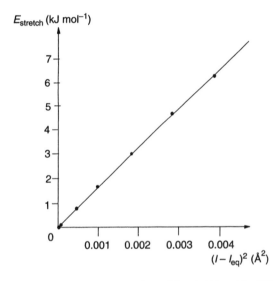

Figure 3.7. Energy vs. the square of the extension of the C–C bond in CH_3–CH_3. The data in Table 3.1 were used.

The slope of the graph is

$$k_{stretch}(\text{C–C}) = 1735 \text{ kJ mol}^{-1}\text{Å}^{-2} \tag{3.11}$$

Similarly, the CH bond of methane was stretched using *ab initio* STO-3G calculations; the results are

$$l_{eq}(\text{C–H}) = 1.083 \text{ Å} \tag{3.12}$$

$$k_{stretch}(\text{C–H}) = 1934 \text{ kJ mol}^{-1}\text{Å}^{-2} \tag{3.13}$$

Parameterizing the angle bending term. From Eq. (3.3), a plot of E_{bend} against $(a - a_{eq})^2$ should be linear with a slope of k_{bend}. From STO-3G calculations on bending the H–C–C angle in ethane we get (cf. Table 3.1 and Fig. 3.7)

$$a_{eq}(HCC) = 110.7° \tag{3.14}$$

$$k_{bend}(HCC) = 0.093 \, kJ \, mol^{-1} \, degree^{-2} \tag{3.15}$$

Calculations on staggered butane gave for the C–C–C angle

$$a_{eq}(CCC) = 112.5° \tag{3.16}$$

$$k_{bend}(CCC) = 0.110 \, kJ \, mol^{-1} \, degree^{-2} \tag{3.17}$$

Parameterizing the torsional term. For the ethane case (Fig. 3.4), the equation for energy as a function of dihedral angle can be deduced fairly simply by adjusting the basic equation $E = \cos\theta$ to give $E = 1/2 E_{max}[1+\cos 3(\theta+60)]$. For butane (Fig. 3.5), using Eq. (3.4) and experimenting with a curve-fitting program shows that a reasonably accurate torsional potential energy function can be created with five parameters, k_0 and $k_1 - k_4$:

$$E_{torsion}(CH_3CH_2-CH_2CH_3) = k_0 + \sum_{r=1}^{4} k_r[1 + \cos(r\theta)] \tag{3.18}$$

The values of the parameters $k_0 - k_5$ are given in Table 3.2. The calculated curve can be made to match the experimental one as closely as desired by using more terms (Fourier analysis).

Parameterizing the nonbonded interactions term. To parameterize Eq. (3.5) we might perform *ab initio* calculations in which the separation of two atoms or groups in different

Table 3.2. The experimental potential energy values for rotation about the central C–C bond of $CH_3CH_2-CH_2CH_3$ can be approximated by

$$E_{torsion}(CH_3CH_2-CH_2CH_3) = k_0 + \sum_{r=1}^{4} k_r[1 + \cos(r\theta)]$$

with $k_0 = 20.1$, $k_1 = -4.7$, $k_2 = 1.91$, $k_3 = -7.75$, $k_4 = 0.58$. Experimental energy values at 30°, 90°, and 150° were interpolated from those at 0°, 60°, 120°, and 180°; energies are in kJ mol^{-1}

θ (degrees)	E (calculated)	E (experimental)
0	0.15	0
30	6.7	7.0
60	14	14
90	8.8	9.0
120	3.5	3.3
150	15	15
180	25	25

molecules (to avoid the complication of concomitant changes in bond lengths and angles) is varied, and fit Eq. (3.5) to the energy vs. distance results. For nonpolar groups this would require quite high-level calculations (chapter 5), as van der Waals or dispersion forces are involved. We shall approximate the nonbonded interactions of methyl groups by the interactions of methane molecules, using experimental values of k_{nb} and σ, derived from studies of the viscosity or the compressibility of methane. The two methods give slightly different values [7b], but we can use the values

$$k_{nb} = 4.7 \, \text{kJ mol}^{-1} \tag{3.19}$$

and

$$\sigma = 3.85 \, \text{Å} \tag{3.20}$$

Summary of the parameterization of the forcefield terms. The four terms of Eq. (3.1) were parameterized to give:

$$E_{stretch}(C–C) = 1735(l - 1.538)^2 \tag{3.21}$$

$$E_{stretch}(C–H) = 1934(l - 1.083)^2 \tag{3.22}$$

$$E_{bend}(HCH) = 0.093(a - 110.7)^2 \tag{3.23}$$

$$E_{bend}(CCC) = 0.110(a - 112.5)^2 \tag{3.24}$$

$$E_{torsion}(CH_3CCCH_3) = k_0 + \sum_{r=1}^{4} k_r[1 + \cos(r\theta)] \tag{3.25}$$

The parameters k of Eq. (3.25) are given in Table 3.2.

$$E_{nonbond}(CH_3/CH_3) = 4.7 \left[\left(\frac{3.85}{r} \right)^{12} - \left(\frac{3.85}{r} \right)^6 \right] \tag{3.26}$$

Note that this parameterization is only illustrative of the principles involved; any really viable forcefield would actually be much more sophisticated. The kind we have developed here might at the very best give crude estimates of the energies of alkanes. An accurate, practical forcefield would be parameterized as a best fit to many experimental and/or calculational results, and would have different parameters for different kinds of bonds, e.g. C–C for acyclic alkanes, for cyclobutane and for cyclopropane. A forcefield able to handle not only hydrocarbons would obviously need parameters involving elements other than hydrogen and carbon. Practical forcefields also have different parameters for various *atom types*, like sp^3 carbon vs. sp^2 carbon, or amine nitrogen vs. amide nitrogen. In other words, a different value would be used for, say, stretching involving an sp^3/sp^3 C–C bond than for an sp^2/sp^2 C–C bond. This is clearly necessary since the force constant of a bond depends on the hybridization of the atoms involved; the IR stretch frequency for the sp^3C/sp^3C bond comes at roughly 1200 cm^{-1}, while that for the sp^2C/sp^2C bond is about 1650 cm^{-1} [8]. Since the vibrational frequency of a bond is proportional to the square root of the force constant, the force constants are in the ratio of about $(1650/1200)^2 = 1.9$; for corresponding atoms, force constants are in fact generally roughly proportional to bond order (double bonds and triple bonds are about two and three times as stiff, respectively, as the corresponding single bonds). Some forcefields account for the variation of bond order with conformation (twisting p orbitals out of alignment reduces their overlap) by performing a simple PPP molecular orbital calculation (chapter 6) to obtain the bond order.

A sophisticated forcefield might also consider H/H nonbonded interactions explicitly, rather than simply subsuming them into methyl/methyl interactions (combining atoms into groups is the feature of a *united atom* forcefield). Furthermore, nonbonding interactions between *polar* groups need to be accounted for in a field not limited to hydrocarbons. These are usually handled by the well-known potential energy/electrostatic charge relationship

$$E = k(q_1 q_2 / r)$$

which has also been used to model hydrogen bonding [11].

A subtler problem with the naive forcefield developed here is that stretching, bending, torsional, and nonbonded terms are not completely independent. For example, the butane torsional potential energy curve (Fig. 3.5) does not apply precisely to all CH_3–C–C–CH_3 systems, because the barrier heights will vary with the length of the central C–C bond, obviously decreasing (other things being equal) as the bond is lengthened, since there will be a decrease in the interactions (whatever causes them) between the CH_3's and H's on one of the carbons of the central C–C and those on the other carbon. This could be accounted for by making the k's of Eq. (3.25) a function of the X–Y length. Actually, partitioning the energy of a molecule into stretching, bending, etc. terms is somewhat formal; e.g. the torsional barrier in butane can be considered to be partly due to nonbonded interactions between the methyl groups. It should be realized that there is no one, right functional form for an MM forcefield (see, e.g. [1b]); accuracy, versatility, and speed of computation are the deciding factors in devising a forcefield.

3.2.3   A calculation using our forcefield

Let us apply the naive forcefield developed here to comparing the energies of two 2,2,3,3-tetramethylbutane ((CH_3)$_3$CC(CH_3)$_3$, i.e. t-Bu-Bu-t) geometries. We compare the energy of structure **1** (Fig. 3.8) with all the bond lengths and angles at our "natural" or standard values (i.e. at the STO-3G values we took as the equilibrium bond lengths and angles in section 3.2.2) with that of structure **2**, where the central C–C bond has been stretched from 1.538 to 1.600 Å, but all other bond lengths, as well as the bond angles and dihedral angles, are unchanged. Figure 3.8 shows the nonbonded distances we need, which would be calculated by the program from bond lengths, angles and dihedrals. Using Eq. (3.1):

$$\left(E = \sum_{bonds} E_{stretch} + \sum_{angles} E_{bend} + \sum_{dihedrals} E_{torsion} + \sum_{pairs} E_{nonbond} \right)$$

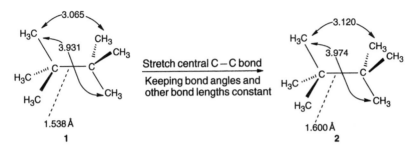

Figure 3.8. Structures for a simple MM "by hand" calculation on the effect of changing the central C–C length of $(CH_3)_3C–C(CH_3)_3$ from 1.538 to 1.600 Å.

For structure **1**

$$\sum_{\text{bonds}} E_{\text{stretch}}(\text{C–C}) = 7 \times 1735(1.538 - 1.538)^2 = 0$$

Bond stretch contribution cf. structure with $l_{\text{eq}} = 1.538$.

$$\sum_{\text{bonds}} E_{\text{stretch}}(\text{C–H}) = 18 \times 1934(1.083 - 1.083)^2 = 0$$

Bond stretch contribution cf. structure with $l_{\text{eq}} = 1.083$.

$$\sum_{\text{angles}} E_{\text{bend}}(\text{HCH}) = 18 \times 0.093(110.7 - 110.7)^2 = 0$$

Bond bend contribution cf. structure with $a_{\text{eq}} = 110.7°$.

$$\sum_{\text{angles}} E_{\text{bend}}(\text{CCC}) = 12 \times 0.110(112.5 - 112.5)^2 = 0$$

Bond bend contribution cf. structure with $a_{\text{eq}} = 112.5°$

$$\sum_{\text{dihedrals}} E_{\text{torsion}}(\text{CH}_3\text{CCCH}_3) = 6 \times 3.5 = 21.0$$

Torsional contribution cf. structure with no gauche–butane interactions

$$\sum_{\text{nonbond}} E_{\text{nonbond}}(\text{anti–CH}_3/\text{CH}_3) + \sum_{\text{nonbond}} E_{\text{nonbond}}(\text{gauche–CH}_3/\text{CH}_3)$$

$$= 3 \times 4.7\left[\left(\frac{3.85}{3.931}\right)^{12} - \left(\frac{3.85}{3.931}\right)^{6}\right] + 6 \times 4.7\left[\left(\frac{3.85}{3.065}\right)^{12} - \left(\frac{3.85}{3.065}\right)^{6}\right]$$

$$= 3 \times (-0.487) + 6 \times (54.05) = -1.463 + 324.3$$

$$= 323 \text{ kJ mol}^{-1}$$

nonbonding contribution cf. structure with noninteracting CH_3s.

Actually, nonbonding interactions are already included in the torsional term (as *gauche*–butane interactions); we might have used an ethane-type torsional function and accounted for CH_3/CH_3 interactions entirely with nonbonded terms. However, in comparing calculated *relative* energies the torsional term will cancel out.

$$E_{total} = E_{stretch} + E_{bend} + E_{torsion} = 0 + 0 + 21.0 + 323\,kJ\,mol^{-1} = 344\,kJ\,mol^{-1}$$

For structure **2**

$$\sum_{bonds} E_{stretch}(C\text{–}C) = 6 \times 1735(1.538 - 1.538)^2 + 1 \times 1735(1.600 - 1.538)^2$$

$$= 0 + 6.67 = 6.67\,kJ\,mol^{-1}$$

Bond stretch contribution cf. structure with $l_{eq} = 1.538$.

$$\sum_{bonds} E_{stretch}(C\text{–}H) = 18 \times 1934(1.083 - 1.083)^2 = 0$$

Bond stretch contribution cf. structure with $l_{eq} = 1.083$.

$$\sum_{angles} E_{bend}(HCH) = 18 \times 0.093(110.7 - 110.7)^2 = 0$$

Bond bend contribution cf. structure with $a_{eq} = 110.7°$

$$\sum_{angles} E_{bend}(CCC) = 12 \times 0.110(112.5 - 112.5)^2 = 0$$

Bond bend contribution cf. structure with $a_{eq} = 112.5°$

$$\sum_{dihedrals} E_{torsion}(CH_3CCCH_3) = 6 \times 3.5 = 21.0$$

Torsional contribution cf. structure with no gauche–butane interactions.

The stretching and bending terms for structure **2** are the same as for structure **1**, except for the contribution of the central C–C bond; strictly speaking, the torsional term should be smaller, since the opposing C(CH₃) groups have been moved apart.

$$\sum_{nonbond} E_{nonbond}(anti\text{–}CH_3/CH_3) + \sum_{nonbond} E_{nonbond}(gauche\text{–}CH_3/CH_3)$$

$$= 3 \times 4.7\left[\left(\frac{3.85}{3.974}\right)^{12} - \left(\frac{3.85}{3.974}\right)^6\right] + 6 \times 4.7\left[\left(\frac{3.85}{3.120}\right)^{12} - \left(\frac{3.85}{3.120}\right)^6\right]$$

$$= 3 \times (-0.673) + 6 \times (41.97) = -2.019 + 251.8$$

$$= 250\,kJ\,mol^{-1}$$

nonbonding contribution cf. structure with noninteracting CH₃s.

$$E_{total} = E_{stretch} + E_{bend} + E_{torsion} = 6.67 + 0 + 21.0 + 250 = 277\,kJ\,mol^{-1}$$

So the relative energies are calculated to be

$$E(\text{structure } \mathbf{2}) - E(\text{structure } \mathbf{1}) = 277 - 344 = -67\,kJ\,mol^{-1}$$

This crude method predicts that stretching the central C/C bond of 2,2,3,3-tetramethylbutane from the approximately normal $sp^3\text{-}C\text{-}sp^3\text{-}C$ length of 1.583 Å (structure **1**) to the quite "unnatural" length of 1.600 Å (structure **2**) will lower the potential energy by 67 kJ mol^{-1}, and indicates that the drop in energy is due very largely to the relief of nonbonded interactions. A calculation using the accurate forcefield MM3 [12] gave an energy difference of 54 kJ mol^{-1} between a "standard" geometry approximately like structure **1**, and a *fully optimized* geometry, which had a central C/C bond length of 1.576 Å. The surprisingly good agreement is largely the result of a fortuitous cancellation of errors, but this does not gainsay the fact that we have used our forcefield to calculate something of chemical interest, namely the relative energy of two molecular geometries. In principle, we could have found the minimum-energy geometry according to this forcefield, i.e. we could have optimized the geometry (chapter 2). Geometry optimization is in fact the main use of MM, and modern programs employ analytical first and second derivatives of the energy with respect to the geometric coordinates for this (chapter 2).

3.3 EXAMPLES OF THE USE OF MM

If we consider the applications of MM from the viewpoint of the goals of those who use it, then the *main* applications have been:

(1) to calculate the geometries (and perhaps energies) of small to medium-sized (i.e. nonpolymeric) molecules, very often in order to a reasonable starting geometry for another type (e.g. *ab initio*) of calculation;

(2) to calculate the geometries and energies of polymers (mainly proteins and nucleic acids);

(3) to calculate the geometries and energies of transition states (infrequent);

(4) as an aid to organic synthesis;

(5) to generate the potential energy function under which molecules move, for molecular dynamics calculations.

These applications are not all independent. For example, a chemist planning a synthesis might use MM to obtain a plausible geometry for an intermediate involved in the synthesis (the use of MM in synthesis is now so common it is likely that this is often not reported in the literature), and a protein or nucleic acid could be studied with molecular dynamics. Examples of these five facets of the use of MM will be given.

3.3.1 Geometries and energies of small- to medium-sized molecules

Molecular mechanics is used mainly to calculate geometries and energies for small- (roughly C_1 to about C_{10}) and medium-sized (roughly C_{11} to C_{100}) organic molecules. It is by no means limited to organic molecules, as forcefields like SYBYL and UFF [5] have been parameterized for most of the periodic table, but the great majority of MM calculations have been done on organics, probably largely because MM was the creation of organic chemists (this is probably because the concept of geometric structure has long been central in organic chemistry). The most frequent use of MM is undoubtedly to obtain reasonable starting structures for *ab initio*, SE, or DFT (chapters 5–7) calculations. Nowadays this is usually done by building the molecule with an interactive builder in a graphical user interface, then optimizing it with MM with the click of a mouse. The resulting structure is then subjected to an *ab initio*, etc. calculation. MM calculations are usually done only for equilibrium structures (i.e. relative minima on the PES), but by constraining geometric parameters one can approximate roughly transition states (below).

The two salient features of MM calculations on small to medium-sized molecules is that they are *fast* and they *can* be *very accurate*. Times required for a geometry optimization of unbranched $C_{20}H_{42}$, of C_{2h} symmetry, with the Merck Molecular Force Field (MMFF), the SE AM1 (chapter 6) and the *ab initio* HF/3-21G (chapter 5) methods, as implemented with the program SPARTAN [13], were 1.2, 16 s, and 57 min, respectively (on an obsolescent machine a few years ago; these times would now by shorter by a factor of at least 2). Clearly as far as speed goes there is no contest between the methods, and the edge in favor of MM increases with the size of the molecule. In fact, MM was till recently the only practical method for calculations on molecules with more than about 100 heavy atoms (in computational chemistry a heavy atom is any atom heavier than helium). Even programs not designed specifically for macromolecules will handle molecules with 1000 or more atoms on machines of modest power (e.g. a good PC).

Molecular mechanics energies can be very accurate *for families of compounds for which the forcefield has been parameterized*. Appropriate parameterization permits calculation of ΔH_f° (heat of formation, enthalpy of formation) in addition to strain energy [1f]. For the MM2 program (see below), for standard hydrocarbons ΔH_f° errors are usually only 0–4 kJ mol^{-1}, which is comparable to experimental error, and for oxygen containing organics the errors are only 0–8 kJ mol^{-1} [14]; the errors in MM conformational energies are often only about 2 kJ mol^{-1} [15]. MM geometries are usually reasonably good for small to medium-sized molecules [4,9a,16]; for the MM3 program (see below) the RMS error in bond lengths for cholesteryl acetate was only about 0.007 Å [4]. "Bond length" is, if unqualified, somewhat imprecise, since different methods of measurement give somewhat different values [4,9a] (section 5.5.1). MM geometries are routinely used as input structures for quantum-mechanical calculations, but in fact the MM geometry and energy are in some cases as good or better than those from a "higher-level calculation" [17]. The benchmark MM programs for small to medium-sized molecules are probably MM2 and MM3, which will presumably be gradually supplanted by MM4 [4]; the MMFF [18] is likely to become very popular too, not least because of its implementation in SPARTAN [13].

3.3.2   Geometries and energies of polymers

Next to generating geometries and energies of small to medium-sized molecules, the main use of MM is to model polymers, mainly biopolymers (proteins, nucleic acids, polysaccharides). Forcefields have been developed specifically for this; two of the most widely-used of these are CHARMM (Chemistry at Harvard using Molecular Mechanics) [19] (the academic version; the commercial version is CHARMm) and the forcefields in the computational package AMBER (Assisted Model Building with Energy Refinement) [20]. CHARMM was designed to deal with biopolymers, mainly proteins, but has been extended to handle a range of small molecules. AMBER is perhaps the most widely used set of programs for biological polymers, being able to model proteins, nucleic acids, and carbohydrates. Programs like AMBER and CHARMM that model large molecules have been augmented with quantum mechanical methods (SE [21] and even *ab initio* [22]) to investigate small regions where treatment of electronic processes like transition state formation may be critical.

An extremely important aspect of the modelling (which is done largely with MM) of biomolecules is designing pharmacologically active molecules that can fit into active sites (the pharmacophores) of biomolecules and serve as useful drugs. For example, a molecule might be designed to bind to the active site of an enzyme and block the undesired reaction of the enzyme with some other molecule. Pharmaceutical chemists computationally craft a molecule that is sterically and electrostatically complementary to the active site, and try to *dock* the potential drug into the active site. The binding energy of various candidates can be compared and the most promising ones can then be synthesized, as the second step on the long road to a possible new drug. The computationally assisted design of new drugs and the study of the relationship of structure to activity (quantitative structure–activity relationships, QSAR) is one of the most active areas of computational chemistry [23].

3.3.3 Geometries and energies of transition states

By far the main use of MM is to find reasonable geometries for the ground states of molecules, but it has also been used to investigate transition states. The calculation of transition states involved in conformational changes is a fairly straightforward application of MM, since "reactions" like the interconversion of butane or cyclohexane conformers do not in involve the deep electronic reorganization that we call bond-making or bond-breaking. The changes in torsional and nonbonded interactions that accompany them are the kinds of processes that MM was designed to model, and so good transition state geometries and energies can be expected for this particular kind of process; transition state *geometries* cannot be (readily) measured, but the MM energies for conformational changes agree well with experiment: indeed, one of the two very first applications of MM [3a,d] was to the rotational barrier in biphenyls (the other was to the S_N2 reaction [3c]). Since MM programs are usually not able to optimize an input geometry toward a saddle point (see below), one normally optimizes to a minimum subject to the symmetry constraint expected for the transition state. Thus for ethane, optimization to a minimum within D_{3h} symmetry (i.e. by constraining the HCCH dihedral to be $0°$, or by starting with a structure of exactly D_{3h} symmetry) will give the transition state, while optimization with D_{3d} symmetry gives the ground-state conformer (Fig. 3.9). Optimizing an input C_{2v} cyclohexane structure (Fig. 3.10) gives the stationary point nearest this input structure, which is the transition state for interconversion of enantiomeric twist cyclohexane conformers.

There are several examples of the application of MM to actual chemical reactions, as distinct from conformational changes; the ones mentioned here are taken from the review by Eksterowicz and Houk [24]. The simplest way to apply MM to transition states is to approximate the transition state by a ground-state molecule. This can sometimes give surprisingly good results. The rates of solvolysis of compounds RX to the

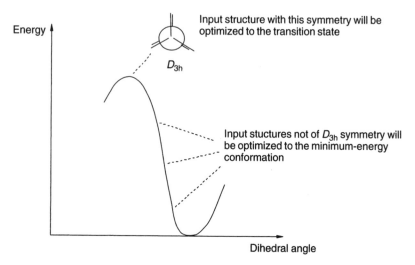

Figure 3.9. Optimizing ethane within D_{3h} symmetry (i.e. by constraining the HCCH dihedral to be $0°$, or by inputting a structure with exact D_{3h} symmetry) will give the transition state, while optimization without requiring D_{3h} symmetry gives the ground-state conformer.

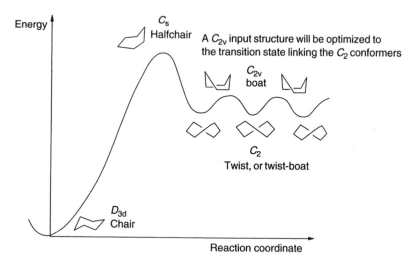

Figure 3.10. Optimizing cyclohexane within C_{2v} symmetry gives a transition state, not one of the minima.

cation correlated well with the energy difference between the hydrocarbon RH, which approximates RX, and the cation R^+, which approximates the transition state leading to this cation. This is not entirely unexpected, as the Hammond postulate [25] suggests that the transition state should resemble the cation. In a similar vein, the activation energy for solvolysis has been approximated as the energy difference between a "methylalkane", with CH_3 corresponding to X in RX, and a ketone, the sp^2 carbon of which corresponds to the incipient cationic carbon of the transition state.

One may wish a more precise approximation to the transition state geometry than is represented by an intermediate or a compound somewhat resembling the transition state. This can sometimes be achieved by optimizing to a minimum subject to the constraint that the bonds being made and broken have lengths believed (e.g. from quantum mechanical calculations on simple systems, or from chemical intuition) to approximate those in the transition state, and with appropriate angles and dihedrals also constrained. With luck this will take the input structure to a point on the potential energy surface near the saddle point. For example, an approximation to the geometry of the transition state for formation of cyclohexene in the Diels–Alder reaction of butadiene with ethene can be achieved (Fig. 3.11) by essentially building a boat conformation of cyclohexene, constraining the two forming C/C bonds to about 2.1 Å, and optimizing, using the CH_2 bridge (later removed) to avoid twisting and to maintain C_s symmetry; optimization with a dihedral constraint removes steric conflict between two hydrogens and gives a reasonable starting structure for, say, an *ab initio* optimization.

The most sophisticated approach to locating a transition state with MM is to use an algorithm that optimizes the input structure to a true saddle point, that is to a geometry characterized by a Hessian with one and only one negative eigenvalue (chapter 2). To do this the MM program must be able not only to calculate second derivatives, but must also be parameterized for the partial bonds in transition states, which is a feature lacking in standard MM forcefields.

MM has been used to study the transition states involved in S_N2 reactions, hydroborations, cycloadditions (mainly the Diels–Alder reaction), the Cope and Claisen rearrangements, hydrogen transfer, esterification, nucleophilic addition to carbonyl groups and electrophilic C/C bonds, radical addition to alkenes, aldol condensations, and various intramolecular reactions [24].

3.3.4   MM in organic synthesis

In the past 15 years or so MM has become widely used by synthetic chemists, thanks to the availability of inexpensive computers (personal computers will easily run MM programs) and user-friendly and relatively inexpensive programs [5]. Since MM can calculate the energies and geometries of ground state molecules and (within the limitations alluded to above) transition states, it can clearly be of great help in planning syntheses. To see which of two or more putative reaction paths should be favored, one might (1) use MM like a hand-held model: examine the substrate molecule for factors like steric hindrance or proximity of reacting groups, or (2) approximate the transition states for alternative reactions using an intermediate or some other plausible proxy (cf. the treatment of solvolysis in the discussion of transition states above), or (3) attempt to calculate the energies of competing transition states (cf. the above discussion of transition state calculations).

The examples given here of the use of MM in synthesis are taken from the review by Lipkowitz and Peterson [26]. In attempts to simulate the metal-binding ability of biological acyclic polyethers, the tricyclic **1** (Fig. 3.12) and a tetracyclic analogue were synthesized, using as a guide the indication from MM that these molecules resemble the cyclic polyether 18-crown-6, which binds the potassium ion; the acyclic compounds were found to be indeed comparable to the crown ether in metal-binding ability.

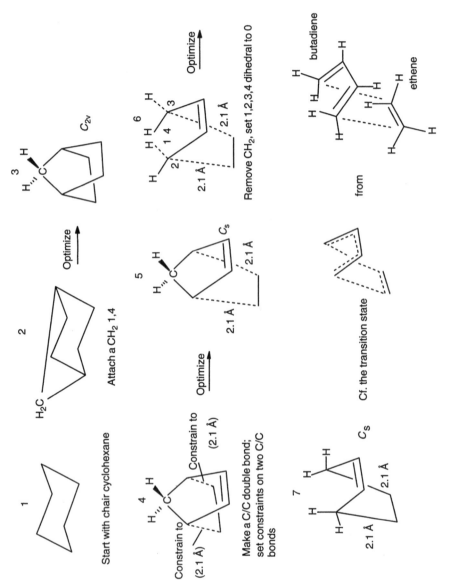

Figure 3.11. Using MM to get the (approximate) transition state for the Diels–Alder reaction of butadiene with ethene.

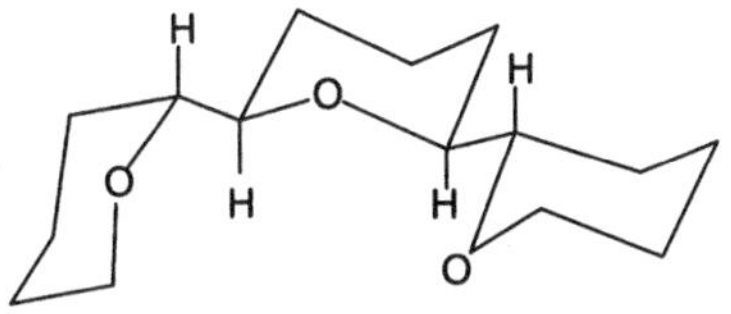

Figure 3.12. Some molecules (**1, 2, 4**) which have been synthesized with the aid of MM.

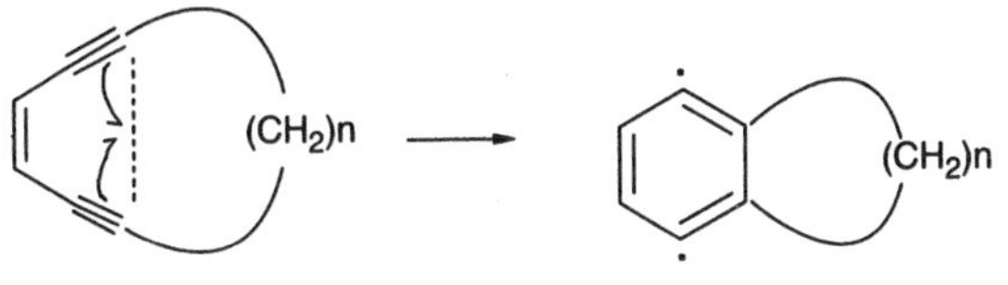

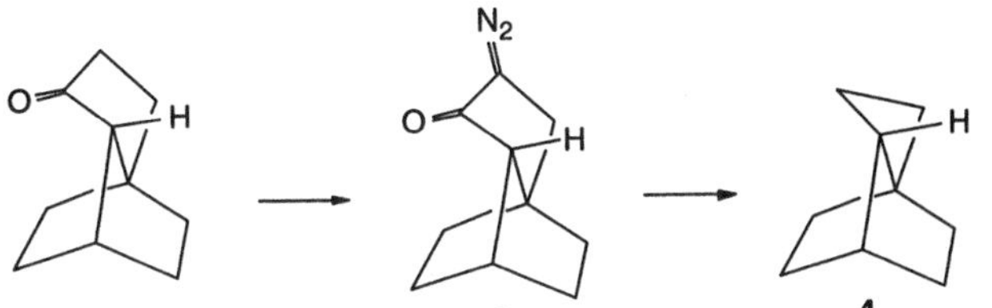

Enediynes like **2** (Fig. 3.12) are able to undergo cyclization to a phenyl-type diradical **3**, which *in vivo* can attack DNA; in molecules with an appropriate triggering mechanism this forms the basis of promising anticancer activity. The effect of the length of the constraining chain (i.e. of *n* in **2**) on the activation energy was studied by MM, aiding the design of compounds (potential drugs) that were found to be more active against tumors than are naturally-occurring enediyne antibiotics.

To synthesize the very strained tricyclic system of **4** (Fig. 3.12), a photochemical Wolff rearrangement was chosen when MM predicted that the skeleton of **4** should be about 109 kJ mol^{-1} less stable than that of the available **5**. Photolysis of the diazoketone **6** gave a high-energy carbene which lay above the carbon skeleton of **4** and so was able to undergo Wolff rearrangement ring contraction to the ketene precursor of **4**.

A remarkable (and apparently still unconfirmed) prediction of MM is the claim that the perhydrofullerene $C_{60}H_{60}$ should be stabler with some hydrogens *inside* the cage [27].

3.3.5 Molecular dynamics and Monte Carlo simulations

Programs like those in AMBER are used not only for calculating geometries and energies, but also for simulating molecular motion, i.e. for molecular dynamics [28], and for calculating the relative populations of various conformations or other geometric arrangements (e.g. solvent molecule distribution around a macromolecule) in Monte Carlo simulations [29]. In molecular dynamics Newton's laws of motion are applied to molecules moving in a MM forcefield, although relatively small parts of the system (system: with biological molecules in particular modelling is often done not on an isolated molecule but on a molecule and its environment of solvent and ions) may be simulated with quantum mechanical methods [21,22]. In Monte Carlo methods random numbers decide how atoms or molecules are moved to generate new conformations or geometric arrangements (states) which are then accepted or rejected according to some filter. Tens of thousands (or more) of states are generated, and the energy of each is calculated by MM, generating a Boltzmann distribution.

3.4   GEOMETRIES CALCULATED BY MM

Figure 3.13 compares geometries calculated with the MMFF with those from a reasonably high-level *ab initio* calculation (MP2(FC)/6-31G*; chapter 5) and from experiment. The MMFF is a popular force field, applicable to a wide variety of molecules. Popular prejudice holds that the *ab initio* method is "higher" than MM and so should give superior geometries. The set of 20 molecules in Fig. 3.13 is also used in chapters 5, 6, and 7, to illustrate the accuracy of *ab initio*, SE, and density functional calculations in obtaining molecular geometries. The data in Fig. 3.13 are analyzed in Table 3.3. Table 3.4 compares dihedral angles for eight molecules, which are also used in chapters 5–7.

This survey suggests that for common organic molecules the MMFF is nearly as good as the *ab initio* MP2(FC)/6-31G* method for calculating geometries. Both methods give good geometries, but while these MM calculations all take effectively about one second, MP2 geometry optimizations on these molecules require typically a few minutes (CH_3COCH_3, 2.4 min; CH_3Cl, 1.4 min; $(CH_3)_2SO$, 3.7 min; 1.5 GHz Pentium 4). For larger molecules where MP2 would need hours, MM calculations might still take only seconds. Note, however, that *ab initio* methods provide information that MM cannot, and are far more reliable for molecules outside those of the kind used in the MM training set (section 3.2.2). The worst MMFF bond length deviation from experiment among the 20 molecules is 0.021 Å (the C=C bond of propene; the MP2 deviation is 0.020 Å); most of the other errors are ca. 0.01 Å or less. The worst bond angle error is 13.6°, for HOF, and for HOCl the deviation is 7.9°, the second worst angle error in the set. This suggests a problem for the MMFF with X–O–Halogen angles, but while for CH_3OF deviation from the MP2 angle (which is likely to be close to experiment) is MMFF−MP2 $= 110.7° − 102.8° = 7.9°$, for CH_3Cl the deviation is only $112.0° − 109.0° = 3.0°$.

MMFF dihedral angles are remarkably good, considering that torsional barriers are believed to arise from subtle quantum mechanical effects. The worst dihedral angle error is 10°, for HOOH, and the second worst, −5.0°, is for the analogous HSSH. The popular *ab initio* HF/3-21G (chapter 5) and SE PM3 (chapter 6) methods also have trouble with HOOH, predicting a dihedral angle of 180°. For those dihedrals not involving OO or SS bonds, (an admittedly small selection), the MMFF errors are only ca. 1°−2°, cf. ca. 2°−6° for MP2.

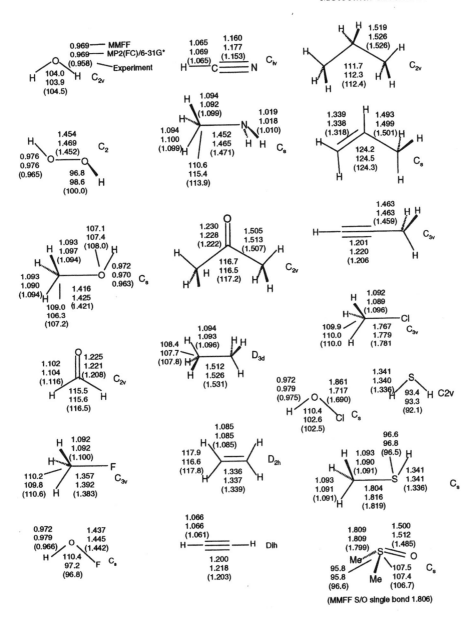

Figure 3.13. A comparison of some MMFF, MP2(FC)/6-31G* and experimental geometries. Calculations are by the author and experimental geometries are from Ref. [30]. Note that all CH bonds are ca. 1 Å, all other bonds range from ca. 1.2–1.8 Å, and all bond angles (except for linear molecules) are ca. 90–120°.

Table 3.3. Error in MMFF MM and MP2(FC)/6-31G* bond lengths and angles, from Fig. 3.13

	Bond length errors, $r - r_{exp}$ (Å)				Bond angle errors, $a - a_{exp}$
C–H	O–H, N–H, S–H	C–C	C–O, N, F, Cl, S		Angles
MeOH −0.001/−0.004 −0.001/0.003	H₂O 0.011/0.011	Me₂CO −0.002/0.006	MeOH −0.005/0.007		H₂O (HOH) −0.5/−0.6
HCHO −0.014/−0.012	H₂O₂ 0.011/0.011	CH₃CH₃ −0.019/−0.005	HCHO 0.017/0.013		H₂O₂ (HOO) −3.2/−1.4
MeF −0.008/−0.008	MeOH 0.009/0.007	CH₂CH₂ −0.013/−0.017/−0.002	MeF −0.005/−0.007		MeOH (HCO)
HCN 0.000/0.004	HOF 0.006/0.013	HCCH −0.003/0.015	HCN 0.007/0.024		HCHO (HCH) −1.0/−0.9
MeNH₂ −0.005/0.001 −0.005/−0.007	MeNH₂ 0.009/0.008	CH₃CH₂CH₃ −0.007/0.000	MeNH₂ −0.019/−0.006		MeF (HCH) −0.4/−0.8
CH₃CH₃ −0.002/−0.003	HOCl −0.003/0.004	CH₂CHCH₃ −0.008/−0.002 0.021/0.020	Me₂CO 0.008/0.006		HOF (HOF) 13.6/0.4
CH₂CH₂ 0.000/0.000	H₂S 0.005/0.004	HCCCH₃ 0.004/0.004 −0.005/0.014	MeCl −0.014/−0.002		MeNH₂ (HCN) −3.3/1.5

CHCH	MeSH		MeSH	Me_2CO (CCC)
0.005/0.005	0.005/0.005		−0.015/−0.003	−0.5/−0.8
MeCl			Me_2SO	CH_3CH_3 (HCH)
−0.004/−0.007			0.010/0.010	0.6/−0.1
MeSH				CH_2CH_2 (HCH)
0.002/0.000				0.1/−1.2
0.002/−0.001				
				$CH_3CH_2CH_3$ (CCC)
				−0.7/−0.1
				CH_2CHCH_3 (CCC)
				−0.1/0.2
				MeCl (HCH)
				−0.1/0.0
				H_2S (HSH)
				1.3/1.2
				MeSH (CSH)
				0.1/0.3
				Me_2SO (CSC)
				−0.8/−0.8
				(CSO)
				0.8/0.7
3+, 8−, two 0	7+, 1−, none 0	2+, 7−, none 0	4+, 5−, none 0	7+, 11−, none 0
4+, 7−, two 0	8+, 0−, none 0	5+, 3−, one 0	6+, 3−, none 0	6+, 11−, one 0
mean of 13:	mean of 8:	mean of 9:	mean of 9:	mean of 18:
0.004/0.004	0.007/0.008	0.009/0.008	0.011/0.009	1.7/0.7

Errors are given as MMFF/MP2. In some cases (e.g. MeOH) errors for two bonds are given, on one line and on the line below. A minus sign means that the calculated value is less than the experimental. The numbers of positive, negative, and zero deviations from experiment are summarized at the bottom of each column. The averages at the bottom of each column are arithmetic means of the absolute values of the errors.

Table 3.4. MMFF, MP2(FC)/6-31G* and experimental dihedral angles (degrees).

| Molecule | Dihedral Angles | | | |
	MMFF	MP2/6-31G*	Exp.	Errors
HOOH	129.4	121.3	119.1[a]	10/2.2
FOOF	90.7	85.8	87.5[b]	3.2/−1.7
FCH$_2$CH$_2$F	72.1	69.0	73[b]	−1.0/−4
(FCCF)				
FCH$_2$CH$_2$OH				
(FCCO)	65.9	60.1	64.0[c]	1.9/−3.9
(HOCC)	53.5	54.1	54.6[c]	−1.1/−0.5
ClCH$_2$CH$_2$OH				
(ClCCO)	65.7	65.0	63.2[b]	2.5/1.8
(HOCC)	56.8	64.3	58.4[b]	−1.6/5.9
ClCH$_2$CH$_2$F				
(ClCCF)	69.8	65.9	68[b]	1.8/−2.1
HSSH	84.2	90.4	90.6[a]	−6.4/−0.2
FSSF	82.9	88.9	87.9[b]	−5.0/1.0
				Deviations: 5+, 5−/4+, 6−
				mean of 10: 3.5/2.3;

Errors are given in the *Errors* column as MMFF/MP2/6-31G*. A minus sign means that the calculated value is less than the experimental. The numbers of positive and negative deviations from experiment and the average errors (arithmetic means of the absolute values of the errors) are summarized at the bottom of the *Errors* column. Calculations are by the author; references to experimental measurements are given for each measurement. The AM1 and PM3 dihedrals vary by a fraction of a degree depending on the input dihedral. Some molecules have calculated minima at other dihedrals in addition to those given here, e.g. FCH$_2$CH$_2$F at FCCF 180°.

[a] Hehre [30], pp. 151, 152.

[b] M. D. Harmony, V. W. Laurie, R. L. Kuczkowski, R. H. Schwenderman, D. A. Ramsay, F. J. Lovas, W. H. Lafferty, A. G. Makai, "Molecular Structures of Gas-Phase Polyatomic Molecules Determined by Spectroscopic Methods," J. Phys. Chem. Ref. Data, 1979, *8*, 619–721.

[c] J. Huang and K. Hedberg, J. Am. Chem. Soc., 1989, *111*, 6909.

3.5   FREQUENCIES CALCULATED BY MM

Any method that can calculate the energy of a molecular geometry can in principle calculate vibrational frequencies, since these can be obtained from the second derivatives of energy with respect to molecular geometry (section 2.5), and the masses of the vibrating atoms. Some commercially available MM programs, for example the MMFF as implemented in SPARTAN [13], can calculate frequencies. Frequencies are useful (section 2.5) (1) for characterizing a species as a minimum (no imaginary frequencies) or a transition state or higher-order saddle point (one or more imaginary frequencies), (2) for obtaining zero-point energies to correct frozen-nuclei energies (section 2.2), and (3) for interpreting or predicting infrared spectra.

(1) *Characterizing a species.* This is not often done with MM, because MM is used mostly to create input structures for other kinds of calculations, and to study known (often biological) molecules. Nevertheless MM can yield information on the curvature of the potential energy surface (as calculated by that particular forcefield, anyway) at the point in question. For example, the MMFF-optimized geometries of D_{3d} (staggered) and D_{3h} (eclipsed) ethane (Fig. 3.3) show, respectively, no imaginary frequencies and one imaginary frequency, the latter corresponding to rotation about the C/C bond. Thus the MMFF (correctly) predicts the staggered conformation to be a minimum, and the eclipsed to be a transition state connecting successive minima along the torsional reaction coordinate. Again, calculations on cyclohexane conformations with the MMFF correctly give the boat an imaginary frequency corresponding to a twisting motion leading to the twist conformation, which latter has no imaginary frequencies (Fig. 3.10). Although helpful for characterizing conformations, particularly hydrocarbon conformations, MM is less appropriate for species in which bonds are being formed and broken. For example, the symmetrical (D_{3h}) species in the $F^- + CH_3 - F$ S_N2 reaction, with equivalent C/F partial bonds, is incorrectly characterized by the MMFF as a minimum rather than a transition state, and the C/C bonds are calculated to be 1.289 Å long, cf. the value of ca. 1.8 Å from methods known to be trustworthy for transition states.

(2) *Obtaining zero-point energies (ZPEs).* ZPEs are essentially the sum of the energies of each normal-mode vibration. They are added to the raw energies (the frozen-nuclei energies, corresponding to the stationary points on a Born-Oppenheimer surface; section 2.3) in accurate calculations of relative energies using ab initio (chapter 5) or DFT (chapter 7) methods. However, the ZPEs used for those corrections are usually obtained from an *ab initio* or DFT calculation.

(3) *Infrared spectra.* The ability to calculate the energies (cm^{-1}) and relative intensities of molecular vibrations amounts to being able to calculate infrared spectra. MM as such cannot calculate the *intensities* of vibrational modes, since these involve changes in dipole moments (section 5.5.3), and dipole moment is related to electron distribution, a concept that lies outside MM. However, approximate intensities can be calculated by assigning dipole moments to bonds or charges to atoms, and such methods have been implemented in MM programs [31], although MM programs that calculate intensities are not yet widely used. Figures 3.14–3.17 compare the IR spectra of acetone, benzene, dichloromethane and methanol, calculated with the MM3 program[6] with the experimental spectra, and with spectra calculated by the *ab initio* MP2(FC)/6-31G* method; the data for Figs 3.14–3.17 are in Tables 3.5–3.8 (in chapters 5–7 spectra for these four molecules, calculated by *ab initio*, SE, and density functional methods, are also given). The MP2 spectra generally match experiment better than the MM3, although the latter method furnishes a rapid way of obtaining approximate IR spectra. For a series of related compounds, MM3 might be a reasonable way to quickly investigate trends in frequencies and intensities. Extensive surveys of MMFF and MM4 frequencies showed that MMFF root-mean-square errors are ca. $60\,cm^{-1}$, and MM4 errors 25–$52\,cm^{-1}$ [5b].

---

[6]The MM3 frequencies and intensities were kindly provided by Dr. J. -H. Lii of the Department of Chemistry of the University of Georgia, Athens, Georgia, USA.

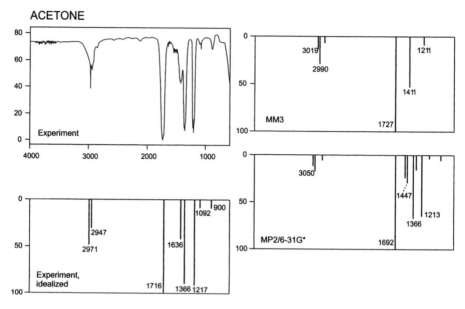

Figure 3.14. Experimental (gas phase) and MM (MM3) and MP2(FC)/6-31G* calculated infra red spectra of acetone. The MM3 spectrum is based on the data in Table 3.5; the MP2 spectrum is that shown in Fig. 5.33.

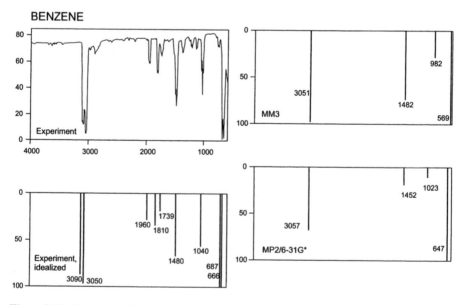

Figure 3.15. Experimental (gas phase) and MM (MM3) and MP2(FC)/6-31G* calculated infra red spectra of benzene. The MM3 spectrum is based on the data in Table 3.6; the MP2 spectrum is that shown in Fig. 5.34.

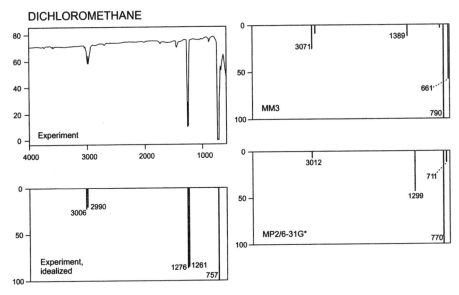

Figure 3.16. Experimental (gas phase) and MM (MM3) and MP2(FC)/6-31G* calculated infra
red spectra of dichloromethane. The MM3 spectrum is based on the data in Table 3.7; the MP2
spectrum is that shown in Fig. 5.35.

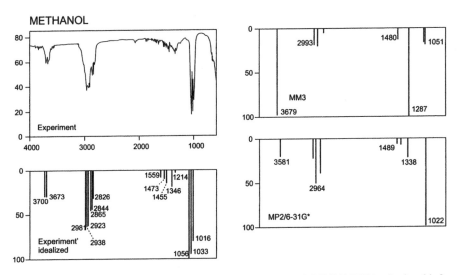

Figure 3.17. Experimental (gas phase) and MM (MM3) and MP2(FC)/6-31G* calculated infra
red spectra of methanol. The MM3 spectrum is based on the data in Table 3.8; the MP2 spectrum
is that shown in Fig. 5.36.

Table 3.5. Calculated (MM3) IR spectrum of acetone

Frequency (cm^{-1})	Relative intensity
3018	17
3016	11
2990	27
2902	9
1727	100
1411	49
1211	7

Intensities have been normalized, and those of less than 2 percent relative intensity have been ignored.

Table 3.6. Calculated (MM3) IR spectrum of benzene

Frequencies	Intensities
3051	96
1482	75
982	29
569	100

The intensities of degenerate frequencies have been combined, intensities have been normalized, and those of less than 2 percent relative intensity have been ignored.

Table 3.7. Calculated (MM3) IR spectrum of dichloromethane

Frequencies	Intensities
3071	25
3003	8
1389	13
813	3
790	100
661	59

Intensities have been normalized, and those of less than 2 percent relative intensity have been ignored.

Table 3.8. Calculated (MM3) IR spectrum of methanol

Frequencies	Intensities
3679	98
2993	17
2930	19
2847	3
1480	12
1287	100
1084	15
1051	19

Intensities have been normalized, and those of less than 2 percent relative intensity have been ignored.

3.6   STRENGTHS AND WEAKNESSES OF MM

3.6.1   Strengths

Molecular mechanics is *fast*, as shown by the times for optimization of $C_{20}H_{42}$ in section 3.3. The speed of MM is not always at the expense of *accuracy*: for the kinds of molecules for which it has been parameterized, it can rival or surpass experiment in the reliability of its results (sections 3.3 and 3.4). MM is *undemanding* in its hardware requirements: except perhaps for work on large biopolymers, MM calculations on moderately well-equipped personal computers are quite practical. The characteristics of speed, (frequent) accuracy and modest computer requirements have given MM a place in many modelling programs.

Because of its speed and the availability of parameters for almost all the elements (section 3.3), MM – even when it does not provide very accurate geometries – can supply reasonably good input geometries for SE, *ab initio* or density functional calculations, and this is one of its main applications. The fairly recent ability of MM programs to calculate IR spectra with some accuracy [16,32] may presage an important application, since frequency calculation by quantum mechanical methods usually requires considerably more time than geometry optimization). Note that MM frequencies should be calculated using the MM geometry – unfortunately, MM cannot be used as a shortcut to obtaining frequencies for a species optimized by a quantum mechanical calculation (*ab initio*, density functional or SE), since frequencies must be calculated using the same method used for the geometry optimization (section 2.5).

3.6.2 Weaknesses

The possible pitfalls in using MM are discussed by Lipkowitz [33]. The weaknesses of MM stem from the fact that it ignores electrons. The philosophy behind MM is to think of a molecule as a collection of atoms subject to forces and to use any practical mathematical treatment of these forces to express the energy in terms of the geometric parameters. By parameterization MM can "calculate" electronic properties; for example, using bond dipoles it can find a dipole moment for a molecule, and using values that have been calculated for various atom types by quantum mechanics it can assign charges to atoms. However, such results are obtained purely by analogy, and their reliability can be negated by unexpected electronic factors to which MM is oblivious. MM cannot provide information about the shapes and energies of molecular orbitals nor about related phenomena such as electronic spectra.

Because of the severely empirical nature of MM, interpreting MM parameters in terms of traditional physical concepts is dangerous; for example, the bond-stretching and angle-bending parameters cannot rigorously be identified with spectroscopic force constants [33]; Lipkowitz suggests that the MM proportionality constants (section 3.2.1) be called *potential constants*. Other dangers in using MM are the following:

(1) *Using an inappropriate forcefield.* A field parameterized for one class of compounds is not likely to perform well for other classes.

(2) *Transferring parameters form one forcefield to another.* This is usually not valid.

(3) *Optimizing to a stationary point that may not really be a minimum* (it could be a "maximum", a transition state), and certainly may not be a *global* minimum (chapter 2). If there is reason to be concerned that a structure is not a minimum, alter it slightly by bond rotation and reoptimize; a transition state should slide down toward a nearby minimum (e.g. eclipsed ethane altered slightly from the D_{3h} geometry and optimized goes to the staggered conformer (Fig. 3.9).

(4) *Being taken in by vendor hype.* MM programs, more so than SE ones and unlike *ab initio* or DFT programs, are ruled by empirical factors (the form of the forcefield and the parameters used in it), and vendors do not usually caution buyers about potential deficiencies.

(5) *Ignoring solvent and nearby ions.* For polar molecules using the *in vacuo* structure can lead to quite wrong geometries and energies. This is particularly important for biomolecules. One way to mitigate this problem is to explicitly add solvent molecules or ions to the system, which can considerably increase the time for a calculation. Another might be to subject various plausible *in vacuo*-optimized conformations to single-point

(no geometry optimization) calculations that simulate the effect of solvent and take the resulting energies as being more reliable than the *in vacuo* ones.

(6) *Lack of caution about comparing energies calculated with MM.* The method calculates the energy of a molecule relative to a hypothetical strainless idealization of the molecule. Using MM to calculate the relative energy of two isomers by comparing their strain energies (the normal MM energies) is dangerous because the two strain energies are not necessarily relative to the same hypothetical unstrained species (strain energies are not an unambiguous observable [34]). This is particularly true for functional group isomers, like $(CH_3)_2O/CH_3CH_2OH$ and $CH_3COCH_3/H_2C=C(OH)CH_3$, which have quite different atom types. For isomers consisting of the same kinds of atoms (alkanes cf. alkanes, say), and especially for conformational isomers and E/Z isomers (geometric isomers), a good MM forcefield should give strain energies which reasonably represent relative enthalpies. For example, the MMFF gives for $CH_3COCH_3/H_2C=C(OH)CH_3$ strain energies of $6.9/-6.6 \, kJ \, mol^{-1}$, i.e. relative energies of $0/-0.3 \, kJ \, mol^{-1}$, but the experimental value is ca. $0/44 \, kJ \, mol^{-1}$, i.e. $H_2C=C(OH)CH_3$ is much the higher-energy molecule. On the other hand, the MMFF yields for *gauche*-butane/*anti*-butane strain energies of $-21.3/-18.0 \, kJ \, mol^{-1}$, i.e. relative energies of $0/3.3 \, kJ \, mol^{-1}$, reasonably close to the experimental value of $0/2.8 \, kJ \, mol^{-1}$. For chair (D_{2d}), twist (D_2), and boat (C_{2v}) cyclohexane, the MMFF strain energies are -14.9, 9.9, and $13.0 \, kJ \, mol^{-1}$, i.e. relative energies of 0, 24.8 and $27.9 \, kJ \, mol^{-1}$, cf. the experimental the estimates of 0, 24 and $29 \, kJ \, mol^{-1}$. MM programs can be parameterized to give, not just strain energy, but enthalpies of formation [1f], and the use of these enthalpies should make possible energy comparisons between isomers of disparate structural kinds.

Although chemists often compare stabilities of isomers using enthalpies, we should remember that equilibria are actually determined by free energies. The lowest-enthalpy isomer is not *necessarily* the one of lowest free energy: a higher-enthalpy molecule may have more vibrational and torsional motion (it may be springier and floppier) and thus possess more entropy and hence have a lower free energy. Free energy has an enthalpy and an entropy component, and to calculate the latter, one needs the vibrational frequencies. Programs that calculate frequencies will usually also provide entropies, and with parameterization for enthalpy this can permit the calculation of free energies. Note that the species of lowest free energy is not always the major one present: one low-energy conformation could be outnumbered by one hundred of higher energy, each demanding its share of the Boltzmann pie.

(7) *Assuming that the major conformation determines the product.* In fact, in a mobile equilibrium the product ratio depends on the relative reactivities, not relative amounts, of the conformers (the Curtin–Hammett principle [35]).

(8) *Failure to exercise judgement: small energy differences* (say up to $10–20 \, kJ \, mol^{-1}$) mean nothing in many cases. The excellent energy results referred to in section 3.3 can be expected only for families of molecules (usually small to medium-sized) for which the forcefield has been parameterized.

Many of the above dangers can be avoided simply by performing test calculations on systems for which the results are known (experimentally, or "known" from high-level quantum mechanical calculations). Such a reality check can have salutary effects on the reliability of one's results, and not only with reference to MM.

3.7 SUMMARY OF CHAPTER 3

This chapter explains the basic principles of MM, which rests on a view of molecules as balls held together by springs. MM began in the 1940s with attempts to analyze the rates of racemization of biphenyls and S_N2 reactions.

The potential energy of a molecule can be written as the sum of terms involving bond stretching, angle bending, dihedral angles and nonbonded interactions. Giving these terms explicit mathematical forms constitutes devising a forcefield, and giving actual numbers to the constants in the forcefield constitutes parameterizing the field. An example is given of the devising and parameterization of an MM forcefield.

MM is used mainly to calculate geometries and energies for small to medium-sized molecules. Such calculations are fast and can be very accurate, provided that the forcefield has been carefully parameterized for the types of molecules under study. Calculations on biomolecules is a very important application of MM; the pharmaceutical industry designs new drugs with the aid of MM: for example, examining how various candidate drugs fit into the active sites of biomolecules (docking) and the related aspect of QSAR are of major importance. MM is of some limited use in calculating the geometries and energies of transition states. Organic synthesis now makes considerable use of MM, which enables chemists to estimate which products are likely to be favored and to devise more realistic routes to a target molecule than was hitherto possible. In molecular dynamics MM is used to generate the forces acting on molecules and hence to calculate their motions, and in Monte Carlo simulations MM is used to calculate the energies of the many randomly generated states.

MM is fast, it can be accurate, it is undemanding of computer power, and it provides reasonable starting geometries for quantum mechanical calculations. MM ignores electrons, and so can provide parameters like dipole moment only by analogy. One must be cautious about the applicability of MM parameters to the problem at hand. Stationary points from MM, even when they are relative minima, may not be global minima. Ignoring solvent effects can give erroneous results for polar molecules. MM gives strain energies, the difference of which for structurally similar isomers represent enthalpy differences; parameterization to give enthalpies of formation is possible. Strictly speaking, relative amounts of isomers depend on free energy differences. The major conformation (even when correctly identified) is not necessarily the reactive one.

REFERENCES

[1] General references to MM: (a) A. K. Rappe, C. L. Casewit, "Molecular mechanics across chemistry," University Science Books, Sausalito, CA, 1997, website http://www.chm.colostate.edu/mmac; (b) A. R. Leach, "Molecular Modelling, Principles and Applications," Addison Wesley Longman, Essex (UK), 1996, chapter 3; (c) U. Burkert, N. L. Allinger, "Molecular Mechanics", ACS Monograph 177, American Chemical Society, Washington, DC, 1982; (d) N. L. Allinger, Calculation of Molecular Structures and Energy by Force Methods", in *Advances in Physical Organic Chemistry*, *13*, V. Gold and D. Bethell, Eds., Academic Press, New York, 1976; (e) T. Clark, "A Handbook of Computational Chemistry," Wiley, New York, 1985; (f) I. N. Levine, "Quantum Chemistry," 4th edn, Prentice Hall, Engelwood Cliffs, New Jersey, 1991, pp. 583–587; (g) Issue

No. 7 of Chem. Rev., 1993, *93*; (h) Conformational energies: I. Pettersson, T. Liljefors, in *Reviews in Computational Chemistry*, 1996, *9*; (i) Inorganic and organometallic compounds: C. R. Landis, D. M. Root, T. Cleveland, in *Reviews in Computational Chemistry*, 1995, *6*; (j) Parameterization: J. P. Bowen, N. L. Allinger, *Reviews in Computational Chemistry*, 1991, *2*.

[2] MM history: (a) Ref. [1]; (b) E. M. Engler, J. D. Andose, P. von R. Schleyer, J. Am. Chem. Soc., 1973, *95*, 8005 and references therein; (c) MM up to the end of 1967 is reviewed in detail in: J. E. Williams, P. J. Stang, P. von R. Schleyer, Annu. Rev. Phys. Chem., 1968, *19*, 531.

[3] (a) F. H. Westheimer, J. E. Mayer, J. Chem. Phys., 1946, *14*, 733; (b) T. L. Hill, J. Chem. Phys., 1946, *14*, 465; (c) I. Dostrovsky, E. D. Hughes, C. K. Ingold, J. Chem. Soc., 1946, 173; (d) F. H. Westheimer, J. Chem. Phys., 1947, *15*, 252.

[4] B. Ma, J-H Lii, K. Chen, N. L. Allinger, J. Am. Chem. Soc., 1997, *119*, 2570 and references therein.

[5] (a) Information on and references to MM programs may be found in Ref. [1] and the website of Ref. [1a]; (b) For papers on the popular MMFF and the MM4 forcefield (and information on some others) see the issue of J. Comp. Chem., 1996, *17*.

[6] The force constant is defined as the proportionality constant in the equation force = $k \times$ extension (of length or angle), so integrating force with respect to extension to get the energy (= force × extension) needed to stretch the bond gives $E = (k/2)(\text{extension})^2$, i.e. $k =$ force constant $= 2k_{stretch}$ (or $2k_{bend}$).

[7] (a) A brief discussion and some parameters: P. W. Atkins, "Physical Chemistry," 5th edn, Freeman New York, 1994, pp.772, 773; it is pointed out here that $e^{-r/\sigma}$ is actually a much better representation of the compressive potential than is r^{-12}; (b) W. J. Moore, "Physical Chemistry," 4th edn, Prentice-Hall, New Jersey, 1972, p. 158 (from J. O. Hirschfelder, C. F. Curtis and R. B. Bird, "Molecular Theory of Gases and Liquids," Wiley, New York, 1954). Note that our k_{nb} is called 4ϵ here and must be multiplied by 8.31/1000 to convert it to our units of kJ mol^{-1}.

[8] Infra red spectroscopy: R. M. Silverstein, G. C. Bassler and T. C. Morrill, "Spectrometric Identification of Organic Compounds," 4th edn, Wiley, New York, 1981, chapter 3.

[9] Different methods of structure determination give somewhat different results; this is discussed in Ref. [4] and in: (a) B. Ma, J-H Lii, H. F. Schaefer, N. L. Allinger, J. Phys. Chem., 1996, *100*, 8763; (b) A. Domenicano, I. Hargittai, Eds., "Accurate Molecular Structures," Oxford Science Publications, New York, 1992.

[10] To properly parameterize a MM force field only high-level *ab initio* calculations (or density functional) calculations would actually be used, but this does not affect the principle being demonstrated.

[11] Ref. [1b, pp. 148–181].

[12] MM3: N. L. Allinger, Y. H. Yuh, J-H Lii, J. Am. Chem Soc., 1989, *111*, 8551. The calculation was performed with MM3 as implemented in Spartan (Ref. [13]).

[13] Spartan SGI Version 4.0.4; Spartan is a comprehensive computational chemistry program with MM, *ab initio*, density functional and SE capability, combined with powerful graphical input and output.

[14] Ref. [1c, pp. 182–184].

[15] K. Gundertofte, T. Liljefors, P.-O. Norby, I. Pettersson, J. Comp. Chem., 1996, *17*, 429.

[16] Comparison of various force fields for geometry (and vibrational frequencies): T. A. Halgren, J. Comp. Chem., 1996, *17*, 553.

[17] The Merck force field (Ref. [18]) often gives geometries that are satisfactory for energy calculations (i.e. for single-point energies) with quantum mechanical methods; this could be very useful for large molecules: W. J. Hehre, J. Yu, P. E. Klunzinger, "A Guide to Molecular Mechanics and Molecular Orbital calculations in Spartan," Wavefunction Inc., Irvine CA, 1997, chapter 4.

[18] T. A. Halgren, J. Comp. Chem., 1996, *17*, 490.

[19] M. C. Nicklaus, J. Comp. Chem., 1997, *18*, 1056; the difference between CHARMM and CHARMm is explained here.

[20] (a) www.amber.ucsf.edu/amber/amber.html; (b) W. D. Cornell, P. Cieplak, C. I. Bayly, I. R. Gould, K. M. Merz, Jr., D. M Ferguson, D. C Sellmeyer, T. Fox, J. W. Caldwell, P. A. Kollman, J. Am. Chem. Soc., 1995, *117*, 5179; (c) V. Barone, G. Capecchi, Y. Brunel, M.-L. D. Andries, R. Subra, J. Comp. Chem., 1997, *18*, 1720.

[21] (a) M. J. Field, P. A. Bash, M. Karplus, J. Comp. Chem., 1990, *11*, 700; (b) P. A. Bash. M. J. Field, M. Karplus, J. Am. Chem. Soc., 1987, *109*, 8092.

[22] U. C. Singh, P. Kollman, J. Comp Chem., 1986, *7*, 718.

[23] (a) Reference [1b, chapter 10]; (b) H.-D. Höltje, G. Folkers "Molecular Modelling, Applications in Medicinal Chemistry," VCH, Weinheim, Germany, 1996; (c) "Computer-Assisted Lead Finding and Optimization," H. van de Waterbeemd, B. Testa, G. Folkers, Eds., VCH, Weinheim, Germany, 1997.

[24] J. E. Eksterowicz and K. N. Houk, Chem. Rev., 1993, *93*, 2439.

[25] M. B. Smith and J. March, "March's Advanced Organic Chemistry," Wiley, New York, 2000, pp. 284–285.

[26] K. B. Lipkowitz, M. A. Peterson, Chem Rev., 1993, *93*, 2463.

[27] M. Saunders, Science, 1991, *253*, 330.

[28] (a) Ref. [1b, chapter 6]; (b) M. Karplus, G. A. Putsch, Nature, 1990, *347*, 631; (c) C. L. Brooks III, D. A. Case, Chem. Rev., 1993, *93*, 2487; (d) M. Eichinger, H. Grubmüller, H. Heller, P. Tavan, J. Comp. Chem., 1997, *18*, 1729; (e) G. E. Marlow, J. S. Perkyns, B. M. Pettitt, Chem. Rev., 1993, *93*, 2503; (f) J. Aqvist, A. Warshel, Chem. Rev., 1993, *93*, 2523.

[29] Ref. [1b, chapter 7].

[30] W. J. Hehre, L. Radom, p. v. R. Schleyer, J. A. Pople, "*Ab initio* Molecular Orbital Theory," Wiley, New York, 1986.

[31] J.-H. Lii and N. L. Allinger, J. Comp. Chem., 1992, *13*, 1138. The MM3 program with the ability to calculate IR intensities can be bought by academic users from QCPE (the Quantum Chemistry Exchange Program of the University of Indiana) and by commercial users from Tripos Associates of St. Louis, Missouri.

[32] N. Nevins, N. L. Allinger, J. Comp. Chem., 1996, *17*, 730 and references therein.

[33] K. B. Lipkowitz, J. Chem. Ed., 1995, *72*, 1070.

[34] (a) K. Wiberg, Angew. Chem., Int. Ed. Engl., 1986, *25*, 312; (b) Issue No. 5 of Chem. Rev., 1989, *89*; (c) S. Inagaki, Y. Ishitani, T. Kakefu, J. Am. Chem. Soc., 1994, *116*, 5954; (d) S. Nagase, Acc. Chem. Res., 1995, *28*, 469; (e) S. Gronert, J. M. Lee, J. Org. Chem., 1995, *60*, 6731; (f) A. Sella, H. Basch, S. Hoz, J. Am. Chem. Soc., 1996, *118*, 416; (g)

S. Grime, J. Am. Chem. Soc., 1996, *118*, 1529; (h) V. Balaji, J. Michl, Pure Appl. Chem., 1988, *60*, 189; (i) K. B. Wiberg, J. W. Ochterski, J. Comp. Chem., 1997, *18*, 108.

[35] J. I. Seeman, Chem. Rev., 1983, *83*, 83.

EASIER QUESTIONS

1. What is the basic idea behind MM?
2. What is a forcefield?
3. What are the two basic approaches to parameterizing a forcefield?
4. Why does parameterizing a forcefield for transition states present special problems?
5. What is the main advantage of MM, generally speaking, over the other methods of calculating molecular geometries and relative energies?
6. Why is it not valid in all cases to obtain the relative energies of isomers by comparing their MM strain energies?
7. What class of problems cannot be dealt with by MM?
8. Give four applications for MM. Which is the most widely used?
9. MM can calculate the values (cm^{-1}) of vibrational frequencies, but without "outside assistance" it can't calculate their intensities. Explain.
10. Why is it not valid to calculate a geometry by some slower (e.g. *ab initio*) method, then use that geometry for a fast MM frequency calculation?

HARDER QUESTIONS

1. One big advantage of MM over other methods of calculating geometries and relative energies is speed. Does it seem likely that continued increases in computer speed could make MM obsolete?
2. Do you think it is possible (in practical terms? In principle?) to develop a forcefield that would accurately calculate the geometry of any kind of molecule?
3. What advantages or disadvantages are there to parameterizing a forcefield with the results of "high-level" calculations rather than the results of experiments?
4. Would you dispute the suggestion that no matter how accurate a set of MM results might be, they cannot provide insight into the factors affecting a chemical problem, because the "ball and springs" model is unphysical?
5. Would you agree that hydrogen bonds (e.g. the attraction between two water molecules) might be modelled in MM as weak covalent bonds, as strong van der Waals or dispersion forces, or as electrostatic attractions? Is any one of these three approaches to be preferred in principle?
6. Replacing small groups by "pseudoatoms" in a forcefield (e.g. CH_3 by an "atom" about as big) obviously speeds up calculations. What disadvantages might accompany this simplification?
7. Why might the development of an accurate and versatile forcefield for inorganic molecules be more of a challenge than for organic molecules?

8. What factor(s) might cause an electronic structure calculation (e.g. *ab initio* or DFT) to give geometries or relative energies very different from those obtained from MM?

9. Compile a list of molecular characteristics/properties that cannot be calculated purely by MM.

10. How many parameters do you think a reasonable forcefield would need to minimize the geometry of 1,2-dichloroethane?

Chapter 4

Introduction to Quantum Mechanics in Computational Chemistry

It is by logic that we prove, but by intuition that we discover.
J. H. Poincaré, ca. 1900.

4.1 PERSPECTIVE

Chapter 1 outlined the tools that computational chemists have at their disposal, chapter 2 set the stage for the application of these tools to the exploration of potential energy surfaces, and chapter 3 introduced one of these tools, molecular mechanics. In this chapter you will be introduced to *quantum mechanics*, and to *quantum chemistry*, the application of quantum mechanics to chemistry. Molecular mechanics is based on *classical* physics, physics before *modern* physics; one of the cornerstones of modern physics is quantum mechanics, and *ab initio* (chapter 5), semiempirical (SE) (chapter 6), and density functional (chapter 7) methods belong to quantum chemistry. This chapter is designed to ease the way to an understanding of the role of quantum mechanics in computational chemistry. The word *quantum* comes from Latin (*quantus*, "how much?", plural *quanta*) and was first used in our sense by Max Planck in 1900, as an adjective and noun, to denote the constrained *quantities* or amounts in which energy can be emitted or absorbed. Although the term *quantum mechanics* was apparently first used by Born (of the Born–Oppenheimer approximation, section 2.3) in 1924, in contrast to classical mechanics, the matrix algebra and differential equation techniques that we now associate with the term were presented in 1925 and 1926 (section 4.2.6).

"Mechanics" as used in physics is traditionally the study of the behavior of bodies under the action of forces like, e.g. gravity (celestial mechanics). Molecules are made of nuclei and electrons, and quantum chemistry deals, fundamentally, with the motion of electrons under the influence of the electromagnetic force exerted by nuclear charges. An understanding of the behavior of electrons in molecules, and thus of the structures and reactions of molecules, rests on quantum mechanics and in particular on that adornment of quantum chemistry, the Schrödinger equation. For that reason we will consider in outline the development of quantum mechanics leading up to the

Schrödinger equation, and then the birth of quantum chemistry with (at least as far as molecules of reasonable size goes) the application of the Schrödinger equation to chemistry by Hückel. This *simple Hückel method* is currently disdained by some theoreticians, but its discussion here is justified by the fact that (1) it continues to be useful in research and (2) it "is immensely useful as a model, today … because it is the model which preserves the ultimate physics, that of nodes in wave functions. It is the model which throws away absolutely everything except the last bit, the only thing that if thrown away would leave nothing. So it provides fundamental understanding."[1] A discussion of a generalization of the simple Hückel method, the extended Hückel method, sets the stage for chapter 5. The historical approach used here, although perforce somewhat superficial, may help to ameliorate the apparent arbitrariness of certain features of quantum chemistry [1,2]. An excellent introduction to quantum chemistry is the text by Levine [3].

Our survey of the factors that led to modern physics and quantum chemistry will follow the sequence:

(1) the origins of quantum theory: blackbody radiation and the photoelectric effect;

(2) radioactivity (brief);

(3) relativity (very brief);

(4) the nuclear atom;

(5) the Bohr atom;

(6) the wave mechanical atom and the Schrödinger equation.

4.2   THE DEVELOPMENT OF QUANTUM MECHANICS (THE SCHRÖDINGER EQUATION)

4.2.1   The origins of quantum theory: blackbody radiation and the photoelectric effect

Three discoveries mark the transition from classical to modern physics: quantum theory, radioactivity, and relativity (Fig. 4.1). Quantum theory had its origin in the study of blackbody radiation and the photoelectric effect.

Blackbody radiation
A blackbody is one that is a perfect absorber of radiation: it absorbs all the radiation falling on it, without reflecting any. More relevant for us, the radiation emitted by a hot blackbody depends (as far as the distribution of energy with wavelength goes) only on the temperature, not on the material the body is made of, and is thus amenable to relatively simple analysis. The sun is approximately a blackbody; in the lab a good source of blackbody radiation is a furnace with blackened insides and a small aperture for the radiation to escape. In the second half of the nineteenth century the distribution of

---

[1] Personal communication from Professor Roald Hoffmann (see the extended Hückel method, section 4.4).

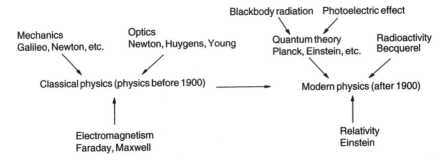

Figure 4.1. The discoveries marking the transition from classical to modern to physics. Although radioactivity was discovered in 1896, its understanding had to wait for relativity and quantum theory.

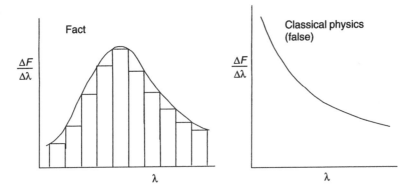

Figure 4.2. In the limit the bar graph becomes a curve, the graph of $f(\lambda)$ vs. λ, where $f(\lambda) = \lim_{\Delta\lambda \to 0} \Delta F/\Delta\lambda = df/dy$, essentially intensity of radiation vs. wavelength. Planck's efforts to find the function $f(\lambda)$ led to the quantum theory.

energy with respect to wavelength that characterizes blackbody radiation was studied, in research that is associated mainly with Lummer and Pringsheim [1]. They plotted the flux ΔF (in modern SI units, $J s^{-1} m^{-2}$) per wavelength emitted by a blackbody over a wavelength range $\Delta\lambda$ vs. the wavelength, for various temperatures (Fig. 4.2): $\Delta F/\Delta\lambda$ vs. λ. The result is a histogram or bar graph in which the area of each rectangle is $(\Delta F/\Delta\lambda)\Delta\lambda = \Delta F$ and represents the flux (energy per second per unit area) emitted in the wavelength range covered by that $\Delta\lambda$; $\Delta F/\Delta\lambda$ can be called the flux density for that particular wavelength range $\Delta\lambda$. The total area of all the rectangles is the total flux emitted over its whole wavelength range by the blackbody. As $\Delta\lambda$ approaches zero (note that for the nonmonochromatic radiation from a blackbody the flux *at* a particular wavelength is essentially zero) the histogram approaches being a smooth curve, the ratio of finite increments approximates a derivative, and we can ask: what is the function (Fig. 4.2) $dF/d\lambda = f(\lambda)$? In the answer to this question lay the beginnings of quantum theory.

Late nineteenth century physics, classical physics at its zenith, predicted that the flux density emitted by a black body should rise without limit as the wavelength decreases. This is because classical physics held that radiation of a particular frequency was emitted by oscillators (atoms or whatever) vibrating with that frequency, and that the average energy of an oscillator was independent of its frequency; since the number of possible frequencies increases without limit, the flux density (energy per second per unit area per wavelength interval) from the blackbody should rise without limit toward higher frequencies or shorter wavelengths, into the ultraviolet, and so the total flux (energy per second per unit area) should be infinite. This is clearly absurd and was recognized as being absurd; in fact, it was called "the ultraviolet catastrophe" [1]. To understand the nature of blackbody radiation and to escape the ultraviolet catastrophe, physicists in the 1890s tried to find the function (Fig. 4.2) $f(\lambda)$.

Without breaking with classical physics, Wien had found a theoretical equation that fit the Lummer–Pringsheim curve at relatively short wavelengths, and Rayleigh and Jeans one that fit at relatively long wavelengths. Max Planck[2] adopted a different approach: he found, in 1900, a purely empirical equation $dF/d\lambda = f(\lambda)$ that fit the facts, and then tried to interpret the equation theoretically. To do this he had to make two assumptions:

(1) the total energy possessed by the oscillators in the frequency range $v + dv$ (v is the Greek letter *nu*, commonly used for frequency, not to be confused with υ, vee, commonly used for velocity) is proportional to the frequency:

$$E_{tot}(v + dv) \propto v \qquad (4.1)$$

(2) the emission or absorption of radiation of frequency v by the collection of oscillators is caused by jumps between energy levels, with loss or gain of a quantity of energy kv

$$\Delta E = kv \qquad (4.2)$$

The constant k, now recognized as a fundamental constant of nature, 6.626×10^{-34} J s particle^{-1}, is called Planck's constant, and is denoted by h, so Eq. (4.2) becomes

$$\Delta E = hv \qquad (4.3)$$

Why the letter h? Evidently because h is sometimes used in mathematics to denote infinitesimals and Planck intended to let this quantity go to zero. In the event, it turned out to be small but finite. Apparently the letter was first used to denote the new constant in a talk given by Planck at a meeting (*Sitzung*) of the German Physical Society in Berlin, on 14 December 1900 [4]. The interpretation of Eq. (4.3), a fundamental equation of quantum theory, as meaning that the energy represented by radiation of frequency v is absorbed and emitted in *quantized* amounts hv (definite, constrained amounts; jerkily rather than continuously) was, ironically, apparently never fully accepted by Planck [5]. Planck's constant is a measure of the graininess of our universe: because it is so small processes involving energy changes often seem to take place smoothly,

---

[2]Max Planck, born Kiel, Germany, 1858. Ph.D. Berlin 1879. Professor, Kiel, Berlin. Nobel prize in physics for quantum theory of blackbody radiation 1918. Died Göttingen, 1947.

but on an ultramicroscopic scale the graininess is there [6]. The constant h is the hallmark of quantum expressions, and its finite value distinguishes our universe from a nonquantum one.

The photoelectric effect
An apparently quite separate (but in science no two phenomena are really ever unrelated) phenomenon that led to Eq. (4.3), which is to say to quantum theory, is the photoelectric effect: the ejection of electrons from a metal surface exposed to light. The first inkling of this phenomenon was due to Hertz,[3] who in 1888 noticed that the potential needed to elicit a spark across two electrodes decreased when ultraviolet light shone on the negative electrode. Beginning in 1902, the photoelectric effect was first studied systematically by Lenard,[4] who showed that the phenomenon observed by Hertz was due to electron emission.

Facts (Fig. 4.3) that classical physics could not explain were the existence of a threshold frequency for electron ejection, that the kinetic energy of the electrons is linearly related to the frequency of the light, and the fact that the electron flux (electrons per unit area per second) is proportional to the intensity of the light. Classical physics predicted that the electron flux should be proportional to the light frequency, decreasing with a decrease in frequency, but without sharply falling to zero below a certain frequency, and that the kinetic energy of the electrons should be proportional to the intensity of the light, not the frequency.

These facts were explained by Einstein[5] in 1905 in a way that now appears very simple, but in fact relies on concepts that were at the time revolutionary. Einstein went beyond Planck and postulated that not only was the *process* of absorption and emission of light quantized, but that light itself was quantized, consisting of in effect of particles of energy

$$E_{\text{particle}} = h\nu \tag{4.4}$$

where ν is the frequency of the light. These particles were given the name *photons* (Arthur Compton, ca. 1923). If the energy of the photon before it removes an electron from the metal is equal to the energy required to tear the electron free of the metal, plus the kinetic energy of the free electron, then

$$h\nu = W + \tfrac{1}{2}m_e v^2 \tag{4.5}$$

where W is the work function of the metal, energy needed to remove an electron (with no energy left over), m_e is the mass of an electron, v is the velocity of electron ejected by

[3] Heinrich Hertz, born Hamburg, Germany, 1857. Ph.D. Berlin, 1880. Professor, Karlsruhe, Bonn. Discoverer of radio waves. Died Bonn, 1894.
[4] Philipp Lenard, German physicist, born Pozsony, Austria–Hungary (now Bratislava, Slovakia), 1862. Ph.D. Heidelberg 1886. Professor, Heidelberg. Nobel prize in physics 1905, for work on cathode rays. Lenard supported the Nazis and rejected Einstein's theory of relativity. Died Messelhausen, Germany, 1947.
[5] Albert Einstein, German–Swiss–American physicist. Born Ulm, Germany, 1879. Ph.D. Zürich 1905. Professor Zürich, Prague, Berlin; Institute for Advanced Studies, Princeton, New Jersey. Nobel Prize in physics 1921 for theory of the photoelectric effect. Best known for the special (1905) and general (1915) theories of relativity. Died Princeton, 1955.

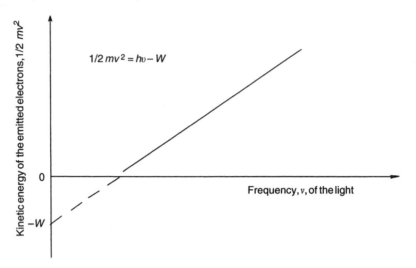

Figure 4.3. The photoelectric effect. Einstein explained the effect by extending to light Planck's idea of the absorption and emission of energy in discrete amounts: he postulated that light itself consisted of discrete particles.

the photon, and $\frac{1}{2}m_e v^2$ is the kinetic energy of the free electron. Rearranging Eq. (4.5):

$$\tfrac{1}{2}m_e v^2 = h\nu - W \tag{4.6}$$

Thus a plot of the kinetic energies of the electrons ($\frac{1}{2}m_e v^2$) vs. the frequency v of the light should be a straight line of positive slope (h; this is one way to find Planck's constant) intersecting the v axis at a positive value ($v = W/h$), as experiment indeed showed (Fig. 4.3).

Planck's explanation of the blackbody radiation curves (1900 [4]) and Einstein's explanation of the facts of the photoelectric effect (1905 [7]) indicated that the flow of energy in physical processes did not take place continuously, as had been believed, but rather jerkily, in discrete jumps, quantum by quantum. The contributions of Planck and Einstein were the signal developments marking the birth of quantum theory and the transition from classical to modern physics.

4.2.2   Radioactivity

Brief mention of radioactivity is in order because it, along with quantum mechanics and relativity, transformed classical into modern physics. Radioactivity was discovered by Becquerel in 1896. However, an understanding of how materials like uranium and radium could emit, over the years, a million times more energy than would be permitted by chemical reactions, had to await Einstein's special theory of relativity (section 4.2.3), which showed that a tiny, unnoticeable decrease in mass represented the release of a large amount of energy.

4.2.3 Relativity

Relativity is relevant to computational chemistry because it must often be explicitly taken into account in accurate calculations involving atoms heavier than about chlorine or bromine (see below) and because, strictly speaking, the Schrödinger equation, the fundamental equation of quantum chemistry, is an approximation to a relativistic equation, the Dirac[6] equation.

Relativity was discovered in by Einstein in 1905, when he formulated the special theory of relativity, which deals with nonaccelerated motion in the absence of significant gravitational fields (general relativity, published by Einstein in 1915, is concerned with accelerated motion and gravitation). Special relativity predicted a relationship between mass and energy, the famous $E = mc^2$ equation and, of more direct relevance to computational chemistry, showed that the mass of a particle increases with its velocity, dramatically so near the velocity of light. In heavier elements the inner electrons are moving at a significant fraction of the speed of light, and the relativistic increase in their masses affects the chemistry of these elements (actually, some physicists do not like to think in terms of rest mass and relativistic mass, but that is a controversy that need not concern us here). In computational chemistry relativistic effects on electrons are usually accounted for by what are called effective core potentials or pseudopotentials (section 5.3.3).

4.2.4 The nuclear atom

The "nuclear atom" is the picture of the atom as a positive nucleus surrounded by negative electrons. Although the idea of atoms in speculative philosophy goes back to at least the time of Democritus,[7] the atom as the basis of a scientifically credible theory emerges only in nineteenth century, with the rationalization by Dalton[8] in 1808 of the law of definite proportions. Nevertheless, atoms were regarded by many scientists of the positivist school of Ernst Mach as being at best a convenient hypothesis, despite the success of the atomistic Maxwell–Boltzmann[9] kinetic theory of gases and it was not until 1908, when Perrin's[10] experiments confirmed Einstein's atomistic analysis of

---

[6]Paul Adrien Maurice Dirac, born Bristol, England, 1902. Ph.D. Cambridge, 1926. Professor, Cambridge, Dublin Institute for Advanced Studies, University of Miami, Florida State University. Nobel prize in physics 1933 (shared with Schrödinger). Known for his mathematical elegance, for connecting relativity with quantum theory, and for predicting the existence of the positron. Died Tallahassee, Florida, 1984.

[7]Democritus, Greek philosopher, born Abdera, Thrace (the eastern Balkans) ca. 470 B.C. Died ca. 370 B.C.

[8]John Dalton, born Eaglesfield, England, 1766. Considered the founder of quantitative chemical atomic theory: law of definite proportions, pioneered determination of atomic weights. Cofounder of British Association for the Advancement of Science. Died Manchester, England, 1844.

[9]Ludwig Boltzmann, born Vienna 1844. Ph.D. Vienna. Professor Graz, Vienna. Developed the kinetic theory of gases independently of Maxwell (viz., Boltzmann constant's, k). Firm supporter of the atomic theory in opposition to Mach and Ostwald, helped develop concept of entropy (S). Died Duino, Austria (now in Italy), 1906 (suicide incurred by depression). Inscribed on gravestone: $S = k \log W$.

[10]Jean Perrin, born Lille, France, 1870. Ph.D. École Normal Supérieure, Paris. Professor University of Paris. Nobel prize in physics 1923. Died New York, 1942.

Brownian motion that the reality of atoms was at last accepted by such eminent holdouts as Boltzmann's opponent Ostwald.[11]

The atom has an internal structure; it is thus not "atomic" in the Greek sense and is more than the mere restless particle of kinetic theories of gases or of Brownian motion. This was shown by two lines of work: the study of the passage of electricity through gases and the behavior of certain solutions. The study of the passage of electricity through gases at low pressure was a very active field of research in the nineteenth century and only a few of the pioneers in what we can now see as the incipient field of subatomic physics will be mentioned here. The observation by Plücker in 1858 of a fluorescent glow near the cathode on the glass walls of a current-carrying evacuated tube was one of the first inklings that particles might be elicited from atoms. That these were indeed particles rather than electromagnetic rays was indicated by Crookes in the 1870s, by showing that they could be deflected by a magnet. Goldstein showed in 1886 the presence of particles of opposite charge to those emitted from the cathode, and christened the latter "cathode rays." That the cathode rays were negative *particles* was proved by Perrin in 1895, when he showed that they imparted a charge to an object on which they fell. Further evidence of the particle nature of cathode rays came at around the same time from Thomson,[12] who showed (1897) that they are deflected in the expected direction by an electric field. Thomson also measured their mass-to-charge ratio and from the smallest possible value of charge in electrochemistry calculated the mass of these particles to be about 1/1837 of the mass of a hydrogen atom. Lorentz later applied the name "electron" to the particle, adopting a term that had been appropriated from the Greek by Stoney for a unit of electric current ($\varepsilon\lambda\varepsilon\kappa\tau\rho\sigma\nu$, amber, which acquires a charge when rubbed). Thomson has been called the discoverer of the electron.

It was perhaps Thomson who first suggested a specific structure for the atom in terms of subatomic particles. His "plum pudding" model (ca. 1900), which placed electrons in a sea of positive charge, like raisins in a pudding, accorded with the then-known facts in evidently permitting electrons to be removed under the influence of an electric potential. The modern picture of the atom as a positive nucleus with extranuclear electrons was proposed by Rutherford[13] in 1911. It arose from experiments in which alpha particles from a radioactive sample were shot through very thin gold foil. Most of the time the particles passed through, but occasionally one bounced back, indicating that the foil was mostly empty space, but that present were particles which were small and, compared to the mass of the electron (which was much too light to stop an alpha particle), massive. From these experiments emerged our picture of the atom as consisting of a small, relatively massive positive nucleus surrounded by electrons: the nuclear atom.

---

[11] Wilhelm Friedrich Ostwald, German chemist, born Riga, Latvia, 1853. Ph.D. Dorpat, Estonia. Professor Riga, Leipzig. A founder of physical chemistry, opponent of the atomic theory till convinced by the work of Einstein and Perrin. Nobel prize in chemistry 1909. Died near Leipzig, 1932.

[12] Sir Joseph John Thomson, born near Manchester, 1856. Professor, Cambridge. Nobel prize in physics 1906. Knighted 1908. Died Cambridge, 1940.

[13] Ernest Rutherford (Baron Rutherford), born near Nelson New Zealand, 1871. Studied at Cambridge under J. J. Thomson. Professor McGill University (Montreal), Manchester, and Cambridge. Nobel prize in chemistry 1908 for work on radioactivity, alpha particles, and atomic structure. Knighted 1914. Died London, 1937.

Rutherford gave the name *proton* (from Greek $\pi\rho\omega\tau o\zeta$, primary or first) to the least massive of these nuclei (the hydrogen nucleus).

There is another thread to the development of the concept of the atom as a composite of subatomic particles. The enhanced effect of electrolytes (solutes that provide an electrically conducting solutions) on boiling and freezing points and on the osmotic pressure of solutions led Arrhenius[14] in 1884 to propose that these substances exist in water as atoms or groups of atoms with an electric charge. Thus sodium chloride in solution would not, as was generally held, exist as NaCl molecules but rather as an a positive sodium "atom" and a negative chlorine "atom"; the presence of two particles instead of the expected one accounted for the enhanced effects. The ability of atoms to lose or gain charge hinted at the existence of some kind of subatomic structure, and although the theory was not warmly received (Arrhenius was almost failed on his Ph.D. exam), the confirmation by Thomson (ca. 1900) that the atom contains electrons made acceptable the concept of charged atoms with chemical properties quite different from those of the neutral ones. Arrhenius' was awarded the Nobel prize for his Ph.D. work.

4.2.5   The Bohr atom

The nuclear atom as formulated by Rutherford faced a serious problem: the electrons orbit the nucleus like planets orbiting the Sun. An object engaged in circular (or elliptical) motion experiences an acceleration because its direction is changing and thus its velocity, which unlike speed is a vector, is also changing. An electron in circular motion about a nucleus would experience an acceleration toward the nucleus, and since from Maxwell's equations of electromagnetism an accelerated electric charge radiates away energy, the electron should lose energy by spiralling in toward the nucleus, ending up there, with no kinetic and potential energy; calculations show this should happen in a fraction of a second [8].

A way out of this dilemma was suggested by Bohr[15] in 1913 [9,10]. He retained the classical picture of electrons orbiting the nucleus in accord with Newton's laws, but subject to the constraint that the angular momentum of an electron must be an integral multiple of $h/2\pi$:

$$mvr = n(h/2\pi), \quad n = 1, 2, 3, \ldots \tag{4.7}$$

where m is the electron mass, v is the electron velocity, r is the radius of electron orbit, and h is the Planck's constant. Equation (4.7) is the Bohr postulate, that electrons can defy Maxwell's laws provided they occupy an orbit whose angular momentum (i.e. an orbit of appropriate radius) satisfies Eq. (4.7). The Bohr postulate is not based on a whim, as most textbooks imply, but rather follows from: (1) the Plank equation Eq. (4.3), $\Delta E = h\nu$ and (2) starting with an orbit of large radius such that the motion is essentially linear and classical physics applies, as no acceleration is involved, then extrapolating

---

[14]Svante Arrhenius, born near Uppsala, Sweden, 1859. Ph.D. University of Stockholm. Nobel prize in chemistry 1903. Professor Stockholm. Died Stockholm 1927.

[15]Niels Bohr, born Copenhagen, 1885. Ph.D. University of Copenhagen. Professor, University of Copenhagen. Nobel prize in physics 1922. Founder of the "Copenhagen school" interpretation of quantum theory. Died Copenhagen, 1962.

to small-radius orbits. The fading of quantum-mechanical equations into their classical analogues as macroscopic conditions are approached is called the *correspondence principle* [11].

Using the postulate of Eq. (4.7) and classical physics, Bohr derived an equation for the energy of an orbiting electron in a one-electron atom (a hydrogen-like atom, H or He$^+$, etc.) in terms of the charge on the nucleus and some constants of nature. Starting with the total energy of the electron as the sum of its kinetic and potential energies:

$$E_t = \frac{1}{2}mv^2 - \frac{Ze^2}{4\pi\varepsilon_0 r} \tag{4.8}$$

where Z is the nuclear charge (1 for H, 2 for He, etc.), e is the charge on the electron, ε_0 is the permitivity of the vacuum. Using force = mass × acceleration:

$$\frac{Ze^2}{4\pi\varepsilon_0 r^2} = \frac{mv^2}{r} \tag{4.9}$$

i.e.

$$\frac{Ze^2}{4\pi\varepsilon_0 r} = mv^2 \tag{4.10}$$

So from Eq. (4.8)

$$E_t = \tfrac{1}{2}mv^2 - mv^2 = -\tfrac{1}{2}mv^2 \tag{4.11}$$

From Eqs (4.7) and (4.10):

$$v = \frac{Ze^2}{2\varepsilon_0 nh} \tag{4.12}$$

So from Eqs (4.11) and (4.12):

$$E_t = -\frac{Z^2 e^4 m}{8\varepsilon_0^2 n^2 h^2} \tag{4.13}$$

Equation (4.13) expresses the total (kinetic plus potential) energy of the electron of a hydrogen-like atom in terms of four fundamental quantities of our universe: electron charge, electron mass, the permittivity of empty space, and Planck's constant. From Eq. (4.13) the energy change involved in emission or absorption of light by a hydrogen-like atom is simply

$$\Delta E = E_{t2} - E_{t1} = \frac{mZ^2 e^4}{8\varepsilon_0^2 h^2}\left(\frac{1}{n_1^2} - \frac{1}{n_2^2}\right) \tag{4.14}$$

where ΔE is the energy of a state characterized by *quantum number n_2*, minus the energy of a state characterized by quantum number n_1. Note that from Eq. (4.13) the total energy increases (becomes less negative) as n increases (= 1, 2, 3, ...), so higher-energy states are associated with higher quantum numbers n and $\Delta E > 0$ corresponds to absorption of energy and $\Delta E < 0$ to emission of energy. The Planck relation between the amount of radiant energy absorbed or emitted and its frequency ($\Delta E = h\nu$, Eq. (4.3)), Eq. (4.14) enables one to calculate the frequencies of spectroscopic absorption and emission lines for hydrogen-like atoms. The agreement with experiment is excellent, and the same is true too for the calculated ionization energies of hydrogen-like atoms (ΔE for $n_2 = \infty$ in Eq. (4.14)).

4.2.6 The wave mechanical atom and the Schrödinger equation

The Bohr approach works well for hydrogen-like atoms, atoms with one electron: hydrogen, singly-ionized helium, doubly-ionized lithium, etc. However, it showed many deficiencies for other atoms, which is to say, almost all atoms of interest other than hydrogen. The problems with the Bohr atom for these cases were described below.

(1) There were lines in the spectra corresponding to transitions other than simply between two n values (cf. Eq. (4.14)). This was rationalized by Sommerfeld in 1915, by the hypothesis of elliptical rather than circular orbits, which essentially introduced a new quantum number k, a measure of the eccentricity of the elliptical orbit. Electrons could have the same n but different ks, increasing the variety of possible electronic transitions; k is related to what we now call the azimuthal quantum number, l; $l = k-1$).

(2) There were lines in the spectra of the alkali metals that were not accounted for by the quantum numbers n and k. In 1925 Goudsmit and Uhlenbeck showed that these could be explained by assuming that the electron spins on an axis; the magnetic field generated by this spin around an axis could reinforce or oppose the field generated by the orbital motion of the electron around the nucleus. Thus for each n and k there are two closely-spaced "magnetic levels," making possible new, closely-spaced spectral lines. The spin quantum number, $m_s = +\frac{1}{2}$ or $-\frac{1}{2}$, was introduced to account for spin.

(3) There were new lines in atomic spectra in the presence of an *external* magnetic field (not to be confused with the fields generated by the electron itself). This Zeeman effect (1896) was accounted for by the hypothesis that the electron orbital plane can take up only a limited number of orientations, each with a different energy, with respect to the external field. Each orientation was associated with a magnetic quantum number m_m (often designated m) $= -l, -(l-1), \ldots, (l-1), l$. Thus in an external magnetic field the numbers n, k (later l) and m_s are insufficient to describe the energy of an electron and new transitions, invoking m_m, are possible.

The only quantum number that flows naturally from the Bohr approach is the principal quantum number, n; the azimuthal quantum number l (a modified k), the spin quantum number m_s and the magnetic quantum number m_m are all *ad hoc*, improvised to meet an experimental reality. Why should electrons move in elliptical orbits that depend on the principal quantum number n? Why should electrons spin, with only two values for this spin? Why should the orbital plane of the electron take up with respect to an external magnetic field only certain orientations, which depend on the azimuthal quantum number? All four quantum numbers should follow naturally from a satisfying theory of the behaviour of electrons in atoms.

The limitations of the Bohr theory arise because it does not reflect a fundamental facet of nature, namely the fact that particles possess wave properties. These limitations were transcended by the *wave mechanics* of Schrödinger,[16] when he devised his famous equation in 1926 [12,13]. Actually, the year before the Schrödinger equation was

---

[16]Erwin Schrödinger, born Vienna, 1887. Ph.D. University of Vienna. Professor Stuttgart, Berlin, Graz (Austria), School for Advanced Studies Dublin, Vienna. Nobel prize in physics 1933 (shared with Dirac). Died Vienna, 1961.

published, Heisenberg[17] published his matrix mechanics approach to calculating atomic (and in principle molecular) properties. The matrix approach is at bottom equivalent to Schrödinger's use of differential equations, but the latter has appealed to chemists more because, like physicists of the time, they were unfamiliar with matrices (section 4.3.3), and because the wave approach lends itself to a physical picture of atoms and molecules while manipulating matrices perhaps tends to resemble numerology. Matrix mechanics and wave mechanics are usually said to mark the birth of quantum *mechanics* (1925, 1926), as distinct from quantum theory (1900). We can think of quantum mechanics as the rules and equations used to calculate the properties of molecules, atoms, and subatomic particles.

Wave mechanics grew from the work of de Broglie,[18] who in 1923 was led to this "wave-particle duality" by his ability to deduce the Wien blackbody equation (section 4.2.1) by treating light as a collection of particles (light quanta) analogous to an ideal gas [14]. This suggested to de Broglie that light (traditionally considered a wave motion) and the atoms of an ideal gas were actually not fundamentally different. He derived a relationship between the wavelength of a particle and its momentum, by using the time-dilation principle of special relativity, and also from an analogy between optics and mechanics. The reasoning below, while less profound than de Broglie's, may be more accessible. From the special theory of relativity, the relation between the energy of a photon and its mass is

$$E_p = mc^2 \qquad (4.15)$$

where c is the velocity of light. From the Planck equation (4.3) for the emission and absorption of radiation, the energy E_p of a photon may be equated with the energy change ΔE of an oscillator, and we may write

$$E_p = h\nu \qquad (4.16)$$

From Eqs (4.15) and (4.16)

$$mc^2 = h\nu \qquad (4.17)$$

Since $\nu = c/\lambda$, Eq. (4.17) can be written

$$mc = h/\lambda \qquad (4.18)$$

and because the product of mass and velocity is momentum, Eq. (4.18) can be written

$$p_p = h/\lambda \qquad (4.19)$$

relating the momentum of a photon (in its particle aspect) to its wavelength (in its wave aspect). If Eq. (4.19) can be generalized to any particle, then we have

$$p = h/\lambda \qquad (4.20)$$

---

[17]Werner Heisenberg, born Würzburg, Germany, 1901. Ph.D. Munich, 1923. Professor, Leipzig University, Max Planck Institute. Nobel Prize 1932 for his famous uncertainty principle of 1927. Director of the German atomic bomb/reactor project 1939–1945. Held various scientific administrative positions in postwar (Western) Germany 1945–1970. Died Munich 1976.

[18]Louis de Broglie, born Dieppe, 1892. Ph.D. University of Paris. Professor Sorbonne, Institut Henri Poincaré (Paris). Nobel prize in physics 1929. Died Paris, 1987.

relating the momentum of a particle to its wavelength; this is the de Broglie equation.

If a particle has wave properties it should describable by somehow combining the de Broglie equation and a classical wave equation. A highly developed nineteenth century mathematical theory of waves was at Schrödinger's disposal, and the union of a classical wave equation with Eq. (4.20) was one of the ways that he derived his wave equation. Actually, it is said that the Schrödinger equation cannot actually be *derived*, but is rather a postulate of quantum mechanics that can only be justified by the fact that it works [15]; this fine philosophical point will not be pursued here. Of his three approaches [15], Schrödinger's simplest is outlined here. A standing wave (one with fixed ends like a vibrating string or a sound wave in a flute) whose amplitude varies with time and with the distance from the ends is described by

$$\frac{d^2 f(x)}{dx^2} = -\frac{4\pi^2}{\lambda^2} f(x) \tag{4.21}$$

where $f(x)$ is the amplitude of the wave, x is the distance from some chosen origin, and λ is the wavelength. From Eq. (4.20):

$$\lambda = h/mv \tag{4.22}$$

where λ is the wavelength of particle of mass m and velocity v. Identifying the wave with a particle and substituting for λ from (4.22) into (4.21):

$$\frac{d^2 f(x)}{dx^2} = -\frac{4\pi^2 m^2 v^2}{h^2} f(x) \tag{4.23}$$

Since the total energy of the particle is the sum of its kinetic and potential energies:

$$E_{kin} = E - E_{pot} = E - V \tag{4.24}$$

where E is the total energy of the particle, and V is the potential energy (the usual symbol), i.e.

$$\tfrac{1}{2}mv^2 = E - V \tag{4.25}$$

Substituting Eq. (4.25) for mv^2 into Eq. (4.23):

$$\frac{d^2 f(x)}{dx^2} = -\frac{8\pi^2 m}{h^2}(E - V)f(x) \tag{4.26}$$

where $f(x)$ is the amplitude of the particle/wave at a distance x from some chosen origin, m is the mass of the particle, E is the total energy (kinetic + potential) of the particle, and V is the potential energy of the particle (possibly a function of x). This is the Schrödinger equation for one-dimensional (1D) motion along the spatial coordinate x. It is usually written

$$\frac{d^2\psi}{dx^2} + \frac{8\pi^2 m}{h^2}(E - V)\psi = 0 \tag{4.27}$$

where ψ is the amplitude of the particle/wave at a distance x from some chosen origin. The 1D Schrödinger equation is easily elevated to 3D status by replacing the 1D operator d^2/dx^2 by its 3D analogue

$$\frac{\partial^2}{\partial x^2} + \frac{\partial^2}{\partial y^2} + \frac{\partial^2}{\partial z^2} = \nabla^2 \tag{4.28}$$

∇^2 is the Laplacian operator "del squared." Replacing d^2/dx^2 by ∇^2, Eq. (4.27) becomes

$$\nabla^2 \psi + \frac{8\pi^2 m}{h^2}(E - V)\psi = 0 \tag{4.29}$$

This is a common way of writing the Schrödinger equation. It relates the amplitude ψ of the particle/wave to the mass m of the particle, its total energy E and its potential energy V. The meaning of ψ itself is unknown [2] but the currently popular interpretation of ψ^2, due to Born (section 2.3) and Pauli[19] is that it is proportional to the probability of finding the particle near a point $P(x, y, z)$ (recall that ψ is a function of x, y, z):

$$\text{Prob}(dx, dy, dz) = \psi^2 \, dx \, dy \, dz \tag{4.30}$$

$$\text{Prob}(V) = \int_V \psi^2 \, dx \, dy \, dz \tag{4.31}$$

The probability of finding the particle in an infinitesimal cube of sides dx, dy, dz is $\psi^2 \, dx \, dy \, dz$, and the probability of finding the particle somewhere in a volume V is the integral over that volume of ψ^2 with respect to dx, dy, dz (a triple integral); ψ^2 is thus a *probability density function*, with units of probability per unit volume. Born's interpretation was in terms of the probability of a particular state, Pauli's the chemist's usual view, that of a particular location.

The Schrödinger equation overcame the limitations of the Bohr approach (see the beginning of section 4.2.6): the quantum numbers follow naturally from it (actually the spin quantum number m_s requires a relativistic form of the Schrödinger equation, the Dirac equation, and electron "spin" is apparently not really due to the particle spinning like a top). The Schrödinger equation can be solved in an exact analytical way only for one-electron systems like the hydrogen atom, the helium monocation and the hydrogen molecule ion, but the mathematical approach is complicated and of no great relevance to the application of this equation to the study of serious molecules. However a brief account of the results for hydrogen-like atoms is in order.

The standard approach to solving the Schrödinger equation for *hydrogen-like atoms* involves transforming it from Cartesian (x, y, z) to polar coordinates (r, θ, ϕ), since these accord more naturally with the spherical symmetry of the system. This makes it possible to separate the equation into three simpler equations, $f(r) = 0$, $f(\theta) = 0$, and $f(\phi) = 0$. Solution of the $f(r)$ equation gives rise to the n quantum number, solution of the $f(\theta)$ equation to the l quantum number, and solution of the $f(\phi)$ equation to the m_m (often simply called m) quantum number. For each specific $n = n'$, $l = l'$ and

---

[19]Wolfgang Pauli, born Vienna, 1900. Ph.D. Munich 1921. Professor Hamburg, Zurich, Princeton, Zurich. Best known for the Pauli exclusion principle. Nobel Prize 1945. Died Zurich 1958.

$m_m = m'_m$ there is a mathematical function obtained by combining the appropriate $f(r)$, $f(\theta)$ and $f(\phi)$:

$$\psi(r, \theta, \phi; n', l', m'_m) = f(r)f(\theta)f(\phi) \tag{4.32}$$

The function $\psi(r, \theta, \phi)$ (clearly ψ could also be expressed in Cartesians), depends *functionally* on r, θ, ϕ and *parametrically* on n, l and m_m: for each particular set (n', l', m'_m) of these numbers there is a particular function with the spatial coordinates variables r, θ, ϕ (or x, y, z). A function like $k \sin x$ is a function of x and depends only parametrically on k. This ψ function is an *orbital* ("quasi-orbit"; the term was invented by Mulliken, section 4.3.4), and you are doubtless familiar with plots of its variation with the spatial coordinates. Plots of the variation of ψ^2 with spatial coordinates indicate variation of the electron density (recall the Born interpretation of the wavefunction) in space due to an electron with quantum numbers n', l' and m'_m. We can think of an orbital as a region of space occupied by an electron with a particular set of quantum numbers, or as a mathematical function ψ describing the energy and the shape of the spatial domain of an electron. For an atom or molecule with more than one electron, the assignment of electrons to orbitals is an (albeit very useful) *approximation*, since orbitals follow from solution of the Schrödinger equation for a hydrogen atom.

The Schrödinger equation that we have been talking about is actually the *time-independent* (and nonrelativistic) Schrödinger equation: the variables in the equation are spatial coordinates, or spatial and spin coordinates (section 5.2.3.1) when electron spin is taken into account. The time-independent equation is the one most widely-used in computational chemistry, but the more general *time-dependent Schrödinger equation*, which we shall not examine, is important in certain applications, like some treatments of the interaction of a molecule with light, since light (radiation) is composed of time-varying electric and magnetic fields. The time-dependent density functional theory (DFT) method of calculating UV spectra (chapter 7) is based on the time-dependent Schrödinger equation.

4.3   THE APPLICATION OF THE SCHRÖDINGER EQUATION TO CHEMISTRY BY HÜCKEL

4.3.1   Introduction

The quantum mechanical methods described in this book are all molecular orbital (MO) methods, or oriented toward the MO approach: *ab initio* and SE methods use the MO method, and density functional methods are oriented toward the MO approach. There is another approach to applying the Schrödinger equation to chemistry, namely the valence bond (VB) method. Basically the MO method allows atomic orbitals (AOs) to interact to create the MO of a molecule, and does not focus on individual bonds as shown in conventional structural formulas. The VB method, on the other hand, takes the molecule, mathematically, as a sum (linear combination) of structures each of which corresponds to a structural formula with a certain pairing of electrons [16]. The MO method explains in a relatively simple way phenomena that can be understood only with difficulty using the VB method, like the triplet nature of dioxygen or the fact that

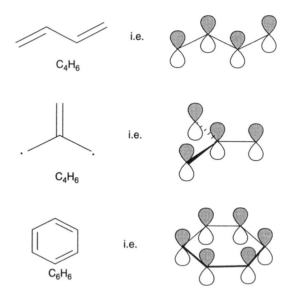

Figure 4.4. The SHM is used mainly for planar arrays of π systems.

benzene is aromatic but cyclobutadiene is not [17]. With the application of computers to quantum chemistry the MO method almost eclipsed the VB approach, but the latter has in recent years made a limited comeback [18].

The first application of quantitative quantum theory to chemical species significantly more complex than the hydrogen atom was the work of Hückel[20] on unsaturated organic compounds, in 1930–1937 [19]. This approach, in its simplest form, focuses on the p electrons of double bonds, aromatic rings and heteroatoms. Although Hückel did not initially explicitly consider orbital hybridization (the concept is usually credited to Pauling,[21] 1931 [20]), the method as it became widely applied [21] confines itself to planar arrays of sp^2-hybridized atoms, usually carbon atoms, and evaluates the consequences of the interactions among the p electrons (Fig. 4.4). Actually, the simple Hückel method has been occasionally applied to nonplanar systems [22]. Because of the importance of the concept of hybridization in the simple Hückel method a brief discussion of this concept is warranted.

4.3.2   Hybridization

Hybridization is the mixing of orbitals on an atom to produce new, "hybridized" (in the spirit of the biological use of the term), AOs. This is done mathematically but can be appreciated pictorially (Fig. 4.5). One way to justify the procedure theoretically is to

[20]Erich Hückel, born Berlin, 1896. Ph.D. Göttingen. Professor, Marburg. Died Marburg, 1980.
[21]Linus Pauling, born Portland, Oregon, 1901. Ph.D. Caltech. Professor, Caltech. Known for work in quantum chemistry and biochemistry, campaign for nuclear disarmament, and controversial views on vitamin C. Nobel prize for chemistry 1954, for peace 1963. Died near Big Sur, CA, 1994.

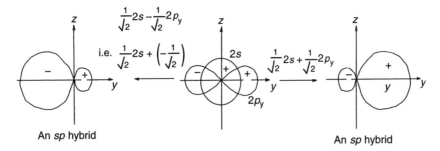

An *sp* hybrid   An *sp* hybrid

Figure 4.5. Hybridization is forming new AOs, on an atom, by mathematically mixing (combining) "original" AOs on that atom. Mixing two orbitals gives two hybrid orbitals, and in general n AOs give n hybrid AOs. Orbitals are mathematical functions and so can be added and subtracted as shown above.

recognize that AOs are vectors in the generalized mathematical sense of being elements of a vector space [23] (if not in the restricted sense of the physicist as physical entities with magnitude and direction); it is therefore permissible to take linear combinations of these vectors to produce new members of the vector space. A good, brief introduction to hybridization in is given by Streitwieser [24].

In a familiar example, a $2s$ orbital can be mixed with three $2p$ orbitals to give four hybrid orbitals; this can be done in an infinite number of ways, such as (from now on ϕ will be used for AOs and ψ for MOs):

$$\phi_1 = \tfrac{1}{2}(s + p_x + p_y + p_z), \quad \phi_2 = \tfrac{1}{2}(s + p_x + p_y - p_z),$$
$$\phi_3 = \tfrac{1}{2}(s + p_x - p_y - p_z), \quad \phi_4 = \tfrac{1}{2}(s + p_x - p_y + p_z) \tag{4.33}$$

or

$$\phi_a = \tfrac{1}{2}(s + p_x + 2^{1/2}p_z), \quad \phi_b = \tfrac{1}{2}(s + p_x - 2^{1/2}p_z),$$
$$\phi_c = \tfrac{1}{2}(s - p_x + 2^{1/2}p_y), \quad \phi_d = \tfrac{1}{2}(s - p_x - 2^{1/2}p_y) \tag{4.34}$$

Both the set (4.33) and the set (4.34) consist of four sp^3 orbitals, since the electron density contributions from the component s and p orbitals to the hybrid is, in each case (considering the *squares* of the coefficients; recall the Born interpretation of the square of a wavefunction, section 4.2.6) in the ratio 1 : 3, i.e. $\tfrac{1}{4} : 3(\tfrac{1}{4})$ and $\tfrac{1}{4} : (\tfrac{1}{4} + \tfrac{2}{4})$, and in each set we have used a total of one s orbital, and one each of the p_x, p_y and p_z orbitals. The total electron density from each component orbital is unity, e.g. for s, $4(\tfrac{1}{4})$.

Hybridization is purely a mathematical procedure, originally invented to reconcile the quantum mechanical picture of electron density in s, p, etc. orbitals with traditional views of directed valence. For example, it is sometimes said that in the absence of hybridization combining a carbon atom with four unpaired electrons with four hydrogen atoms would give a methane molecule with three equivalent, mutually perpendicular bonds and a fourth, different, bond (Fig. 4.6). Actually, this is incorrect: the $2s$ and three $2p$ orbitals of an unhybridized carbon along with the four $1s$ orbitals of four hydrogen atoms provide, without invoking hybridization, a tetrahedrally symmetrical

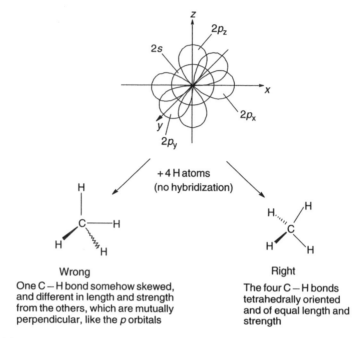

Figure 4.6. Hybridization is not needed to explain bonding, e.g. the tetrahedral geometry of methane.

valence electron distribution that leads to tetrahedral methane with four equivalent bonds (Fig. 4.6). In fact, it has been said "It is sometimes convenient to combine AOs to form hybrid orbitals that have well defined directional character and then to form MOs by combining these hybrid orbitals. This recombination of AOs to form hybrids is *never* necessary..." [25]. Interestingly, the MOs accommodating the four highest-energy electron pairs of methane (the eight valence electrons) are *not* equivalent in *energy* (not *degenerate*). This is an experimental fact that can be shown by photoelectron spectroscopy [26]. Instead of four orbitals of the same energy we have three degenerate orbitals and one lower in energy (and of course the almost undisturbed $1s$ core orbital of carbon). This surprising arrangement is a consequence of the fact that symmetry requires one combination (i.e. one MO) of carbon and hydrogen orbitals (essentially a weighted sum of the C$2s$ and the four H$1s$ orbitals) to be unique and the other three AO combinations (the other three MOs) to be degenerate (they involve the C$2p$ and the H$1s$ orbitals) [26,27]. It must be emphasized that although the methane valence orbitals are *energetically* different, the electron and nuclear distribution *is* tetrahedrally symmetrical – the molecule indeed has T_d (section 2.6) symmetry. The four MOs formed directly from AOs are the *canonical* MOs. They are *delocalized* (spread out over the molecule), and do not correspond to the familiar four bonding Csp^3/H$1s$ MOs, each of which is localized between the carbon nucleus and a hydrogen nucleus. However, the canonical MOs can be mathematically manipulated to give the familiar localized MOs (section 5.2.3.1).

Another example illustrates a situation somewhat similar to that we saw with methane, and what was until fairly recently a serious controversy: the best way to represent the carbon/carbon double bond [28]. The currently popular way to conceptualize the C=C bond has it resulting from the union of two sp^2-hybridized carbon atoms (Fig. 4.7); the sp^2 orbitals on each carbon overlap end-on forming a σ bond and the p orbitals on each carbon overlap sideways forming a π bond. Note that the usual depiction of a carbon p orbital is unrealistically spindle-shaped, necessitating depicting overlap with connecting lines as in Fig. 4.7. Figure 4.8 shows a picture in better accord with the calculated electron density in the p orbital, i.e. corresponding to the square of the wavefunction. The two leftover sp^2 orbitals can be used to bond to, say, hydrogen atoms, as shown. From this viewpoint the double bond is thus composed of a σ bond and a π bond. However, this is not the only way to represent the C=C bond. One can, for example, mathematically construct a carbon atom with two sp^2 orbitals and two sp^5 orbitals; the union of two such carbons gives a double bond formed from two sp^5/sp^5 bonds (Fig. 4.9), rather than from a σ bond and a π bond. Which is right? They are only different ways of viewing the same thing: the electron density in the C=C bond decreases smoothly from the central C/C axis in both models (Fig. 4.10), and the experimental ^{13}C/H NMR coupling constant for the C–H bond would, in both models, be predicted to correspond to about 33% s character in the orbital used by carbon to bond to hydrogen [29]. The ability of the hybridization concept to correlate and rationalize acidities of hydrocarbons in terms of the s character of the carbon orbital in a C–H bond [29] is an example of the usefulness of this idea. Most of the systems studied by the simple Hückel method are essentially flat, as expected for sp^2 arrays, and many

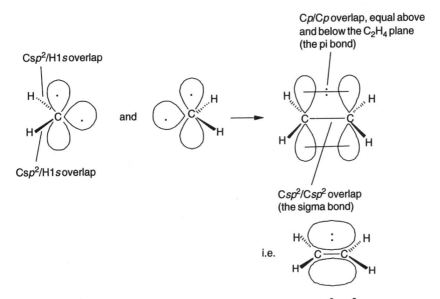

Figure 4.7. The currently popular view of the C/C double bond: an $sp^2/sp^2\sigma$ bond and a p/p π bond. Compare this with Figs 4.8 and 4.9.

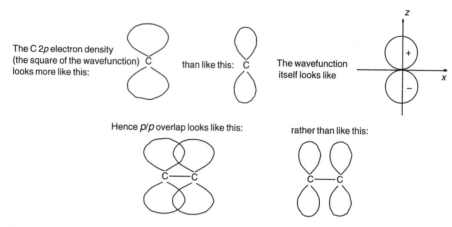

Figure 4.8. The electron density is represented by the *square* of the mathematical function we call the orbital. A carbon $2p$ orbital is actually more buxom than its conventional representation, and two $2p$ orbitals overlap better than the usual picture, e.g. Fig. 4.7, suggests.

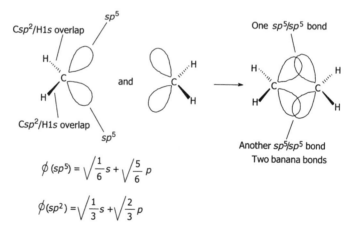

$$\phi(sp^5) = \sqrt{\frac{1}{6}}s + \sqrt{\frac{5}{6}}p$$

$$\phi(sp^2) = \sqrt{\frac{1}{3}}s + \sqrt{\frac{2}{3}}p$$

Figure 4.9. The C/C double bond can be built from two sp^5 orbitals. The result is the same as using a σ bond and a π bond (Fig. 4.7): see Fig. 4.10.

properties of these molecules can be at least qualitatively understood by considering the in-plane σ electrons of the overlapping sp^2 orbitals to simply represent a framework that holds the perpendicular p orbitals, in which we are interested, in an orientation allowing neighboring p orbitals to overlap.

Before moving on to Hückel theory we take a look at matrices, since matrix algebra is the simplest and most elegant way to handle the linear equations that arise when MO theory is applied to chemistry.

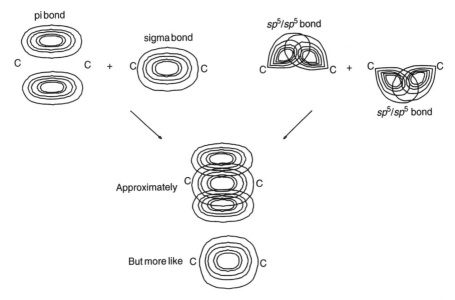

Figure 4.10. The model of a C/C double bond as a σ/π bond is at bottom really equivalent to the $sp^5/sp^5 + sp^5/sp^5$ model: both result in the same electron distribution, which is the physically significant thing. There are no gaps in electron density between the carbons: as the contribution to density from the σ bond (or one of the sp^5/sp^5 bonds) falls off, the contribution from the π bond (or the other sp^5/sp^5 bond) increases. The electron density falls off smoothly with distance from the C/C axis. For some purposes one of the models, σ/π or bent (banana) bonds, may be more useful.

4.3.3   Matrices and determinants

Matrix algebra was invented by Cayley[22] as a systematic way of dealing with systems of linear equations. The single equation in one unknown

$$ax = b$$

has the solution

$$x = a^{-1}b$$

Consider next the system of two equations in two unknowns

$$(1) \quad a_{11}x + a_{12}y = c_1$$

$$(2) \quad a_{21}x + a_{22}y = c_2$$

---

[22] Arthur Cayley, lawyer and mathematician, born Richmond, England, 1821. Graduated Cambridge. Professor, Cambridge. After Euler and Gauss, history's most prolific author of mathematical papers. Died Cambridge, 1895.

The subscripts of the unknowns coefficients a indicate row 1, column 1, row 1, column 2, etc. We will see that using matrices the solutions (the values of x and y) can be expressed in a way analogous to that for the equation $ax = b$.

A matrix is a rectangular array of "elements" (numbers, derivatives, etc.) that obeys certain rules of addition, subtraction, and multiplication. Curved or angular brackets are used to denote a matrix:

$$\begin{pmatrix} 1 & 2 \\ 7 & 2 \end{pmatrix} \qquad \begin{pmatrix} 5 \\ 2 \\ 0 \end{pmatrix} \qquad (0 \ \ 0 \ \ 7 \ \ 4)$$

$$2 \times 2 \text{ matrix} \quad 3 \times 1 \text{ matrix} \quad 1 \times 4 \text{ matrix}$$

or

$$\begin{bmatrix} 1 & 2 \\ 7 & 2 \end{bmatrix} \qquad \begin{bmatrix} 5 \\ 2 \\ 0 \end{bmatrix} \qquad [0 \ \ 0 \ \ 7 \ \ 4]$$

Do not confuse matrices with determinants (below), which are written with straight lines, e.g.

$$\begin{vmatrix} 1 & 2 \\ 7 & 3 \end{vmatrix}$$

is a determinant, not a matrix. This determinant represents the number $1 \times 3 - 2 \times 7 = 3 - 14 = -11$. In contrast to a determinant, a matrix is not a number, but rather an operator (although some would consider matrices to be generalizations of numbers, with e.g. the 1×1 matrix $(3) = 3$). An operator *acts on* a function (or a vector) to give a new function, e.g. d/dx acts on (differentiates) $f(x)$ to give $f'(x)$:

$$\frac{d}{dx} f(x) = \frac{df(x)}{dx} = f'(x)$$

and the square root operator acts on y^2 to give y. When we have done matrix multiplication you will see that a matrix can act on a vector and rotate it through an angle to give a new vector.

Let us look at matrix addition, subtraction, multiplication by scalars, and matrix multiplication (multiplication of a matrix by a matrix).

Addition and subtraction

Matrices of the same size (2×2 and 2×2, 3×1 and 3×1, etc.) can be added just by adding corresponding elements:

$$\begin{pmatrix} 2 & 1 \\ 7 & 4 \end{pmatrix} + \begin{pmatrix} 1 & 3 \\ 5 & 6 \end{pmatrix} = \begin{pmatrix} 2+1 & 1+3 \\ 7+5 & 4+6 \end{pmatrix} = \begin{pmatrix} 3 & 4 \\ 12 & 10 \end{pmatrix}$$

$$\begin{pmatrix} 7 \\ 0 \\ 3 \end{pmatrix} + \begin{pmatrix} 4 \\ 4 \\ 1 \end{pmatrix} = \begin{pmatrix} 7+4 \\ 0+4 \\ 3+1 \end{pmatrix} = \begin{pmatrix} 11 \\ 4 \\ 4 \end{pmatrix}$$

Subtraction is similar:

$$\begin{pmatrix} 2 & 1 \\ 7 & 4 \end{pmatrix} - \begin{pmatrix} 1 & 3 \\ 5 & 6 \end{pmatrix} = \begin{pmatrix} 2-1 & 1-3 \\ 7-5 & 4-6 \end{pmatrix} = \begin{pmatrix} 1 & -2 \\ 2 & -2 \end{pmatrix}$$

Multiplication by a scalar

A scalar is an ordinary number (in contrast to a vector or an operator), e.g. 1, 2, $\sqrt{2}$, 1.714, π, etc. To multiply a matrix by a scalar we just multiply every element by the number:

$$2 \begin{pmatrix} 2 & 1 \\ 7 & 4 \end{pmatrix} = \begin{pmatrix} 2 \times 2 & 2 \times 1 \\ 2 \times 7 & 2 \times 4 \end{pmatrix} = \begin{pmatrix} 4 & 2 \\ 14 & 8 \end{pmatrix}$$

Matrix multiplication

We could define matrix multiplication to be analogous to addition: simply multiplying corresponding elements. After all, in mathematics any rules are permitted, as long as they do not lead to contradictions. However, as we shall see in a moment, for matrices to be useful in dealing with simultaneous equations we must adopt a slightly more complex multiplication rule. The easiest way to understand matrix multiplication is to first define series multiplication. If

$$\text{series } a = S_a = a_1\, a_2\, a_3 \cdots, \quad \text{and} \quad \text{series } b = S_b = b_1\, b_2\, b_3 \cdots$$

then we define the series product as

$$S_a\, S_b = a_1 b_1 + a_2 b_2 + a_3 b_3 + \cdots$$

So for example, if $S_a = 5\ 2\ 1$ and $S_b = 3\ 6\ 2$, then

$$S_a\, S_b = 5(3) + 2(6) + 1(2) = 15 + 12 + 2 = 29$$

Now it is easy to understand matrix multiplication: if $\mathbf{AB} = \mathbf{C}$, where $\mathbf{A}$, $\mathbf{B}$, and $\mathbf{C}$ are matrices, then element i, j of the product matrix $\mathbf{C}$ is the series product of row i of $\mathbf{A}$ and column j of $\mathbf{B}$. For example,

$$\mathbf{AB} = \begin{pmatrix} 1 & 3 \\ 7 & 2 \end{pmatrix} \begin{pmatrix} 2 & 4 \\ 5 & 6 \end{pmatrix} = \begin{pmatrix} 1(2) + 3(5) & 1(4) + 3(6) \\ 7(2) + 2(5) & 7(4) + 2(6) \end{pmatrix} = \begin{pmatrix} 17 & 22 \\ 24 & 40 \end{pmatrix}$$

(With practice, you can multiply simple matrices in your head.) Note that matrix multiplication is not *commutative*: $\mathbf{AB}$ is not *necessarily* $\mathbf{BA}$, e.g.

$$\mathbf{BA} = \begin{pmatrix} 2 & 4 \\ 5 & 6 \end{pmatrix} \begin{pmatrix} 1 & 3 \\ 7 & 2 \end{pmatrix} = \begin{pmatrix} 2(1) + 4(7) & 2(3) + 4(2) \\ 5(1) + 6(7) & 5(3) + 6(2) \end{pmatrix} = \begin{pmatrix} 30 & 14 \\ 47 & 27 \end{pmatrix}$$

(Two matrices are identical if and only if their corresponding elements are the same.) Note that two matrices may be multiplied together only if the number of columns of the first equals the number of rows of the second. Thus we can multiply $\mathbf{A}(2 \times 2)\mathbf{B}(2 \times 2)$, $\mathbf{A}(2 \times 2)\mathbf{B}(2 \times 3)$, $\mathbf{A}(3 \times 1)\mathbf{B}(1 \times 3)$, and so on. A useful mnemonic is $(a \times b)(b \times c) = (a \times c)$, meaning, for example that $\mathbf{A}(2 \times 1)$ times $\mathbf{B}(1 \times 2)$ gives $\mathbf{C}(2 \times 2)$:

$$\begin{pmatrix} 5 \\ 2 \end{pmatrix} \begin{pmatrix} 0 & 3 \end{pmatrix} = \begin{pmatrix} 5(0) & 5(3) \\ 2(0) & 2(3) \end{pmatrix} = \begin{pmatrix} 0 & 15 \\ 0 & 6 \end{pmatrix}$$

It is helpful to know beforehand the size, i.e. (2×2), (3×3), whatever, of the matrix you will get on multiplication.

To get an idea of why matrices are useful in dealing with systems of linear equations, let us go back to our system of equations

$$(1) \quad a_{11}x + a_{12}y = c_1$$

$$(2) \quad a_{21}x + a_{22}y = c_2$$

Provided certain conditions are met this can be solved for x and y, e.g. by solving (1) for x in terms of y then substituting for x in (2) etc. Now consider the equations from the matrix viewpoint. Since

$$\mathbf{AB} = \begin{pmatrix} a_{11} & a_{12} \\ a_{21} & a_{22} \end{pmatrix} \begin{pmatrix} x \\ y \end{pmatrix} = \begin{pmatrix} a_{11}x + a_{12}y \\ a_{21}x + a_{22}y \end{pmatrix}$$

clearly **AB** corresponds to the left-hand side of the system, and the system can be written

$$\mathbf{AB} = \mathbf{C}, \quad \text{where } \mathbf{C} = \begin{pmatrix} c_1 \\ c_2 \end{pmatrix}$$

A is the coefficients matrix, **B** is the unknowns matrix, and **C** is the constants matrix. Now, if we can find a matrix $\mathbf{A}^{-1}$ such that $\mathbf{A}^{-1}\mathbf{AB} = \mathbf{B}$ (analogous to the numbers $a^{-1}ab = b$) then

$$\mathbf{A}^{-1}\mathbf{AB} = \mathbf{A}^{-1}\mathbf{C}, \quad \text{i.e. } \mathbf{B} = \mathbf{A}^{-1}\mathbf{C}$$

Thus the unknowns matrix is simply the inverse of the coefficients matrix times the constants matrix. Note that we multiplied by $\mathbf{A}^{-1}$ on the left ($\mathbf{A}^{-1}\mathbf{AB} = \mathbf{A}^{-1}\mathbf{C}$), which is not the same as multiplying on the right, which would give $\mathbf{ABA}^{-1} = \mathbf{CA}^{-1}$; this is not necessarily the same as **B**.

To see that a matrix can act as an operator consider the vector from the origin to the point $P(3, 4)$. This can be written as a column matrix, and multiplying it by the rotation matrix shown transforms it (rotates it) into another matrix:

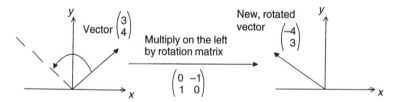

Some important kinds of matrices
These matrices are particularly important in computational chemistry:

(1) the zero matrix (the null matrix),
(2) diagonal matrices,
(3) the unit matrix (the identity matrix),
(4) the inverse of another matrix,
(5) symmetric matrices,

(6) the transpose of another matrix,

(7) orthogonal matrices.

(1) The zero matrix or null matrix, **0**, is any matrix with all its elements zero. Examples:

$$\begin{pmatrix} 0 & 0 \\ 0 & 0 \end{pmatrix} \quad \begin{pmatrix} 0 & 0 & 0 \\ 0 & 0 & 0 \end{pmatrix} \quad (0 \ 0 \ 0 \ 0)$$

Clearly, multiplication by the zero matrix (when the $(a \times b)(b \times c)$ mnemonic permits multiplication) gives a zero matrix.

(2) A diagonal matrix is a *square* matrix that has all its off-diagonal elements zero; the (principal) diagonal runs from the upper left to the lower right. Examples:

$$\begin{pmatrix} 2 & 0 \\ 0 & 4 \end{pmatrix} \quad \begin{pmatrix} 3 & 0 & 0 \\ 0 & 6 & 0 \\ 0 & 0 & 1 \end{pmatrix} \quad \begin{pmatrix} 0 & 0 & 0 \\ 0 & 0 & 0 \\ 0 & 0 & 0 \end{pmatrix}$$

(3) the unit matrix or identity matrix **1** or **I** is a diagonal matrix whose diagonal elements are all unity. Examples:

$$\begin{pmatrix} 1 & 0 \\ 0 & 1 \end{pmatrix} \quad \begin{pmatrix} 1 & 0 & 0 \\ 0 & 1 & 0 \\ 0 & 0 & 1 \end{pmatrix} \quad (1)$$

Since diagonal matrices are square, unit matrices must be square (but zero matrices can be any size). Clearly, multiplication (when permitted) by the unit matrix leaves the other matrix unchanged: $\mathbf{1A} = \mathbf{A1} = \mathbf{A}$.

(4) The inverse $\mathbf{A}^{-1}$ of another matrix **A** is the matrix that, multiplied **A**, on the left or right, gives the unit matrix: $\mathbf{A}^{-1}\mathbf{A} = \mathbf{AA}^{-1} = \mathbf{1}$. Example:

$$\text{If } \mathbf{A} = \begin{pmatrix} 1 & 2 \\ 3 & 4 \end{pmatrix}, \quad \text{then } \mathbf{A}^{-1} = \begin{pmatrix} -2 & 1 \\ \frac{3}{2} & \frac{-1}{2} \end{pmatrix}$$

Check it out.

(5) A symmetric matrix is a square matrix for which $a_{ij} = a_{ji}$ for each element. Examples:

$$\begin{pmatrix} 1 & 4 \\ 4 & 3 \end{pmatrix} a_{12} = a_{21} = 4, \quad \begin{pmatrix} 2 & 7 & 1 \\ 7 & 3 & 5 \\ 1 & 5 & 4 \end{pmatrix} a_{12} = a_{21} = 7, \text{ etc.}$$

Note that a symmetric matrix is unchanged by rotation about its principal diagonal. The complex-number analogue of a symmetric matrix is a *Hermitian matrix* (after the mathematician Charles Hermite); this has $a_{ij} = a_{ji}^*$, e.g. if element $(2, 3) = a + bi$, then element $(3, 2) = a - bi$, the complex conjugate of element $(2, 3)$; $i = \sqrt{-1}$. Since all the matrices we will use are *real* rather than complex, attention has been focussed on real matrices here.

(6) The transpose ($\mathbf{A}^T$ or $\tilde{\mathbf{A}}$) of a matrix $\mathbf{A}$ is made by exchanging rows and columns. Examples:

$$\text{If } \mathbf{A} = \begin{pmatrix} 2 & 3 \\ 4 & 7 \end{pmatrix}, \quad \text{then } \mathbf{A}^T = \begin{pmatrix} 2 & 4 \\ 3 & 7 \end{pmatrix}$$

$$\text{If } \mathbf{A} = \begin{pmatrix} 2 & 1 & 6 \\ 1 & 7 & 2 \end{pmatrix}, \quad \text{then } \mathbf{A}^T = \begin{pmatrix} 2 & 1 \\ 1 & 7 \\ 6 & 2 \end{pmatrix}$$

Note that the transpose arises from twisting the matrix around to interchange rows and columns. Clearly the transpose of a symmetric matrix $\mathbf{A}$ is the same matrix $\mathbf{A}$. For complex-number matrices, the analogue of the transpose is the *conjugate transpose* $\mathbf{A}^\dagger$; to get this form $\mathbf{A}^*$, the complex conjugate of $\mathbf{A}$, by converting each complex number element $a + bi$ in $\mathbf{A}$ to its complex conjugate $a - bi$, then switch the rows and columns of $\mathbf{A}^*$ to get $(\mathbf{A}^*)^T = \mathbf{A}^\dagger$. Physicists call $\mathbf{A}^\dagger$ the *adjoint* of $\mathbf{A}$, but mathematicians use adjoint to mean something else.

(7) An orthogonal matrix is a square matrix whose inverse is its transpose: if $\mathbf{A}^{-1} = \mathbf{A}^T$ then $\mathbf{A}$ is orthogonal. Examples:

$$\mathbf{A}_1 = \begin{pmatrix} \frac{1}{\sqrt{2}} & -\frac{1}{\sqrt{2}} \\ \frac{1}{\sqrt{2}} & \frac{1}{\sqrt{2}} \end{pmatrix}, \quad \mathbf{A}_2 = \begin{pmatrix} \frac{1}{\sqrt{6}} & -\frac{1}{\sqrt{2}} & -\frac{1}{\sqrt{3}} \\ \frac{2}{\sqrt{6}} & 0 & \frac{1}{\sqrt{3}} \\ \frac{1}{\sqrt{6}} & \frac{1}{\sqrt{2}} & -\frac{1}{\sqrt{3}} \end{pmatrix}$$

We saw that for the inverse of a matrix, $\mathbf{A}^{-1}\mathbf{A} = \mathbf{A}\mathbf{A}^{-1} = \mathbf{1}$, so for an orthogonal matrix $\mathbf{A}^T\mathbf{A} = \mathbf{A}\mathbf{A}^T = \mathbf{1}$, since here the transpose is the inverse. Check this out for the matrices shown. The complex analogue of an orthogonal matrix is a *unitary matrix*; its inverse is its conjugate transpose.

The columns of an orthonormal matrix are orthonormal vectors. This means that if we let each column represent a vector, then these vectors are mutually orthogonal and each one is normalized. Two or more vectors are orthogonal if they are mutually perpendicular (i.e. at right angles), and a vector is normalized if it is of unit length. Consider the matrix $\mathbf{A}_1$ above. If column 1 represents the vector $\mathbf{v}_1$ and column 2 the vector $\mathbf{v}_2$, then we can picture these vectors like this (the long side of a right triangle is of unit length if the squares of the other sides sum to 1):

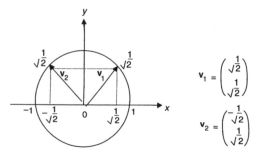

The two vectors are orthogonal: from the diagram the angle between them is clearly $90°$ since the angle each makes with, say, the x-axis is $45°$. Alternatively, the angle can be calculated from vector algebra: the dot product (scalar product) is

$$\mathbf{v}_1 \cdot \mathbf{v}_2 = |\mathbf{v}_1||\mathbf{v}_2| \cos \theta$$

where $|\mathbf{v}|$ ("mod v") is the absolute value of the vector, i.e. its length:

$$|\mathbf{v}| = (v_x^2 + v_y^2)^{1/2} \quad \text{(or } (v_x^2 + v_y^2 + v_z^2)^{1/2} \text{ for a 3D vector).}$$

Each vector is normalized, i.e. $|\mathbf{v}_1| = |\mathbf{v}_2| = (\frac{1}{2} + \frac{1}{2})^2 = 1$.
    The dot product is also

$$\mathbf{v}_1 \cdot \mathbf{v}_2 = v_{1x} v_{2x} + v_{1y} v_{2y} \quad \text{(with an obvious extension to 3D space)}$$

i.e.

$$\cos \theta = (v_{1x} v_{2x} + v_{1y} v_{2y})/|\mathbf{v}_1||\mathbf{v}_2|$$

$$= \left[\left(\tfrac{1}{\sqrt{2}} \right) \left(-\tfrac{1}{\sqrt{2}} \right) + \left(\tfrac{1}{\sqrt{2}} \right) \left(\tfrac{1}{\sqrt{2}} \right) \right] /(1)(1) = 0$$

and so

$$\theta = 90°$$

Likewise, the three columns of the matrix $\mathbf{A}_2$ above represent three mutually perpendicular, normalized vectors in 3D space. A better name for an orthogonal matrix would be an orthonormal matrix. Orthogonal matrices are important in computational chemistry because MOs can be regarded as orthonormal vectors in a generalized n-dimensional space (Hilbert space, after the mathematician David Hilbert). We extract information about MOs from matrices with the aid of *matrix diagonalization*.

Matrix diagonalization
Modern computer programs use matrix diagonalization to calculate the energies (*eigenvalues*) of MOs and the sets of coefficients (*eigenvectors*) that help define their size and shape. We met these terms, and matrix diagonalization, briefly in section 2.5; "eigen" means suitable or appropriate, and we want solutions of the Schrödinger equation that are appropriate to our particular problem. If a matrix $\mathbf{A}$ can be written $\mathbf{A} = \mathbf{PDP}^{-1}$, where $\mathbf{D}$ is a diagonal matrix (you could call $\mathbf{P}$ and $\mathbf{P}^{-1}$ pre- and postmultiplying matrices), then we say that $\mathbf{A}$ is *diagonalizable* (can be diagonalized). The process of finding $\mathbf{P}$ and $\mathbf{D}$ (getting $\mathbf{P}^{-1}$ from $\mathbf{P}$ is simple for the matrices of computational chemistry – see below) is *matrix diagonalization*. For example,

$$\text{if } \mathbf{A} = \begin{pmatrix} 4 & -2 \\ 1 & 1 \end{pmatrix},$$

$$\text{then} \quad \mathbf{P} = \begin{pmatrix} 1 & 2 \\ 1 & 1 \end{pmatrix}, \quad \mathbf{D} = \begin{pmatrix} 2 & 0 \\ 0 & 3 \end{pmatrix}, \quad \text{and } \mathbf{P}^{-1} = \begin{pmatrix} -1 & 2 \\ 1 & -2 \end{pmatrix}$$

Check it out. Linear algebra texts describe an analytical procedure using determinants, but computational chemistry employs a numerical iterative procedure called Jacobi matrix diagonalization, or some related method, in which the off-diagonal elements are made to approach zero.

Now, it can be proved that if and only if **A** is a symmetric matrix (or more generally, if we are using complex numbers, a Hermitian matrix – see symmetric matrices, above), then **P** is orthogonal (or more generally, unitary – see orthogonal matrices, above) and so the inverse $\mathbf{P}^{-1}$ of the premultiplying matrix **P** is simply the transpose of **P**, $\mathbf{P}^T$ (or more generally, what computational chemists call the *conjugate transpose* $\mathbf{A}^\dagger$ – see transpose, above). Thus

if $\mathbf{A} = \begin{pmatrix} 0 & 1 \\ 1 & 0 \end{pmatrix}$,

then $\mathbf{P} = \begin{pmatrix} 0.707 & 0.707 \\ 0.707 & -0.707 \end{pmatrix}$, $\mathbf{D} = \begin{pmatrix} 1 & 0 \\ 0 & -1 \end{pmatrix}$, $\mathbf{P}^{-1} = \begin{pmatrix} 0.707 & 0.707 \\ 0.707 & -0.707 \end{pmatrix}$

(In this simple example the transpose of **P** happens to be identical with **P**.) In the spirit of numerical methods 0.707 is used instead of $\frac{1}{\sqrt{2}}$. A matrix like **A** above, for which the premultiplying matrix **P** is orthogonal (and so for which $\mathbf{P}^{-1} = \mathbf{P}^T$) is said to be *orthogonally diagonalizable*. The matrices we will use to get MO eigenvalues and eigenvectors are orthogonally diagonalizable. A matrix is orthogonally diagonalizable if and only if it is symmetric; this has been described as "one of the most amazing theorems in linear algebra" (see S. Roman, "An Introduction to Linear Algebra with Applications", Harcourt Brace, 1988, p. 408) because the concept of orthogonal diagonalizability is not simple, but that of a symmetric matrix is very simple.

Determinants

A determinant is a *square* array of elements that is a shorthand way of writing a sum of products; if the elements are numbers, then the determinant is a number. Examples:

$$\begin{vmatrix} a_{11} & a_{12} \\ a_{21} & a_{22} \end{vmatrix} = a_{11}a_{22} - a_{12}a_{21}, \qquad \begin{vmatrix} 5 & 2 \\ 4 & 3 \end{vmatrix} = 5(3) - 2(4) = 7$$

As shown here, a 2×2 determinant can be expanded to show the sum it represents by "cross multiplication." A higher-order determinant can be expanded by reducing it to smaller determinants until we reach 2×2 determinants; this is done like this:

$$\begin{vmatrix} 2 & 1 & 3 & 0 \\ 1 & 7 & 3 & 5 \\ 3 & 4 & 6 & 1 \\ 1 & 8 & 2 & -2 \end{vmatrix} = 2 \begin{vmatrix} 7 & 3 & 5 \\ 4 & 6 & 1 \\ 8 & 2 & -2 \end{vmatrix} - 1 \begin{vmatrix} 1 & 3 & 5 \\ 3 & 6 & 1 \\ 1 & 2 & -2 \end{vmatrix}$$

$$+ 3 \begin{vmatrix} 1 & 7 & 5 \\ 3 & 4 & 1 \\ 1 & 8 & -2 \end{vmatrix} - 0 \begin{vmatrix} 1 & 7 & 3 \\ 3 & 4 & 6 \\ 1 & 8 & 2 \end{vmatrix}$$

Here we started with element (1, 1) and moved across the first row. The first of the above four terms is 2 times the determinant formed by striking out the row and column in which 2 lies, the second term is *minus* 1 times the determinant formed by striking out the row and column in which 1 lies, the third term is *plus* 3 times the determinant formed by striking out the row and column in which 3 lies, and the fourth term is *minus* 0 times the determinant formed by striking out the row and column in which 0 lies; thus starting with the element of row 1, column 1, we move along the row and multiply by +1,

$-1, +1, -1$. It is also possible to start at, say element $(2, 1)$, the number 1, and move across the second row $(-, +, -, +)$, or to start at element $(1,2)$ and go down the column $(-, +, -, +)$, etc. One would likely choose to work along a row or column with the most zeroes. The $(n - 1) \times (n - 1)$ determinants formed in expanding an $n \times n$ determinant are called *minors*, and a minor with its appropriate $+$ or $-$ sign is a *cofactor*. Expansion of determinants using minors/cofactors is called Lagrange expansion (Joseph Louis Lagrange, 1773). There are also other approaches to expanding determinants, such as manipulating them to make all the elements but one of a row or column zero; see any text on matrices and determinants. The third-order determinants in the example above can be reduced to second-order ones and so the fourth-order determinant can be evaluated as a single number. Obviously every determinant has a corresponding square matrix and every square matrix has a corresponding determinant, but a determinant is not a matrix; it is a *function* of a matrix, a rule that tells us how to take the set of numbers in a matrix and get a new number. Approaches to the study of determinants were made by Seki in Japan and Leibnitz in Europe, both in 1683. The word "determinant" was first used in our sense by Cauchy (1812), who also wrote the first definitive treatment of the topic.

Some properties of determinants
These are stated in terms of rows, but also hold for columns; D is "the determinant."

(1) If each element of a row is zero, D is zero (obvious from Lagrange expansion).

(2) Multiplying each element of a row by k multiplies D by k (obvious from Lagrange expansion).

(3) Switching two rows changes the sign of D (since this changes the sign of each term in the expansion).

(4) If two rows are the same D is zero. (follows from 3, since if $n = -n$, n must be zero.

(5) If the elements of one row are a multiple of those of another, D is zero (follows from 2 and 4).

(6) Multiplying a row by k and adding it (adding corresponding elements) to another row leaves D unchanged (in the Laplace expansion the terms with k cancel).

(7) A determinant A can be written as the sum of two determinants B and C which differ only in row i in accordance with this rule: if row i of A is $b_{i1} + c_{i1}$ $b_{i2} + c_{i2} \cdots$ then row i of B is b_{i1} $b_{i2} \cdots$ and row i of C is c_{i1} $c_{i2} \cdots$ An example makes this clear; with row i = row 3:

$$\begin{vmatrix} 1 & 3 & 6 \\ 5 & 4 & 2 \\ 8 & 11 & 9 \end{vmatrix} = \begin{vmatrix} 1 & 3 & 6 \\ 5 & 4 & 2 \\ 5+3 & 7+4 & 4+5 \end{vmatrix} = \begin{vmatrix} 1 & 3 & 6 \\ 5 & 4 & 2 \\ 5 & 7 & 4 \end{vmatrix} + \begin{vmatrix} 1 & 3 & 6 \\ 5 & 4 & 2 \\ 3 & 4 & 5 \end{vmatrix}$$

4.3.4 The simple Hückel method – theory

The derivation of the Hückel method (SHM, or simple Hückel theory, SHT; also called Hückel MO method, HMO method) given here is not rigorous and has been strongly criticized [30]; nevertheless it has the advantage of showing how with simple arguments one can use the Schrödinger equation to develop, more by a plausibility argument than a proof, a method that gives useful results and which can be extended to more powerful methods with the retention of many useful concepts from the simple approach.

The Schrödinger equation (section 4.2.6, Eq. (4.29))

$$\nabla^2\psi + \frac{8\pi^2 m}{h^2}(E - V)\psi = 0$$

can after very simple algebraic manipulation be rewritten

$$\left(-\frac{h^2}{8\pi^2 m}\nabla^2 + V\right)\psi = E\psi \tag{4.35}$$

This can be abbreviated to the seductively simple-looking form

$$\hat{H}\psi = E\psi \tag{4.36}$$

where

$$\hat{H} = \left(-\frac{h^2}{8\pi^2 m}\nabla^2 + V\right) \tag{4.37}$$

The symbol $\hat{H}$ ("H hat" or "H peak") is an *operator* (section 4.3.3): it specifies that an operation is to be performed on ψ, and Eq. (4.36) says that the result of the operation will be E multiplied by ψ. The operation to be performed on ψ (i.e. $\psi(x, y, z)$) is "differentiate it twice with respect to x, to y and to z, add the partial derivatives, and multiply the sum by $-h^2/8\pi^2 m$; then add this result to V times ψ" (now you can see why symbols replaced words in mathematical discourse). The notation $\hat{H}\psi$ means $\hat{H}$ *of* ψ, not $\hat{H}$ *times* ψ.

Eq. (4.36) says that an operator ($\hat{H}$) acting on a function (ψ) equals a constant (E) times the function (H hat of psi equals E psi). Such an equation

$$\hat{O}f = kf, \quad \hat{O} = \text{operator} \tag{4.38}$$

is called an *eigenvalue equation*. The functions f and constants k that satisfy Eq. (4.38) are eigenfunctions and eigenvalues, respectively, of the operator O. The operator $\hat{H}$ is called the Hamiltonian operator, or simply the Hamiltonian. The term is named after the mathematician Sir William Rowan Hamilton, who formulated Newton's equations of motion in a manner analogous to the quantum mechanical equation (4.36). Eigenvalue equations are very important in quantum mechanics, and we shall again meet eigenfunctions and eigenvalues.

The eigenvalue formulation of the Schrödinger equation is the starting point for our derivation of the Hückel method. We will apply Eq. (4.36) to molecules, so in this context $\hat{H}$ and ψ are the molecular Hamiltonian and wavefunction, respectively. From

$$\hat{H}\psi = E\psi$$

we get

$$\psi\hat{H}\psi = E\psi^2 \tag{4.39}$$

Note that this is not the same as $\hat{H}\psi^2 = E\psi^2$, just as $x\, df(x)/dx$, say, is not the same as $dxf(x)/dx$. Integrating and rearranging we get

$$E = \frac{\int \psi\hat{H}\psi\, dv}{\int \psi^2\, dv} \tag{4.40}$$

The integration variable dv indicates integration with respect to spatial coordinates (x, y, z in a Cartesian coordinate system), and integration over all of space is implied, since that is the domain of an electron in a molecule, and thus the domain of the variables of the function ψ. One might wonder why not simply use $E = \hat{H}\psi/\psi$; the problems with this function are that it goes to infinity as ψ approaches zero, and it is not well-behaved with regard to finding a minimum by differentiation.

Next we approximate the molecular wavefunction ψ as a linear combination of AOs (LCAO). The MO concept as a tool in interpreting electronic spectra was formalized by Mulliken[23] starting in 1932 and building on earlier (1926) work by Hund[24] [31] (recall that Mulliken coined the word orbital). The postulate behind the LCAO approach is that an MO can be "synthesized" by combining simpler functions, now called *basis functions*; these functions comprise a *basis set*. This way of calculating MOs is based on suggestions of Pauling (1928) [32] and Lennard-Jones[25] (1929) [33]. Perhaps the most important early applications of the LCAO method were the SHM (1931) [19], in which p AOs orbitals are combined to give π AOs (probably the first time that the MOs of relatively big molecules were represented as a weighted sum of AOs with optimized coefficients), and the treatment of all the lower electronic states of the hydrogen molecule by Coulson[26] and Fischer (1949) [34]. The basis functions are usually located on the atoms of the molecule, and may or may not (see the discussion of basis functions in section 5.3) be conventional AOs. The wavefunction can in principle be approximated as accurately as desired by using enough suitable basis functions. In this simplified derivation of the Hückel method we at first consider a molecule with just two atoms, with each atom contributing one basis function to the MO. Combining basis functions on different atoms to give MOs spread over the molecule is somewhat analogous to combining AOs on the same atom to give hybrid AOs (section 4.3.2) [27]. The combination of n basis functions always gives n MOs, as indicated in Fig. 4.11, and we expect two MOs for the two-AO diatomic molecule we are using here.

Using the LCAO approximation

$$\psi = c_1\phi_1 + c_2\phi_2 \tag{4.41}$$

where ϕ_1 and ϕ_2 are basis functions on atoms 1 and 2, and c_1 and c_2 are weighting coefficients to be adjusted to get the best ψ, and substituting into Eq. (4.40) we get

$$E = \frac{\int (c_1\phi_1 + c_2\phi_2)\hat{H}(c_1\phi_1 + c_2\phi_2)\,dv}{\int (c_1\phi_1 + c_2\phi_2)^2\,dv} \tag{4.42}$$

---

[23] Robert Mulliken, born Newburyport, Massachusetts, 1896. Ph.D. University of Chicago. Professor New York University, University of Chicago, Florida State University. Nobel prize in chemistry 1966, for the MO method. Died Arlington, Virginia, 1986.

[24] Friedrich Hund, born Karlsruhe, Germany, 1896. Ph.D. Marburg, 1925, Professor Rostock, Leipzig, Jena, Frankfurt, Göttingen. Died Göttingen, 1997.

[25] John Edward Lennard-Jones, born Leigh, Lancaster, England, 1894. Ph.D. Cambridge, 1924. Professor Bristol. Best known for the Lennard-Jones potential function for nonbonded atoms. Died Stoke-on-Trent, England, 1954.

[26] Charles A. Coulson, born Worcestershire, England, 1910. Ph.D. Cambridge, 1935. Professor of theoretical physics, King's College, London; professor of mathematics, Oxford; professor of theoretical chemistry, Oxford. Died Oxford, 1974. Best known for his book "Valence" (the 1st Ed., 1952).

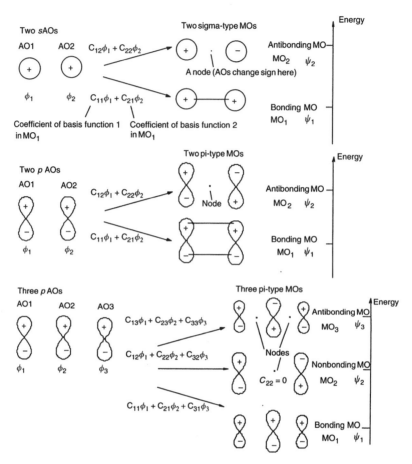

Figure 4.11. Linear combination of *n* AOs (or, more generally, basis functions) gives *n* MOs. The coefficients *c* are weighting factors that determine the magnitude and the sign of the contribution from each basis function. The functions contributing to the MO change sign at a node (actually a *nodal plane*) and the energy of the MOs increases with the number of nodes.

If we multiply out the terms in Eq. (4.42) we get

$$E = \frac{c_1^2 H_{11} + 2c_1 c_2 H_{12} + c_2^2 H_{22}}{c_1^2 S_{11} + 2c_1 c_2 S_{12} + c_2^2 S_{22}} \tag{4.43}$$

where

$$\int \phi_1 \hat{H} \phi_1 \, dv = H_{11}, \quad \int \phi_1 \hat{H} \phi_2 \, dv = H_{12} = \int \phi_2 \hat{H} \phi_1 \, dv = H_{21},$$

$$\int \phi_2 \hat{H} \phi_2 \, dv = H_{22}, \quad \int \phi_1^2 \, dv = S_{11}, \tag{4.44}$$

$$\int \phi_1 \phi_2 \, dv = S_{12} = \int \phi_2 \phi_1 \, dv = S_{21}, \quad \int \phi_2^2 \, dv = S_{22}$$

Note that in Eqs (4.43) and (4.44) the H_{ij} are not operators hence are not given hats; they are *integrals* involving $\hat{H}$ and basis functions ϕ.

For any particular molecular geometry (i.e. nuclear configuration: section 2.3, the Born–Oppenheimer approximation) the energy of the ground electronic state is the minimum energy possible for that particular nuclear arrangement and the collection of electrons that goes with it. Our objective now is to minimize the energy with respect to the basis set coefficients. We want to find the c's corresponding to the minimum on an energy vs. c's potential energy surface. To do this we follow a standard calculus procedure: set $\partial E/\partial c_1$ equal to zero, explore the consequences, then repeat for $\partial E/\partial c_2$. In theory, setting the first derivatives equal to zero guarantees only that we will find in "MO space" (an abstract space defined by an energy axis and two or more coefficient axes) a stationary point (cf. section 2.2), but examining the second derivatives shows that the procedure gives an energy minimum if all or most of the electrons are in bonding MOs, which is the case for most real molecules [35]. Write Eq. (4.43) as

$$E(c_1^2 S_{11} + 2c_1 c_2 S_{12} + c_2^2 S_{22}) = c_1^2 H_{11} + 2c_1 c_2 H_{12} + c_2^2 H_{22} \qquad (4.45)$$

and differentiate with respect to c_1:

$$\left(\frac{\partial E}{\partial c_1}\right)(c_1^2 S_{11} + 2c_1 c_2 S_{12} + c_2^2 S_{22}) + E(2c_1 S_{11} + 2c_2 S_{22}) = 2c_1 H_{11} + 2c_2 H_{12}$$

Set $\partial E/\partial c_1 = 0$:

$$E(2c_1 S_{11} + 2c_2 S_{22}) = 2c_1 H_{11} + 2c_2 H_{12}$$

This can be written

$$(H_{11} - ES_{11})c_1 + (H_{12} - ES_{12})c_2 = 0 \qquad (4.46)$$

The analogous procedure, beginning with Eq. (4.45) and differentiating with respect to c_2 leads to

$$(H_{12} - ES_{12})c_1 + (H_{22} - ES_{22})c_2 = 0 \qquad (4.47)$$

Equation (4.47) can be written as Eq. (4.48):

$$(H_{21} - ES_{21})c_1 + (H_{22} - ES_{22})c_2 = 0 \qquad (4.48)$$

since as shown in Eqs (4.44) $H_{12} = H_{21}$ and $S_{12} = S_{21}$, and the form used in Eq. (4.48) is preferable because it makes it easy to remember the pattern for the two-basis function system examined here and for the generalization (see below) to n basis functions. Equations (4.46) and (4.48) form a system of simultaneous linear equations:

$$
\begin{aligned}
(H_{11} - ES_{11})c_1 + (H_{12} - ES_{12})c_2 &= 0 \\
(H_{21} - ES_{21})c_1 + (H_{22} - ES_{22})c_2 &= 0
\end{aligned}
\qquad (4.49)
$$

The pattern is that the subscripts correspond to the row and column in which they lie; this is literally true for the matrices and determinants we will consider later, but even for the system of equations (4.49) we note that in the first equation (row 1), the coefficient of c_1 has the subscripts 11 (row 1, column 1) and the coefficient of c_2 has the subscripts 12 (row 1, column 2), while in the second equation (row 2) the coefficient of c_1 has the subscripts 21 (row 2, column 1) and the coefficient of c_2 has the subscripts 22 (row 2, column 2).

The system of equations (4.49) are called *secular equations*, because of a supposed resemblance to certain equations in astronomy that treat the long-term motion of the planets; from the Latin *saeculum*, a long period of time (not to be confused with secular meaning worldly as opposed to religious, which is from the Latin *secularis*, worldly, temporal). From the secular equations we can find the basis function coefficients c_1 and c_2, and thus the MOs ψ, since the c's and the basis functions ϕ make up the MOs (Eq. (4.41)). The simplest, most elegant and most powerful way to get the coefficients and energies of the MOs from the secular equations is to use matrix algebra (section 4.3.3). The following exposition may seem a little involved, but it must be emphasized that in practice the matrix method is implemented automatically on a computer, to which it is highly suited.

The secular equations (4.49) are equivalent to the single matrix equation

$$\begin{pmatrix} H_{11} - E S_{11} & H_{12} - E S_{12} \\ H_{21} - E S_{21} & H_{22} - E S_{22} \end{pmatrix} \begin{pmatrix} c_1 \\ c_2 \end{pmatrix} = \begin{pmatrix} 0 \\ 0 \end{pmatrix} \tag{4.50}$$

Since the $H - ES$ matrix is an H matrix minus an ES matrix, and since the ES matrix is the product of an S matrix and the scalar E, Eq. (4.50) can be written:

$$\left[\begin{pmatrix} H_{11} & H_{12} \\ H_{21} & H_{22} \end{pmatrix} - \begin{pmatrix} S_{11} & S_{12} \\ S_{21} & S_{22} \end{pmatrix} E \right] \begin{pmatrix} c_1 \\ c_2 \end{pmatrix} = \begin{pmatrix} 0 \\ 0 \end{pmatrix} \tag{4.51}$$

which can be more concisely rendered as

$$[\mathbf{H} - S E]\mathbf{c} = \mathbf{0} \tag{4.52}$$

and Eq. (4.52) can be written

$$\mathbf{H}\mathbf{c} = S E \mathbf{c} \tag{4.53}$$

$\mathbf{H}$ and $\mathbf{S}$ are square matrices and $\mathbf{c}$ and $\mathbf{0}$ are column matrices (Eq. (4.51)), and E is a scalar (an ordinary number). We have been developing these equations for a system of two basis functions, so there should be two MOs, each with its own energy and its own pair of c's (Fig. 4.11). We need two energy values and four c's: we want to be able to calculate c_{11} and c_{21} of ψ_1 (MO$_1$, energy level 1) and c_{12} and c_{22} of ψ_2 (MO$_2$, 0 energy level; in keeping with common practice the energies of the MOs are designated ε_1 and ε_2. Eq. (4.53) can be extended (our simple derivation shortchanges us here) [36] to encompass the four c's and two ε's; the result is

$$\mathbf{H}\mathbf{C} = \mathbf{S}\mathbf{C}\varepsilon \tag{4.54}$$

We now have only square matrices; in Eq. (4.53) $\mathbf{c}$ was a column matrix and E was not a matrix, but rather a scalar – an ordinary number. The matrices are:

$$\mathbf{H} = \begin{pmatrix} H_{11} & H_{12} \\ H_{21} & H_{22} \end{pmatrix}, \quad \mathbf{C} = \begin{pmatrix} c_{11} & c_{12} \\ c_{21} & c_{22} \end{pmatrix},$$

$$\mathbf{S} = \begin{pmatrix} S_{11} & S_{12} \\ S_{21} & S_{22} \end{pmatrix}, \quad \boldsymbol{\varepsilon} = \begin{pmatrix} \varepsilon_1 & 0 \\ 0 & \varepsilon_2 \end{pmatrix} \tag{4.55}$$

The $\mathbf{H}$ matrix is an energy-elements matrix, the *Fock*[27] *matrix*, whose elements are integrals H_{ij} (Eqs (4.44). Fock actually pointed out the need to take electron spin into account in more elaborate calculations than simple Hückel method; we will meet "real" Fock matrices in chapter 5. For now, we just note that in the simple (and extended) Hückel methods as an *ad hoc* prescription at most two electrons, paired, are allowed in each MO. Each H_{ij} represents some kind of energy term, since $\hat{H}$ is an energy operator (section 4.3.3). The meaning of the H_{ij}'s is discussed later in this section.

The $\mathbf{C}$ matrix is the *coefficient matrix*, whose elements are the weighting factors c_{ij} that determine to what extent each basis function ϕ (roughly, each atomic orbital on an atom) contributes to each MO ψ. Thus c_{11} is the coefficient of ϕ_1 in ψ_1, c_{21} the coefficient of ϕ_2 in ψ_1, etc., with the first subscript indicating the basis function and the second subscript the MO (Fig. 4.11). In each column of $\mathbf{C}$ the c's belong to the same MO.

The $\mathbf{S}$ matrix is the *overlap matrix*, whose elements are *overlap integrals* S_{ij} which are a measure of how well pairs of basis functions (roughly, AOs) overlap. Perfect overlap, between identical functions on the same atom, corresponds to $S_{ii} = 1$, while zero overlap, between different functions on the same atom or well-separated functions on different atoms, corresponds to $S_{ij} = 0$.

The diagonal $\boldsymbol{\varepsilon}$ matrix is an energy-levels matrix, whose diagonal elements are MO energy levels ε_i, corresponding to the MOs ψ_i. Each ε_i is ideally the negative of the energy needed to remove an electron from that orbital, i.e. the negative of the ionization energy for that orbital. Thus it is ideally the energy of an electron attracted to the nuclei and repelled by the other electrons, relative to the energy of that electron and the corresponding ionized molecule, infinitely separated from one another. This is seen by the fact that photoelectron spectra correlate well with the energies of the occupied orbitals, in more elaborate (*ab initio*) calculations [26]. In simple Hückel calculations, however, the quantitative correlation is largely lost.

Now suppose that the basis functions ϕ had these properties (the H and S integrals, involving ϕ, are defined in Eqs (4.44)):

$$S_{11} = 1, \quad S_{12} = S_{21} = 0, \quad S_{22} = 1 \tag{4.56}$$

More succinctly, suppose that

$$S_{ij} = \delta_{ij} \tag{4.57}$$

---

[27]Vladimer Fock, born St. Petersburg, 1898. Ph.D. Petrograd University, 1934. Professor Leningrad University, also worked at various institutes in Moscow. Worked on quantum mechanics and relativity, e.g. the Klein–Fock equation for particles with spin in an electromagnetic field. Died Leningrad, 1974.

where δ_{ij} is the Kronecker delta (Leopold Kronecker, German mathematician, ca. 1860) which has the property of being 1 or 0 depending on whether i and j are the same or different. Then the **S** matrix (Eqs (4.55)) would be

$$\mathbf{S} = \begin{pmatrix} 1 & 0 \\ 0 & 1 \end{pmatrix} \tag{4.58}$$

Since this is a unit matrix Eq. (4.54) would become

$$\mathbf{HC} = \mathbf{C}\varepsilon \tag{4.59}$$

and by multiplying on the right by the inverse of **C** we get

$$\mathbf{H} = \mathbf{C}\varepsilon\mathbf{C}^{-1} \tag{4.60}$$

So from the definition of matrix diagonalization, diagonalization of the **H** matrix will give the **C** and the ε matrices, i.e. will give the coefficients c and the MO energies ε (Eqs (4.55)), *if* $S_{ij} = \delta_{ij}$ (Eq. (4.57)). This is a big if, and in fact it is not true. $S_{ij} = \delta_{ij}$ would mean that the basis functions are both orthogonal and normalized, i.e. orthonormal. *Orthogonal* atomic (or molecular) orbitals or functions ϕ have zero net overlap (Fig. 4.12), corresponding to $\int \phi_i \phi_j \, dv = 0$. A *normalized* orbital or function ϕ has the property $\int \phi\phi \, dv = 1$. We can indeed use a set of normalized basis functions: a suitable normalization constant k applied to an unnormalized basis function ϕ' will ensure normalization ($\phi = k\phi'$). However, we cannot choose a set of *orthogonal* atom-centered basis functions, because orthogonality implies zero overlap between the two functions in question, and in a molecule the overlap between pairs of basis functions will depend on the geometry of the molecule (Fig. 4.12). (However, as we will see later, the basis functions can be manipulated mathematically to give *combinations* of the original functions which *are* orthonormal).

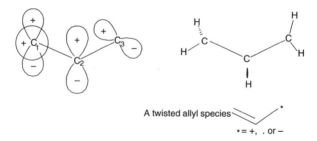

Figure 4.12. We cannot simply choose a set of orthonormal basis functions, because in a typical molecule many pairs of basis functions will not be orthogonal, i.e. will not have zero overlap. In the allyl species shown, the $2s$ and the $2p$ functions (i.e. AOs) on C_1 are orthogonal (the + part of the p orbital cancels the − part in overlap with the s orbital; in general AOs on the same atom are orthogonal), and the $2p$ functions on C_2 and C_3 are also orthogonal, if their axes are at right angles. However, the $C_1(2s)/C_2(2p)$ and the $C_1(2p)/C_2(2p)$ pairs are not orthogonal.

The assumption of basis function orthonormality is a drastic approximation, but it greatly simplifies the Hückel method, and in the present context it enables us to reduce Eq. (4.54) to Eq. (4.59), and thus to obtain the coefficients and energy levels by diagonalizing the Fock matrix. Later we will see that in the absence of the orthogonality assumption the set of basis functions can be mathematically transformed so that a modified Fock matrix can be diagonalized; in the simple Hückel method we are spared this transformation. In the matrix approach to the Hückel method, then, we must diagonalize the Fock matrix $\mathbf{H}$; to do this we have to assign numbers to the matrix elements H_{ij}, and this brings us to other simplifying assumptions of the SHM, concerning the H_{ij}.

In the SHM the energy integrals H_{ij} are approximated as just three quantities (the units are, e.g. kJ mol^{-1}): α, the *coulomb integral*

$$\int \phi_i \hat{H} \phi_i \, dv = H_{ii} = \alpha \quad \text{i.e. basis functions on the same atom,} \tag{4.61a}$$

β, the *bond integral* or *resonance integral*

$$\int \phi_i \hat{H} \phi_j \, dv = H_{ij} = \int \phi_j \hat{H} \phi_i \, dv = H_{ji} = \beta \tag{4.61b}$$

for basis functions on adjacent atoms, and finally

$$\int \phi_i \hat{H} \phi_j \, dv = H_{ij} = \int \phi_j \hat{H} \phi_i \, dv = H_{ji} = 0 \tag{4.61c}$$

for basis functions neither on the same or on adjacent atoms.

To give these approximations some physical significance, we must realize that in quantum mechanical calculations the zero of energy is normally taken as corresponding to infinite separation of the particles of a system. In the simplest view, α, the coulomb integral, is the energy of the molecule relative to a zero of energy taken as the electron and basis function (i.e. AO; in the SHM, ϕ is usually a carbon p AO) at infinite separation. Since the energy of the system actually *decreases* as the electron falls from infinity into the orbital, α is negative (Fig. 4.13). The negative of α, in this view, is the ionization energy (a positive quantity) of the orbital (the ionization energy of the orbital is defined as the energy needed to remove an electron from the orbital to an infinite distance away).

The quantity β, the bond integral or resonance integral, is, in the simplest view, the energy of an electron in the overlap region (roughly, a two-center MO) of adjacent p orbitals relative to a zero of energy taken as the electron and two-center MO at infinite separation. Like α, β is negative energy quantity. A rough, naive estimate of the value of β would be the negative of the average of the ionization energies (a positive quantity) of the two adjacent AOs, multiplied by some fraction to allow for the fact that the two orbitals do not coincide but are actually separated. These views of α and β are oversimplifications [30].

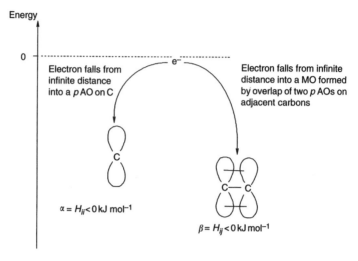

Figure 4.13. The coulomb integral α is most simply (but not too accurately) viewed as the energy of an electron in a carbon $2p$ orbital, relative to its energy an infinite distance away. The bond integral (resonance integral) β is most simply (but not too accurately) viewed as the energy of an electron in an MO formed by adjacent $2p$ orbitals, relative to its energy an infinite distance away.

We derived the 2×2 matrices of Eqs (4.55) starting with a two-orbital system. These results can be generalized to n orbitals:

$$\mathbf{H} = \begin{pmatrix} H_{11} & H_{12} & \cdots & H_{1n} \\ H_{21} & H_{22} & \cdots & H_{2n} \\ \vdots & \vdots & \cdots & \vdots \\ H_{n1} & H_{n2} & \cdots & H_{nn} \end{pmatrix} \tag{4.62}$$

The H elements of Eq. (4.62) become α, β, or 0 according to the rules of Eqs (4.61). This will be clear from the examples in Fig. 4.14.

The computer algorithms for matrix diagonalization use some version of the Jacobi rotation method [37], which proceeds by successive numerical approximations (textbooks describe a diagonalization method based on expanding the determinant corresponding to the matrix; this is not used in computational chemistry). Therefore in order to diagonalize our Fock matrices we need numbers in place of α and β. In methods more advanced than the SHM, like the extended Hückel method (EHM), other SE methods, and *ab initio* methods, the H_{ij} integrals are calculated to give numerical (in energy units) values. In the SHM we simply use energy values in $|\beta|$ units relative to α (recall that β is a negative quantity: Fig. 4.13). The matrix of Fig. 4.14(a) then becomes

$$\mathbf{H} = \begin{pmatrix} \alpha & \beta \\ \beta & \alpha \end{pmatrix} = \begin{pmatrix} 0 & -1 \\ -1 & 0 \end{pmatrix} \tag{4.63}$$

An electron in an MO represented by a 1,2-type interaction is lower in energy than one in a p orbital (1,1-type interaction) by one $|\beta|$ energy unit. Similarly, the **H** matrix of

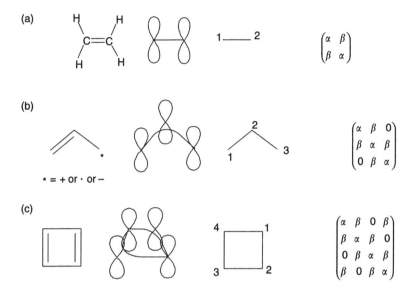

Figure 4.14. Some conjugated molecules, their p orbital arrays, simplified representations of the molecules, and their simple Hückel–Fock matrices. Same-atom interactions are α, adjacent-atom interactions are β, and all other interactions are 0. To diagonalize the matrices, we use $\alpha = 0$ and $\beta = -1$.

Fig. 4.14(b) becomes

$$\mathbf{H} = \begin{pmatrix} 0 & -1 & 0 \\ -1 & 0 & -1 \\ 0 & -1 & 0 \end{pmatrix} \tag{4.64}$$

and the $\mathbf{H}$ matrix of Fig. 4.14(c) becomes

$$\mathbf{H} = \begin{pmatrix} 0 & -1 & 0 & -1 \\ -1 & 0 & -1 & 0 \\ 0 & -1 & 0 & -1 \\ -1 & 0 & -1 & 0 \end{pmatrix} \tag{4.65}$$

The $\mathbf{H}$ matrices can be written down simply by setting all i, i-type interactions equal to 0, and all i, j-type interactions equal to -1 where i and j refer to atoms that are bonded together, and equal to 0 when i and j refer to atoms that are not bonded together.

Diagonalization of the two-basis-function matrix of Eq. (4.63) gives

$$\mathbf{H} = \begin{pmatrix} 0 & -1 \\ -1 & 0 \end{pmatrix} = \begin{pmatrix} 0.707 & 0.707 \\ 0.707 & -0.707 \end{pmatrix} \begin{pmatrix} 1 & 0 \\ 0 & -1 \end{pmatrix} \begin{pmatrix} 0.707 & 0.707 \\ 0.707 & -0.707 \end{pmatrix} \tag{4.66}$$

$$\phantom{\mathbf{H} = }\quad\;\; \mathbf{C} \qquad\qquad\quad \boldsymbol{\varepsilon} \qquad\qquad \mathbf{C}^{-1}$$

Comparing Eq. (4.66) with Eq. (4.60), we see that we have obtained the matrices we want: the coefficients matrix C and the MO energy levels matrix ε. The columns of C are eigenvectors, and the diagonal elements of ε are eigenvalues; cf. Eq. (4.38) and the associated discussion of eigenfunctions and eigenvalues. The result of Eq. (4.66) is readily checked by actually multiplying the matrices (multiplication here is aided by knowing that an analytical rather than numerical diagonalization shows that ± 0.707 are approximations to $\frac{1}{\sqrt{2}}$). Note that $CC^{-1} = 1$, and that C^{-1} is the transpose of C. The first eigenvector of C, the left-hand column, corresponds to the first eigenvalue of ε, the top left element; the second eigenvector corresponds to the second eigenvalue. The individual eigenvectors, v_1 and v_2, are column matrices:

$$\underbrace{\begin{pmatrix} 0.707 \\ 0.707 \end{pmatrix}}_{v_1} \equiv 1 \quad \text{and} \quad \underbrace{\begin{pmatrix} 0.707 \\ -0.707 \end{pmatrix}}_{v_2} \equiv -1 \tag{4.67}$$

Figure 4.15 shows a common way of depicting the results for this two-orbital calculation. Since the coefficients are weighting factors for the contributions of the basis functions to the MOs (Fig. 4.11 and associated discussion), the c's of eigenvector v_1 combine with the basis functions to give MO_1 (ψ_1) and the c's of eigenvector v_2 combine with these same basis functions to give MO_2 (ψ_2). MOs below α are bonding and MOs above α are antibonding. The ε matrix translates into an energy level diagram with ψ_1 of energy $\alpha + \beta$ and ψ_2 of energy $\alpha - \beta$, i.e. the MOs lie one $|\beta|$ unit below and one $|\beta|$ above the nonbonding α level. Since β, like α, is negative, the $\alpha + \beta$ and $\alpha - \beta$ levels are of lower and higher energy, respectively, than the nonbonding α level. Diagonalization of the three-basis function matrix of Eq. (4.64) gives

$$\begin{pmatrix} 0 & -1 & 0 \\ -1 & 0 & -1 \\ 0 & -1 & 0 \end{pmatrix}$$

$$= \underbrace{\begin{pmatrix} 0.500 & 0.707 & 0.500 \\ 0.707 & 0 & -0.707 \\ 0.500 & -0.707 & 0.500 \end{pmatrix}}_{\substack{v_1 \quad v_2 \quad v_3 \\ C}} \underbrace{\begin{pmatrix} 1.414 & 0 & 0 \\ 0 & 0 & 0 \\ 0 & 0 & -1.414 \end{pmatrix}}_{\substack{\varepsilon_1, \ 0, \ 0 \\ 0, \ \varepsilon_2, \ 0 \\ 0, \ 0, \ \varepsilon_3 \\ \varepsilon}} \underbrace{\begin{pmatrix} 0.500 & 0.707 & 0.500 \\ 0.707 & 0 & -0.707 \\ 0.500 & -0.707 & 0.500 \end{pmatrix}}_{C^{-1}} \tag{4.68}$$

The energy levels and MOs corresponding to these results are shown in Fig. 4.16.

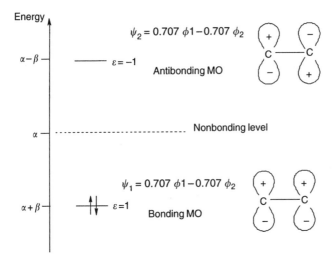

Figure 4.15. The π MOs and π energy levels for a two-p-orbital system in the SHM. The MOs are composed of the basis functions (two p AOs) and the eigenvectors, while the energies of the MOs follow from the eigenvalues (Eq. (4.66)). The paired arrows represent a pair of electrons of opposite spin (in the electronic ground state of the neutral ethene molecule ψ_1 is occupied and ψ_2 is empty).

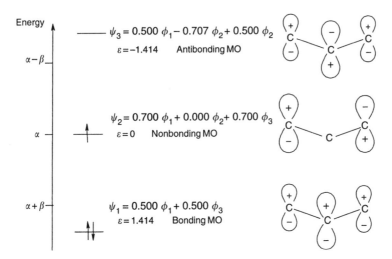

Figure 4.16. The π MOs and π energy levels for an acyclic three-p-orbital system in the simple Hückel method. The MOs are composed of the basis functions (three p AOs) and the eigenvectors (the c's), while the energies of the MOs follow from the eigenvalues (Eq. (4.68)). In the drawings of the MOs, the relative sizes of the AOs in each MO suggest the relative contribution of each AO to that MO. This diagram is for the propenyl radical. The paired arrows represent a pair of electrons of opposite spin, in the fully-occupied lowest MO, ψ_1, and the single arrow represents an unpaired electron in the nonbonding MO, ψ_2; the highest π MO, ψ_3, is empty in the radical.

Diagonalization of the four-basis-function matrix of Eq. (4.65) gives

$$
\begin{pmatrix}
0 & -1 & 0 & -1 \\
-1 & 0 & -1 & 0 \\
0 & -1 & 0 & -1 \\
-1 & 0 & -1 & 0
\end{pmatrix}
$$

$$
= \begin{pmatrix}
0.500 & 0.500 & 0.500 & 0.500 \\
0.500 & -0.500 & 0.500 & -0.500 \\
0.500 & -0.500 & -0.500 & 0.500 \\
0.500 & 0.500 & -0.500 & -0.500
\end{pmatrix}
\begin{pmatrix}
2 & 0 & 0 & 0 \\
0 & 0 & 0 & 0 \\
0 & 0 & 0 & 0 \\
0 & 0 & 0 & -2
\end{pmatrix}
$$

$$
\times \begin{pmatrix}
0.500 & 0.500 & 0.500 & 0.500 \\
0.500 & -0.500 & -0.500 & 0.500 \\
0.500 & 0.500 & -0.500 & -0.500 \\
0.500 & -0.500 & 0.500 & -0.500
\end{pmatrix}
\tag{4.69}
$$

$$
\begin{array}{cccc}
v_1 & v_2 & v_3 & v_4
\end{array}
\begin{pmatrix}
\varepsilon_1 & 0 & 0 & 0 \\
0 & \varepsilon_2 & 0 & 0 \\
0 & 0 & \varepsilon_3 & 0 \\
0 & 0 & 0 & \varepsilon_4
\end{pmatrix}
$$

$$
\mathbf{C} \qquad\qquad \varepsilon \qquad\qquad \mathbf{C}^{-1}
$$

The energy levels and MOs from these results are shown in Fig. 4.17. Note that all these matrix diagonalizations yield orthonormal eigenvectors: $v_i \cdot v_i = 1$ and $v_i \cdot v_j = 0$, as required the fact that the Fock matrices are symmetric (see the discussion of matrix diagonalization in section 4.3.3).

4.3.5   The simple Hückel method – applications

Applications of the SHM are discussed in great detail in several books [21]; here we will deal only with those applications which are needed to appreciate the utility of the method and to smooth the way for the discussion of certain topics (like bond orders and atomic charges) in later chapters. We will discuss: the nodal properties of the MOs; stability as indicated by energy levels and aromaticity (the $4n + 2$ rule); resonance energies; and bond orders and atomic charges.

The nodal properties of the MOs
A node of an MO is a plane at which, as we proceed along the sequence of basis functions, the sign of the wavefunction changes (Figs 4.15–4.17). For a given molecule, the number of nodes in the π orbitals increases with the energy. In the two-orbital system (Fig. 4.15), ψ_1 has zero nodes and ψ_2 has one node. In the three-orbital system (Fig. 4.16), ψ_1, ψ_2 and ψ_3 have zero, one and two nodes, respectively. In the cyclic four-orbital system (Fig. 4.17), ψ_1 has zero nodes, ψ_2 and ψ_3, which are degenerate (of the same energy) each have one node (one nodal plane), and ψ_4 has two nodes. In a given molecule, the energy of the MOs increases with the number of nodes. The nodal properties of the SHM π orbitals form the basis of one of the simplest ways of

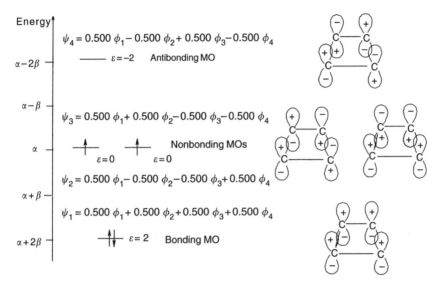

Figure 4.17. The π MOs and π energy levels for a cyclic four-p-orbital system in the SHM. The MOs are composed of the basis functions (four p AOs) and the eigenvectors, while the energies of the MOs follow from the eigenvalues (Eq. (4.69)). This particular diagram is for the square cyclobutadiene molecule. The paired arrows represent a pair of electrons of opposite spin, in the fully-occupied lowest MO, ψ_1, and the single arrows represents unpaired electrons of the same spin, one in each of the two nonbonding MOs, ψ_2 and ψ_3; the highest π MO, ψ_4, is empty in the neutral molecule.

understanding the predictions of the Woodward–Hoffmann orbital symmetry rules [38]. For example, the thermal conrotatary and disrotatary ring closure/opening of polyenes can be rationalized very simply in terms of the symmetry of the highest occupied π MO of the open-chain species. That the highest π MO should dominate the course of this kind of reaction is indicated by more detailed considerations (including extended Hückel calculations) [38]. Figure 4.18 shows the situation for the ring closure of a 1,3-butadiene to a cyclobutene. The phase ($+$ or $-$) of the π HOMO (ψ_2) at the end carbons (the atoms that bond) is opposite on each face, because this orbital has one node in the middle of the C_4 chain. You can see this by sketching the MO as the four AOs contributing to it, or even – remembering the node – drawing just the end AOs. For the electrons in ψ_2 to bond, the end groups must rotate in the same sense (*conrotation*) to bring orbital lobes of the same phase together. Remember that plus and minus phase has nothing to do with electric charge, but is a consequence of the wave nature of electrons (section 4.2.6): two electron waves can reinforce one another and form a bonding pair if they are "vibrating in phase"; an out-of-phase interaction represents an antibonding situation. Rotation in opposite senses (*disrotation*) would bring opposite-phase lobes together, an antibonding situation. The mechanism of the reverse reaction is simply the forward mechanism in reverse, so the fact that the thermodynamically favored process is the *ring-opening* of a cyclobutene simply means that the cyclobutene shown would open to the butadiene shown on heating. Photochemical processes can also be

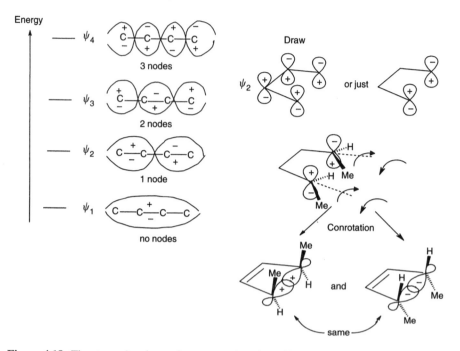

Figure 4.18. The stereochemistry of many reactions is easily predicted from the symmetry of MOs, usually the highest occupied π MO (π HOMO). In the ring closure of 1,3-butadiene to cyclobutene the phase (+ or $-$) of the HOMO (ψ_2) at the end carbons (the atoms that bond) is such that closure must occur in a conrotatory sense, giving a definite stereochemical outcome. In the example above there is only one product. The *reverse* process is actually thermodynamically favored, and the *cis* dimethyl cyclobutene opens to the *cis, trans* diene. No attempt is made here to show quantitatively the positions of the energy levels or to size the AOs according to their contributions to the MOs.

accommodated by the Woodward–Hoffmann orbital symmetry rules if we realize that absorption of a photon creates an electronically excited molecule in which the previous lowest unoccupied MO (LUMO) is now the HOMO. For more about orbital symmetry and chemical reactions see, e.g. the book by Woodward and Hoffmann [38].

Stability as indicated by energy levels and aromaticity
The MO energy levels obtained from an SHM calculation must be filled with electrons according to the species under consideration. For example, the neutral ethene molecule has two π electrons, so the diagrams of Fig. 4.19(a) (cf. Fig. 4.15) with one, two and three π electrons, would refer to the cation, the neutral and the anion. We might expect the neutral, with its bonding π orbital ψ_1 full and its antibonding π orbital ψ_2, empty, to be resistant to oxidation (which would require removing electronic charge from the low-energy ψ_1) and to reduction (which would require adding electronic charge to the high-energy ψ_2).

The propenyl (allyl) system has two, three or four π electrons, depending on whether we are considering the cation, radical or anion (Fig. 4.19(b); cf. Fig. 4.16). The cation

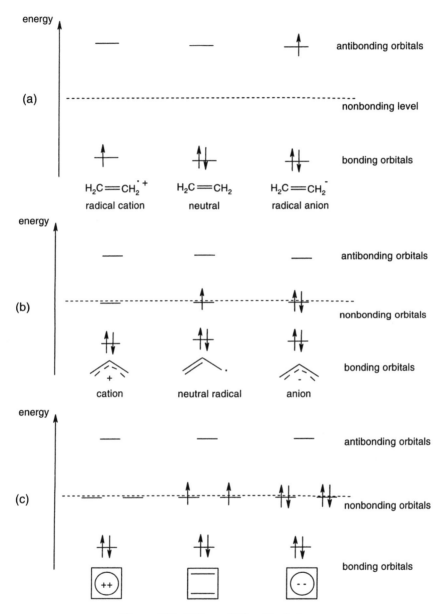

Figure 4.19. Filling π MOs with electrons.

might be expected to be resistant to oxidation, which requires removing an electron from a low-lying π orbital (ψ_1) and to be moderately readily reduced, as this involves adding an electron to the nonbonding π orbital ψ_2, a process that should not be strongly favorable or unfavorable. The radical should be easier to oxidize than the cation, for this requires removing an electron from a nonbonding, rather than a lower-lying bonding, orbital, and the ease of reduction of the radical should be roughly comparable to that of the cation, as both can accommodate an electron in a nonbonding orbital. The anion should be oxidized with an ease comparable to that of the radical (removal of an electron from the nonbonding ψ_2, but be harder to reduce (addition of an electron to the antibonding ψ_3).

The cyclobutadiene system (Fig. 4.19(c); cf. Fig. 4.17) can be envisaged with, amongst others, two (the dication), four (the neutral molecule) and six π (the dianion) electrons. The predictions one might make for the behavior of these three species toward redox reactions are comparable to those just outlined for the propenyl cation, radical and anion, respectively (note the analogous occupancy of bonding, nonbonding and antibonding orbitals). The neutral cyclobutadiene molecule is, however, predicted by the SHM to have an unusual electronic arrangement for a diene: in filling the π orbitals, from the lowest-energy one up, one puts electrons of the same spin into the degenerate ψ_2 and ψ_3 in accordance with Hund's rule of maximum multiplicity. Thus the SHM predicts that cyclobutadiene will be a diradical, with two unpaired electrons of like spin. Actually, more advanced calculations [39] indicate, and experiment confirms, that cyclobutadiene is a singlet molecule with two single and two double C/C bonds. A square cyclobutadiene diradical with four 1.5 C/C bonds would distort to a rectangular, closed-shell (i.e. no unpaired electrons) molecule with two single and two double bonds (Fig. 4.20). This could have been predicted by augmenting the SHM result with a knowledge of the phenomenon known as the Jahn–Teller effect [40]: cyclic systems (and certain others) with an odd number of electrons in degenerate (equal-energy) MOs will distort to remove the degeneracy.

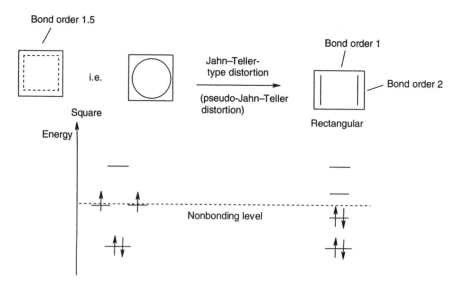

Figure 4.20. Cyclic systems with degenerate energy levels tend to undergo a geometric distortion to remove the degeneracy, a consequence of the Jahn–Teller theorem.

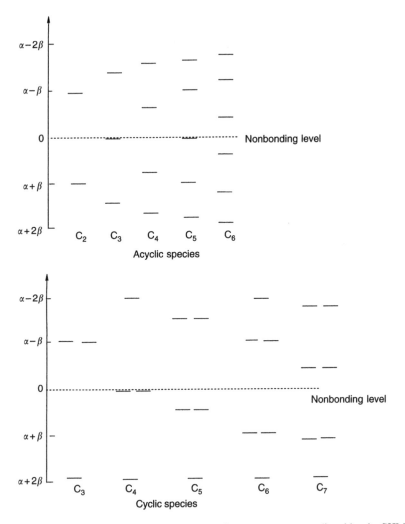

Figure 4.21. The MO pattern for acyclic and cyclic π systems, as predicted by the SHM.

What general pattern of MOs emerges from the SHM? Acyclic π systems (ethene, the propenyl system, 1,3-butadiene, etc.), have MOs distributed singly and evenly on each side of the nonbonding level; the odd-AO systems also have one nonbonding MO (Fig. 4.21). Cyclic π systems (the cyclopropenyl system, cyclobutadiene, the cyclopentadienyl system, benzene, etc.) have a lowest MO and pairs of degenerate MOs, ending with one highest or a pair of highest MOs, depending on whether the number of MOs is even or odd. The total number of MOs is always equal to the number of basis functions, which in the SHM is, for organic polyenes, the number of p orbitals (Fig. 4.21). The pattern for cyclic systems can be predicted qualitatively simply by sketching the polygon, with one vertex down, inside a circle (Fig. 4.22). If the circle is of radius $2|\beta|$ the energies can even be calculated by trigonometry [41]. It follows from this

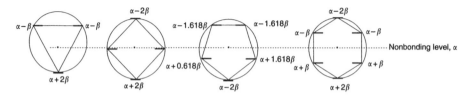

Figure 4.22. A useful mnemonic for getting the SHM pattern for cyclic π systems. Setting the radius of the circle at $2|\beta|$, the energy separations from the nonbonding level can even be calculated by trigonometry.

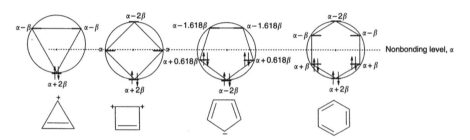

Figure 4.23. Hückel's rule says that cyclic π systems with $4n + 2\pi$ electrons ($n = 0, 1, 2, \ldots$; $4n + 2 = 2, 6, 10, \ldots$) should be especially stable, since they have all bonding levels full and all antibonding levels empty. The special stability is usually equated with aromaticity. Shown here are the cyclopropenyl cation, the cyclobutadiene dication, the cyclopentadienyl anion, and benzene; *formal* structures are given for these species – the actual molecules do not have single and double bonds, but rather electron delocalization makes all C/C bonds the same.

pattern that cyclic species (not necessarily neutral) with 2, 6, 10, ... π electrons have filled π MOs and might be expected to show particular stability, analogously to the filled AOs of the unreactive noble gases (Fig. 4.23). The archetype of such molecules is, of course, benzene, and the stability is associated with the general collection of properties called aromaticity [17]. These results, which were first perceived by Hückel [19] (1931–1937), are summarized in a rule called the $4n + 2$ rule or Hückel's rule, although the $4n + 2$ formulation was evidently actually due to Doering and Knox (1954) [42]. This says that cyclic arrays of sp^2-hybridized atoms with $4n+2\pi$ electrons are characteristic of aromatic molecules; the canonical aromatic molecule benzene with six π electrons corresponds to $n = 1$. For neutral molecules with formally fully conjugated perimeters this amounts to saying that those with an odd number of C/C double bonds are aromatic and those with an even number are antiaromatic (see *resonance energies*, below).

Hückel's rule has been abundantly verified [17] notwithstanding the fact that the SHM, when applied without regard to considerations like the Jahn–Teller effect (see above) incorrectly predicts $4n$ species like cyclobutadiene to be triplet diradicals. The Hückel rule also applies to ions; for example, the cyclopropenyl with system two π electrons, the cyclopropenyl cation, corresponds to $n = 0$, and is strongly aromatic.

Other aromatic species are the cyclopentadienyl anion (six π electrons, $n = 1$; Hückel predicted the enhanced acidity of cyclopentadiene) and the cycloheptatrienyl cation. Only reasonably planar species can be expected to provide the AO overlap need for cyclic electron delocalization and aromaticity, and care is needed in applying the rule.

Resonance energies

The SHM permits the calculation of a kind of stabilizing energy, or, more accurately, an energy that reflects the stability of molecules. This energy is calculated by comparing the total electronic energy of the molecule in question with that of a reference compound, as shown below for the propenyl systems, cyclobutadiene, and the cyclobutadiene dication.

The propenyl cation, Fig. 4.19(b); cf. Fig. 4.16. If we take the total π electronic energy of a molecule to be simply the number of electrons in a π MO times the energy level of the orbital, summed over occupied orbitals (a gross approximation, as it ignores interelectronic repulsion), then for the propenyl cation

$$E_\pi \text{ (prop. cation)} = 2(\alpha + 1.414\beta) = 2\alpha + 2.828\beta$$

We want to compare this energy with that of two electrons in a normal molecule with no special features (the propenyl cation has the special feature of an empty p orbital adjacent to the formal C/C double bond), and we choose neutral ethene for our reference energy (Fig. 4.15)

$$E_\pi \text{ (reference)} = 2(\alpha + \beta) = 2\alpha + 2\beta$$

The stabilization energy is then

$$E \text{ (stab, cation)} = E_\pi \text{ (prop. cation)} - E_\pi \text{ (reference)}$$
$$= (2\alpha + 2.828\beta) - (2\alpha + 2\beta) = 0.828\beta$$

Since β is negative, the π-electronic energy of the propenyl cation is calculated to be below that of ethene: providing an extra, empty p orbital for the electron pair causes the energy to drop. Actually, resonance energy is usually presented as a positive quantity, e.g. "$100\,kJ\,mol^{-1}$." We can interpret this as $100\,kJ\,mol^{-1}$ below a reference system. To avoid a negative quantity in SHM calculations like these, we can use $|\beta|$ instead of β.

The propenyl radical, Fig. 4.16. The total π electronic energy by the SHM is

$$E_\pi \text{ (prop. radical)} = 2(\alpha + 1.414\beta) + \alpha = 3\alpha + 2.828\beta$$

For the reference energy we use one ethene molecule and one nonbonding p electron (like the electron in a methyl radical):

$$E_\pi \text{ (reference)} = (2\alpha + 2\beta) + \alpha = 3\alpha + 2\beta$$

The stabilization energy is then

$$E \text{ (stab, radical)} = E_\pi \text{ (prop. radical)} - E_\pi \text{ (reference)}$$
$$= (3\alpha + 2.828\beta) - (3\alpha + 2\beta) = 0.828\beta$$

The propenyl anion. An analogous calculation (cf. Fig. 4.16, with four electrons for the anion) gives

$$E(\text{stab, anion}) = E_\pi(\text{prop. anion}) - E_\pi(\text{reference})$$

$$= (4\alpha + 2.828\beta) - (4\alpha + 2\beta) = 0.828\beta$$

Thus the SHM predicts that all three propenyl species will be lower in energy than if the π electrons were localized in the formal double bond and (for the radical and anion) in one p orbital. Because this lower energy is associated with the ability of the electrons to spread or be delocalized over the whole π system, what we have called $E(\text{stab})$ is often denoted as the delocalization energy, and designated E_D. Note that E_R (or E_D) is always some multiple of β (or is zero). Since electron delocalization can be indicated by the familiar resonance symbolism the Hückel delocalization energy is often equated with resonance energy, and designated E_R. The accord between calculated delocalization and the ability to draw resonance structures is not perfect, as indicated by the next example.

Cyclobutadiene (Fig. 4.17). The total π electronic energy is

$$E_\pi(\text{cyclobutadiene}) = 2(\alpha + 2\beta) + 2\alpha = 4\alpha + 4\beta$$

Using two ethene molecules as our reference system:

$$E_\pi(\text{reference}) = 2\alpha + 2\beta$$

and so for $E(\text{stab})$ ($= E_D$ or E_R) we get

$$E(\text{stab, cyclobutadiene}) = E_\pi(\text{cyclobutadiene}) - E_\pi(\text{reference})$$

$$= (4\alpha + 4\beta) - (4\alpha + 4\beta) = 0$$

Cyclobutadiene is predicted by this calculation to have no resonance energy, although we can readily draw two "resonance structures" exactly analogous to the Kekulé structures of benzene. The SHM predicts a resonance energy of 2β for benzene. Equating $2|\beta|$ with the commonly-quoted resonance energy of $150\,\text{kJ}\,\text{mol}^{-1}$ ($36\,\text{kcal}\,\text{mol}^{-1}$) for benzene gives a value of $75\,\text{kJ}\,\text{mol}^{-1}$ for $|\beta|$, but this should be taken with more than a grain of salt, for outside a closely related series of molecules, β has little or no quantitative meaning [43]. However, in contrast to the failure of simple resonance theory in predicting aromatic stabilization (and other chemical phenomena) [44], the SHM is quite successful.

The cyclobutadiene dication (cf. Fig. 4.17). The total π electronic energy is

$$E_\pi(\text{dication}) = 2(\alpha + 2\beta) = 2\alpha + 4\beta$$

Using one ethene molecule as the reference:

$$E_\pi(\text{reference}) = 2\alpha + 2\beta$$

and so

$$E(\text{stab, dication}) = E_\pi(\text{dication}) - E_\pi(\text{reference})$$

$$= (2\alpha + 4\beta) - (2\alpha + 2\beta) = 2\beta$$

Thus the stabilization energy calculation agrees with the deduction from the disposition of filled MOs (i.e. with the $4n + 2$ rule) that the cyclobutadiene dication should be stabilized by electron delocalization, which is in some agreement with experiment [45].

More sophisticated calculations indicate that cyclic $4n$ systems like cyclobutadiene (where planar; cyclooctatetraene, for example, is buckled by steric factors and is simply an ordinary polyene) are actually *destabilized* by π electronic effects: their resonance energy is not just zero, as predicted by the SHM, but less than zero. Such systems are *antiaromatic* [17,46].

Bond orders

The meaning of this term is easy to grasp in a qualitative, intuitive way: an ideal single bond has a bond order of one, and ideal double and triple bonds have bond orders of two and three, respectively. Invoking Lewis electron-dot structures, one might say that the order of a bond is the number of electron pairs being shared between the two bonded atoms. Calculated quantum mechanical bond orders should be more widely applicable than those from the Lewis picture, because electron pairs are not localized between atoms in a clean pairwise manner; thus a weak bond, like a hydrogen bond or a long single bond, might be expected to have a bond order of less than one. However, there is no unique definition of bond order in computational chemistry, because there seems to be no single, correct method to assign electrons to particular atoms or pairs of atoms [47]. Various quantum mechanical definitions of bond order can be devised [48], based on basis-set coefficients. Intuitively, these coefficients for a pair of atoms should be relevant to calculating a bond order, since the bigger the contribution two atoms make to the wavefunction (whose square is a measure of the electron density; section 4.2.6), the bigger should be the electron density between them. In the SHM the order of a bond between two atoms A_i and B_j is defined as

$$B_{i,j} = 1 + \sum_{\text{all occ}} n c_i c_j \qquad (4.70)$$

Here the 1 denotes the single bond of the ubiquitous spectator σ bond framework, which is taken as always contributing a σ bond order of unity. The other term is the π bond order; its value is obtained by summing over all the occupied MOs the number of electrons n in each of these MOs times the product of the c's of the two atoms for each MO. This is illustrated in the following examples.

Ethene. The occupied orbital is ψ_1, which has 2 electrons), and the coefficients of c_1 and c_2 for this orbital are 0.707, 0.707 (Eq. (4.66)). Thus

$$B_{i,j} = 1 + \sum_{\text{all occ}} n c_i c = 1 + 2(0.707)0.707 = 1 + 1.000 = 2.000$$

which is reasonable for a double bond. The order of the σ bond is 1 and that of the π bond is 1.

The ethene radical anion. The occupied orbitals are ψ_1, which has 2 electrons, and ψ_2, which has 1 electron; the coefficients of c_1 and c_2 for ψ_1 are 0.707, 0.707 and for

ψ_2, 0.707, −0.707 (Eq. (4.66)). Thus

$$B_{i,j} = 1 + \sum_{\text{all occ}} nc_i c = 1 + 2(0.707)0.707 + 1(0.707)(-0.707)$$

$$= 1 + 1 - 0.500 = 1.500$$

The π bond order of 0.500 (1.500 − σ bond order) accords with two electrons in the bonding MO and one electron in the antibonding orbital.

Atomic charges

In an intuitive way, the charge on an atom might be thought to be a measure of the extent to which the atom repels or attracts a charged probe near it, and to be measurable from the energy it takes to bring a probe charge from infinity up to near the atom. However, this would tell us the charge at a point *outside* the atom, for example a point on the van der Waals surface of the molecule, and the repulsive or attractive forces on the probe charge would be due to the molecule as a whole. Although atomic charges are generally considered to be experimentally unmeasurable, chemists find the concept very useful (thus calculated charges are used to parameterize molecular mechanics force fields – chapter 3), and much effort has gone into designing various definitions of atomic charge [47,48]. Intuitively, the charge on an atom should be related to the basis set coefficients of the atom, since the more the atom contributes to a multicenter wavefunction (one with contributions from basis functions on several atoms), the more it might be expected to lose electronic charge by delocalization into the rest of the molecule (cf. the discussion of bond order above). In the SHM the charge on an atom A_i is defined as (cf. Eq. (4.70))

$$q_i = 1 - \sum_{\text{all occ}} nc_i^2 \tag{4.71}$$

The summation term is the *charge density*, and is a measure of the electronic charge on the molecule due to the π electrons. For example, having no π electrons (an empty p orbital, formally a cationic carbon) would mean a π electron charge density of zero; subtracting this from unity gives a charge on the atom of +1. Again, having two π electrons in a p orbital would mean a π electron charge density of 2 on the atom; subtracting this from unity gives a charge on the atom of −1 (a filled p orbital, formally an anionic carbon). The application of Eq. (4.71) will be illustrated using methylenecyclopropene (Fig. 4.24).

Methylenecyclopropene (Fig. 4.24).

$$q_1 = 1 - \sum_{\text{all occ}} nc_1^2 = 1 - [2(0.282)^2 + 2(0.815)^2] = 1 - 1.487 = -0.487$$

$$q_2 = 1 - \sum_{\text{all occ}} nc_2^2 = 1 - [2(0.612)^2 + 2(0.254)^2] = 1 - 0.878 = 0.122$$

$$q_3 = q_4 = 1 - \sum_{\text{all occ}} nc_3^2 = 1 - [2(0.523)^2 + 2(-0.368)^2] = 1 - 0.817 = 0.182$$

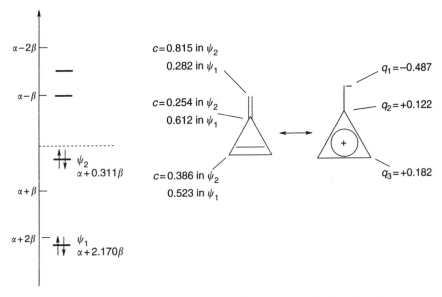

Figure 4.24. The SHM charges on the atoms of a molecule can be calculated from the number of electrons in each occupied MO and the coefficients of these MOs. The predicted dipolar nature of methylenecyclopropene has been ascribed to a cyclopropenyl-cation-like resonance contributor.

The results of this charge calculation are summarized in Fig. 4.24; the negative charge on the exocyclic carbon and the positive charges on the ring carbons are in accord with the resonance picture (Fig. 4.24), which invokes a contribution from the aromatic cyclopropenyl cation [49]. Note that the charges sum to (essentially) zero, as they must for a neutral molecule (the hydrogens, which actually also carry charges, have been excluded from consideration here). A high-level calculation places a total charge (carbon plus hydrogen) – albeit defined in a different way – of -0.37 on the CH_2 group and $+0.37$ on the ring (cf. -0.487 and $+0.487$ for the exocyclic carbon and the ring carbons in the SHM calculation).

4.3.6 Strengths and weaknesses of the SHM

Strengths
The SHM has been extensively used to correlate, rationalize, and predict many chemical phenomena, having been applied with surprising success to dipole moments, esr spectra, bond lengths, redox potentials, ionization potentials, UV and IR spectra, aromaticity, acidity/basicity, and reactivity, and specialized books on the SHM should be consulted for details [21]. The method will probably give some insight into any phenomenon that involves predominantly the π electron systems of conjugated molecules. The SHM may have been underrated [50] and reports of its death are probably exaggerated. However, the SHM is not used very much in research nowadays, partly because more

sophisticated π electron approaches like the PPP method (section 6.2.2) are available, but mainly because of the phenomenal success of all-valence-electron SE methods (chapter 6), which are applicable to quite large molecules, and of the increasing power of all-electron *ab initio* (chapter 5) and DFT (chapter 7) methods.

Weaknesses

The defects of the SHM arise from the fact that it treats only π electrons, and these only very approximately. The basic Hückel method described here has been augmented in an attempt to handle non-π substituents, e.g. alkyl groups, halogen groups, etc., and heteroatoms instead of carbon. This has been done by treating the substituents as π centers and embodying empirically altered values of α and β, so that in the Fock matrix values other than -1 and 0 appear. However, the values of these modified parameters that have been employed vary considerably [51], which tends to diminish one's confidence in their reliability.

The approximations in the SHM are its peremptory treatment of the overlap integrals S (section 4.3.4, discussion in connection with Eqs (4.55)), its drastic truncation of the possible values of the Fock matrix elements into just α, β and 0 (section 4.3.4, discussion in connection with Eqs (4.61)), its complete neglect of electron spin, and its glossing over (although not exactly ignoring) interelectronic repulsion by incorporating this into the α and β parameters.

The overlap integrals S are divided into just two classes:

$$\int \phi_i \phi_j \, dv = S_{ij} = 1 \text{ or } 0$$

depending on whether the orbitals on the atoms i and j are on the same or different atoms. This approximation, as explained earlier, reduces the matrix form of the secular equations to standard eigenvalue form $\mathbf{HC} = \mathbf{C}\varepsilon$ (Eq. (4.59)), so that the Fock matrix can (after giving its elements numerical values) be diagonalized without further ado (the ado is explained in section 4.3.7, in connection with the extended Hückel method). In the older determinant, as opposed to matrix, treatment (section 4.3.7), the approximation greatly simplifies the determinants. In fact, however, the overlap integral between adjacent carbon p orbitals is ca. 0.24 [52].

Setting the Fock matrix elements equal to just α, β and 0. Setting

$$\int \phi_i \hat{H} \phi_j \, dv = H_{ij} = \alpha, \beta \text{ or } 0$$

depending on whether the orbitals on the atoms i and j are on the same, adjacent or further-removed atoms is an approximation, because all the H_{ii} terms are not the same, and all the adjacent-atom H_{ij} terms are not the same either; these energies depend on the environment of the atom in the molecule; for example, atoms in the middle of a conjugated chain should have different H_{ii} and H_{ij} parameters than ones at the end of the chain. Of course, this approximation simplifies the Fock matrix (or the determinant in the old determinant method, section 4.3.7).

The neglect of electron spin and the deficient treatment of interelectronic repulsion is obvious. In the usual derivation (section 4.3.4): in Eq. (4.40) the integration is carried out with respect to only spatial coordinates (ignoring spin coordinates; contrast *ab initio* theory, section 5.2), and in calculating π energies (section 4.3.5) we simply took the sum of the number of electrons in each occupied MO times the energy level of the MO. However, the energy of an MO is the energy of an electron in the MO moving in the force field of the nuclei and all the other electrons (as pointed out in section 4.3.4, in explaining the matrices of Eqs (4.55)). If we calculate the total electronic energy by simply summing MO energies times occupancy numbers, we are assuming, wrongly, that the electron energies are independent of one another, i.e. that the electrons do not interact. An energy calculated in this way is said to be a sum of one-electron energies. The resonance energies calculated by the SHM can thus be only very rough, unless the errors tend to cancel in the subtraction step, which in fact probably occurs to some extent (this is presumably why the method of Hess and Schaad for calculating resonance energies works so well [50]). The neglect of electron repulsion and spin in the usual derivation of the SHM is discussed in Ref. [30].

4.3.7 The determinant method of calculating the Hückel c's and energy levels

An older method of obtaining the coefficients and energy levels from the secular equations (Eqs (4.49) for a two-basis-function system) utilizes determinants rather than matrices. The method is much more cumbersome than the matrix diagonalization approach of section 4.3.4, but in the absence of cheap, readily-available computers (matrix diagonalization is easily handled by a personal computer) its erstwhile employment may be forgiven. It is outlined here because traditional presentations of the SHM [21] use it.

Consider again the secular equations (4.49):

$$(H_{11} - ES_{11})c_1 + (H_{12} - ES_{12})c_2 = 0$$
$$(H_{21} - ES_{21})c_1 + (H_{22} - ES_{22})c_2 = 0$$

By considering the requirements for nonzero values of c_1 and c_2 we can find how to calculate the c's and the MO energies (since the coefficients are weighting factors that determine how much each basis function contributes to the MO, zero c's would mean no contributions from the basis functions and hence no MOs; that would not be much of a molecule). Consider the system of linear equations

$$A_{11}x_1 + A_{12}x_2 = b_1, \quad A_{21}x_1 + A_{22}x_2 = b_1$$

Using determinants:

$$x_1 = \frac{\begin{vmatrix} b_1 & A_{12} \\ b_2 & A_{22} \end{vmatrix}}{D}, \quad x_2 = \frac{\begin{vmatrix} A_{11} & b_1 \\ A_{21} & b_2 \end{vmatrix}}{D}, \quad D = \begin{vmatrix} A_{11} & A_{12} \\ A_{21} & A_{22} \end{vmatrix}$$

where D is the *determinant of the system*. If $b_1 = b_2 = 0$ (the situation in the secular equations), then in the equations for x_1 and x_2 the numerator is zero, and so $x_1 = 0/D$

and $x_2 = 0/D$. The only way that x_1 and x_2 can be nonzero in this case is that the determinant of the system be zero, i.e.

$$D = 0$$

for then $x_1 = 0/0$ and $x_2 = 0/0$, and $0/0$ can have any finite value; mathematicians call it indeterminate. This is easy to see:

$$\text{Let} \quad \frac{0}{0} = a, \quad \text{then } a \times 0 = 0$$

which is true for any finite value of a.

So for the secular equations the requirement that the c's be nonzero is that the determinant of the system be zero:

$$D = \begin{vmatrix} H_{11} - E S_{11} & H_{12} - E S_{12} \\ H_{21} - E S_{21} & H_{22} - E S_{22} \end{vmatrix} = 0 \tag{4.72}$$

Equation (4.72) can be generalized to n basis functions (cf. the matrix of Eq. (4.62)):

$$\begin{vmatrix} H_{11} - E S_{11} & H_{12} - E S_{12} & \cdots & H_{1n} - E S_{1n} \\ H_{21} - E S_{21} & H_{22} - E S_{22} & \cdots & H_{2n} - E S_{2n} \\ \vdots & \vdots & \cdots & \vdots \\ H_{n1} - E S_{n1} & H_{n2} - E S_{n2} & \cdots & H_{nn} - E S_{nn} \end{vmatrix} = 0 \tag{4.73}$$

If we invoke the SHM simplification of orthogonality of the S integrals (pp. 37–39), then $S_{ii} = 1$ and $S_{ij} = 0$ and Eq. (4.75) becomes

$$\begin{vmatrix} H_{11} - E & H_{12} & \cdots & H_{1n} \\ H_{21} & H_{22} - E & \cdots & H_{2n} \\ \vdots & \vdots & \cdots & \vdots \\ H_{n1} & H_{n2} & \cdots & H_{nn} - E \end{vmatrix} = 0 \tag{4.74}$$

Substituting α, β and 0 for the appropriate H's (p. 39) we get

$$\begin{vmatrix} \alpha - E & \beta & \cdots & 0 \\ \beta & \alpha - E & \cdots & 0 \\ \vdots & \vdots & \cdots & \vdots \\ 0 & 0 & \cdots & \alpha - E \end{vmatrix} = 0 \tag{4.75}$$

The diagonal terms will always be $\alpha - E$, but the placement of β and 0 will depend on which i, j terms are adjacent and which are further-removed, which depends on the numbering system chosen (see below). Since multiplying or dividing a determinant by a number is equivalent to multiplying or dividing the elements of one row or column by that number (section 4.3.3), multiplying both sides of Eq. (4.75) by $1/\beta$ n times, i.e. by $(1/\beta)^n$ gives

$$\begin{vmatrix} (\alpha - E)/\beta & 1 & \cdots & 0 \\ 1 & (\alpha - E)/\beta & \cdots & 0 \\ \vdots & \vdots & \cdots & \vdots \\ 0 & 0 & \cdots & (\alpha - E)/\beta \end{vmatrix} = 0 \qquad (4.76)$$

Finally, if we define $(\alpha - E)/\beta = x$, we get

$$\begin{vmatrix} x & 1 & \cdots & 0 \\ 1 & x & \cdots & 0 \\ \vdots & \vdots & \cdots & \vdots \\ 0 & 0 & \cdots & x \end{vmatrix} = 0 \qquad (4.77)$$

The diagonal terms are always x but the off-diagonal terms, 1 for adjacent and 0 for nonadjacent orbital pairs, depend on the numbering (which does not affect the results: Fig. 4.25). Any specific determinant of the type in Eq. (4.77) can be expanded into a polynomial of order n (where the determinant is of order $n \times n$), making Eq. (4.77) yield the polynomial equation:

$$x^n + a_1 x^{n-1} + a_2 x^{n-2} + \cdots + a_n = 0 \qquad (4.78)$$

The polynomial can be solved for x and then the energy levels can be found from $(\alpha - E)/\beta = x$, i.e. from

$$E = \alpha - \beta x \qquad (4.79)$$

The coefficients can then be calculated from the energy levels by substituting the E's into one of the secular equations, finding the ratio of the c's, and normalizing to get the actual c's. An example will indicate how the determinant method can be implemented.

Consider the propenyl system. In the secular determinant the i, i-type interactions will be represented by x, adjacent i, j-type interactions by 1, and nonadjacent i, j-type interactions by 0. For the determinantal equation we can write (Fig. 4.25)

$$\begin{vmatrix} x & 1 & 0 \\ 1 & x & 1 \\ 0 & 1 & x \end{vmatrix} = 0 \qquad (4.80)$$

(Compare this with the Fock matrix for the propenyl system). Solving this equation (see section 4.3.3):

$$\begin{vmatrix} x & 1 & 0 \\ 1 & x & 1 \\ 0 & 1 & x \end{vmatrix} = x \begin{vmatrix} x & 1 \\ 1 & x \end{vmatrix} - 1 \begin{vmatrix} 1 & 1 \\ 0 & x \end{vmatrix} + 0 \begin{vmatrix} 1 & x \\ 0 & 1 \end{vmatrix}$$

$$= x(x^2 - 1) - (x - 0) + 0 = x^3 - x - x = x^3 - 2x = 0 \qquad (4.81)$$

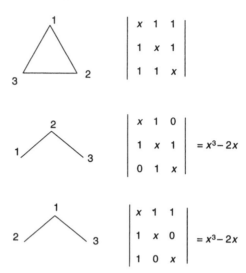

Figure 4.25. The determinants corresponding to different numbering patterns can seem to differ, but on expansion they give the same polynomial.

This cubic can be factored (but in general polynomial equations require numerical approximation methods):

$$x(x^2 - 2) = 0 \quad \text{so } x = 0 \text{ and } x^2 - 2 = 0 \text{ or } x = \pm\sqrt{2}$$

From $(\alpha - E)/\beta = x$, $E = \alpha - x\beta$ and

$$x = 0 \qquad \text{leads to } E = \alpha$$
$$x = +\sqrt{2} \quad \text{leads to } E = \alpha - \sqrt{2}\beta$$
$$x = -\sqrt{2} \quad \text{leads to } E = \alpha + \sqrt{2}\beta$$

So we get the same energy levels as from matrix diagonalization ($\sqrt{2} = 1.414$).

To find the coefficients we substitute the energy levels into the secular equations; for the propenyl system these are, projecting from the secular equations for a two-orbital system, Eqs (4.49):

$$(H_{11} - ES_{11})c_1 + (H_{12} - ES_{12})c_2 + (H_{13} - ES_{13})c_3 = 0$$
$$(H_{21} - ES_{21})c_1 + (H_{22} - ES_{22})c_2 + (H_{23} - ES_{23})c_3 = 0 \qquad (4.82)$$
$$(H_{31} - ES_{31})c_1 + (H_{32} - ES_{32})c_2 + (H_{33} - ES_{33})c_3 = 0$$

These can be simplified (Eqs (4.57) and (4.61)) to

$$(\alpha - E)c_1 + \beta c_2 + 0c_3 = 0, \quad \beta c_1 + (\alpha - E)c_2 + \beta c_3 = 0, \quad 0c_1 + \beta c_2 + (\alpha - E)c_3 = 0$$
$$(4.83)$$

For the energy level $E = \alpha + \sqrt{2}\beta$ (MO level 1, ψ_1), substituting into the first secular equation we get

$$-\sqrt{2}\beta c_{11} + \beta c_{21} = 0, \quad \text{so } c_{21}/c_{11} = \sqrt{2}$$

(Recall the c_{ij} notation; c_{11} is the coefficient for atom 1 in ψ_1, c_{21} is the coefficient for atom 2 in ψ_1, etc.) Substituting $E = \alpha + \sqrt{2}\beta$ into the second secular equation we get

$$-\beta c_{11} + \beta c_{31} = 0, \quad \text{so } c_{11} = c_{31}$$

We now have the relative values of the c's:

$$c_{11}/c_{11} = 1, \quad c_{21}/c_{11} = \sqrt{2}, \quad c_{31}/c_{11} = 1 \qquad (4.84)$$

To find the actual values of the c's, we utilize the fact that the MO (we are talking about MO level 1, ψ_1) must be normalized:

$$\int \psi_1^2 \, dv = 1 \qquad (4.85)$$

Now, from the LCAO method

$$\psi_1 = c_{11}\phi_1 + c_{21}\phi_2 + c_{31}\phi_3 \qquad (4.86)$$

Therefore

$$\psi_1^2 = c_{11}^2\phi_1^2 + c_{21}^2\phi_2^2 + c_{31}^2\phi_3^2 + 2c_{11}c_{21}\phi_1\phi_2$$
$$+ 2c_{11}c_{31}\phi_1\phi_3 + 2c_{21}c_{31}\phi_2\phi_3 \qquad (4.87)$$

So from Eq. (4.87), and recalling that in the SHM we pretend that the basis functions ϕ are orthonormal, i.e. that $S_{ij} = \delta_{ij}$, we get

$$\int \psi_1^2 \, dv = c_{11}^2 + c_{21}^2 + c_{31}^2 = 1 \qquad (4.88)$$

Using the ratios of the c's from Eq. (4.84):

$$\frac{c_{11}^2}{c_{11}^2} + \frac{c_{21}^2}{c_{11}^2} + \frac{c_{31}^2}{c_{11}^2} = \frac{1}{c_{11}^2}$$

i.e.

$$1^2 + (\sqrt{2})^2 + 1^2 = \frac{1}{c_{11}^2}$$

and so

$$c_{11} = \tfrac{1}{2}, \quad c_{21} = (\sqrt{2})c_{11} = \tfrac{1}{\sqrt{2}}, \quad c_{31} = c_{11} = 1$$

By substituting into the secular equations (4.83) the E values for ψ_2 and ψ_3 we could find the ratios of the c's for ψ_2 and ψ_3 and with the aid of the orthonormalization equation analogous to Eq. (4.88) we could get the actual values of c_{12}, c_{22}, c_{32} and c_{13}, c_{23}, and c_{33}. Although this somewhat clumsy way of finding the c's from the energy levels was streamlined (see, e.g. [21d]), the determinant method has been replaced by matrix diagonalization implemented in a computer program.

4.4   THE EXTENDED HÜCKEL METHOD

4.4.1   Theory

In the SHM, as in all modern MO methods, a Fock matrix is diagonalized to give coefficients (which with the basis set give the wavefunctions of the MOs) and energy levels (i.e. MO energies). The SHM and the extended Hückel method (EHM, extended Hückel theory, EHT) differ in how the elements of the Fock matrix are obtained and how the overlap matrix is treated. The EHM was popularized and widely applied by Hoffmann[28] [53], although earlier work using the approach had been done by Wolfsberg and Helmholz [54]. We now compare point by point the SHM and the EHM.

Simple Hückel method

(1) *Basis set is limited to p orbitals.* Each element of the Fock matrix **H** is an integral that represents an interaction between two orbitals. The orbitals are in almost all cases a set of p orbitals (usually carbon $2p$) supplied by an sp^2 framework, with the p orbital axes parallel to one another and perpendicular to the plane of the framework. In other words, the set of basis orbitals – the *basis set* – is limited (in the great majority of cases) to p_z orbitals (taking the framework plane, i.e. the molecular plane, to be the xy plane).

(2) *Orbital interaction energies are limited to α, β and 0.* The Fock matrix orbital interactions are limited to α, β and 0, depending on whether the H_{ij} interaction is, respectively, i, i, adjacent, or further-removed. The value of β does not vary smoothly with the separation of the orbitals, although logically it should decrease continuously to zero as the separation increases.

(3) *Fock matrix elements are not actually calculated.* The Fock matrix elements are not any definite physical quantities, but rather energy levels relative to α in units of $|\beta|$, making them 0 or -1. One can try to estimate α and β, but the SHM does not define them quantitatively.

(4) *Overlap integrals are limited to 1 or 0.* We pretend that the overlap matrix **S** is a unit matrix, by setting $S_{ij} = \delta_{ij}$. This enables us to simplify $\mathbf{HC} = \mathbf{SC}\varepsilon$ (Eq. (4.54)) to the standard eigenvalue form $\mathbf{HC} = \mathbf{C}\varepsilon$ (Eq. (4.59)) and so $\mathbf{H} = \mathbf{C}\varepsilon\mathbf{C}^{-1}$, which is the same as saying that the SHM Fock matrix is directly diagonalized to give the c's and ε's.

Now compare these four points with the corresponding features of the EHM.

Extended Hückel method

(1) *All valence s and p orbitals are used in the basis set.* As in the SHM each element of the Fock matrix is an integral representing an interaction between two orbitals; however, in the EHM the basis set is not just a set of $2p_z$ orbitals but rather the set of valence-shell orbitals of each atom in the molecule (the derivation of the secular equations

---

[28]Roald Hoffmann, born Zloczow, Poland, 1937. Ph.D. Harvard, 1962, Professor, Cornell. Nobel prize 1981(shared with Kenichi Fukui; section 7.3.5) for work with organic chemist Robert B. Woodward, showing how the symmetry of molecular orbitals influences the course of chemical reactions (the Woodward–Hoffmann rules or the conservation of orbital symmetry). Main exponent of the extended Hückel method. He has written poetry, and several popular books on chemistry.

says nothing about what kinds of orbitals we are considering). Thus each hydrogen atom contributes a $1s$ orbital to the basis set and each carbon atom a $2s$ and three $2p$ orbitals. Lithium and beryllium, although they have no $2p$ electrons, are assigned a $2s$ and three $2p$ orbitals (experience shows that this works better than omitting these basis functions) so the atoms from lithium to fluorine each contribute a $2s$ and three $2p$ orbitals. A basis set like this, which uses the normal valence orbitals of atoms, is called a *minimal valence basis set.*

(2) *Orbital interaction energies are calculated and vary smoothly with geometry.* The EHM Fock matrix orbital interactions H_{ij} are calculated in a way that depends on the distance apart of the orbitals, so their values vary smoothly with orbital separation.

(3) *Fock matrix elements are actually calculated.* The EHM Fock matrix elements are calculated from well-defined physical quantities (ionization energies) with the aid of well-defined mathematical functions (overlap integrals), and so are closely related to ionization energies and have definite quantitative values.

(4) *Overlap integrals are actually calculated.* We do not in effect ignore the overlap matrix, i.e. we do not set it equal to a unit matrix. Instead, the elements of the overlap matrix are calculated, each S_{ij} depending on the distance apart of the atoms i and j, which has the important consequence that the S values depend on the geometry of the molecule. Since **S** is not taken as a unit matrix, we cannot go directly from $\mathbf{HC} = \mathbf{SC}\varepsilon$ to $\mathbf{HC} = \mathbf{C}\varepsilon$ and thus we cannot simply diagonalize the EHM Fock to get the c's and ε's.

These four points are elaborated on below.

(1) *Use of a minimal valence basis set* in the EHM is more realistic than treating just the $2p_z$ orbitals, since all the valence electrons in a molecule are likely to be involved in determining its properties. Further, the SHM is largely limited to π systems, i.e. to alkenes and aromatics and derivatives of these with attached π electron groups, but the EHM, in contrast, can in principle be applied to any molecule. The use of a minimal valence basis set makes the Fock matrix much larger than in the "corresponding" SHM calculation. For example in an SHM calculation on ethene, only two orbitals are used, the $2p_z$ on C_1 and the $2p_z$ on C_2, and the SHM Fock matrix is (using the compact Dirac notation $\langle \phi_i | \hat{H} | \phi_j \rangle = \int \phi_i \hat{H} \phi_j \, dv$)

$$\mathbf{H}(\text{SHM}) = \begin{pmatrix} \langle C_1(2p_z)|\hat{H}|C_1(2p_z)\rangle & \langle C_1(2p_z)|\hat{H}|C_2(2p_z)\rangle \\ \langle C_2(2p_z)|\hat{H}|C_1(2p_z)\rangle & \langle C_2(2p_z)|\hat{H}|C_2(2p_z)\rangle \end{pmatrix} = \begin{pmatrix} 0 & -1 \\ -1 & 0 \end{pmatrix}$$

$$2 \times 2 \text{ matrix} \qquad\qquad (4.89)$$

To write down the EHM Fock matrix, let us label the valence orbitals like this:

$$
\begin{array}{lllll}
H_1(1s)\phi_1 & C_1(2s)\phi_5 & C_1(2p_x)\phi_7 & C_1(2p_y)\phi_9 & C_1(2p_z)\phi_{11} \\
H_2(1s)\phi_2 & C_2(2s)\phi_6 & C_2(2p_x)\phi_8 & C_2(2p_y)\phi_{10} & C_2(2p_z)\phi_{12} \\
H_3(1s)\phi_3 & & & & \\
H_4(1s)\phi_4 & & & &
\end{array}
$$

The SHM basis set for ethene.
Each carbon has one $2p$ basis function.
C_2H_4 has two basis functions

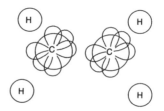

The EHM basis set for ethene.
Each carbon has one $2s$ and three $2p$ basis functions.
Each H has one $1s$ basis function.
C_2H_4 has 12 basis functions.

Figure 4.26. The SHM normally uses only one basis function per "heavy atom": only one $2p$ orbital on each carbon, oxygen, nitrogen, etc., ignoring the hydrogens. The EHM uses for each carbon, oxygen, nitrogen, etc., a $2s$ and three $2p$ orbitals, and for each hydrogen a $1s$ orbital. This is called a minimal valence basis set.

Then

$$\mathbf{H}(\text{EHM}) = \begin{pmatrix} \langle\phi_1|\hat{H}|\phi_1\rangle & \langle\phi_1|\hat{H}|\phi_2\rangle & \cdots & \langle\phi_1|\hat{H}|\phi_{12}\rangle \\ \langle\phi_2|\hat{H}|\phi_1\rangle & \langle\phi_2|\hat{H}|\phi_2\rangle & \cdots & \langle\phi_2|\hat{H}|\phi_{12}\rangle \\ \vdots & \vdots & \cdots & \vdots \\ \langle\phi_{12}|\hat{H}|\phi_1\rangle & \langle\phi_{12}|\hat{H}|\phi_2\rangle & \cdots & \langle\phi_{12}|\hat{H}|\phi_{12}\rangle \end{pmatrix} \quad (4.90)$$

12×12 matrix

The SHM and EHM basis sets are shown in Fig. 4.26.

(2) *The EHM Fock matrix interactions i, j do not have just two values (α or β) as in the SHM, but are functions of the orbitals (the basis functions) ϕ_i and ϕ_j and of the separation of these orbitals*, as explained in (3) below.

(3) *The EHM matrix elements $\langle\phi_i\hat{H}\phi_i\rangle$ and $\langle\phi_i\hat{H}\phi_j\rangle$ are calculated* (rather than set equal to 0 or −1), although the calculation is a simple one using overlap integrals and experimental ionization energies; in *ab initio* calculations (chapter 5) and more advanced SE calculations (chapter 6), the mathematical form of $\hat{H}$ taken into account. The i,i-type interactions are taken as being proportional to the negative of the ionization energy [55] of the orbital ϕ_i and the i,j-type interactions as being proportional to the overlap integral between ϕ_i and ϕ_j and the negative of the average of the ionization energies I_i and I_j of ϕ_i and ϕ_j (the negative of the orbital ionization energy is the energy of an electron in the orbital, compared to the zero of energy of the electron and the ionized species infinitely separated and at rest):

$$\langle\phi_i\hat{H}\phi_i\rangle = -I_i \quad (4.91)$$

$$\langle\phi_i\hat{H}\phi_j\rangle = -\tfrac{1}{2}K S_{ij}(I_i + I_j) \quad (4.92)$$

A proportionality constant K of about 2 is commonly used.

For H($1s$), C($2s$) and C($2p$), experiment shows

$$I(\text{H}(1s)) = 13.6\,\text{eV}, \quad I(\text{C}(2s)) = 20.8\,\text{eV}, \quad I(\text{C}(2p)) = 11.3\,\text{eV} \quad (4.93)$$

The overlap integrals are calculated using Slater-type (section 5.3.2) functions for the basis functions, e.g.

$$\phi(1s) = \left(\frac{\zeta_1^3}{\pi}\right)^{1/2} \exp(-\zeta_1|\mathbf{r} - \mathbf{R}_{1s}|) \tag{4.94}$$

$$\phi(2s) = \left(\frac{\zeta_2^5}{96\pi}\right)^{1/2} |\mathbf{r} - \mathbf{R}_{2s}| \exp\left(\frac{-\zeta_2|\mathbf{r} - \mathbf{R}_{2s1}|}{2}\right) \tag{4.95}$$

where the parameters ζ depend on the particular atom (H, C, etc.) and orbital (1s, 2s, etc. The variable $\mathbf{r} - \mathbf{R}$ is the distance of the electron from the atomic nucleus on which the function is centered; $\mathbf{r}$ is the vector from the origin of the Cartesian coordinate system to the electron, and $\mathbf{R}$ is the vector from the origin to the nucleus on which the basis function is centered:

$$|\mathbf{r} - \mathbf{R}_A| = [(x - x_A)^2 + (y - y_A)^2 + (z - z_a)^2]^{1/2} \tag{4.96}$$

where (x_A, y_A, z_A) are the coordinates of the nucleus bearing the Slater function. The Slater function is thus a function of three variables x,y,z and depends parametrically on the location (x_A, y_A, z_A) of the nucleus A on which it is centered. The Fock matrix elements are thus calculated with the aid of overlap integrals whose values depend the location of the basis functions; this means that the MOs and their energies will depend on the actual *geometry* used in the input, whereas in a simple Hückel calculation, the MOs and their energies depend only on the *connectivity* of the molecule).

(4) *The overlap matrix* **S** in the EHM is not simply treated as a unit matrix, in effect ignoring it, for the purpose of diagonalizing the Fock matrix. Rather, the overlap integrals are actually evaluated, not only to help calculate the Fock elements, but also to reduce the equation $\mathbf{HC} = \mathbf{SC\varepsilon}$ to the standard eigenvalue form $\mathbf{HC} = \mathbf{C\varepsilon}$. This is done in the following way. Suppose the original set of basis functions $\{\phi_i\}$ could be transformed by some process into an *orthonormal* set $\{\phi_i'\}$ (since atom-centered basis functions cannot be orthogonal, as explained in section 4.3.4, the new set must be delocalized over several centers and is in fact a linear combination of the atom-centered set) such that with a new set of coefficients c' we have LCAO MOs with the same energy levels as before, i.e.

$$S_{ij}' = \int \phi_i' \phi_j' \, dv = \delta_{ij} \tag{4.97}$$

where δ_{ij} is the Kronecker delta (Eq. (4.57)). The result of the process referred to above is

$$\mathbf{HC} = \mathbf{SC\varepsilon} \xrightarrow{\text{Process}} \mathbf{H'C'} = \mathbf{S'C'\varepsilon} \tag{4.98}$$

(ε, not ε', as the energy will not depend on manipulation of a given set of basis functions) where the matrices **H**, **C**, **S** and ε were defined in section 4.3.4 (Eqs (4.55)) and **H'** and **S'** are analogous to **H** and **S** with ϕ' in place of ϕ and **C'** is the matrix of coefficients c' that satisfies the equation with the energy levels ε (the elements of ε) being the same as in the original equation $\mathbf{HC} = \mathbf{SC\varepsilon}$. Since from Eq. (4.97) $\mathbf{S'} = 1$, the unit matrix (section 4.3.3), Eq. (4.98) simplifies to

$$\mathbf{HC} = \mathbf{SC\varepsilon} \xrightarrow{\text{Process}} \mathbf{H'C'} = \mathbf{C'\varepsilon} \tag{4.99}$$

The *Process* that effects the transformation is called *orthogonalization*, since the result is to make the basis functions orthogonal. The favored orthogonalization procedure in computational chemistry, which I will now describe, is Löwdin orthogonalization (after the quantum chemist Per-Olov Löwdin).

Define a matrix $\mathbf{C}'$ such that

$$\mathbf{C}' = \mathbf{S}^{1/2}\mathbf{C}, \quad \text{i.e.} \quad \mathbf{C} = \mathbf{S}^{-1/2}\mathbf{C}' \tag{4.100}$$

(By multiplying on the left by $\mathbf{S}^{-1/2}$ and noting that $\mathbf{S}^{-1/2}\mathbf{S}^{1/2} = \mathbf{S}^0 = 1$.)

Substituting (4.100) into $\mathbf{HC} = \mathbf{SC}\varepsilon$ and multiplying on the left by $\mathbf{S}^{-1/2}$ we get

$$\mathbf{S}^{-1/2}\mathbf{H}\mathbf{S}^{-1/2}\mathbf{C}' = \mathbf{S}^{-1/2}\mathbf{S}\mathbf{S}^{-1/2}\mathbf{C}'\varepsilon \tag{4.101}$$

Let

$$\mathbf{S}^{-1/2}\mathbf{H}\mathbf{S}^{-1/2} = \mathbf{H}' \tag{4.102}$$

and note that $\mathbf{S}^{-1/2}\mathbf{S}\mathbf{S}^{-1/2} = \mathbf{S}^{1/2}\mathbf{S}^{-1/2} = 1$ then we have from (4.101) and (4.102)

$$\mathbf{H}'\mathbf{C}' = 1\mathbf{C}'\varepsilon$$

i.e.

$$\mathbf{H}'\mathbf{C}' = \mathbf{C}'\varepsilon \tag{4.103}$$

Thus the orthogonalizing *Process* of (4.99) (or rather one possible orthogonalization process, Löwdin orthogonalization) is the use of an *orthogonalizing matrix* $\mathbf{S}^{-1/2}$ to transform $\mathbf{H}$ by pre- and postmultiplication (Eq. 102) into $\mathbf{H}'$. $\mathbf{H}'$ satisfies the standard eigenvalue equation (Eq. (4.103)), so

$$\mathbf{H}' = \mathbf{C}'\varepsilon\mathbf{C}'^{-1} \tag{4.104}$$

In other words, using $\mathbf{S}^{-1/2}$ we transform the original Fock matrix $\mathbf{H}$, which is not directly diagonalizable to eigenvector and eigenvalue matrices $\mathbf{C}$ and ε, into a related matrix $\mathbf{H}'$ which *is* diagonalizable to eigenvector and eigenvalue matrices $\mathbf{C}'$ and ε. The matrix $\mathbf{C}'$ is then transformed to the desired $\mathbf{C}$ by multiplying by $\mathbf{S}^{-1/2}$ (Eq. (4.100)). So without using the drastic $\mathbf{S} = 1$ approximation we can use matrix diagonalization to get the coefficients and energy levels from the Fock matrix.

The orthogonalizing matrix $\mathbf{S}^{-1/2}$ is calculated from $\mathbf{S}$: the integrals S are calculated and assembled into $\mathbf{S}$, which is then diagonalized:

$$\mathbf{S} = \mathbf{PDP}^{-1} \tag{4.105}$$

Now it can be shown that any function of a matrix $\mathbf{A}$ can be obtained by taking the same function of its corresponding diagonal alter ego and pre- and postmultiplying by the diagonalizing matrix $\mathbf{P}$ and its inverse $\mathbf{P}^{-1}$:

$$f(\mathbf{A}) = \mathbf{P}f(\mathbf{D})\mathbf{P}^{-1} \tag{4.106}$$

and diagonal matrices have the nice property that $f(\mathbf{D})$ is the diagonal matrix whose diagonal element $i, j = f(\text{element } i, j \text{ of } \mathbf{D})$. So the inverse square root of $\mathbf{D}$ is the matrix whose elements are the inverse square roots of the corresponding elements of **D**. Therefore

$$\mathbf{S}^{-1/2} = \mathbf{P}\mathbf{D}^{-1/2}\mathbf{P}^{-1} \tag{4.107}$$

and to find $\mathbf{D}^{-1/2}$ we (or rather the computer) simply take the inverse square root of the diagonal (i.e. the nonzero) elements of **D**. To summarize: **S** is diagonalized to give **P**, $\mathbf{P}^{-1}$ and **D**, **D** is used to calculate $\mathbf{D}^{-1/2}$, then the orthogonalizing matrix $\mathbf{S}^{-1/2}$ is calculated (Eq. (4.107)) from **P**, $\mathbf{D}^{-1/2}$ and $\mathbf{P}^{-1}$. The orthogonalizing matrix is then used to convert **H** to **H**′, which can be diagonalized to give the eigenvalues and the eigenvectors (section 4.4.2).

Review of the EHM procedure
The EHM procedure for calculating eigenvectors and eigenvalues, i.e. coefficients (or in effect MOs – the c's along with the basis functions comprise the MOs) and energy levels, bears several important resemblances to that used in more advanced methods (chapters 5 and 6) and so is worth reviewing.

(1) An input structure (a molecular geometry) must be specified and submitted to calculation. The geometry can be specified in Cartesian coordinates (probably the usual way nowadays) or as bond lengths, angles and dihedrals (internal coordinates), depending on the program. In practice a virtual molecule would likely be created with an interactive model-building program (usually by clicking together groups and atoms) which would then supply the EHM program with either Cartesian or internal coordinates.

(2) The EHM program calculates the overlap integrals S and assembles the overlap matrix **S**.

(3) The program calculates the Fock matrix elements $H_{ij} = \langle \phi_i | \hat{H} | \phi_j \rangle$ (Eqs (4.91) and (4.92)) using stored values of ionization energies I, the overlap integrals S, and the proportionality constant K of that particular program. The matrix elements are assembled into the Fock matrix **H**.

(4) The overlap matrix is diagonalized to give **P**, **D** and $\mathbf{P}^{-1}$ (Eq. (4.105)) and $\mathbf{D}^{-1/2}$ is then calculated by finding the inverse square roots of the diagonal elements of **D**. The orthogonalizing matrix $\mathbf{S}^{-1/2}$ is then calculated from **P**, $\mathbf{D}^{-1/2}$ and $\mathbf{P}^{-1}$ (Eq. (4.107)).

(5) The Fock matrix **H** in the atom-centered nonorthogonal basis $\{\phi\}$ is transformed into the matrix **H**′ in the delocalized, linear combination orthogonal basis $\{\phi'\}$ by pre- and postmultiplying **H** by the orthogonalizing matrix $\mathbf{S}^{-1/2}$ (Eq. (4.102)).

(6) **H**′ is diagonalized to give **C**′, ε and $\mathbf{C}'^{-1}$ (Eq. (4.104)). We now have the energy levels ε (the diagonal elements of the $\dot{\varepsilon}$ matrix).

(7) **C**′ must be transformed to give the coefficients c of the original, atom-centered set of basis functions $\{\phi\}$ in the MOs (i.e. to convert the elements c' to c). To get the c's in the MOs $\psi_j = c_{1j}\phi_1 + c_{2j}\phi_2 + \cdots$, we transform **C**′ to **C** by premultiplying by $\mathbf{S}^{-1/2}$ (Eq. (4.100)).

Molecular energy and geometry optimization in the EHM
Steps (1)–(7) take an input geometry and calculate its energy levels (the elements of $\boldsymbol{\varepsilon}$) and their MOs or wavefunctions (the ψ's; from the c's, the elements of **C**, and the basis functions ϕ). Now, clearly any method in which the energy of a molecule depends on its

geometry can in principle be used to find minima and transition states (see chapter 2). This brings us to the matter of how the EHM calculates the energy of a molecule. The energy of a molecule, that is, the energy of a particular nuclear configuration on the potential energy surface, is the sum of the electronic energies and the internuclear repulsions ($E_{electronic} + V_{NN}$).

In comparing the energies of isomers, or of two geometries of the same molecule, one should, strictly, compare $E^{total} = E_{electronic} + V_{NN}$. The electronic energy is the sum of kinetic energy and potential energy (electron–electron repulsion and electron–nucleus attraction) terms. The internuclear repulsion, due to all pairs of interacting nuclei and trivial to calculate, is usually represented by V, a symbol for potential energy. The EHM ignores V_{NN}. Furthermore, the method calculates electronic energy simply as the sum of one-electron energies (section 4.3.5), ignoring electron–electron repulsion. Hoffmann's tentative justification [53a] for ignoring internuclear repulsion and using a simple sum of one-electron energies was that when the *relative* energies of isomers are calculated, by subtracting two values of E^{total}, the electron repulsion and nuclear repulsion terms approximately cancel, i.e. that changes in energy that accompany changes in geometry are due mainly to alterations of the MO energy levels. Actually, it seems that the (quite limited) success of the EHM in predicting molecular geometry is due to the fact that E^{total} is approximately *proportional* to the sum of the occupied MO energies; thus although the EHM energy difference is not equal to the difference in total energies, it is (or tends to be) approximately proportional to this difference [56]. In any case, the real strength of the EHM lies in the ability of this fast and widely applicable method to assist chemical intuition, if *provided with* a reasonable molecular geometry.

4.4.2   An illustration of the EHM: the protonated helium molecule

Protonation of a helium atom gives $He-H^+$, the helium hydride cation, the simplest heteronuclear molecule [57]. Conceptually, of course, this can also be formed by the union of a helium dication and a hydride ion, or a helium cation and a hydrogen atom:

$$He : +H^+ \rightarrow He : H^+ \text{ or } He^{++} + : H^- \rightarrow He : H^+ \text{ or } He^{+\cdot} + {}^\bullet H \rightarrow He : H^+$$

Its lower symmetry makes this molecule better than H_2 for illustrating molecular quantum mechanical calculations (most molecules have little or no symmetry). Following the prescription in points (1)–(7), we calculate the following results.

(1) Input structure
We choose a plausible bond length: 0.800 Å (the H–H bond length is 0.742 Å and the H–X bond length is ca. 1.0 Å, where X is a "first-row" element (in quantum chemistry, first-row means Li to F, not H and He). The Cartesian coordinates could be written $H_1(0, 0, 0)$, $He_2(0, 0, 0.800)$.

(2) Overlap integrals and overlap matrix
The minimal valence basis set here consists of the hydrogen 1s orbital (ϕ_1) and the helium 1s orbital (ϕ_2). The needed integrals are $S_{11} = S_{22}$ and $S_{12} = S_{21}$, where

$S_{ij} = \int \phi_i \phi_j \, dv$. The Slater functions for ϕ_1 and ϕ_2 are [58]

$$\phi_1(\text{H}1s) = \left(\frac{\zeta_H^3}{\pi}\right)^{1/2} e^{-\zeta_H|\mathbf{r}-\mathbf{R}_H|} \tag{4.108}$$

and

$$\phi_2(\text{He}1s) = \left(\frac{\zeta_{He}^3}{\pi}\right)^{1/2} e^{-\zeta_H|\mathbf{r}-\mathbf{R}_{He}|} \tag{4.109}$$

Reasonable values [57] are $\zeta_H = 1.24\,\text{Bohr}^{-1}$ and $\zeta_{He} = 2.0925\,\text{Bohr}^{-1}$, if r is in atomic units, a.u. (see section 5.2.2); 1 a.u. $= 0.5292\,\text{Å}$. The overlap integrals are $S_{11} = S_{22} = 1$ (as must be the case if ϕ_1 and ϕ_2 are normalized and $S_{12} = S_{21} = 0.435$ (for all well-behaved functions $\int f_1 f_2 \, dq = \int f_2 f_1 \, dq$).
   The overlap matrix is thus

$$\mathbf{S} = \begin{pmatrix} 1 & 0.435 \\ 0.435 & 1 \end{pmatrix} \tag{4.112}$$

(3) Fock matrix
We need the matrix elements $H_{11} = H_{22}$ and $H_{12} = H_{21}$, where the integrals $H_{ij} = \langle \phi_i | \hat{H} | \phi_j \rangle$ are not actually calculated from first principles but rather are estimated, with the aid of overlap integrals and orbital ionization energies:

$$\langle \phi_i \hat{H} \phi_j \rangle = -I_i$$

$$\langle \phi_i \hat{H} \phi_j \rangle = -\tfrac{1}{2} K S_{ij}(I_i + I_j)$$

Using simply the ionization energies (cf. [55]):

$$I(\text{H}) = I_1 = 13.6\,\text{eV}, \quad I(\text{He}) = I_2 = 24.6\,\text{eV}$$

Hoffmann used in his initial calculations [53a] $K = 1.75$. So

$$H_{11} = -13.6\,\text{eV}$$
$$H_{12} = H_{21} = -\tfrac{1}{2}(1.75)(0.435)(13.6 + 24.6) = -14.5$$
$$H_{22} = -24.6$$

And the Fock matrix is

$$\mathbf{H} = \begin{pmatrix} -13.6 & -14.5 \\ -14.5 & -24.6 \end{pmatrix} \tag{4.113}$$

(4) Orthogonalizing matrix

As explained above, we (a) diagonalize **S**, (b) calculate $\mathbf{D}^{-1/2}$, then (c) calculate the orthogonalizing matrix $\mathbf{S}^{-1/2}$:

(a) Diagonalize **S**

$$\mathbf{S} = \begin{pmatrix} 1 & 0.435 \\ 0.435 & 1 \end{pmatrix}$$

$$= \underbrace{\begin{pmatrix} 0.707 & 0.707 \\ 0.707 & -0.707 \end{pmatrix}}_{\mathbf{P}} \underbrace{\begin{pmatrix} 1.435 & 0 \\ 0 & 0.565 \end{pmatrix}}_{\mathbf{D}} \underbrace{\begin{pmatrix} 0.707 & 0.707 \\ 0.707 & -0.707 \end{pmatrix}}_{\mathbf{P}^{-1}} \tag{4.114}$$

(b) Calculate $\mathbf{D}^{-1/2}$

$$\mathbf{D}^{-1/2} = \begin{pmatrix} 1.435^{-1/2} & 0 \\ 0 & 0.565^{-1/2} \end{pmatrix} = \begin{pmatrix} 0.835 & 0 \\ 0 & 1.330 \end{pmatrix} \tag{4.115}$$

(c) Calculate the orthogonalizing matrix $\mathbf{S}^{-1/2}$

$$\mathbf{S}^{-1/2} = \underbrace{\begin{pmatrix} 0.707 & 0.707 \\ 0.707 & -0.707 \end{pmatrix}}_{\mathbf{P}} \underbrace{\begin{pmatrix} 0.835 & 0 \\ 0 & 1.330 \end{pmatrix}}_{\mathbf{D}^{-1/2}} \underbrace{\begin{pmatrix} 0.707 & 0.707 \\ 0.707 & -0.707 \end{pmatrix}}_{\mathbf{P}^{-1}}$$

$$= \begin{pmatrix} 1.083 & -0.248 \\ -0.248 & 1.083 \end{pmatrix} \tag{4.116}$$

*(5) Transformation of the original Fock matrix **H** to **H'***

Using Eq. (102):

$$\mathbf{H'} = \underbrace{\begin{pmatrix} 1.083 & -0.248 \\ -0.248 & 1.083 \end{pmatrix}}_{\mathbf{S}^{-1/2}} \underbrace{\begin{pmatrix} -13.6 & -14.5 \\ -14.5 & -24.6 \end{pmatrix}}_{\mathbf{H}} \underbrace{\begin{pmatrix} 1.083 & -0.248 \\ -0.248 & 1.083 \end{pmatrix}}_{\mathbf{S}^{-1/2}}$$

$$= \begin{pmatrix} -9.67 & -7.65 \\ -7.68 & -21.74 \end{pmatrix} \tag{4.117}$$

*(6) Diagonalization of **H'***

From Eq. (4.104) ($\mathbf{H'} = \mathbf{C'}\boldsymbol{\varepsilon}\mathbf{C'}^{-1}$), diagonalization of **H'** gives an eigenvector matrix **C'** and the eigenvalue matrix $\boldsymbol{\varepsilon}$; the columns of **C'** are the coefficients of the transformed, orthonormal basis functions:

$$\mathbf{H'} = \begin{pmatrix} -9.67 & -7.65 \\ -7.68 & -21.74 \end{pmatrix}$$

$$= \underbrace{\begin{pmatrix} 0.436 & 0.899 \\ 0.900 & -0.437 \end{pmatrix}}_{\mathbf{C'}} \underbrace{\begin{pmatrix} -25.5 & 0 \\ 0 & -5.95 \end{pmatrix}}_{\boldsymbol{\varepsilon}} \underbrace{\begin{pmatrix} 0.436 & 0.900 \\ 0.899 & -0.437 \end{pmatrix}}_{\mathbf{C'}^{-1}} \tag{4.118}$$

We now have the energy levels (-25.5 and -5.95 eV), but the eigenvectors of $\mathbf{C}'$ must be transformed to give us the coefficients of the original, nonorthogonal basis functions.

(7) Transformation of $\mathbf{C}'$ to $\mathbf{C}$

Using Eq. (4.102), ($\mathbf{C} = \mathbf{S}^{-1/2}\mathbf{C}'$):

$$\mathbf{C} = \underbrace{\begin{pmatrix} 1.083 & -0.248 \\ -0.248 & 1.083 \end{pmatrix}}_{\mathbf{S}^{-1/2}} \underbrace{\begin{pmatrix} 0.436 & 0.899 \\ 0.900 & -0.437 \end{pmatrix}}_{\mathbf{C}'}$$

$$= \begin{pmatrix} 0.249 & 1.082 \\ 0.867 & -0.696 \end{pmatrix} \tag{4.119}$$

Note that unlike the case in the SHM, the sum of the squares of the c's for an MO does not equal 1, since overlap integrals S_{ij} for basis functions on different atoms are not set equal to 0; in other words, the basis functions are not assumed to be orthogonal, and the overlap matrix is not a unit matrix. Thus for ψ_1:

$$\psi_1 = c_1\phi_1 + c_2\phi_2, \quad \text{so}$$

$$\int \psi_1^2 dv = \int (c_1^2\phi_1^2 + 2c_1c_2\phi_1\phi_2 + c_2^2\phi_2^2)\, dv = 1$$

since the probability of finding an electron in ψ_1 somewhere in space is 1. The basis functions ϕ are normalized, so

$$c_1^2 + 2c_1c_2S_{12} + c_2^2 = 1, \quad \text{i.e.} \quad c_1^2 + c_2^2 = 1 - 2c_1c_2S_{12}$$

4.4.3 The extended Hückel method – applications

The EHM was initially applied to the geometries (including conformations) and relative energies of hydrocarbons [53a], but the calculation of these two basic chemical parameters is now much better handled by SE methods like AM1 and PM3 (chapter 6) and by *ab initio* (chapter 5) methods. The main use of the EHM nowadays is to study large, extended systems [59] like polymers, solids and surfaces. Indeed, of four papers by Hoffmann and coworkers in the J. Am. Chem. Soc. in 1995, using the EHM, three applied it to such polymeric systems [60]. The ability of the method to illuminate problems in solid-state science makes it useful to physicists. Even when not applied to polymeric systems, the EHM is frequently used to study large, heavy-metal-containing molecules [61] that might not be very amenable to *ab initio* or to other SE approaches.

4.4.4 Strengths and weaknesses of the EHM

Strengths
One big advantage of the EHM over *ab initio* methods (chapter 5), more elaborate SE methods (chapter 6), and DFT methods (chapter 7), is that the EHM can be applied to

very large systems, and can treat almost any element since the only element-specific parameter needed is an ionization energy, which is usually available. In contrast, more elaborate SE methods have not been parameterized for most elements (although recent parameterizations of PM3 and MNDO for transition metals make these much more generally useful than hitherto – chapter 6, section 6.2.6.7). For *ab initio* and DFT methods, basis sets may not be available for elements of interest, and besides, *ab initio* and even DFT methods are hundreds of times slower than the EHM and thus limited to much smaller systems. The applicability of the EHM to large systems and a wide variety of elements is one reason why it has been extensively applied to polymeric and solid-state structures. The EHM is faster than more elaborate SE methods because calculation of the Fock matrix elements is so simple and because this matrix needs to be diagonalized only *once* to yield the eigenvalues and eigenvectors; in contrast, SE methods like AM1 and PM3 (chapter 6), as well as *ab initio* calculations, require repeated matrix diagonalization because the Fock matrix must be iteratively refined in the SCF procedure (section 5.2.3.6).

The spartan reliance of the EHM on empirical parameters helps to make it relatively easy (in the right hands) to interpret its results, which depend, in the last analysis, only on geometry (which affects overlap integrals) and ionization energies. With a strong dose of chemical intuition this has enabled the method to yield powerful insights, such as counterintuitive orbital mixing [62], and the very powerful Woodward–Hoffmann rules [38].

The applicability to large systems, including polymers and solids, containing almost any kind of atom, and the relative transparency of the physical basis of the results, are the main advantages of the EHM.

Surprisingly for such a conceptually simple method, the EHM has a theoretically-based advantage over otherwise more elaborate SE methods like AM1 and PM3, in that it treats orbital overlap properly: those other methods use the "neglect of differential overlap" or NDO approximation (section 6.2), meaning that they take $S_{ij} = \delta_{ij}$, as in the SHM. This can lead to superior results from the EHM [63].

The EHM is a very valuable teaching tool because it follows straightforwardly from the SHM yet uses overlap integrals and matrix orthogonalization in the same fashion as the mathematically more elaborate *ab initio* method.

Finally, the EHM, albeit more elaborately parameterized than in its original incarnation, has recently been shown to offer some promise as a serious competitor to the very useful and popular SE AM1 method (section 6.2.6.4) for calculating molecular geometries [64].

Weaknesses

The weaknesses of the standard EHM probably arise at least in part from the fact that it does not (contrast the *ab initio* method, chapter 5) take into account electron spin or electron–electron repulsion, ignores the fact that molecular geometry is partly determined by internuclear repulsion, and makes no attempt to overcome these defects by parameterization (unlike the recent variation which, with the aid of careful parameterization, evidently gives good geometries [64]).

The standard EHM gives, by and large, poor geometries and energies. Although it predicts a C–H bond length of ca. 1.0 Å, it yields C/C bond lengths of 1.92, 1.47 and

0.85 Å for ethane, ethene and ethyne, respectively, cf. the actual values of 1.53, 1.33 and 1.21Å, and although the favored conformation of an alkane is usually correctly identified, the energy barriers and differences are generally at best in only modest agreement with experiment. Because of this inability to reliably calculate geometries, EHM calculations are usually not used for geometry optimizations, but rather utilize experimental geometries.

4.5 SUMMARY OF CHAPTER 4

This chapter introduces the application of quantum mechanics (QM) to computational chemistry by outlining the development of QM up to the Schrödinger equation and then showing how this equation led to the SHM, from which the EHM followed.

Quantum mechanics teaches, basically, that energy is *quantized*: absorbed and emitted in discrete packets (*quanta*) of magnitude $h\nu$, where h is Planck's constant and ν (Greek *nu*) is the frequency associated with the energy. QM grew out of studies of blackbody radiation and of the photoelectric effect. Besides QM, radioactivity and relativity contributed to the transition from classical to modern physics. The classical Rutherford nuclear atom suffered from the deficiency that Maxwell's electromagnetic theory demanded that its orbiting electrons radiate away energy and swiftly fall into the nucleus. This problem was countered by Bohr's quantum atom, in which an electron could orbit stably if its angular momentum was an integral multiple of $h/2\pi$. However, the Bohr model contained several *ad hoc* fixes and worked only for the hydrogen atom. The deficiencies of the Bohr atom were surmounted by Schrödinger's wave mechanical atom; this was based on a combination of classical wave theory and the de Broglie postulate that any particle is associated with a wavelength $\lambda = h/p$, where p is the momentum. The four quantum numbers follow naturally from the wave mechanical treatment and the model does not break down for atoms beyond hydrogen.

Hückel was the first to apply QM to species significantly more complex than the hydrogen atom. The Hückel approach is treated nowadays within the framework of the concept of hybridization: the π electrons in p orbitals are taken into account and the σ electrons in an sp^2 framework are ignored. Hybridization is a purely mathematical convenience, a procedure in which atomic (or molecular) orbitals are combined to give new orbitals; it is analogous to the combination of simple vectors to give new vectors (an orbital is actually a kind of vector).

The SHM (SHT, HMO method) starts with the Schrödinger equation in the form $\hat{H}\psi = E\psi$ where $\hat{H}$ is a Hamiltonian operator, ψ is a MO wavefunction and E is the energy of the system (atom or molecule). By expressing ψ as a LCAO and minimizing E with respect to the LCAO coefficients one obtains a set of simultaneous equations, the secular equations. These are equivalent to a single matrix equation, $\mathbf{HC} = \mathbf{SC\varepsilon}$; $\mathbf{H}$ is an energy matrix, the Fock matrix, $\mathbf{C}$ is the matrix of the LCAO coefficients, $\mathbf{S}$ is the overlap matrix and ε is a diagonal matrix whose nonzero, i.e. diagonal, elements are the MO energy levels. The columns of $\mathbf{C}$ are called eigenvectors and the diagonal elements of ε are called eigenvalues. By the drastic approximation $\mathbf{S} = \mathbf{1}$ ($\mathbf{1}$ is the unit matrix), the matrix equation becomes $\mathbf{HC} = \mathbf{C\varepsilon}$, i.e. $\mathbf{H} = \mathbf{C\varepsilon C}^{-1}$ which is the same as saying that diagonalization of $\mathbf{H}$ gives $\mathbf{C}$ and ε, i.e gives the MO coefficients in the

LCAO, and the MO energies. To get numbers for $\mathbf{H}$ the SHM reduces all the Fock matrix elements to α (the coulomb integral, for AOs on the same atom) and β (the bond integral or resonance integral, for AOs not on the same atom; for nonadjacent atoms β is set $= 0$). To get actual numbers for the Fock elements, α and β are defined as energies relative to α, in units of $|\beta|$; this makes the Fock matrix consist of just 0s and -1s, where the 0s represent same-atom interactions and nonadjacent-atom interactions, and the -1s represent adjacent-atom interactions. The use of just two Fock elements is a big approximation. The SHM Fock matrix is easily written down just by looking at the way the atoms in the molecule are connected. Applications of the SHM include predicting:

(a) The nodal properties of the MOs, very useful in applying the Woodward–Hoffmann rules.

(b) The stability of a molecule based on its filled and empty MOs, and its delocalization energy or resonance energy, based on a comparison of its total π-energy with that of a reference system. The pattern of filled and empty MOs led to Hückel's rule (the $4n + 2$ rule) which says that planar molecules with completely conjugated p orbitals containing $4n + 2$ electrons should be aromatic.

(c) Bond orders and atom charges, which are calculated from the AO coefficients of the occupied MOs (in the SHM LCAO treatment, p AOs are basis functions that make up the MOs).

The strengths of the SHM lie in the qualitative insights it gives into the effect of molecular structure on π orbitals. Its main triumph in this regard was its spectacularly successful prediction of the requirements for aromaticity (the Hückel $4n + 2$ rule).

The weaknesses of the SHM arise from the fact that it treats only π electrons (limiting its applicability largely to planar sp^2 arrays), its all-or-nothing treatment of overlap integrals, the use of just two values for the Fock integrals, and its neglect of electron spin and interelectronic repulsion. Because of these approximations it is not used for geometry optimizations and its quantitative predictions are sometimes viewed with suspicion. For obtaining eigenvectors and eigenvalues from the secular equations an older and inelegant alternative to matrix diagonalization is the use of determinants.

The EHM (EHT) follows from the SHM by using a basis set that consists not just of p orbitals, but rather of all the valence AOs (a minimal valence basis set), by calculating (albeit very empirically) the Fock matrix integrals, and by explicitly calculating the overlap matrix $\mathbf{S}$ (whose elements are also used in calculating the Fock integrals). Because $\mathbf{S}$ is not taken as a unit matrix, the equation $\mathbf{HC} = \mathbf{SC}\varepsilon$ must be transformed to one without $\mathbf{S}$ before matrix diagonalization can be applied. This is done by a matrix multiplication process called orthogonalization, involving $\mathbf{S}^{-1/2}$, which converts the original Fock matrix $\mathbf{H}$, based on nonorthogonal atom-centered basis functions, into a Fock matrix $\mathbf{H}'$, based on orthogonal linear combinations of the original basis functions. With these new basis functions, $\mathbf{H}'\mathbf{C}' = \mathbf{C}'\varepsilon$, i.e $\mathbf{H}' = \mathbf{C}'\varepsilon\mathbf{C}'^{-1}$, so that diagonalization of $\mathbf{H}'$ gives the eigenvectors (of the new basis functions, which are transformed back to those corresponding to the original set: $\mathbf{C}' \rightarrow \mathbf{C}$) and eigenvalues of $\mathbf{H}$.

Because the overlap integrals needed by the EHM depend on molecular geometry, the method can in principle be used for geometry optimization, although for the conventional EHM the results are generally poor, so known geometries are used as input.

Applications of the EHM involve largely the study of big molecules and polymeric systems, often containing heavy metals.

The strengths of the EHM derive from its simplicity: it is very fast and so can be applied to large systems; the only empirical parameters needed are (valence-state) ionization energies, which are available for a wide range of elements; the results of calculations lend themselves to intuitive interpretation since they depend only on geometry and ionization energies, and on occasion the proper treatment of overlap integrals even gives better results than those from more elaborate SE methods. The fact that the EHM is conceptually simple yet incorporates several features of more sophisticated methods enables it to serve as an excellent introduction to quantum mechanical computational methods.

The weaknesses of the EHM are due largely to its neglect of electron spin and electron–electron repulsion and the fact that it bases the energy of a molecule simply on the sum of the one-electron energies of the occupied orbitals, which ignores electron–electron repulsion and internuclear repulsion; this is at least partly the reason it usually gives poor geometries.

REFERENCES

[1] For general accounts of the development of quantum theory see: J. Mehra and H. Rechenberg, "The Historical Development of Quantum Theory", Springer-Verlag, New York, 1982; T. S. Kuhn, "Black-body Theory and the Quantum Discontinuity 1894–1912", Oxford University Press, Oxford, 1978. (b) An excellent historical and scientific exposition, at a somewhat advanced level: M. S. Longair, "Theoretical Concepts in Physics", Cambridge University Press, Cambridge, 1983, chapters 8–12.

[2] A great deal has been written speculating on the meaning of quantum theory, some of it serious science, some philosophy, some mysticism. Some leading references are: (a) A. Whitaker, "Einstein, Bohr and the Quantum Dilemma", Cambridge University Press, 1996; (b) V. J. Stenger, "The Unconscious Quantum", Prometheus, Amherst, NY, 1995; (c) P. Yam, Scientific American, June 1997, p. 124; (d) D. Z. Albert, Scientific American, May 1994, p. 58; (e) D. Z. Albert, "Quantum Mechanics and Experience", Harvard University Press, Cambridge, MA, 1992; (f) D. Bohm and H. B. Hiley, "The Undivided Universe", Routledge, New York, 1992; (g) J. Baggott, "The Meaning of Quantum Theory", Oxford University Press, New York, 1992; (h) M. Jammer, "The Philosophy of Quantum Mechanics", Wiley, New York, 1974.

[3] I. N. Levine, "Quantum Chemistry", 5th Ed., Prentice Hall, Upper Saddle River, NJ, 2000.

[4] Sitzung der Deutschen Physikalischen Gesellschaft, 14 December 1900, Verhandlung 2, p. 237. This presentation and one of October leading up to it (Verhandlung 2, p. 202) were combined in: M. Planck, Annalen. Phys., 1901, *4*(4), 553.

[5] M. J. Klein, Physics Today, 1966, *19*, 23.

[6] For a good and amusing account of quantum strangeness (and relativity effects) and how things might be if Planck's constant had a considerably different value, see G. Gamov and R. Stannard, "The New World of Mr Tompkins", Cambridge University Press, Cambridge, 1999. This is based on the classics by George Gamow, "Mr Tompkins in Wonderland" (1940) and "Mr Tompkins Explores the Atom" (1944), which were united in "Mr Tompkins in Paperback," Cambridge University Press, Cambridge, 1965.

[7] A. Einstein, Ann. Phys., 1905, *17*, 132. Actually, the measurements are very difficult to do accurately, and the Einstein linear relationship may have been more a prediction than an explanation of established facts.

[8] (a) For an elementary treatment of Maxwell's equations and the loss of energy by an accelerated electric charge, see R. K. Adair, "Concepts in Physics," Academic Press, New York, 1969, chapter 21. (b) For a brief historical introduction to Maxwell's equations see M. S. Longair, "Theoretical Concepts in Physics", Cambridge University Press, Cambridge, 1983, chapter 3. For a rigorous treatment of the loss of energy by an accelerated electric charge see Longair, chapter 9.

[9] N. Bohr, Phil. Mag., 1913, *26*, 1.

[10] For example, S. T. Thornton and A. Rex, "Modern Physics for Scientists and Engineers", Saunders, Orlando, FL., 1993, pp. 155–164.

[11] See, e.g. Ref. [2a], *loc. cit.*

[12] E. Schrödinger, Ann. Phys., 1926, *79*, 361. This first Schrödinger equation paper, a nonrelativistic treatment of the hydrogen atom, has been described as "one of the greatest achievements of twentieth-century physics" (Ref. [13, p. 205]).

[13] W. Moore, "Schrödinger. Life and thought," Cambridge University Press, Cambridge, 1989.

[14] L. de Broglie, "Recherche sur la Theorie des Quanta", thesis presented to the faculty of sciences of the University of Paris, 1924.

[15] Ref. [13, chapter 6].

[16] For example, Ref. [3, pp. 410–419, 604–613].

[17] V. I. Minkin, M. N. Glukhovtsev, and B. Ya. Simkin, "Aromaticity and Antiaromaticity: Electronic and Structural Aspects," Wiley, New York, 1994.

[18] (a) Generalized VB method: R. A. Friesner, R. B. Murphy, M. D. Beachy, M. N. Ringnalda, W. T. Pollard, B. D. Dunietz, Y. Cao, J. Phys. Chem. A, 1999, *103*, 1913, and references therein; (b) J. G. Hamilton and W. E. Palke, J. Am. Chem. Soc., 1993, *115*, 4159.

[19] The pioneering benzene paper: E. Hückel, Z. Physik, 1931, *70*, 204. Other papers by Hückel, on the double bond and on unsaturated molecules, are listed in his autobiography, "Ein Gelehrtenleben. Ernst und Satire," Verlag Chemie, Weinheim, 1975, pp. 178–179.

[20] L. Pauling, "The Nature of the Chemical Bond," 3 Ed., Cornell University Press, Ithaca, NY, 1960, pp. 111–126.

[21] (a) A compact but quite thorough treatment of the SHM see Ref. [3], pp. 629–649. (b) A good, brief introduction to the SHM is: J. D. Roberts, "Notes on Molecular Orbital Calculations", Benjamin, New York, 1962. (c) A detailed treatment: A. Streitweiser, "Molecular Orbital Theory for Organic Chemists," Wiley, New York, 1961. (d) The SHM and its atomic orbital and molecular orbital background are treated in considerable depth in "H. E. Zimmerman, "Quantum Mechanics for Organic Chemists," Academic Press, New York, 1975. (e) Perhaps the definitive presentation of the SHM is E. Heilbronner and H. Bock, "Das HMO Modell und seine Anwendung," Verlag Chemie, Weinheim, Germany, vol. 1 (basics and implementation), 1968; vol. 2, (examples and solutions), 1970; vol. 3 (tables of experimental and calculated quantities), 1970. An English translation of vol. 1 is available: "The HMO Model and its Application. Basics and Manipulation", Verlag Chemie, 1976.

[22] For example Ref. [21b, pp. 87–90; [21c, pp. 380–391] and references therein; Ref. [21d, chapter 4].

[23] See any introductory book on linear algebra.

[24] Ref. [21c, chapter 1].

[25] J. Simons and J. Nichols, "Quantum Mechanics in Chemistry," Oxford University Press, New York, 1997, p. 133.

[26] See, e.g. F. A. Carey and R. J. Sundberg, "Advanced Organic Chemistry. Part A," 3rd Ed., Plenum, New York, 1990, pp. 30–34.

[27] Y. Jean and F. Volatron, "An Introduction to Molecular Orbitals," Oxford University Press, New York, 1993, pp. 143–144.

[28] (a) P. A. Schultz and R. P. Messmer, J. Am. Chem. Soc., 1993, *115*, 10925. (b) P. B. Karadakov, J. Gerratt, D. L. Cooper, and M. Raimondi, J. Am. Chem. Soc., 1993, *115*, 6863.

[29] (a) T. H. Lowry and K. S. Richardson, "Mechanism and Theory in Organic Chemistry", Harper and Row, New York, 1981, pp. 26, 270. (b) A. Streiwieser, R. A. Caldwell and G. R. Ziegler, J. Am. Chem. Soc., 1969, *91*, 5081, and references therein.

[30] M. J. S. Dewar, "The Molecular Orbital Theory of Organic Chemistry," McGraw-Hill, New York, 1969, pp. 92–98.

[31] (a) For a short review of the state of MO theory in its early days see R. S. Mulliken, J. Chem Phys., 1935, 3, 375. (b) A personal account of the development of MO theory: R. S. Mulliken, "Life of a Scientist: An Autobiographical Account of the Development of Molecular Orbital Theory with an Introductory Memoir by Friedrich Hund," Springer-Verlag, New York, 1989. (c) For an account of the "tension" between the MO approach of Mulliken and the VB approach of Pauling see A. Simões and K. Gavroglu in "Conceptual Perspectives in Quantum Chemistry," J.-L. Calais and E. Kryachko, Eds., Kluwer Academic Publishers, London, 1997.

[32] L. Pauling, Chem. Rev., 1928, *5*, 173.

[33] J. E. Lennard-Jones, Trans. Faraday Soc., 1929, *25*, 668.

[34] C. A. Coulson and I. Fischer, Philos. Mag., 1949, *40*, 386.

[35] Ref. [21d, pp. 52–53].

[36] As Dewar points out in Ref. [30], this derivation is not really satisfactory. A rigorous approach is a simplified version of the derivation of the Hartree–Fock equations (section 5.2.3). It starts with the total molecular wavefunction expressed as a determinant, writes the energy in terms of this wavefunction and the Hamiltonian and finds the condition for minimum energy subject to the molecular orbitals being orthonormal (cf. orthogonal matrices, section 4.3.3). The procedure is explained in some detail in section 5.2.3).

[37] See, e.g. D. W. Rogers, "Computational Chemistry Using the PC," VCH, New York, 1990, pp. 92–94.

[38] R. B. Woodward and R. Hoffmann, "The Conservation of Orbital Symmetry," Verlag Chemie, Weinheim, Germany, 1970.

[39] (a) For a nice review of the cyclobutadiene problem see B. K. Carpenter in "Advances in Molecular Modelling," JAI Press, Greenwich, Connecticut, 1988. (b) Calculations on the degenerate interconversion of the rectangular geometries: J. C. Santo-García, A. J. Pérez-Jiménez, and F. Moscardó, Chem. Phys. Letter, 2000, *317*, 245.

[40] (a) Strictly speaking, cyclobutadiene exhibits a pseudo-Jahn–Teller effect: D. W. Kohn and P. Chen, J. Am. Chem. Soc., 1993, *115*, 2844. (b) For "A beautiful example of the Jahn–Teller effect" (MnF$_3$) see M. Hargittai, J. Am. Chem. Soc., 1997, *119*, 9042. (c) Review: T. A. Miller, Angew. Chem. Int. Ed., 1994, *33*, 962.

[41] A. A. Frost and B. Musulin, J. Chem. Phys., 1953, *21*, 572.

[42] W. E. Doering and L. H. Knox, J. Am. Chem. Soc., 1954, *76*, 3203.

[43] M. J. S. Dewar, "The Molecular Orbital Theory of Organic Chemistry," McGraw-Hill, New York, 1969, pp. 95–98.

[44] M. J. S. Dewar, "The Molecular Orbital Theory of Organic Chemistry," McGraw-Hill, New York, 1969, pp. 236–241.

[45] (a) Ref. [17, pp. 157–161]. (b) K. Krogh-Jespersen, P. von R. Schleyer, J. A. Pople, and D. Cremer, J. Am. Chem. Soc., 1978, *100*, 4301. (c) The cyclobutadiene dianion, another potentially aromatic system, has recently been prepared: K. Ishii, N. Kobayashi, T. Matsuo, M. Tanaka, A. Sekiguchi, J. Am. Chem. Soc., 2001, *123*, 5356.

[46] S. Zilberg and Y. Haas, J. Phys. Chem. A, 1998, *102*, 10843, 10851.

[47] The most rigorous approach to assigning electron density to atoms and bonds within molecules is the atoms-in molecules (AIM) method of Bader and coworkers: R. F. W. Bader, "Atoms in Molecules," Clarendon Press, Oxford, 1990.

[48] Various approaches to defining bond order and atom charges are discussed in F. Jensen, "Introduction to Computational Chemistry," Wiley, New York, 1999, chapter 9.

[49] Ref. [17, pp. 177–180].

[50] For leading references see: (a) B. A. Hess and L. J. Schaad, J. Chem. Educ., 1974, *51*, 640; (b) B. A. Hess and L. J. Schaad, Pure and Appl. Chem., 1980, *52*, 1471.

[51] See, e.g. Ref. [21c, chapters 4 and 5].

[52] See, e.g. Ref. [21c, pp. 13, 16].

[53] (a) R. Hoffmann, J. Chem. Phys., 1963, *39*, 1397; (b) Hoffmann, J. Chem. Phys, 1964, *40*, 2474; (c) R. Hoffmann, J. Chem. Phys., 1964, 40, 2480; (d) R. Hoffmann, J. Chem. Phys., 1964, *40*, 2745; (e) R. Hoffmann, Tetrahedron, 1966, *22*, 521; (f) R. Hoffmann, Tetrahedron, 1966, *22*, 539; (g) P. J. Hay, J. C. Thibeault, R. Hoffmann J. Am. Chem. Soc., 1975, *97*, 4884.

[54] M. Wolfsberg and L. Helmholz, J. Chem. Phys., 1952, *20*, 837.

[55] Actually, *valence state* ionization energies are usually used; see H. O. Pritchard, H. A. Skinner, Chem. Rev., 1955, *55*, 745; J. Hinze, H. H. Jaffe, J. Am. Chem. Soc., 1962, *84*, 540; A. Stockis, R. Hoffmann J. Am. Chem. Soc., 1980, *102*, 2952.

[56] F. L. Pilar, "Elementary Quantum Chemistry," McGraw-Hill, New York, 1990, pp. 493–494.

[57] A. Szabo and N. S. Ostlund, "Modern Quantum Chemistry," McGraw-Hill, 1989, pp. 168–179. This describes an *ab initio* (chapter 5) calculation on HeH$^+$, but gives information relevant to our EHM calculation.

[58] C. C. J. Roothaan, J. Chem. Phys., 1951, *19*, 1445.

[59] R. Hoffmann, "Solids and Surfaces: A Chemist's View of Bonding in Extended Structures." VCH publishers, 1988.

[60] (a) A polymeric rhenium compound: H. S. Genin, K. A. Lawlwr, R. Hoffmann, W. A. Hermann, R. W. Fischer, and W. Scherer, J. Am. Chem. Soc., 1995, *117*, 3244.

(b) Chemisorption of ethyne on silicon: Q. Liu and R. Hoffmann, J. Am. Chem. Soc., 1995, *117*, 4082. (c) A carbon/sulfur polymer: H. Genin and R. Hoffmann. J. Am. Chem. Soc., 1995, *117*, 12328.

[61] (a) IrH$_2$(SC$_5$H$_5$N)$_2$(PH$_3$)$_2$: Q Liu, and R Hoffmann, J. Am,, Chem Soc., 1995, *117*, 10108. (b) [Ni(SH)$_2$]$_6$: P. Alemany and R. Hoffmann. J. Am, Chem, Soc., 1993, *115*, 8290; Mn clusters: D. M. Proserpio, R. Hoffmann, G. C. Dismukes, J. Am, Chem Soc., 1992, *114*, 4374.

[62] J. H. Ammeter, H.-B. Bürgi, J C. Thibeault, R. Hoffmann, J. Am, Chem Soc., 1978, *100*, 3686.

[63] Superior results from EHM compared to MINDO/3 and MNDO, for nonplanarity of certain C/C double bonds: J. Spanget-Larsen and R Gleiter, Tetrahedron, 1983, *39*, 3345.

[64] EHM modified to give good geometries: S. L Dixon and P. C. Jurs, J. Comp. Chem., 1994, *15*, 733.

4.6   EASIER QUESTIONS

1. What do you understand by the term *quantum mechanics*?

2. Outline the experimental results that led to quantum mechanics.

3. What approximations are used in the SHM?

4. How could the SHM Fock matrix for 1,3-butadiene be modified in an attempt to recognize the fact that the molecule has, formally anyway, two double bonds and one single bond?

5. What are the most important kinds of results that can be obtained from Hückel calculations?

6. Write down the simple Hückel Fock matrices (in each case using α, β and 0, and 0, -1 and 0) for: (1) the pentadienyl radical (2) the cyclopentadienyl radical (3) trimethylenemethane, C(CH$_2$)$_3$ (4) trimethylenecyclopropane (5) 3-methylene-1,4-pentadiene.

7. The SHM predicts the propenyl cation, radical and anion to have the same resonance energy (stabilization energy). Actually, we expect the resonance energy to decrease as we add π electrons; why should this be the case?

8. What molecular feature cannot be obtained at all from the simple Hückel method? Why?

9. List the differences between the underlying theory of the SHM and the EHM.

10. A 400 × 400 matrix is easily diagonalized. How many carbons would an alkane have for its EHM Fock matrix to be 400 × 400 (or just under this size)? How many carbons would a (fully) conjugated polyene have if its SHM Fock matrix were 400 × 400?

4.7   HARDER QUESTIONS

1. Do you think it is reasonable to describe the Schrödinger equation as a postulate of quantum mechanics? What is a postulate?

2. What is the probability of finding a particle *at* a point?

3. Suppose we tried to simplify the SHM even further, by ignoring *all* interactions i, j; $i \neq j$ (ignoring adjacent interactions instead of setting them $= \beta$). What effect would this have on energy levels? Can you see the answer without looking at a matrix or determinant?

4. How might the i,j-type interactions in the simple Hückel Fock matrix be made to assume values other than just -1 and 0?

5. What is the result of using as a reference system for calculating the resonance energy of cyclobutadiene, not two ethene molecules, but 1,3-butadiene? What does this have to do with antiaromaticity? Is there any way to decide if one reference system is better than another?

6. What is the problem with unambiguously defining the charge on an atom in a molecule?

7. It has been reported that the extended Hückel method can be parameterized to give good geometries. Do you think this might be possible for the simple Hückel method? Why or why not?

8. Give the references to a journal paper that used the SHM, and one that used the EHM, within the last decade. Give an abstract of each paper.

9. The ionization energies usually used to parameterize the EHM are not ordinary atomic ionization energies, but rather *valence-state AO ionization energies* (VSAO ionization energies). What does the term "valence state" mean here? Should the VSAO ionization energies of the orbitals of an atom depend somewhat on the hybridization of the atom? In what way?

10. Which should require more empirical parameters: a molecular mechanics force field (chapter 3) or an EHM program? Explain.

Chapter 5

Ab initio calculations

"I could have done it in a much more complicated way", said the
Red Queen, immensely proud.

Lewis Carroll, ca. 1870.

5.1 PERSPECTIVE

Chapter 4 showed how quantum mechanics was first applied to molecules of real chemical interest (*pace* chemical physics) by Erich Hückel, and how the extension of the simple Hückel method (SHM) by Hoffmann gave a technique of considerable usefulness and generality, the extended Hückel method (EHM). The SHM and EHM are both based on the Schrödinger equation, and this makes them quantum mechanical methods. Both depend on reference to experimental quantities (i.e. on parameterization against experiment) to give actual values of calculated parameters: the SHM gives energy levels in terms of a parameter β which we could try to assign a value by comparison with experiment (actually the results of SHM calculations are usually left in terms of β), while the EHM needs experimental valence ionization potentials to calculate the Fock matrix elements. The need for parameterization against experiment makes the SHM and the EHM *semiempirical* ("semiexperimental") theories. In this chapter, we deal with a quantum mechanical approach that does not rely on calibration against measured chemical parameters and is therefore called *ab initio* [1] meaning "from the first", from first principles (it is true that *ab initio* calculations give results in terms of fundamental physical constants – Planck's constant, the speed of light, the charge of the electron – that must be measured to obtain their actual numerical values, but a chemical theory could hardly be expected to calculate the fundamental physical parameters of our universe).

5.2   THE BASIC PRINCIPLES OF THE *AB INITIO* METHOD

5.2.1   Preliminaries

In chapter 4, we saw that wavefunctions and energy levels could be obtained by diagonalizing a Fock matrix: the equation

$$\mathbf{H} = \mathbf{C}\boldsymbol{\varepsilon}\mathbf{C}^{-1} \tag{5.1}$$

is just another way of saying that diagonalization of $\mathbf{H}$ gives the coefficients or eigenvectors (the columns of $\mathbf{C}$ that, combined with the basis functions, yield the wavefunctions of the molecular orbitals) and the energy levels or eigenvalues (the diagonal elements of $\boldsymbol{\varepsilon}$). Equation (5.1) followed from

$$\mathbf{HC} = \mathbf{SC}\boldsymbol{\varepsilon} \tag{5.2}$$

which gives Eq. (5.1) when $\mathbf{S}$ is approximated as a unit matrix (simple Hückel method, section 4.3.4) or when the original Fock matrix is transformed into $\mathbf{H}$ (into $\mathbf{H}'$ in the notation of section 4.4.1) using an orthogonalizing matrix calculated from $\mathbf{S}$ (EHM section 4.4.1). To do a simple or an extended Hückel calculation the algorithm assembles the Fock matrix $\mathbf{H}$ and diagonalizes it. This is also how an *ab initio* calculation is done; the essential difference compared to the Hückel methods lies in the *evaluation of the matrix elements*.

In the simple Hückel method the Fock matrix elements H_{ij} are not calculated, but are instead set equal to 0 or -1 according to simple rules based on atomic connectivity (section 4.3.3); in the EHM the H_{ij} are calculated from the relative positions (through S_{ij}) of the orbitals or basis functions and the ionization potentials of these orbitals (section 4.4.1); in neither case is H_{ij} calculated from first principles. Section 4.3.3, Eqs (44) imply that H_{ij} is:

$$H_{ij} = \int \phi_i \hat{H} \phi_j \, dv \tag{5.3}$$

In *ab initio* calculations H_{ij} is calculated from Eq. (5.3) by actually performing the integration using explicit mathematical expressions for the basis functions ϕ_i and ϕ_j and the Hamiltonian operator $\hat{H}$; of course the integration is done by a computer following a detailed algorithm. How this algorithm works will now be outlined.

5.2.2   The Hartree SCF method

The simplest kind of *ab initio* calculation is a Hartree–Fock (HF) calculation. Modern molecular HF calculations grew out of calculations first performed on atoms by Hartree[1] in 1928 [2]. The problem that Hartree addressed arises from the fact that for any atom (or molecule) with more than *one* electron an exact analytic solution of the Schrödinger equation (section 4.3.2) is not possible, because of the electron–electron

---

[1] Douglas Hartree, born Cambridge, England, 1897. Ph.D. Cambridge, 1926. Professor applied mathematics, theoretical physics, Manchester, Cambridge. Died Cambridge, 1958.

repulsion term(s). Thus, for the helium atom the Schrödinger equation (cf. section 4.3.3, Eqs (4.36) and (4.37)) is, in SI units

$$\left[-\frac{h^2}{8\pi^2 m}(\nabla_1^2 + \nabla_2^2) - \frac{Ze^2}{4\pi\varepsilon_0 r_1} - \frac{Ze^2}{4\pi\varepsilon_0 r_2} + \frac{e^2}{4\pi\varepsilon_0 r_{12}}\right]\Psi = E\Psi \qquad (5.4)$$

Here m is the mass (kg) of the electron, e is the charge (coulombs, positive) of the proton (= minus the charge on the electron), the variables r_1, r_2, and r_{12} are the distances (meters) of electrons 1 and 2 from the nucleus, and from each other, $Z = 2$ is the number of protons in the nucleus, and ε_0 is something called the permitivity of empty space; the factor $4\pi\varepsilon_0$ is needed to make SI units consistent.

Hamiltonians can be written much more simply by using *atomic units*. Let's take Planck's constant, the electron mass, the proton charge, and the permitivity of space as the building blocks of a system of units in which $h/2\pi$, m, e, and $4\pi\varepsilon_0$ are numerically equal to 1 (i.e. $h = 2\pi$, $m = 1$, $e = 1$, and $\varepsilon_0 = 1/4\pi$; the numerical values of physical constants are always dependent on our system of units). These ($h/2\pi$, m, e, and $4\pi\varepsilon_0$) are the units of angular momentum, mass, charge, and permitivity in the system of atomic units. In this system, Eq. (5.4) becomes

$$\left(-\frac{1}{2}\nabla_1^2 - \frac{1}{2}\nabla_2^2 - \frac{Z}{r_1} - \frac{Z}{r_2} + \frac{1}{r_{12}}\right)\Psi = E\Psi \qquad (5.5)$$

Using atomic units simplifies writing quantum-mechanical expressions, and also means that the numerical (in these units) results of calculations are independent of the currently accepted values of physical constants in terms of kg, coulombs, meters, and seconds (of course, when we convert from atomic to SI units we must use accepted SI values of m, e, etc.). The atomic units of energy and length are particularly important to us. We can get the atomic unit of a quantity by combining $h/2\pi$, m, e, and $4\pi\varepsilon_0$ to give the expression with the required dimensions. The atomic units of length and energy, the bohr and the hartree, turn out to be:

Length 1 bohr $= a_0 = 4\pi\varepsilon_0(h/2\pi)^2/me^2 = \varepsilon_0 h^2/\pi me^2 = 0.05292\,\text{nm} = 0.5292\,\text{Å}$

Energy 1 hartree $= E_h$ (or h) $= e^2/4\pi\varepsilon_0 a_0$; 1 hartree/particle $= 2625.5\,\text{kJ mol}^{-1}$

The bohr is the radius of a hydrogen atom in the Bohr model (section 4.2.5), or the most probable distance of the electron from the nucleus in the fuzzier Schrödinger picture (section 4.2.5). The hartree is the energy needed to move a stationary electron one bohr distant from a proton away to infinity. The energy of a hydrogen atom, relative to infinite proton/electron separation as zero, is $-\frac{1}{2}$ hartree: the potential energy is -1 h and the kinetic energy (always positive) is $\frac{1}{2}$ h. Note that atomic units derived by starting with the old Gaussian system (cm, grams, statcoulombs) differ by a $4\pi\varepsilon_0$ factor from the SI-derived ones.

The Hamiltonian

$$\hat{H} = -\frac{1}{2}\nabla_1^2 - \frac{1}{2}\nabla_2^2 - \frac{Z}{r_1} - \frac{Z}{r_2} + \frac{1}{r_{12}} \qquad (5.6)$$

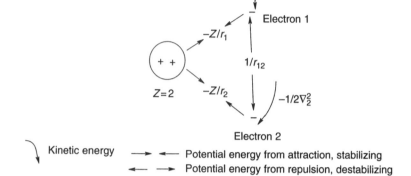

Figure 5.1. The terms in the helium atom Hamiltonian,

$$\hat{H} = -\frac{1}{2}\nabla_1^2 - \frac{1}{2}\nabla_2^2 - \frac{Z}{r_1} - \frac{Z}{r_2} + \frac{1}{r_{12}}$$

consists of five terms, signifying (Fig. 5.1) from left to right: the kinetic energy of electron 1, the kinetic energy of electron 2, the potential energy of the attraction of the nucleus (charge $Z = 2$) for electron 1, the potential energy of the attraction of the nucleus for electron 2, and the potential energy of the repulsion between electrons 1 and 2. Actually this is not the *exact* Hamiltonian, for it neglects effects due to relativity and to magnetic interactions such as spin–orbit coupling [3]; these effects are rarely important in calculations involving lighter atoms, say those in the first two full rows of the periodic table (up to about chlorine). Relativistic quantum chemical calculations will be briefly discussed later. The wavefunction ψ is the "total", overall wavefunction of the atom and can be approximated, as we will see later for molecular HF calculations, as a combination of wavefunctions for various energy levels. The problem with solving Eq. (5.5) exactly arises from the $1/r_{12}$ term. This makes it impossible to separate the Schrödinger equation for helium into two one-electron equations which, like the hydrogen atom equation, could be solved exactly [4]. This problem arises in any system with three or more interacting moving objects, and in fact the *many-body problem* is an old one even in classical mechanics, going back to eighteenth century studies in celestial mechanics. The impossibility of an analytic solution to polyelectronic systems prompted Hartree's approach to calculating wavefunctions and energy levels for atoms.

Hartree's method was to write a plausible approximate polyelectronic wavefunction (a "guess") for an atom as the product of one-electron wavefunctions:

$$\Psi_0 = \psi_0(1)\psi_0(2)\psi_0(3)\cdots\psi_0(n) \tag{5.7}$$

This function is called a Hartree product. Here ψ_0 is a function of the coordinates of all the electrons in the atom, $\psi_0(1)$ is a function of the coordinates of electron 1, $\psi_0(2)$ is a function of the coordinates of electron 2, etc.; the one-electron functions $\psi_0(1)$, $\psi_0(2)$, etc. are called atomic orbitals (molecular orbitals if we were dealing with a molecule). The initial guess, ψ_0, is our zeroth approximation to the true total wavefunction ψ, zeroth because we have not yet started to refine it with the Hartree process; it is based on the zeroth approximations $\psi_0(1)$, $\psi_0(2)$, etc. To apply the Hartree process we first solve for electron one a *one-electron* Schrödinger equation in which the electron–electron repulsion comes from electron one and an average, smeared-out electrostatic field calculated from $\psi_0(2)$, $\psi_0(3)$, ..., $\psi_0(n)$, due to all the *other* electrons. The only moving particle in this equation is electron one. Solving this equation gives $\psi_1(1)$, an improved version of $\psi_0(1)$. We next solve for electron 2 a one-electron Schrödinger equation with electron two moving in an average field due to the electrons of $\psi_1(1)$, $\psi_0(3)$, ..., $\psi_0(n)$, continuing to electron n moving in a field due to $\psi_1(1)$, $\psi_1(2)$, ..., $\psi_1(n-1)$. This completes the first cycle of calculations and gives

$$\Psi_1 = \psi_1(1)\psi_1(2)\psi_1(3)\cdots\psi_1(n) \tag{5.8}$$

Repetition of the cycle gives

$$\Psi_2 = \psi_2(1)\psi_2(2)\psi_2(3)\cdots\psi_2(n) \tag{5.9}$$

The process is continued for k cycles till we have a wavefunction ψ_k and/or an energy calculated from ψ_k that are essentially the same (according to some reasonable criterion) as the wavefunction and/or energy from the previous cycle. This happens when the functions $\psi(1)$, $\psi(2)$, ..., $\psi(n)$ are changing so little from one cycle to the next that the smeared-out electrostatic field used for the electron–electron potential has (essentially) ceased to change. At this stage the field of cycle k is essentially the same as that of cycle $k-1$, i.e. it is "consistent with" this previous field, and so the Hartree procedure is called the *self-consistent-field-procedure*, which is usually abbreviated as the *SCF* procedure.

There are two problems with the Hartree product of Eq. (5.7). Electrons have a property called spin, among the consequences of which is that not more than two electrons can occupy one atomic or molecular orbital (this is one statement of the Pauli (section 4.2.6) exclusion principle). In the Hartree approach we acknowledge this only in an *ad hoc* way, simply by not placing more than two electrons in any of the component orbitals ψ that make up our (approximate) total wavefunction ψ. Another problem comes from the fact that electrons are indistinguishable. If we have a wavefunction of the coordinates of two or more indistinguishable particles, then switching the positions of two of the particles, i.e. exchanging their coordinates, must either leave the function unchanged or change its sign. This is because all physical manifestations of the wavefunction must be unchanged on switching indistinguishable particles, and these manifestations depend only on its *square* (more strictly on the square of its absolute value, i.e. on $|\psi|^2$, to allow for the fact that ψ may be a complex, as distinct from a real, function). This should be clear from the equations below for a two-particle function:

If $\Psi_a = f(x_1, y_1, z_1; x_2, y_2, z_2)$ and $\Psi_b = f(x_2, y_2, z_2; x_1, y_1, z_1)$

then $|\Psi_a|^2 = |\Psi_b|^2$

if and only if $\Psi_b = \Psi_a$ or $\Psi_b = -\Psi_a$

If switching the coordinates of two of the particles leaves the function unchanged, it is said to be symmetric with respect to particle exchange, while if the function changes sign it is said to be antisymmetric with respect to particle exchange. Comparing the predictions of theory with the results of experiment shows [5] that electronic wavefunctions are actually antisymmetric with respect to exchange (such particles are called fermions, after the physicist Enrico Fermi; particles like photons whose wavefunctions are exchange-symmetric are called bosons, after the physicist S. Bose). Any rigorous attempt to approximate the wavefunction ψ should use an antisymmetric function of the coordinates of the electrons $1, 2, \ldots, n$, but the Hartree product is symmetric rather than antisymmetric; e.g. if we approximate a helium atom wavefunction as the product of two hydrogen atom $1s$ orbitals, then if $\psi_a = 1s(x_1, y_1, z_1)1s(x_2, y_2, z_2)$ and $\psi_b = 1s(x_2, y_2, z_2)1s(x_1, y_1, z_1)$, then $\psi_a = \psi_b$.

These defects of the Hartree SCF method were corrected by Fock (section 4.3.4) and by Slater[2] in 1930 [6], and Slater devised a simple way to construct a total wavefunction ψ from one-electron functions (i.e. orbitals) such that ψ will be antisymmetric to electron switching. Hartree's iterative, average-field approach supplemented with electron spin and antisymmetry leads to the HF equations.

5.2.3   The HF equations

5.2.3.1   Slater determinants

The Hartree wavefunction (above) is a product of one-electron functions called orbitals, or, more precisely, *spatial* orbitals: these are functions of the usual space coordinates x, y, z. The Slater wavefunction is composed, not just of spatial orbitals, but of *spin orbitals*. A spin orbital ψ(spin) is the product of a spatial orbital and a spin function, α or β: The spin orbitals corresponding to a given spatial orbital are

$$\psi(\text{spin } \alpha) = \psi(\text{spatial})\alpha = \psi(x, y, z)\alpha \quad \text{and}$$

$$\psi(\text{spin } \beta) = \psi(\text{spatial})\beta = \psi(x, y, z)\beta$$

As the function ψ(spatial) has as its variables the coordinates x, y, z, so the spin functions α and β have as *their* variables a spin coordinate, sometimes denoted ξ (Greek letter *kzi* or *zi*) or ω (Greek *omega*). We know that a wavefunction ψ fits in with an operator and eigenvalues, say the energy operator and energy eigenvalues, according to the equation $\hat{H}\psi = E\psi$. Analogously, the spin functions α and β are associated with the spin operator $\hat{S}_z$ according to $\hat{S}_z\alpha = \frac{1}{2}(h/2\pi)\alpha$ and $\hat{S}_z\beta = -\frac{1}{2}(h/2\pi)\beta$. Unlike most other functions, then, α and β each have only one eigenvalue, $\frac{1}{2}(h/2\pi)$ and $-\frac{1}{2}(h/2\pi)$, respectively. A spin function has the peculiar property that it is zero unless $\xi = \frac{1}{2}(\alpha$ spin function) or $\xi = -\frac{1}{2}(\beta$ spin function). A function that is zero everywhere except at one value of its variable, where it spikes sharply, is a *delta function*

---

[2]John Slater, born Oak Park Illinois, 1900. Ph.D. Harvard, 1923. Professor of physics, Harvard, 1924–1930; MIT 1930–1966; University of Florida at Gainesville, 1966–1976. Author of 14 textbooks, contributed to solid-state physics and quantum chemistry, developed X-alpha method (early density functional theory method). Died Sanibel Island, Florida, 1976.

(invented by Dirac – section 4.2.3). Since the spin function ψ(spin α or β) describing an electron exists only when the spin variable $\xi = \pm\frac{1}{2}$, these two values can be considered the allowed values of the spin quantum number m_s mentioned in section 4.2.6. Sometimes an electron with spin quantum number $\frac{1}{2}$ ("an electron with spin $\frac{1}{2}$") is called an α electron, and said to have *up* spin, and an electron with spin $-\frac{1}{2}$ is called a β electron, and said to have *down* spin. Up and down electrons are often denoted by arrows $\uparrow$ and $\downarrow$, respectively. A nice, brief treatment of the delta function and of the mathematical treatment of the spin functions is given by Levine [7].

The Slater wavefunction differs from the Hartree function not only in being composed of spin orbitals rather than just spatial orbitals, but also in the fact that it is not a simple product of one-electron functions, but rather a determinant (section 4.3.3) whose elements are these functions. To construct a Slater wavefunction (Slater determinant) for a closed-shell species (the only kind we consider in any detail here), we use each of the occupied spatial orbitals to make two spin orbitals, by multiplying the spatial orbital by α and, separately, by β. The spin orbitals are then filled with the available electrons. An example should make the procedure clear (Fig. 5.2). Suppose we wish to write a Slater determinant for a four-electron closed-shell system. We need two spatial molecular orbitals, since each can hold a maximum of two electrons; each spatial orbital ψ(spatial) is used to make two spin orbitals, ψ(spatial) α and ψ(spatial)β (alternatively, each spatial orbital could be thought of as a composite of two spin orbitals, which we are separating and using to build the determinant). Along the first (top) row of a determinant we write successively the first α spin orbital, the first β spin orbital, the second α spin orbital, and the second β spin orbital, using up our occupied spatial (and thus spin) orbitals. Electron one is then assigned to all four spin orbitals of the first row – in a sense it is allowed to roam among these four spin orbitals [8]. The second row of

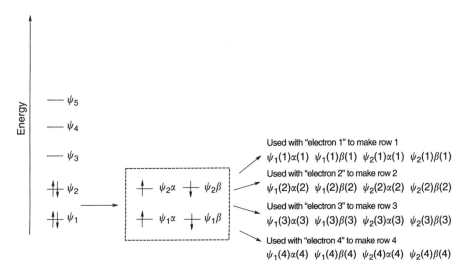

Figure 5.2. A Slater determinant is made from spin orbitals derived from the occupied spatial molecular orbitals and two spin functions, α and β.

the determinant is the same as the first, except that it refers to electron two rather than electron one; likewise the third and fourth rows refer to electrons three and four, respectively. The result is the determinant of Eq. (5.10).

$$\Psi = \frac{1}{\sqrt{4!}} \begin{vmatrix} \psi_1(1)\alpha(1) & \psi_1(1)\beta(1) & \psi_2(1)\alpha(1) & \psi_2(1)\beta(1) \\ \psi_1(2)\alpha(2) & \psi_1(2)\beta(2) & \psi_2(2)\alpha(2) & \psi_2(2)\beta(2) \\ \psi_1(3)\alpha(3) & \psi_1(3)\beta(3) & \psi_2(3)\alpha(3) & \psi_2(3)\beta(3) \\ \psi_1(4)\alpha(4) & \psi_1(4)\beta(4) & \psi_2(4)\alpha(4) & \psi_2(4)\beta(4) \end{vmatrix} \qquad (5.10)$$

(The $1/\sqrt{(4!)}$ factor ensures that the wavefunction is normalized, i.e. that $|\Psi|^2$ integrated over all space= 1 [9].) This *Slater determinant* ensures that there are no more than two electrons in each spatial orbital, since for each spatial orbital there are only two one-electron spin functions, and it ensures that Ψ is antisymmetric since switching two electrons amounts to exchanging two rows of the determinant, and this changes its sign (section 4.3.3). Note that instead of assigning the electrons successively to row 1, row 2, etc., we could have placed them in column 1, column 2, etc.: Ψ' of Eq. (5.11) = Ψ of Eq. (5.10). Some authors use the row format for the electrons, others the column format.

$$\Psi' = \frac{1}{\sqrt{4!}} \begin{vmatrix} \psi_1(1)\alpha(1) & \psi_1(2)\alpha(2) & \psi_1(3)\alpha(3) & \psi_1(4)\alpha(4) \\ \psi_1(1)\beta(1) & \psi_1(2)\beta(2) & \psi_1(3)\beta(3) & \psi_1(4)\beta(4) \\ \psi_2(1)\alpha(1) & \psi_2(2)\alpha(2) & \psi_2(3)\alpha(3) & \psi_2(4)\alpha(4) \\ \psi_2(1)\beta(1) & \psi_2(2)\beta(2) & \psi_2(3)\beta(3) & \psi_2(4)\beta(4) \end{vmatrix} \qquad (5.11)$$

Slater determinants enforce the Pauli exclusion principle, which forbids any two electrons in a system to have all quantum numbers the same. This is readily seen for an atom: if the three quantum numbers n, l and m_m of $\psi(x, y, z)$ (section 4.2.6) and the spin quantum number m_s of α or β were all the same for any electron, two rows (or columns, in the alternative formulation) would be identical and the determinant, hence the wavefunction, would vanish (section 4.3.3).

For $2n$ electrons (we are limiting ourselves for now to *even*-electron species, as the theory for these is simpler) the general form of a Slater determinant is clearly the $2n \times 2n$ determinant

$$\Psi_{2n} = \frac{1}{\sqrt{(2n)!}}$$

$$\times \begin{vmatrix} \psi_1(1)\alpha(1) & \psi_1(1)\beta(1) & \psi_2(1)\alpha(1) & \psi_2(1)\beta(1) & \cdots & \psi_n(1)\beta(1) \\ \psi_1(2)\alpha(2) & \psi_1(2)\beta(2) & \psi_2(2)\alpha(2) & \psi_2(2)\beta(2) & \cdots & \psi_n(2)\beta(2) \\ \vdots & \vdots & \vdots & \vdots & & \\ \psi_1(2n)\alpha(2n) & \psi_1(2n)\beta(2n) & \psi_2(2n)\alpha(2n) & \psi_2(2n)\beta(2n) & \cdots & \psi_n(2n)\beta(2n) \end{vmatrix}$$

$$(5.12)$$

The Slater determinant for the total wavefunction Ψ of a $2n$-electron atom or molecule is a $2n \times 2n$ determinant with $2n$ rows due to the $2n$ electrons and $2n$ columns due to the $2n$ spin orbitals (you can interchange the row/column format); since these are closed-shell species, the number of spatial orbitals ψ is half the number of electrons. We use the lowest n occupied spatial orbitals (the lowest $2n$ spin orbitals) to make the determinant.

The determinant (= total molecular wavefunction Ψ) just described will lead to (remainder of section 5.2) n occupied, and a number of unoccupied, component spatial molecular orbitals ψ. These orbitals ψ from the straightforward Slater determinant are called *canonical* (in mathematics the word means "in simplest or standard form") molecular orbitals. Since each occupied spatial ψ can be thought of as a region of space which accommodates a pair of electrons, we might expect that when the shapes of these orbitals are displayed ("visualized"; section 5.5.6) each one would look like a bond or a lone pair. However, this is often not the case; e.g. we do not find that one of the canonical MOs of water connects the O with one H, and another canonical MO connects the O with another H. Instead most of these MOs are spread over much of a molecule – delocalized (lone pairs, unlike conventional bonds, do tend to stand out). However, it is possible to combine the canonical MOs to get localized MOs which look like our conventional bonds and lone pairs. This is done by using the columns (or rows) of the Slater Ψ to create a Ψ with modified columns (or rows): if a column/row of a determinant is multiplied by k and added to another column/row, the determinant is unchanged (section 4.3.3). We see that if this is applied to the Slater determinant, we will get a "new" determinant corresponding to exactly the same total wavefunction, i.e. to the same molecule, but built up from different component occupied MOs ψ. The new Ψ and the new ψ's are no less or more correct than the previous ones, but by appropriate manipulation of the columns/rows the ψ's can be made to correspond to our ideas of bonds and lone pairs. These localized MOs are sometimes useful.

5.2.3.2   Calculating the atomic or molecular energy

The next step in deriving the HF equations is to express the energy of the molecule or atom in terms of the total wavefunction Ψ; the energy will then be minimized with respect to each of the component molecular (or atomic; an atom is a special case of a molecule) spin orbitals $\psi\alpha$ and $\psi\beta$ (cf. the minimization of energy with respect to basis function coefficients in section 4.3.3). The derivation of these equations involves considerable algebraic manipulation, which is at times hard to follow without actually writing out the intermediate expressions. The procedure has been summarized by Pople and Beveridge [10], and a less condensed account is given by Lowe [11].

It follows from the Schrödinger equation that the energy of a system is given by

$$E = \frac{\int \Psi^* \hat{H} \Psi \, d\tau}{\int \Psi^* \Psi \, d\tau} \tag{5.13}$$

This is similar to Eq. (4.40) in chapter 4, but here the total wavefunction Ψ has been specified, and allowance has been made for the possibility of Ψ being a complex function by utilizing its complex conjugate Ψ^*; this ensures that E, the energy of the atom or molecule, will be real. If Ψ is complex then $\Psi^2 \, d\tau$ will not be a real number, while $\Psi^* \Psi \, d\tau = |\Psi|^2 \, d\tau$ will, as must be the case for a probability. Integration is with respect to three spatial coordinates and one spin coordinate, for each electron. This is symbolized by $d\tau$ (τ = Greek *tau*), which means $dx\,dy\,dz\,d\xi$, so for a $2n$-electron system these integrals are actually $4 \times 2n$-fold, each electron having its set of four coordinates. We assume the use of orthonormal functions (section 4.3.4), since this makes several integrals disappear in the derivation of the energy. Working with the usual normalized wavefunctions makes the denominator unity, and Eq. (5.13) can then be written as

$$E = \int \Psi^* \hat{H} \Psi \, d\tau$$

or using the more compact Dirac notation for integrals (section 4.4.1)

$$E = \langle \Psi | \hat{H} | \Psi \rangle \tag{5.14}$$

In Eq. (5.14) it is understood that the first Ψ is actually Ψ^*, and that the integration variables are the space and spin coordinates. The vertical bars are only to visually separate the operator from the two functions, for clarity.

We next substitute into Eq. (5.14) the Slater determinant for Ψ (and Ψ^*) and the explicit expression for the Hamiltonian. A simple extension of the helium Hamiltonian of Eq. (5.5) to a molecule with $2n$ electrons and μ atomic nuclei (the μth nucleus has charge Z_μ) gives

$$\hat{H} = \sum_{i=1}^{2n} -\frac{1}{2}\nabla_i^2 - \sum_{\text{all } \mu, i} \frac{Z_\mu}{r_{\mu i}} + \sum_{\text{all } i, j} \frac{1}{r_{ij}} \tag{5.15}$$

Just like the helium Hamiltonian, the molecular Hamiltonian $\hat{H}$ in Eq. (5.15) is composed (from left to right) of electron kinetic energy terms, nucleus-electron attraction potential energy terms, and electron–electron repulsion potential energy terms (cf. Fig. 5.1). This is actually the *electronic* Hamiltonian, since nucleus-nucleus repulsion potential energy terms have been omitted; from the Born–Oppenheimer approximation (section 2.3) these can simply be added to the electronic energy after this has been calculated, giving the total molecular energy for a molecule with "frozen nuclei" (calculation of the vibrational energy, the zero-point energy (ZPE), is discussed later). Calculation of the internuclear potential energy is trivial:

$$V_{NN} = \sum_{\text{all } \mu, v} \frac{Z_\mu Z_v}{r_{\mu v}} \tag{5.16}$$

Substituting into Eq. (5.14) the Slater determinant and the molecular Hamiltonian gives, after much algebraic manipulation

$$E = 2\sum_{i=1}^{n} H_{ii} + \sum_{i=1}^{n}\sum_{j=1}^{n}(2J_{ij} - K_{ij}) \tag{5.17}$$

for the electronic energy of a $2n$-electron molecule (the sums are over the n occupied spatial orbitals ψ). The terms in Eq. (5.17) have these meanings:

$$H_{ii} \quad H_{ii} = \int \psi_i^*(1)\hat{H}^{\text{core}}(1)\psi_i^*(1)\,dv \tag{5.18}$$

where

$$\hat{H}^{\text{core}}(1) = -\frac{1}{2}\nabla_1^2 - \sum_{\text{all } \mu} \frac{Z_\mu}{r_{\mu 1}} \tag{5.19}$$

The operator $\hat{H}^{\text{core}}$ is so called because it leads to H_{ii}, the electronic energy of a single electron moving simply under the attraction of a nuclear "core", with all the other electrons stripped away; H_{ii} is the electronic energy of, for example, H, He$^+$, H$_2^+$, or CH$_4^{9+}$ (of course, it is different for these various species). Note that $\hat{H}^{\text{core}}(1)$ represents the kinetic energy of electron 1 plus the potential energy of attraction of that electron to each of the nuclei μ; the 1 in parentheses in these equations is just a label showing that the same electron is being considered in ψ_i^*, ψ_i and $\hat{H}^{\text{core}}$ (we could have used, say, 2 instead). The integration in Eq. (5.18) is respect to spatial coordinates only, ($dv = dx\,dy\,dz$, not $d\tau$) because spin coordinates have been "integrated out": on integration, i.e. summation over the discrete spin variable, these give 0 or 1 [7,10,12]. We are left with the three integration variables (x, y, z) and so the integral is threefold.

$$J_{ij} \quad J_{ij} = \int \psi_i^*(1)\psi_i(1) \left(\frac{1}{r_{12}}\right) \psi_j^*(2)\psi_j(2)\,dv_1\,dv_2 \qquad (5.20)$$

J is called a coulomb integral; it represents the electrostatic (i.e. coulombic) repulsion between an electron in ψ_i and one in ψ_j, i.e. between the charge clouds of orbitals ψ_i and ψ_j. This may be clearer if one considers the integral as a sum of potential energy terms involving repulsion between infinitesimal volume elements dv (Fig. 5.3). The 1 and 2 are just labels showing we are considering two electrons. The integrals J and K allow each electron to experience the *average electrostatic repulsion of a charge cloud due to all the other electrons*. Since J represents potential energy corresponding to a destabilizing electrostatic repulsion, it is positive. As for H_{ii} in Eq. (5.18), the integration is with respect to spatial coordinates because the spin coordinates have been integrated out. There are six integration variables, x, y, z for electron 1 (dv_1) and x, y, z for electron 2 (dv_2), and so the integral is sixfold. Note that the *ab initio* coulomb

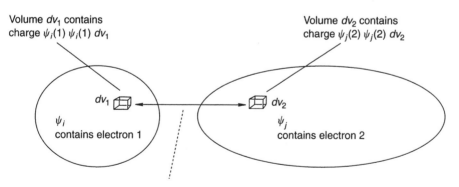

Potential energy between dv_1 and dv_2 is $\psi_i(1)\,\psi_i(1)\,dv_1\,\dfrac{1}{r_{12}}\,\psi_j(2)\,\psi_j(2)\,dv_2$
(product of the charges divided by their distance apart)

Figure 5.3. The coulomb integral (J integral) represents the electrostatic repulsion between two charge clouds, due to electron 1 in orbital ψ_i and electron 2 in orbital ψ_j.

$$J_{ij} = \int \psi_i^*(1)\psi_i(1) \left(\frac{1}{r_{12}}\right) \psi_j^*(2)\psi_j(2)\,dv_1\,dv_2$$

integral J is not the same as what we called a coulomb integral in simple Hückel theory; that was $\alpha = \int \phi_i \hat{H} \phi_i \, dv$ (Eq. (4.61)) and represents at least crudely the energy of an electron in the p orbital ϕ_i (section 4.3.4). The *ab initio* coulomb integral can also be written

$$J_{ij} = \int \psi_i^*(1)\psi_j^*(2) \left(\frac{1}{r_{12}}\right) \psi_i(1)\psi_j(2) \, dv_1 \, dv_2 \qquad (5.21)$$

but unlike (5.20) this does not notationally emphasize the repulsion (invoked by the $1/r_{12}$ operator) between electrons 1 and 2 (on the left and right, respectively, of $1/r_{12}$ in Eq. (5.20)).

$$K_{ij} \quad K_{ij} = \int \psi_i^*(1)\psi_j^*(2) \left(\frac{1}{r_{12}}\right) \psi_i(2)\psi_j(1) \, dv_1 \, dv_2 \qquad (5.22)$$

K is called an exchange integral; mathematically, it arises from Slater determinant expansion terms that differ only in exchange of electrons. Note that the terms on either side of $1/r_{12}$ differ by exchange of electrons. It is often said to have no simple physical interpretation, but looking at Eq. (5.17), we can regard K as a kind of correction to J, reducing the effect of J (both J and K are positive, with K smaller), i.e. reducing the electrostatic potential energy due to the mutual ψ_i, ψ_j charge cloud repulsion referred to in connection with J. This reduction in repulsion arises because as particles with an antisymmetric wavefunction, two electrons cannot occupy the same spin orbital (roughly, cannot be at the same point in space), and can occupy the same spatial orbital only if they have opposite spins, so two electrons *of the same spin* avoid each other more assiduously than expected only from the coulombic repulsion taken into account by J. We could consider the summed $2J - K$ terms of Eq. (5.17) to be the true coulombic repulsion, corrected for electron spin, i.e. corrected for the Pauli exclusion principle effect. The J and K interactions are shown in Fig. 5.4 for a four-electron molecule,

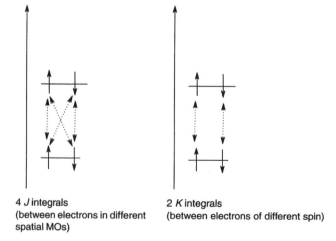

4 J integrals
(between electrons in different spatial MOs)

2 K integrals
(between electrons of different spin)

Figure 5.4. The J integrals represent interactions between electrons in different spatial orbitals; the K integrals represent interactions between electrons of the same spin.

the smallest closed-shell system in which K integrals arise. A detailed exposition of the significance of the HF integrals is given by Dewar [13]. Note that outside the nucleus the only significant forces in atoms and molecules are electrostatic; there are no vague "quantum-mechanical forces" in chemistry [14]. Chemical reactions involve the shuffling of atomic nuclei under the influence of the electromagnetic force.

5.2.3.3 The variation theorem (variation principle)

The energy calculated from Eq. (5.14) is the *expectation value* of the energy operator $\hat{H}$, i.e. the expectation value of the Hamiltonian operator. In quantum mechanics an integral of a wavefunction "over" an operator, as E in Eq. (5.14) is an integral of Ψ over $\hat{H}$, is the expectation value of that operator. The expectation value is the value (strictly, the quantum-mechanical average value) of the physical quantity represented by the operator. Every "observable", i.e. every measurable property of a system, has a quantum mechanical operator from which the property could be calculated, at least in principle, by integrating the wavefunction over the operator. The expectation value of the energy operator $\hat{H}$ is the energy of the molecule (or atom). Of course this energy will be the exact, true energy of the molecule only if the wavefunction Ψ and the Hamiltonian $\hat{H}$ are exact. The variation theorem states that *the energy calculated from Eq. (5.14) must be greater than or equal to the true ground-state energy of the molecule.* The theorem [15] (it can be stated more rigorously, specifying that $\hat{H}$ must be time-independent and Ψ must be normalized and well-behaved) is very important in quantum chemistry: it assures us that any ground state (we examine electronic ground states much more frequently than we do excited states) energy we calculate "variationally" (i.e. using Eq. (5.14)) must be greater than or equal to the true energy of the molecule. In practice, any molecular wavefunction we insert into Eq. (5.14) is always only an approximation to the true wavefunction and so the variationally calculated molecular energy will always be greater than the true energy. The HF energy is variational (the method starts with Eq. (5.14)) so the variation theorem gives us at least some indication of the true energy and of how good our wavefunction is: the correct energy always lies below any calculated by the HF method, and the better the wavefunction, the lower the calculated energy. The HF energy actually levels off at a value *above* the true energy as the HF wavefunction, based on a Slater determinant, is improved; this is discussed in section 5.5, in connection with post-HF methods.

5.2.3.4 Minimizing the energy; the HF equations

The HF equations are obtained from Eq. (5.17) by minimizing the energy with respect to the atomic or molecular orbitals ψ. The minimization is carried out with the constraint that the orbitals remain orthonormal, for orthonormality was imposed in deriving Eq. (5.17). Minimizing a function subject to a constraint can be done using the method of undetermined Lagrangian multipliers [16]. For orthonormality the overlap integrals S must be constants ($=\delta_{ij}$, i.e. 0 or 1) and at the minimum the energy is constant ($=E_{\min}$). Thus at $E_{\min}$ any linear combination of E and S_{ij} is constant:

$$E + \sum_{i=1}^{n} \sum_{j=1}^{n} l_{ij} S_{ij} = constant \qquad (5.23)$$

where l_{ij} are the Lagrangian multipliers; we do not know what they are, physically, yet (they are "undetermined"). Differentiating with respect to the ψ's of the S's:

$$dE + d\sum_{i=1}^{n}\sum_{j=1}^{n} l_{ij} S_{ij} = 0 \qquad (5.24)$$

Substituting the expression for E from Eqs (5.17) into (5.24) we get

$$2\sum_{i=1}^{n} dH_{ii} + \sum_{i=1}^{n}\sum_{j=1}^{n}(2dJ_{ij} - dK_{ij}) + \sum_{i=1}^{n}\sum_{j=1}^{n} l_{ij} dS_{ij} = 0 \qquad (5.25)$$

Note that this procedure of minimizing the energy with respect to the *molecular orbitals* ψ is somewhat analogous to the minimization of energy with respect to the *atomic orbital coefficients c* in the less rigorous procedure which gave the Hückel secular equations in section 4.3.3. It is also somewhat similar to finding a relative minimum on a PES (section 2.4), but with energy in that case being varied with respect to geometry. Since the procedure starts with Eq. (5.14) and varies the MO's to find the minimum value of E, it is called the variation method; the variation theorem (section 5.2.3.3) assures us that the energy we calculate from the results will be greater than or equal to the true energy.

From the definitions of H_{ii}, J_{ij}, K_{ij} and S_{ij} we get

$$dH_{ii} = \int d\psi_i^*(1)\hat{H}^{\text{core}}(1)\psi_i(1)\,dv_1 + \int \psi_i^*(1)\hat{H}^{\text{core}}(1)\,d\psi_i(1)\,dv_1 \qquad (5.26)$$

$$dJ_{ij} = \int d\psi_i^*(1)\hat{J}_j(1)\psi_i(1)\,dv_1 + \int d\psi_j^*(1)\hat{J}_i(1)\psi_j(1)\,dv_1 + complex\ conjugate \qquad (5.27)$$

$$dK_{ij} = \int d\psi_i^*(1)\hat{K}_j(1)\psi_i(1)\,dv_1 + \int d\psi_j^*(1)\hat{K}_i(1)\psi_j(1)\,dv_1 + complex\ conjugate \qquad (5.28)$$

where

$$\hat{J}_i(1) = \int \psi_i^*(2)\left(\frac{1}{r_{12}}\right)\psi_i(2)\,dv_2 \qquad (5.29)$$

and

$$\hat{K}_i(1)\psi_j(1) = \psi_i(1)\int \psi_i^*(2)\left(\frac{1}{r_{12}}\right)\psi_j(2)\,dv_2 \qquad (5.30)$$

and similarly for $\hat{J}_j$ and $\hat{K}_j$.

$$dS_{ij} = \int d\psi_i^*(1)\psi_j(1)\,dv_1 + \psi_i^*(1)\,d\psi_j(1)v_1 \qquad (5.31)$$

Using for dH, dJ, dK and dS the expressions in Eqs (5.26), (5.27), (5.28) and (5.31), Eq. (5.25) becomes

$$2\sum_{i=1}^{n}\int d\psi_i^*(1)\left[\hat{H}^{\text{core}}(1)\psi_i(1) + \sum_{j=1}^{n}(2\hat{J}_j(1) - \hat{K}_j(1))\psi_i(1) + \frac{1}{2}\sum_{j=1}^{n/2}l_{ij}\psi_j(1)\right]dv$$

$$+ \text{ complex conjugate } = 0 \qquad (5.32)$$

Since the MOs can be varied independently, and the expression on the left side is zero, both parts of Eq. (5.32) (the part shown and the complex conjugate) equal zero. It can be shown that a consequence of

$$2\sum_{i=1}^{n}\int d\psi_i^*(1)\left[\hat{H}^{\text{core}}(1)\psi_i(1) + \sum_{j=1}^{n}(2\hat{J}_j(1) - \hat{K}_j(1))\psi_i(1)\right.$$

$$\left. + \frac{1}{2}\sum_{j=1}^{n}l_{ij}\psi_j(1)\right]dv = 0 \qquad (5.33)$$

is that

$$\hat{H}^{\text{core}}(1)\psi_i(1) + \sum_{j=1}^{n}(2\hat{J}_j(1) - \hat{K}_j(1))\psi_i(1) + \frac{1}{2}\sum_{j=1}^{n}l_{ij}\psi_j(1)\,dv = 0$$

i.e.

$$\left[\hat{H}^{\text{core}}(1) + \sum_{j=1}^{n}(2\hat{J}_j(1) - \hat{K}_j(1))\right]\psi_i(1) = -\frac{1}{2}\sum_{j=1}^{n}l_{ij}\psi_j(1) \qquad (5.34)$$

Equation (5.34) can be written as:

$$\hat{F}\psi_i(1) = -\frac{1}{2}\sum_{j=1}^{n}l_{ij}\psi_j(1) \qquad (5.35)$$

where $\hat{F}$ is the *Fock operator*:

$$\hat{F} = \hat{H}^{\text{core}}(1) + \sum_{j=1}^{n}(2\hat{J}_j(1) - \hat{K}_j(1)) \qquad (5.36)$$

We want an eigenvalue equation because (cf. section 4.3.3) we hope to be able to use the matrix form of a series of such equations to invoke matrix diagonalization to get eigenvalues and eigenvectors. Equation (5.35) is not quite an eigenvalue equation, because it is not of the form Operation on function = $k \times$ function, but rather Operation on function = sum of ($k \times$ functions). However, by transforming the molecular orbitals

ψ to a new set the equation can be put in eigenvalue form (with a caveat, as we shall see). Equation (5.35) represents a system of equations

$$\hat{F}\psi_1(1) = -\frac{1}{2}[l_{11}\psi_1(1) + l_{12}\psi_2(1) + l_{13}\psi_3(1) + \cdots + l_{1n}\psi_n(1)] \quad i = 2$$

$$\hat{F}\psi_2(1) = -\frac{1}{2}[l_{21}\psi_1(1) + l_{22}\psi_2(1) + l_{23}\psi_3(1) + \cdots + l_{2n}\psi_n(1)] \quad i = 2$$

$$\vdots \tag{5.37}$$

$$\hat{F}\psi_n(1) = -\frac{1}{2}[l_{n1}\psi_1(1) + l_{n2}\psi_2(1) + l_{n3}\psi_3(1) + \cdots + l_{nn}\psi_n(1)] \quad i = n$$

There are n spatial orbitals ψ since we are considering a system of $2n$ electrons and each orbital holds two electrons. The 1 in parentheses on each orbital emphasizes that each of these n equations is a *one*-electron equation, dealing with the same electron (we could have used a 2 or a 3, etc.), i.e. the Fock operator (Eq. 5.36) is a one-electron operator, unlike the general electronic Hamiltonian operator of Eq. (5.15), which is a multi-electron operator (a $2n$ electron operator for our specific case). The Fock operator acts on a total of n spatial orbitals, the $\psi_1, \psi_2, \ldots, \psi_n$ in Eq. (5.35).

The series of equations (5.37) can be written as the single matrix equation (cf. Eq. (4.50))

$$\hat{F} \begin{pmatrix} \psi_1(1) \\ \psi_2(1) \\ \psi_3(1) \\ \vdots \\ \psi_n(1) \end{pmatrix} = -\frac{1}{2} \begin{pmatrix} l_{11} & l_{12} & l_{13} & \cdots & l_{1n} \\ l_{21} & l_{22} & l_{23} & \cdots & l_{2n} \\ \vdots & \vdots & & \cdots & \vdots \\ l_{n1} & l_{n2} & l_{n3} & \cdots & l_{nn} \end{pmatrix} \begin{pmatrix} \psi_1(1) \\ \psi_2(1) \\ \psi_3(1) \\ \vdots \\ \psi_n(1) \end{pmatrix} \tag{5.38}$$

i.e.

$$\hat{F}\psi = -\frac{1}{2}\mathbf{L}\psi \tag{5.39}$$

In Eqs (5.37), each equation will be of the form $\hat{F}\psi_i = k\psi_i$, which is what we want, if all the $l_{ij} = 0$ except for $i = j$ (e.g. in the first equation $\hat{F}\psi_1(1) = -(1/2)l_{11}\psi_1(1)$ if the only nonzero l is l_{11}). This will be the case if in Eq. (5.39) $\mathbf{L}$ is a diagonal matrix. It can be shown that $\mathbf{L}$ is diagonalizable (section 4.3.3), i.e. there exist matrices $\mathbf{P}$, $\mathbf{P}^{-1}$ and a diagonal matrix $\mathbf{L}'$ such that

$$\mathbf{L} = \mathbf{P}\mathbf{L}'\mathbf{P}^{-1} \tag{5.40}$$

Substituting $\mathbf{L}$ from Eqs (5.40) into (5.39):

$$\hat{F}\psi = -\frac{1}{2}\mathbf{P}\mathbf{L}'\mathbf{P}^{-1}\psi \tag{5.41}$$

Multiplying on the left by $\mathbf{P}^{-1}$ and on the right by $\mathbf{P}$ we get

$$\hat{F}\mathbf{P}^{-1}\psi\mathbf{P} = -\frac{1}{2}(\mathbf{P}^{-1}\mathbf{P})\mathbf{L}'(\mathbf{P}^{-1}\psi\mathbf{P})$$

which, since $\mathbf{P}^{-1}\mathbf{P} = 1$ can be written

$$\hat{F}\psi' = -\frac{1}{2}\mathbf{L}'\psi' \tag{5.42}$$

where

$$\psi' = \mathbf{P}^{-1}\psi\mathbf{P} \tag{5.43}$$

We may as well remove the $-\frac{1}{2}$ factor by incorporating it into $\mathbf{L}'$, and we can omit the prime from ψ (we could have *started* the derivation using primes on the ψ's then written $\psi = \mathbf{P}^{-1}\psi'\mathbf{P}$ for Eq. (5.43)). Equation (5.42) then becomes (anticipating the soon-to-be-apparent fact that the diagonal matrix is an energy-level matrix)

$$\hat{F}\psi = \boldsymbol{\varepsilon}\psi \tag{5.44}$$

where

$$\boldsymbol{\varepsilon} = \begin{pmatrix} (-1/2)l_{11} & 0 & 0 & \cdots & 0 \\ 0 & (-1/2)l_{22} & 0 & \cdots & 0 \\ \vdots & \vdots & & \cdots & \vdots \\ 0 & 0 & 0 & \cdots & (-1/2)l_{nn} \end{pmatrix} \tag{5.45}$$

Equation (5.44) is the compact form of (cf. Eq. (5.38)). Thus

$$\hat{F}\begin{pmatrix} \psi_1(1) \\ \psi_2(1) \\ \psi_3(1) \\ \vdots \\ \psi_n(1) \end{pmatrix} = \begin{pmatrix} \varepsilon_1 & 0 & 0 & \cdots & 0 \\ 0 & \varepsilon_2 & 0 & \cdots & 0 \\ \vdots & \vdots & \vdots & \cdots & \vdots \\ 0 & 0 & 0 & \cdots & \varepsilon_n \end{pmatrix}\begin{pmatrix} \psi_1(1) \\ \psi_2(1) \\ \psi_3(1) \\ \vdots \\ \psi_n(1) \end{pmatrix} \tag{5.46}$$

where the superfluous double subscripts on the ε's have been replaced by single ones. Equations (5.44/5.46) are the matrix form of the system of equations

$$\hat{F}\psi_1(1) = \varepsilon_1\psi_1(1)$$

$$\hat{F}\psi_2(1) = \varepsilon_2\psi_2(1)$$

$$\hat{F}\psi_3(1) = \varepsilon_3\psi_3(1) \tag{5.47}$$

$$\vdots$$

$$\hat{F}\psi_n(1) = \varepsilon_n\psi_n(1)$$

These Eqs (5.47) are the HF equations (the matrix form is Eqs (5.44) or (5.46)). By analogy with the Schrödinger equation $\hat{H}\psi = E\psi$, we see that they show that the Fock operator acting on a one-electron wavefunction (an atomic or molecular orbital) generates an energy value times the wavefunction. Thus the Lagrangian multipliers l_{ii} turned out to be (with the $-\frac{1}{2}$ factor) the energy values associated with the orbitals ψ_i. Unlike the Schrödinger equation the HF equations are not quite eigenvalue equations (although they are closer to this ideal than is Eq. (5.35)), because in $\hat{F}\psi_i = k\psi_i$ the Fock operator $\hat{F}$ is itself dependent on ψ_i; in a true eigenvalue equation the operator can be written down without reference to the function on which it acts. The significance of the HF equations is discussed in the next section.

5.2.3.5   The meaning of the HF equations

The HF equations (5.47) (in matrix form Eqs (5.44) and (5.46)) are pseudoeigenvalue equations asserting that the Fock operator $\hat{F}$ acts on a wavefunction ψ_i to generate an energy value ε_i, times ψ_i. *Pseudo*eigenvalue because, as stated above, in a true eigenvalue equation the operator is not dependent on the function on which it acts; in the HF equations $\hat{F}$ depends on ψ because (Eq. (5.36)) the operator contains $\hat{J}$ and $\hat{K}$, which in turn depend (Eqs (5.29) and (5.30)) on ψ. Each of the equations in the set (5.47) is for a single electron ("electron one" is indicated, but any ordinal number could be used), so the HF operator $\hat{F}$ is a one-electron operator, and each spatial molecular orbital ψ is a one-electron function (of the coordinates of the electron). Two electrons can be placed in a spatial orbital because the *full* description of each of these electrons requires a *spin function* α or β (section 5.2.3.1) and each electron "moves in" a different spin orbital. The result is that the two electrons in the spatial orbital ψ do not have all four quantum numbers the same (for an atomic $1s$ orbital, e.g. one electron has quantum numbers $n = 1, l = 0, m = 0$ and $s = \frac{1}{2}$, while the other has $n = 1, l = 0, m = 0$ and $s = -\frac{1}{2}$), and so the Pauli exclusion principle is not violated.

The *functions* ψ are the spatial molecular (or atomic) orbitals or wavefunctions that (along with the spin functions) make up the overall or total molecular (or atomic) wavefunction ψ, which can be written as a Slater determinant (Eq. (5.12)). Concerning the *energies* ε_i, from the fact that

$$\varepsilon_i = \int \psi_i \hat{F} \psi_i \, dv \tag{5.48}$$

(this follows simply from multiplying both sides of a HF equation by ψ_i and integrating, noting that ψ_i is normalized) and the definition of $\hat{F}$ (Eq. (5.36)) we get

$$\varepsilon_i = \int \psi_i(1) \hat{H}^{\text{core}}(1) \psi_i(1) \, dv + \sum_{j=1}^{n} (2J_{ij}(1) - K_{ij}(1)) \tag{5.49}$$

i.e.

$$\varepsilon_i = H_{ii}^{\text{core}} + \sum_{j=1}^{n} (2J_{ij}(1) - K_{ij}(1)) \tag{5.50}$$

(the *operators* $\hat{J}$ and $\hat{K}$ in Eq. (5.36) have been transformed by integration into the *integrals* J and K in Eq. 5.49)). Equation (5.50) shows that ε_i is the energy of an electron in ψ_i subject to interaction with all the other electrons in the molecule: H_{ii}^{core} Eq. (5.19) is the energy of the electron due only to its motion (kinetic energy) and to the attraction of the nuclear core (electron-nucleus potential energy), while the sum of $2J - K$ terms represents the exchange-corrected (via K) coulombic repulsion (through J) energy resulting from the interaction of the electron with all the other electrons in the molecule or atom [17].

In principle the Eqs (5.47) allow us to calculate the molecular orbitals (MOs) ψ and the energy levels ε. We could start with "guesses" (actually obtained by intuition or analogy) of the MOs (the zeroth approximation to these) and use these to construct the operator $\hat{F}$ (Eq. (5.36)), then allow $\hat{F}$ to operate on the guesses to yield energy levels (the first approximation to the ε_i) and new, improved [18] functions (the first calculated approximations to the ψ_i). Using the improved functions in $\hat{F}$ and operating on these gives the second approximations to the ψ_i and ε_i, and the process is continued until ψ_i and ε_i no longer change (within preset limits), which occurs when the smeared-out electrostatic field represented in Eq. (5.17) by $\sum \sum (2J - K)$ (cf. Fig. 5.3) ceases to change appreciably – is consistent from one iteration cycle to the next, i.e. is self-consistent. This is, of course, in exactly the same spirit as the procedure described in section 5.2.2 using the Hartree product as our total or overall wavefunction Ψ. The main difference between the two methods is that the HF method represents Ψ as a Slater determinant of component spin MOs rather than as a simple product of spatial MOs, and a consequence of this is that the calculation of the average coulombic field in the Hartree method involves only the coulomb integral J, but in the HF modification we need the coulomb integral J *and* the exchange integral K, which arises from Slater determinant terms that differ in exchange of electrons. Because K acts as a kind of "Pauli correction" to the classical electrostatic repulsion, reminding the electrons that two of them of the same spin cannot occupy the same spatial orbital, electron–electron repulsion is less in the HF method than if a simple Hartree product were used. Of course K does not arise in calculations involving no electrons of like spin, as in H_2 or (section 4.4.1; also section 5.4.3.6e) HHe^+, which have only two, paired-spin, electrons. At the end of the iterative procedure we have the MO's ψ_i and their corresponding energy levels ε_i, and the total wavefunction Ψ, the Slater determinant of the ψ_i's. The ε_i can be used to calculate the total electronic energy of the molecule, and the MO's ψ_i are useful heuristic approximations to the electron distribution, while the total wavefunction Ψ can in principle be used to calculate anything about the molecule. Applications of the energy levels and the MO's are given in section 5.4.

5.2.3.6 Basis functions and the Roothaan–Hall equations

5.2.3.6a Deriving the Roothaan–Hall equations
As they stand, the HF equations (5.44), (5.46) or (5.47) are not very useful for molecular calculations, mainly because (1) they do not prescribe a mathematically viable procedure getting the initial guesses for the MO wavefunctions ψ_i, which we need to initiate the iterative process (section 5.2.3.5), and (2) the wavefunctions may be so complicated that they contribute nothing to a qualitative understanding of the electron distribution. For calculations on *atoms*, which obviously have much simpler orbitals than molecules, we could use for the ψ's atomic orbital wavefunctions based on the solution of the Schrödinger equation for the hydrogen atom (taking into account the increase of atomic number and the screening effect of inner electrons on outer ones [19]). This yields the atomic wavefunctions as tables of ψ at various distances from the nucleus. This is not a suitable approach for molecules because among molecules there is no prototype species occupying a place analogous to that of the hydrogen atom in the hierarchy of atoms, and as indicated above it does not readily lend itself to an interpretation of how molecular properties arise from the nature of the constituent atoms.

In 1951 Roothaan and Hall independently pointed out [20] that these problems can be solved by representing MOs as linear combinations of basis functions (just as in the SHM, in chapter 4, the π MOs are constructed from atomic p orbitals). For a basis-function expansion of MOs we write

$$\psi_1 = c_{11}\phi_1 + c_{21}\phi_2 + c_{31}\phi_3 + \cdots + c_{m1}\phi_m$$
$$\psi_2 = c_{12}\phi_1 + c_{22}\phi_2 + c_{32}\phi_3 + \cdots + c_{m2}\phi_m$$
$$\psi_3 = c_{13}\phi_1 + c_{23}\phi_2 + c_{33}\phi_3 + \cdots + c_{m3}\phi_m \qquad (5.51)$$
$$\vdots$$
$$\psi_m = c_{1m}\phi_1 + c_{2m}\phi_2 + c_{3m}\phi_3 + \cdots + c_{mm}\phi_m$$

In devising a more compact notation for this set of equations it is very helpful, particularly when we come to the matrix treatment in section 5.2.3.6c, to use different subscripts to denote the MOs ψ and the basis functions ϕ. Conventionally, Roman letters have been used for the ψ's and Greek letters for the ϕ's, or $i, j, k, l, \ldots$ for the ψ's and $r, s, t, u, \ldots$ for the ϕ's. The latter convention will be adopted here, and we can write the Eqs (5.51) as

$$\underset{\substack{| \\ i\text{th MO}}}{\psi_i} = \overset{\substack{m \text{ basis functions} \\ | \\ m}}{\underset{s=1}{\sum}} \underset{\substack{\\ c \text{ of the } s\text{th basis function of } i\text{th MO}}}{c_{si}} \overset{\substack{s\text{th basis function} \\ \diagup}}{\phi_s} \qquad i = 1, 2, 3, \ldots, m\,(m \text{ MOs}) \qquad (5.52)$$

We are expanding each MO ψ in terms of m basis functions. The basis functions are usually (but not necessarily) located on atoms, i.e. for the function $\phi(x, y, z)$, where x, y, z are the coordinates of the electron being treated by this one-electron function, the distance of the electron from the nucleus is:

$$r = [(x - x_0)^2 + (y - y_0)^2 + (z - z_0)^2]^{1/2} \qquad (5.53)$$

where x_0, y_0, z_0 are the coordinates of the atomic nucleus in the coordinate system used to define the geometry of the molecule. Because each basis function may usually be regarded (at least vaguely) as some kind of atomic orbital, this linear combination of basis functions approach is commonly called a linear combination of atomic orbitals (LCAO) representation of the MOs, as in the SHM and EHM (sections 4.3.3 and 4.4.1). The set of basis functions used for a particular calculation is called the *basis set*.

We need at least enough spatial MOs ψ to accommodate all the electrons in the molecule, i.e. we need at least n ψ's for the $2n$ electrons (recall that we are dealing with closed-shell molecules). This is ensured because even the smallest basis sets used in *ab initio* calculations have for each atom at least one basis function corresponding to each orbital conventionally used to describe the chemistry of the atom, and the number of basis functions ϕ is equal to the number of (spatial) MOs ψ (section 4.3.4). An example will make this clear: for an *ab initio* calculation on CH_4, the smallest basis set would specify for C: $\phi(C, 1s)$, $\phi(C, 2s)$, $\phi(C, 2p_x)$, $\phi(C, 2p_y)$, $\phi(C, 2p_z)$ and for each H: $\phi(H, 1s)$. These nine basis functions ϕ (5 on C and $4 \times 1 = 4$ on H) create nine spatial MO's ψ, which could hold 18 electrons; for the 10 electrons of CH_4 we need only 5 spatial MO's. There is no *upper* limit to the size of a basis set: there are commonly many more basis functions, and hence MOs, than are needed to hold all the electrons, so that there are usually many unoccupied MO's. In other words, the number of basis functions m in the expansions (5.52) can be much bigger than the number n of pairs of electrons in the molecule, although only the n occupied spatial orbitals are used to construct the Slater determinant which represents the HF wavefunction (section 5.2.3.1). This point, and basis sets, are discussed further in section 5.3.

To continue with the Roothaan–Hall approach, we substitute the expansion (5.52) for the ψ's into the HF equations (5.47), getting (we will work with m, not n, HF equations since there is one such equation for each MO, and our m basis functions will generate m MOs):

$$\sum_{s=1}^{m} c_{s1} \hat{F} \phi_{sj} = \varepsilon_1 \sum_{sj=1}^{m} c_{s1} \phi_s$$

$$\sum_{s=1}^{m} c_{s2} \hat{F} \phi_s = \varepsilon_2 \sum_{s=1}^{m} c_{s2} \phi_s \qquad (5.54)$$

$$\vdots$$

$$\sum_{s=1}^{m} c_{sm} \hat{F} \phi_s = \varepsilon_m \sum_{s=1}^{m} c_{sm} \hat{F} \phi_s$$

($\hat{F}$ operates on the functions ϕ, not on the c's, which have no variables x, y, z). Multiplying each of these m equations by $\phi_1, \phi_2, \ldots, \phi_m$ (or ϕ_1^* etc. if the ϕ's are complex functions, as is occasionally the case) and integrating, we get m sets of equations (one for each of the basis functions ϕ). Basis function ϕ_1 gives

$$\sum_{s=1}^{m} c_{s1} F_{1s} = \varepsilon_1 \sum_{s=1}^{m} c_{s1} S_{1s}$$

$$\sum_{sj=1}^{m} c_{s2} F_{1s} = \varepsilon_2 \sum_{s=1}^{m} c_{s2} S_{1s} \qquad (5.54\text{-}1)$$

$$\vdots$$

$$\sum_{s=1}^{m} c_{sm} F_{1s} = \varepsilon_m \sum_{s=1}^{m} c_{sm} S_{1s}$$

where

$$F_{rs} = \int \phi_r \hat{F} \phi_s \, dv \quad \text{and} \quad S_{rs} = \int \phi_r \phi_s \, dv \qquad (5.55)$$

Basis function ϕ_2 gives

$$\sum_{s=1}^{m} c_{s1} F_{2s} = \varepsilon_1 \sum_{s=1}^{m} c_{s1} S_{2s}$$

$$\sum_{s=1}^{m} c_{s2} F_{2s} = \varepsilon_2 \sum_{s=1}^{m} c_{s2} S_{2s} \qquad (5.54\text{-}2)$$

$$\vdots$$

$$\sum_{s=1}^{m} c_{sm} F_{2s} = \varepsilon_m \sum_{s=1}^{m} c_{sm} S_{2s}$$

Finally, basis function ϕ_m gives

$$\sum_{s=1}^{m} c_{s1} F_{ms} = \varepsilon_1 \sum_{s=1}^{m} c_{s1} S_{ms}$$

$$\sum_{s=1}^{m} c_{s2} F_{ms} = \varepsilon_2 \sum_{s=1}^{m} c_{s2} S_{ms} \qquad (5.54\text{-}m)$$

$$\vdots$$

$$\sum_{s=1}^{m} c_{sm} F_{ms} = \varepsilon_m \sum_{s=1}^{m} c_{sm} S_{ms}$$

In the m sets of equations (5.54-1) to (5.54-m) each set itself contains m equations (the subscript of ε, for example, runs from 1 to m), for a total of $m \times m$ equations. These equations are the Roothaan-Hall version of the HF equations; they were obtained by substituting for the MOs ψ in the HF equations a linear combination of basis functions (ϕ's weighted by c's). The Roothaan–Hall equations are usually written more compactly, as

$$\sum_{s=1}^{m} F_{rs} c_{si} = \sum_{s=1}^{m} S_{rs} c_{si} \varepsilon_i \quad r = 1, 2, 3, \ldots, m \text{ (for each } i = 1, 2, 3, \ldots, m) \quad (5.56)$$

We have $m \times m$ equations because each of the m spatial MO's ψ we used (recall that there is one HF equation for each ψ, Eqs (5.47)) is expanded with m basis functions. The Roothaan–Hall equations connect the basis functions ϕ (contained in the integrals F and S, Eqs (5.55)), the coefficients c, and the MO energy levels ε. Given a basis set $\{\phi_s, s = 1, 2, 3, \ldots, m\}$ they can be used to calculate the c's, and thus the MOs ψ (Eq. (5.52)) and the MO energy levels ε. The overall electron distribution in the molecule can be calculated from the total wavefunction Ψ, which can be written as a Slater determinant of the "component" spatial wavefunctions ψ (by including spin functions), and in principle anyway, any property of a molecule can be calculated

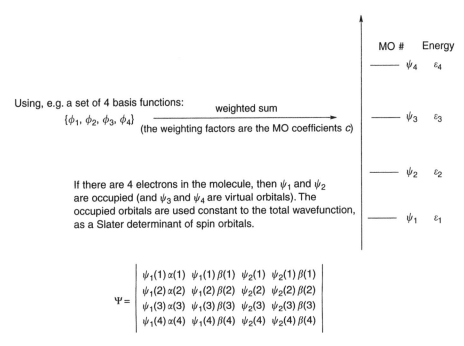

Using, e.g. a set of 4 basis functions:
$\{\phi_1, \phi_2, \phi_3, \phi_4\}$ $\xrightarrow{\text{weighted sum}}$
(the weighting factors are the MO coefficients c)

If there are 4 electrons in the molecule, then ψ_1 and ψ_2 are occupied (and ψ_3 and ψ_4 are virtual orbitals). The occupied orbitals are used constant to the total wavefunction, as a Slater determinant of spin orbitals.

$$\Psi = \begin{vmatrix} \psi_1(1)\,\alpha(1) & \psi_1(1)\,\beta(1) & \psi_2(1) & \psi_2(1)\,\beta(1) \\ \psi_1(2)\,\alpha(2) & \psi_1(2)\,\beta(2) & \psi_2(2) & \psi_2(2)\,\beta(2) \\ \psi_1(3)\,\alpha(3) & \psi_1(3)\,\beta(3) & \psi_2(3) & \psi_2(3)\,\beta(3) \\ \psi_1(4)\,\alpha(4) & \psi_1(4)\,\beta(4) & \psi_2(4) & \psi_2(4)\,\beta(4) \end{vmatrix}$$

MO #   Energy

—— ψ_4 ε_4

—— ψ_3 ε_3

—— ψ_2 ε_2

—— ψ_1 ε_1

Figure 5.5. Pictorial representation of basis functions, MO's, total wavefunction, and energy levels.

from Ψ. The component wavefunctions ψ and their energy levels ε are extremely useful, as chemists rely heavily on concepts like the shape and energies of, for example, the HOMO and LUMO of a molecule (MO concepts are reviewed in chapter 4). The energy levels enable (with a correction term) the total energy of a molecule to be calculated, and so the energies of molecules can be compared and reaction energies and activation energies can be calculated. The Roothaan–Hall equations, then, are a cornerstone of modern *ab initio* calculations, and the procedure for solving them is outlined next. These ideas are summarized pictorially in Fig. 5.5.

The fact that the Roothaan–Hall equations (5.56) are actually a total of $m \times m$ equations suggests that they might be expressible as a single matrix equation, since the single matrix equation $\mathbf{AB} = \mathbf{0}$, where $\mathbf{A}$ and $\mathbf{B}$ are $m \times m$ matrices, represents $m \times m$ "simple" equations, one for each element of the product matrix $\mathbf{AB}$ (work it out for two 2×2 matrices). A single matrix equation would be easier to work with than m^2 equations and might allow us to invoke matrix diagonalization as in the case of the simple and extended Hückel methods (sections 4.3.3 and 4.4.1). To subsume the sets of Eqs (5.54-1) to (5.54-m), i.e. Eqs (5.56), into one matrix equation, we might (eschewing a rigorous deductive approach) suspect that the matrix form is the rather obvious possibility

$$\mathbf{FC} = \mathbf{SC}\varepsilon \tag{5.57}$$

Here $\mathbf{F}$, $\mathbf{C}$ and $\mathbf{S}$ would have to be $m \times m$ matrices, since there are m^2 F's, c's and S's, and ε would be an $m \times m$ diagonal matrix with the nonzero elements $\varepsilon_1, \varepsilon_2, \ldots, \varepsilon_m$, since ε must contain only m elements, but has to be $m \times m$ to make the right-hand side matrix product the same size as that on the left.

This is easily checked: the left-hand side of Eq. (5.57) is

$$
\mathbf{FC} = \begin{pmatrix} F_{11} & F_{12} & F_{13} & \cdots & F_{1m} \\ F_{21} & F_{22} & F_{23} & \cdots & F_{2m} \\ \vdots & \vdots & \ddots & & \vdots \\ F_{m1} & F_{m2} & F_{m3} & \cdots & F_{mm} \end{pmatrix} \begin{pmatrix} c_{11} & c_{12} & c_{13} & \cdots & c_{1m} \\ c_{21} & c_{22} & c_{23} & \cdots & c_{2m} \\ \vdots & \vdots & \ddots & & \vdots \\ c_{m1} & c_{m2} & c_{m3} & \cdots & c_{mm} \end{pmatrix}
$$

$$
= \begin{pmatrix} F_{11}c_{11} + F_{12}c_{21} + F_{13}c_{31} & \cdots & F_{11}c_{12} + F_{12}c_{22} + F_{13}c_{32} & \cdots & \cdots \\ F_{21}c_{11} + F_{22}c_{21} + F_{23}c_{31} & \cdots & F_{21}c_{12} + F_{22}c_{22} + F_{23}c_{33} & \cdots & \cdots \\ & & \vdots & & \end{pmatrix}
$$

$$(5.58)$$

The right-hand side of Eq. (5.57) is

$$
\mathbf{SC\varepsilon} = \begin{pmatrix} S_{11} & S_{12} & \cdots & S_{1m} \\ S_{21} & S_{22} & \cdots & S_{2m} \\ \vdots & \vdots & \ddots & \vdots \\ S_{m1} & S_{m2} & \cdots & S_{mm} \end{pmatrix} \begin{pmatrix} c_{11} & c_{12} & \cdots & c_{1m} \\ c_{21} & c_{22} & \cdots & c_{2m} \\ \vdots & \vdots & \ddots & \vdots \\ c_{m1} & c_{m2} & \cdots & c_{mm} \end{pmatrix} \begin{pmatrix} \varepsilon_{11} & 0 & \cdots & 0 \\ 0 & \varepsilon_{22} & \cdots & 0 \\ \vdots & \vdots & \ddots & \vdots \\ 0 & 0 & \cdots & \varepsilon_{mm} \end{pmatrix}
$$

$$
= \begin{pmatrix} S_{11}c_{11} + S_{12}c_{21} + S_{13}c_{31} & \cdots & S_{11}c_{12} + S_{12}c_{22} + S_{13}c_{32} & \cdots & \cdots \\ S_{21}c_{11} + S_{22}c_{21} + S_{23}c_{31} & \cdots & S_{21}c_{12} + S_{22}c_{22} + S_{23}c_{33} & \cdots & \cdots \\ & & \vdots & & \end{pmatrix} \varepsilon
$$

$$
= \begin{pmatrix} \varepsilon_1(S_{11}c_{11} + S_{12}c_{21} + S_{13}c_{31} \cdots) & \varepsilon_2(S_{11}c_{12} + S_{12}c_{22} + S_{13}c_{32} \cdots) & \cdots \\ \varepsilon_1(S_{21}c_{11} + S_{22}c_{21} + S_{23}c_{31} \cdots) & \varepsilon_2(S_{21}c_{12} + S_{22}c_{22} + S_{23}c_{33} \cdots) & \cdots \\ & \vdots & \end{pmatrix}
$$

$$(5.59)$$

Now compare **FC** (5.58) and **SCε** (5.59). Comparing element a_{11} of **FC** (multiplied out to give a single matrix as shown in (5.58)) with element a_{11} of **SCε** (multiplied out to give a single matrix as shown in (5.59)) we see that *if* **FC** = **SCε**, i.e. if (5.57) is true, then

$$
F_{11}c_{11} + F_{12}c_{21} + F_{13}c_{31} + \cdots = \varepsilon(S_{11}c_{11} + S_{12}c_{21} + S_{13}c_{31} + \cdots)
$$

i.e.

$$
\sum_{s=1}^{m} c_{si} F_{rs} = \varepsilon \sum_{s=1}^{m} c_{si} S_{rs} \tag{5.60}
$$

But this is the first equation of the set (5.53-1). Continuing in this way we see that matching each element of (the multiplied-out) matrix **FC** (5.58) with the corresponding element of (the multiplied-out) matrix **SCε** (5.59) gives one of the equations of the set (5.54-1) to (5.54-m), i.e. of the set (5.56). This can be so only if **FC** = **SCε**, so this matrix equation is indeed equivalent to the set of Eqs (5.56).

Now we have $\mathbf{FC} = \mathbf{SC}\varepsilon$ (5.57), the matrix form of the Roothaan–Hall equations. These equations are sometimes called the Hartree–Fock–Roothaan equations, and, often, the Roothaan equations, as Roothaan's exposition was the more detailed and addresses itself more clearly to a general treatment of molecules. Before showing how they are used to do *ab initio* calculations, a brief review of how we got these equations is in order.

Summary of the derivation of the Roothaan–Hall equations:

1. The total wavefunction Ψ of an atom or molecule was expressed as a Slater determinant of spin MOs ψ(spatial) α and ψ(spatial) β, Eq. (5.12).

2. From the Schrödinger equation we got an expression for the electronic energy of the atom or molecule, $E = \langle \Psi | \hat{H} | \Psi \rangle$, Eq. (5.14).

3. Substituting the Slater determinant for ψ and the explicit form of the Hamiltonian operator $\hat{H}$ into (5.14) gave the energy in terms of the spatial MO's ψ (Eq. (5.17):

$$E = 2 \sum_{i=1}^{n} H_{ii} + \sum_{i=1}^{n} \sum_{j=1}^{n} (2J_{ij} - K_{ij})$$

4. Minimizing E in (5.17) with respect to the ψ's (to find the best ψ's) gave the HF equations $\hat{F}\psi = \varepsilon\psi$ (5.44).

5. Substituting into the HF equations $\hat{F}\psi = \varepsilon\psi$ (5.44) the Roothaan–Hall linear combination of basis functions (LCAO) expansions $\psi_i = \sum c_{si}\phi_s$ (5.52) for the MO's ψ gave the Roothaan–Hall equations (Eqs (5.56)), which can be written compactly as $\mathbf{FC} = \mathbf{SC}\varepsilon$ (Eqs (5.57)).

5.2.3.6b   Using the Roothaan–Hall equations to do ab initio calculations – the SCF procedure

The Roothaan–Hall equations $\mathbf{FC} = \mathbf{SC}\varepsilon$ (Eqs (5.57)) ($\mathbf{F}$, $\mathbf{C}$, $\mathbf{S}$ and ε are defined in connection with Eqs (5.58) and (5.59); the matrix elements F and S are defined by Eqs (5.54) and (5.55)) are of the same matrix form as Eq. (4.54), $\mathbf{HC} = \mathbf{SC}\varepsilon$, in the simple Hückel method (section 4.3.3) and the extended Hückel (section 4.4.1) method. Here, however, we have seen (in outline) how the equation may be rigorously derived. Also, unlike the case in the Hückel methods the Fock matrix elements are rigorously defined theoretically: from Eqs (5.55)

$$F_{rs} = \int \phi_r \hat{F} \phi_s \, dv \qquad (5.61 = 4.54)$$

and Eq. (5.36)

$$\hat{F} = \hat{H}^{\text{core}}(1) + \sum_{j=1}^{n} (2\hat{J}_j(1) - \hat{K}_j(1)) \qquad (5.62 = 5.36)$$

it follows that

$$F_{rs} = \int \phi_r \left[\hat{H}^{\text{core}}(1) + \sum_{j=1}^{n} (2\hat{J}_j(1) - \hat{K}_j(1)) \right] \phi_s \, dv \qquad (5.63)$$

where

$$\hat{H}^{\text{core}}(1) = -\frac{1}{2}\nabla_1^2 - \sum_{\text{all } \mu} \frac{Z_\mu}{r_{\mu 1}} \qquad (5.65 = 5.19)$$

$$\hat{J}_j(1) = \int \psi_j^*(2) \left(\frac{1}{r_{12}}\right) \psi_j(2)\, dv_2 \qquad (5.65 = 5.29)$$

and

$$\hat{K}_i(1)\psi_j(1) = \psi_i(1) \int \psi_i^*(2) \left(\frac{1}{r_{12}}\right) \psi_j(2)\, dv_2 \qquad (5.66 = 5.30)$$

To use The Roothaan–Hall equations we want them in standard eigenvalue-like form so that we can diagonalize the Fock matrix $\mathbf{F}$ of Eq. (5.57) to get the coefficients c and the energy levels ε, just as we did in connection with the extended Hückel method (section 4.4.1). The procedure for diagonalizing $\mathbf{F}$ and extracting the c's and ε's and is exactly the same as that explained for the extended Hückel method (although here the cycle is iterative, i.e. repetitive, see below):

1. The overlap matrix $\mathbf{S}$ is calculated and used to calculate an orthogonalizing matrix $\mathbf{S}^{-1/2}$, as in Eqs (4.107) and (4.108):

$$\mathbf{S} --> \mathbf{D} --> \mathbf{S}^{-1/2} \qquad (5.67)$$

2. $\mathbf{S}^{-1/2}$ is used to convert $\mathbf{F}$ to $\mathbf{F}'$ (cf. (4.104)):

$$\mathbf{F}' = \mathbf{S}^{-1/2}\mathbf{F}\mathbf{S}^{-1/2} \qquad (5.68)$$

The transformed Fock matrix $\mathbf{F}'$ satisfies

$$\mathbf{F}' = \mathbf{C}'\boldsymbol{\varepsilon}\mathbf{C}'^{-1} \qquad (5.69)$$

(cf. Eq. (4.106)). The overlap matrix $\mathbf{S}$ is readily calculated, so if $\mathbf{F}$ can be calculated it can be transformed to $\mathbf{F}'$, which can be diagonalized to give $\mathbf{C}'$ and $\boldsymbol{\varepsilon}$, which latter yields the MO energy levels ε_i.

3. Transformation of $\mathbf{C}'$ to $\mathbf{C}$ (Eq. (4.102)) gives the coefficients c_{si} in the expansion of the MO's ψ in terms of basis functions ϕ:

$$\mathbf{C} = \mathbf{S}^{-1/2}\mathbf{C}' \qquad (5.70)$$

Equations (5.63)–(5.66) show that to calculate $\mathbf{F}$, i.e. each of the matrix elements F, we need the wavefunctions ψ_i, because $\hat{J}$ and $\hat{K}$, the coulomb and exchange operators (Eqs (5.65) and (5.66)), are defined in terms of the ψ's. It looks like we are faced with a dilemma: the point of calculating $\mathbf{F}$ is to get (besides the ε's) the ψ's (the c's with the chosen basis set $\{\phi\}$ make up the ψ's), but to get $\mathbf{F}$ we need the ψ's. The way out of this is to start with a set of approximate c's, e.g. from an extended Hückel calculation, which needs no c's to begin with because the extended Hückel "Fock" matrix elements are calculated from experimental ionization potentials (section 4.4.1). These c's, the *initial guess*, are used with the basis functions ϕ to in effect (section 5.2.3.6c) calculate initial MO wavefunctions ψ, which are used to calculate the $\mathbf{F}$ elements F_{rs}. Transformation of F to F' and diagonalization gives a "first-cycle" set of ε's and (after transformation of C' to C) a first-cycle set of c's. These c's are used to calculate new F_{rs}, i.e. a new F, and this gives a second-cycle set of ε's and c's. The process is continued until things – the ε's, the c's (as the density matrix – section 5.2.3.6c), the energy, or, more usually, some combination of these – stop changing within certain pre-defined limits, i.e. until the cycles have essentially converged on the limiting ε's and c's. Typically, about ten cycles are needed to achieve convergence. It is because the operator $\hat{F}$ depends on the functions ϕ on which it acts, making an iterative approach necessary, that the Roothaan–Hall equations, like the HF equations, are called *pseudo*eigenvalue (see end of section 5.2.3.4 and start of 5.2.3.5).

Now, in the HF method (the Roothaan–Hall equations represent one implementation of the HF method) each electron moves in an average *field* due to all the other electrons (see the discussion in connection with Fig. 5.3, section 5.2.3.2). As the c's are refined the MO wavefunctions improve and so this average field that each electron feels improves (since J and K, although not explicitly calculated (section 5.2.3.6c) improve with the ψ's). When the c's no longer change the field represented by this last set of c's is (practically) the same as that of the previous cycle, i.e. the two fields are "consistent" with one another, i.e. "self-consistent". This Roothaan–Hall–Hartree–Fock iterative process (initial guess, first F, first-cycle c's, second F, second-cycle c's, third F, etc.) is therefore a *self-consistent-field-procedure* or *SCF procedure*, like the HF procedure of section 5.2.2. The terms "HF calculations/method" and "SCF calculations/method" are in practice synonymous. *The key point to the iterative nature of the SCF procedure is that to get the c's (for the MOs ψ) and the MO ε's we diagonalize a Fock matrix F, but to calculate F we need an initial guess for the c's and we then improve the c's by repeatedly recalculating and diagonalizing F.* The procedure is summarized in Fig. 5.6. Note that in the simple and extended Hückel methods we do not need the c's to calculate F, and there is no iterative refinement of the c's, so these are not SCF methods; other semiempirical procedures, however (chapter 6) do use the SCF approach.

5.2.3.6c   Using the Roothaan–Hall equations to do ab initio *calculations –*
the equations in terms of the c's and ϕ's of the LCAO expansion
The key process in the HF *ab initio* calculation of energies and wavefunctions is calculation of the Fock matrix, i.e. of the matrix elements F_{rs} (section 5.2.3.6b). Equation (5.63) expresses these in terms of the basis functions ϕ and the operators $\hat{H}^{\text{core}}$, $\hat{J}$ and $\hat{K}$, but the $\hat{J}$ and $\hat{K}$ operators (Eqs (5.28) and (5.31)) are themselves functions of the MO's ψ and therefore of the c's and the basis functions ϕ. Obviously the F_{rs} can be written explicitly in terms of the c's and ϕ's; such a formulation enables the Fock matrix to be efficiently calculated from the coefficients and the basis functions without *explicitly* evaluating the operators $\hat{J}$ and $\hat{K}$ after each iteration. This more explicit (in terms of the Roothaan–Hall LCAO approach) formulation of the Fock matrix will now be explained.

To see more clearly what is required, write Eq. (5.63) as

$$F_{rs} = \langle \phi_r(1)|\hat{H}^{\text{core}}(1)|\phi_s(1)\rangle + \sum_{j=1}^{n}[2\langle \phi_r(1)|\hat{J}_j(1)\phi_s(1)\rangle - \langle \phi_r(1)|\hat{K}_j(1)|\phi_s(1)\rangle]$$

$$(5.71)$$

using the compact Dirac notation. The operator $\hat{H}^{core}(1)$ involves only the Laplacian differentiation operator, atomic numbers and electron coordinates, so we do not have to consider substituting the Roothaan–Hall c's and ϕ's into $\hat{H}^{core}$. The operators $\hat{J}$ and $\hat{K}$, however, invoke the integrals $\langle \phi_r(1)|\hat{J}(1)\phi_s(1)\rangle$ and $\langle \phi_r(1)|\hat{K}(1)\phi_s(1)\rangle$. We now examine these two integrals.

$\hat{J} = \langle \phi_r(1)|\hat{J}(1)\phi_s(1)\rangle$: from Eq. (5.65),

$$\hat{J}_j(1)\phi_s(1) = \phi_s(1) \int \frac{\psi_j^*(2)\psi_j(2)}{r_{12}} \, dv_2$$

Substituting for $\psi_j^*(2)$ the basis function expansion $\sum c_{tj}^* \phi_t^*(2)$ and for $\psi_j(2)$ the expansion $\sum c_{uj}\phi_u(2)$ (cf. Eq. (5.52)):

$$\hat{J}_j(1)\phi_s(1) = \phi_s(1) \sum_{t=1}^{m} \sum_{u=1}^{m} c_{tj}^* c_{uj} \int \frac{\phi_t^*(2)\phi_u(2)}{r_{12}} \, dv_2$$

where the double sum arises because we multiply the ψ^* sum by the ψ sum. To get the desired expression for $\langle \phi_r(1)|\hat{J}(1)\phi_s(1)\rangle$ we multiply this by $\phi_r^*(1)$ and integrate with respect to the coordinates of electron 1, getting:

$$\langle \phi_r(1)|\hat{J}_j(1)\phi_s(1)\rangle = \sum_{t=1}^{m} \sum_{u=1}^{m} c_{tj}^* c_{uj} \iint \frac{\phi_r^*(1)\phi_s(1)\phi_t^*(2)\phi_u(2)}{r_{12}} \, dv_1 \, dv_2$$

Note that this is really a sixfold integral, since there are three variables (x_1, y_1, z_1) for electron 1, and three (x_2, y_2, z_2) for electron 2, represented by dv_1 and dv_2 respectively. This equation can be written more compactly as

$$\langle \phi_r(1)|\hat{J}_j(1)\phi_s(1)\rangle = \sum_{t=1}^{m} \sum_{u=1}^{m} c_{tj}^* c_{uj}(rs|tu) \tag{5.72}$$

The notation

$$(rs|tu) = \iint \frac{\phi_r^*(1)\phi_s(1)\phi_t^*(2)\phi_u(2)}{r_{12}} \, dv_1 \, dv_2 \tag{5.73}$$

is a common shorthand for this kind of integral, which is called a *two-electron repulsion integral* (or two-electron integral, or electron repulsion integral); the physical significance of these is outlined in section 5.2.3.6d). This parentheses notation should not be confused with the Dirac bra-ket notation, $\langle|$ (a bra) and $|\rangle$ a ket: by definition

$$\langle f|g\rangle = \int f^*(q)g(q) \, dq \tag{5.74}$$

so

$$\langle rs|tu\rangle = \int (\phi_r(1)\phi_s(1))^* \phi_t(1)\phi_u(1) \, dv_1 \tag{5.75}$$

Actually, several notations have been used for the integrals of Eq. (5.73) and for other integrals; make sure to ascertain which symbolism a particular author is using.

$\hat{K} = \langle \phi_r(1) | \hat{K}(1) \phi_s(1) \rangle$: from Eq. (5.66):

$$\hat{K}_j(1)\phi_s(1) = \psi_j(1) \int \frac{\psi_j^*(2)\phi_s(2)}{r_{12}} \, dv_2$$

Substituting for $\psi_j(1)$ the basis function expansion $\sum c_{uj}\phi_u(1)$ and for $\psi_j^*(2)$ the expansion $\sum c_{tj}^*\phi_t^*(2)$ (cf. Eq. (5.52)):

$$\hat{K}_j(1)\phi_s(1) = \phi_u(1) \sum_{t=1}^{m} \sum_{u=1}^{m} c_{tj}^* c_{uj} \int \frac{\phi_t^*(2)\phi_s(2)}{r_{12}} \, dv_2$$

To get the desired expression for $\langle \phi_r(1) | \hat{K}(1) \phi_s(1) \rangle$ we multiply this by $\phi_r^*(1)$ and integrate with respect to the coordinates of electron 1:

$$\langle \phi_r(1) | \hat{K}_j(1)\phi_s(1) \rangle = \sum_{t=1}^{m} \sum_{u=1}^{m} c_{tj}^* c_{uj} \iint \frac{\phi_r^*(1)\phi_u(1)\phi_t^*(2)\phi_s(2)}{r_{12}} \, dv_1 \, dv_2$$

which can be written more compactly as

$$\langle \phi_r(1) | \hat{K}_j(1)\phi_s(1) \rangle = \sum_{t=1}^{m} \sum_{u=1}^{m} c_{tj}^* c_{uj} (ru|ts) \tag{5.76}$$

where of course (cf. (5.73))

$$(ru|ts) = \iint \frac{\phi_r^*(1)\phi_u(1)\phi_t^*(2)\phi_s(2)}{r_{12}} \, dv_1 \, dv_2 \tag{5.77}$$

Substituting Eqs (5.72) and (5.76) for $\langle \phi_r(1) | \hat{J}(1)\phi_s(1) \rangle$ and $\langle \phi_r(1) | \hat{K}(1)\phi_s(1) \rangle$ into Eq. (5.71) for F_{rs}, we get

$$F_{rs} = \langle \phi_r(1) | \hat{H}^{\text{core}}(1) | \phi_s(1) \rangle + \sum_{j=1}^{n} \left[2 \sum_{t=1}^{m} \sum_{u=1}^{m} c_{tj}^* c_{uj} (rs|tu) - \sum_{t=1}^{m} \sum_{u=1}^{m} c_{tj}^* c_{uj} (ru|ts) \right]$$

i.e.

$$F_{rs} = H_{rs}^{\text{core}}(1) + \sum_{t=1}^{m} \sum_{u=1}^{m} \sum_{j=1}^{n} c_{tj}^* c_{uj} [2(rs|tu) - (ru|ts)] \tag{5.78}$$

where the integral of the operator $\hat{H}^{\text{core}}$ over the basis functions has been written as:

$$H_{rs}^{\text{core}}(1) = \langle \phi_r(1) | \hat{H}^{\text{core}}(1) | \phi_s(1) \rangle \tag{5.79}$$

with $\hat{H}^{\text{core}}$ defined by Eq. (5.64).

Equation (5.78), with its ancillary definitions Eqs (5.73), (5.77), and (5.79), is what we wanted: the Fock matrix elements in terms of the basis functions ϕ and their weighting coefficients c, for a closed-shell molecule; m is the number of basis functions and $2n$ is the number of electrons. We can use Eq. (5.78) to calculate MO's and energy levels

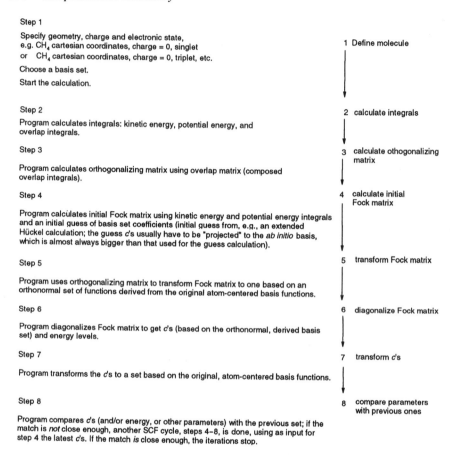

Step 1

Specify geometry, charge and electronic state,
e.g. CH₄ cartesian coordinates, charge = 0, singlet
or CH₄ cartesian coordinates, charge = 0, triplet, etc.

Choose a basis set.

Start the calculation.

1 Define molecule

Step 2

Program calculates integrals: kinetic energy, potential energy, and
overlap integrals.

2 calculate integrals

Step 3

Program calculates orthogonalizing matrix using overlap matrix (composed
overlap integrals).

3 calculate othogonalizing
   matrix

Step 4

Program calculates initial Fock matrix using kinetic energy and potential energy integrals
and an initial guess of basis set coefficients (initial guess from, e.g., an extended
Hückel calculation; the guess *c*'s usually have to be "projected" to the *ab initio* basis,
which is almost always bigger than that used for the guess calculation).

4 calculate initial
   Fock matrix

Step 5

Program uses orthogonalizing matrix to transform Fock matrix to one based on an
orthonormal set of functions derived from the original atom-centered basis functions.

5 transform Fock matrix

Step 6

Program diagonalizes Fock matrix to get *c*'s (based on the orthonormal, derived basis
set) and energy levels.

6 diagonalize Fock matrix

Step 7

Program transforms the *c*'s to a set based on the original, atom-centered basis functions.

7 transform *c*'s

Step 8

Program compares *c*'s (and/or energy, or other parameters) with the previous set; if the
match is *not* close enough, another SCF cycle, steps 4–8, is done, using as input for
step 4 the latest *c*'s. If the match *is* close enough, the iterations stop.

8 compare parameters
   with previous ones

Figure 5.6. Summary of the steps in the Hartree–Fock–Roothaan–Hall SCF procedure.

(section 5.2.3.6b). Given a basis set and molecular geometry (the integrals depend on molecular geometry, as will be illustrated) and starting with an initial guess at the c's, one (or rather the computer algorithm) calculates the matrix elements F_{rs}, assembles them into the Fock matrix $\mathbf{F}$, etc. (section 5.2.3.6b and Fig. 5.6) Let us now examine certain details connected with Eq. (5.78) and this procedure.

5.2.3.6d   Using the Roothaan–Hall equations to do ab initio
calculations – some details

Equation (5.78) is normally modified by subsuming the c's into P_{tu}, the elements of the density matrix $\mathbf{P}$:

$$\mathbf{P} = \begin{pmatrix} P_{11} & P_{12} & P_{13} & \cdots & P_{1m} \\ P_{21} & P_{22} & P_{23} & \cdots & P_{2m} \\ \vdots & \vdots & \ddots & & \vdots \\ P_{m1} & P_{m2} & P_{m3} & \cdots & P_{mm} \end{pmatrix} \tag{5.80}$$

where the density matrix elements are

$$P_{tu} = 2 \sum_{j=1}^{n} c_{tj}^* c_{uj} \quad t = 1, 2, \ldots, m \text{ and } u = 1, 2, \ldots, m \qquad (5.81)$$

(sometimes P is defined as $\sum c^* c$). From Eqs (5.78) and (5.81):

$$F_{rs} = H_{rs}^{\text{core}}(1) + \sum_{t=1}^{m} \sum_{u=1}^{m} P_{tu} \left[(rs|tu) - \frac{1}{2}(ru|ts) \right] \qquad (5.82)$$

Equation (5.82), a slight modification of Eq. (5.78), is the key equation in calculating the *ab initio* Fock matrix. Each density matrix element P_{tu} represents the coefficients c for a particular pair of basis functions ϕ_t and ϕ_u, summed over all the occupied MO's $\psi_i (i = 1, 2, \ldots, n)$. We use the density matrix here just as a convenient way to express the Fock matrix elements, and to formulate the calculation of properties arising from electron distribution (section 5.5.4), although there is far more to the density matrix concept [21]. Equation (5.82) enables the MO wavefunctions ψ (which are linear combinations of the c's and ϕ's) and their energy levels ε to be calculated by iterative diagonalization of the Fock matrix.

Equation (5.17) ($E = 2 \sum H + \sum \sum (2J - K)$) gives one expression for the molecular electronic energy E. If we wish to calculate E from the energy levels, we must note that in the HF method E is not simply twice the sum of the energies of the n occupied energy levels, i.e. it is not the sum of the one-electron energies (as we take it to be in the simple and extended Hückel methods). This is because the MO energy level value ε represents the energy of one electron *subject to interaction with all the other electrons*. The energy of an electron is thus its kinetic energy plus its electron–nuclear attractive potential energy (H^{core}), plus, courtesy of the J and K integrals (section 5.2.3.5 and Eqs (5.48)–(5.50 = 5.83)), the potential energy from repulsion of all the other electrons:

$$\varepsilon_i = H_{ii}^{\text{core}} + \sum_{j=1}^{n} (2J_{ij}(1) - K_{ij}(1)) \qquad (5.83 = 5.50)$$

If we add the energies of electrons 1 and 2, say, we are adding, besides the kinetic energies of these electrons, the repulsion energy of electron 1 on electron 2, 3, 4, ..., and the repulsion energy of electron 2 on electron 1, 3, 4, ... – in other words, we are counting each repulsion twice. The simple sum thus represents properly the total kinetic and electron–nuclear attraction potential energy, but overcounts the electron–electron repulsion potential energy (recall that we are working with $2n$ electrons and thus n filled MOs):

$$E(\text{overestimated}) = 2 \sum_{i=1}^{n} \varepsilon_i \qquad (5.84)$$

Note that we cannot just take half of this simple sum, because only the electron–electron energy terms, not all the terms, have been doubly-counted. The solution is to subtract from $2 \sum \varepsilon$ the superfluous repulsion energy; from our discussion of Eq. (5.50) in section 5.2.3.5 we saw that the sum $\sum (2J - K)$ over n represents the repulsion energy of one electron interacting with all the other electrons, so to remove the superfluous interactions we subtract $\sum\sum(2J - K)$, the sum over n of the repulsion energy sum, to get [13]

$$E_{\text{HF}} = 2\sum_{i=1}^{n}\varepsilon_i - \sum_{i=1}^{n}\sum_{j=1}^{n}(2J_{ij}(1) - K_{ij}(1)) \qquad (5.85)$$

E_{HF} is the HF electronic energy: the sum of one-electron energies corrected (within the average-field HF approximation) for electron–electron repulsion. We can get rid of the integrals J and K over MOs ψ and obtain an equation for E_{HF} in terms of c's and ϕ's. From (5.83),

$$\sum_{j=1}^{n}\sum_{j=1}^{n}(2J_{ij}(1) - K_{ij}(1)) = \sum_{i=1}^{n}\varepsilon_i - \sum_{i=1}^{n}H_{ii}^{\text{core}}$$

and from this and (5.85) we get

$$E_{\text{HF}} = \sum_{i=1}^{n}\varepsilon_i + \sum_{i=1}^{n}H_{ii}^{\text{core}} \qquad (5.86)$$

From the definition of H_{ii}^{core} in Eqs (5.49) and (5.50), i.e. from

$$H_{ii}^{\text{core}} = \langle\psi_i(1)|\hat{H}^{\text{core}}|\psi_i\rangle \qquad (5.87)$$

and the LCAO expansion (5.52)

$$\psi_i = \sum_{s=1}^{m}c_{si}\phi_s \qquad (5.88 = 5.52)$$

we get from Eq. (5.86)

$$E_{\text{HF}} = \sum_{i=1}^{n}\varepsilon_i + \sum_{r=1}^{m}\sum_{s=1}^{m}\sum_{i=1}^{n}c_{ri}^*c_{si}H_{rs}^{\text{core}} \qquad (5.89)$$

Using Eq. (5.81), Eq. (5.89) can be written in terms of the density matrix elements P:

$$E_{\text{HF}} = \sum_{i=1}^{n}\varepsilon_i + \frac{1}{2}\sum_{r=1}^{m}\sum_{s=1}^{m}P_{rs}H_{rs}^{\text{core}} \qquad (5.90)$$

This is the key equation for calculating the HF electronic energy of a molecule. It can be used when self-consistency has been reached, or after each SCF cycle employing the ε's and c's yielded by that particular iteration, and H_{rs}^{core}, which latter does not change from iteration to iteration, since it is composed only of the fixed basis functions and an operator which does not contain ε's or c's: from Eqs (5.64=5.19) and (5.79)

$$H_{rs}^{\text{core}} = \left\langle\phi_r\left|-\frac{1}{2}\nabla_i^2 - \sum_{\text{all }\mu}\frac{Z_\mu}{r_{\mu i}}\right|\phi_s\right\rangle \qquad (5.91)$$

H_{rs}^{core} does not change because the SCF procedure refines the electron–electron repulsion (till the field each electron feels is "consistent" with the previous one), but H_{rs}^{core} in contrast represents only the contribution to the kinetic energy plus electron–nucleus attraction of the electron density associated with each pair of basis functions ϕ_r and ϕ_s.

Equation (5.90) gives the HF electronic energy of the molecule or atom – the energy of the electrons due to their motion (their kinetic energy) plus their energy due to electron–nucleus attraction and (within the HF approximation) to electron–electron repulsion (their potential energy). The *total* energy of the molecule, however, involves not just the electrons but also the nuclei, which contribute potential energy due to internuclear repulsion and kinetic energy due to nuclear motion. This motion persists even at 0 K, because the molecule vibrates even at this temperature; this unavoidable vibrational energy is called the *zero-point vibrational energy* or *zero-point energy* (ZPVE or ZPE; section 2.5, Fig. 2.20 and associated discussion). Calculation of the internuclear repulsion energy is trivial, as this is just the sum of all coulombic repulsions between the nuclei:

$$V_{NN} = \sum_{\text{all } \mu, \nu} \frac{Z_\mu Z_\nu}{r_{\mu\nu}} \qquad (5.92 = 5.16)$$

Calculation of the ZPE is more involved; it requires calculating the frequencies (i.e. the normal-mode vibrations – section 2.5) and summing the energies of each mode [22] (all this is done by standard programs, which print out the ZPE after the frequencies). Adding the HF electronic energy and the internuclear repulsion gives what we might call E_{HF}^{total}, the total "frozen-nuclei" (no ZPE) energy:

$$E_{HF}^{total} = E_{HF} + V_{NN} = \sum_{i=1}^{n} \varepsilon_i + \frac{1}{2} \sum_{r=1}^{m} \sum_{s=1}^{m} P_{rs} H_{rs}^{core} + V_{NN} \qquad (5.93)$$

from (5.90) and (5.92). E_{HF}^{total}, the energy usually displayed at the end of a HF calculation is, in ordinary parlance, "the HF energy". An aggregate of such energies, plotted against various geometries, represents an HF Born–Oppenheimer PES (section 2.3). The zero of energy for the Schrödinger equation for an atom or molecule is normally taken as the energy of the electrons and nuclei at rest at infinite separation. The HF energy (any *ab initio* energy, in fact) of a species is thus relative to the energy of the electrons and nuclei at rest at infinite separation, i.e. it is the negative of the minimum energy required to break up the molecule or atom and separate the electrons and nuclei to infinity. We are normally interested in *relative* energies, *differences* in absolute *ab initio* energies. *Ab initio* energies are discussed in section 5.5.2.

In a geometry optimization (section 2.4) a series of single-point calculations (calculations at a single point on the potential energy surface, i.e. at a single geometry) is done, each of which requires the calculation of E_{HF}^{total}, and the geometry is changed systematically until a stationary point is reached (one where the potential energy surface is flat; *ideally* E_{HF}^{total} should fall monotonically in the case of optimization to a minimum). The ZPE calculation, which is valid only for a stationary point on the potential energy surface (section 2.5; discussion in connection with Fig. 2.19), can be used to correct E_{HF}^{total} of the optimized structure for vibrational energy; adding the ZPE gives the total internal energy of the molecule at 0 K, which we could call E_{0K}^{total}:

$$E_{0K}^{total} = E_{HF}^{total} + ZPE \qquad (5.94)$$

The relative energies of isomers may be calculated by comparing E_{HF}^{total}, but for accurate work the ZPE should be taken into account, even though the required frequency calculations usually take considerably longer than the geometry optimization (sometimes five to ten times as long – see section 5.3.3, Table 5.3). Fortunately, it is valid to correct E_{HF}^{total} with a ZPE from a lower-level optimization-plus-frequency job (not a lower-level frequency job on the higher-level geometry). Figure 2.19 in section 2.5 compares energies for the species in the isomerization of HNC to HCN. The relative energies with/without the ZPE correction for HCN, transition state, and HNC are 0/0, 202/219, and 49.7/52.2 kJ mol^{-1}. The ZPE's of isomers tend to be roughly equal and so to cancel when relative energies are calculated (less so where transition states are involved), but, as implied above, in accurate work it is usual to compare the ZPE-corrected energies E_{0K}^{total}.

5.2.3.6e   Using the Roothaan–Hall equations to do ab initio calculations – an example

The application of the HF method to an actual calculation will now be illustrated in detail with protonated helium, H–He$^+$, the simplest closed-shell heteronuclear molecule. This species was also used to illustrate the details of the EHM in section 4.4.1b. In this simple example all the steps were done with a pocket calculator, except for the evaluation of the integrals (this was done with the *ab initio* program Gaussian 92 [23]) and the matrix multiplication and diagonalization steps (done with the program Mathcad [24]).

Step 1   Specifying the geometry, basis set and MO orbital occupancy

We start by specifying a geometry and a basis set. We will use same geometry as with the EHM, 0.800 Å, i.e. 1.5117 a.u. (bohr). In *ab initio* calculations on molecules, the basis functions are almost always *Gaussian functions* (basis functions are discussed in section 5.3). Gaussian functions differ from the Slater functions we used in the EHM in chapter 4 in that the exponent involves the *square* of the distance of the electron from the point (usually an atomic nucleus) on which the function is centered:

An s-type Slater function

$$\phi = a \exp(-br) \qquad (5.95)$$

An *s*-type Gaussian function

$$\phi = a \exp(-br^2) \qquad (5.96)$$

In *ab initio* calculations the mathematically more tractable Gaussians are used to approximate the physically more realistic Slater functions (see section 5.3). We use here the simplest possible Gaussian basis set: a 1s atomic orbital on each of the two atoms, each 1s orbital being approximated by one Gaussian function. This is called an STO-1G basis set, meaning Slater-type orbitals-one Gaussian, because we are approximating a Slater-type 1s orbital with a Gaussian function. The best STO-1G approximations to

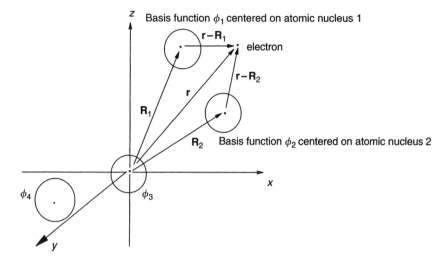

Figure 5.7. A four-atom molecule in a coordinate system. Only one of possibly many electrons is shown. The basis functions ϕ are one-electron functions, usually centered on atomic nuclei. R_1, R_2, etc., are vectors representing the x, y, z coordinates (conveniently as 3×1 column matrices; section 4.3.3) of the nuclei ("of the atoms"), and r is a vector representing the x, y, z coordinates of an electron. The distances of the electron from the centers of the various basis functions are the absolute values of the various vector differences: $|r - R_1|, |r - R_2|$, etc. For a particular molecular geometry, R_1, R_2, etc. are fixed and enter the functions ϕ_1, ϕ_2, etc., only parametrically, i.e. to denote where the ϕ's are centered; r is the variable in these functions, which are thus $\phi(x, y, z)$. Several basis functions may be centered on each nucleus.

the hydrogen and helium $1s$ orbitals in a molecular environment [25] are

$$\phi(\text{H}) = \phi_1 = 0.3696 \exp(-0.4166|r - R_1|^2) \qquad (5.97)$$

$$\phi(\text{He}) = \phi_2 = 0.5881 \exp(-0.7739|r - R_2|^2) \qquad (5.98)$$

where $|r - R_i|$ is the distance of the electron in ϕ_i (ϕ is a one-electron function) from nucleus i on which ϕ_i is centered (Fig. 5.7). The larger constant in the helium exponent as compared to that of hydrogen (0.7739 vs. 0.4166) reflects the intuitively reasonable fact that since an electron in ϕ_2 is bound more tightly to its doubly-charged nucleus than is an electron in ϕ_1 is to its singly-charged nucleus, electron density around the helium nucleus falls off more quickly with distance than does that around the hydrogen nucleus (Fig. 5.8).

We have a geometry and a basis set, and wish to do an SCF calculation on HHe^+ with both electrons in the lowest MO, ψ_1, i.e. on the *singlet* ground state. In general, SCF calculations proceed from specification of geometry, basis set, charge and multiplicity. The multiplicity is a way of specifying the number of unpaired electrons:

$$\text{multiplicity} = S = 2s + 1 \qquad (5.99)$$

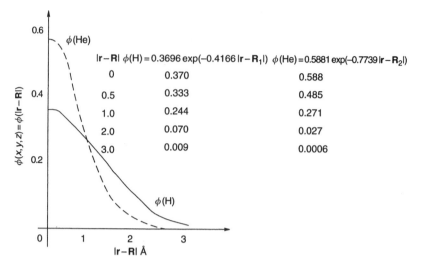

Figure 5.8. Electron density around the helium nucleus falls off more quickly than electron density around the lower-charge hydrogen nucleus.

where s = total number of unpaired electron spins (each electron has a spin of $\pm\frac{1}{2}$, taking each unpaired spin as $+\frac{1}{2}$. Figure 5.9 shows some examples of the specification of charge and multiplicity. By default an SCF calculation is performed on the *ground state* of specified multiplicity, i.e. the MO's are filled from ψ_1 up to give the lowest-energy state of that multiplicity.

Step 2   Calculating the integrals
Having specified a HF calculation on singlet HHe$^+$, with H–He $= 0.800$ Å (1.5117 bohr), using an STO-1G basis set, the most straightforward way to proceed is to now calculate all the integrals, and the orthogonalizing matrix $\mathbf{S}^{-1/2}$ that will be used to transform the Fock matrix $\mathbf{F}$ to $\mathbf{F}'$ and to convert the transformed coefficient matrix $\mathbf{C}'$ to $\mathbf{C}$ (Eqs (5.67)–(5.70)). The integrals are those required for H^{core}, the one-electron part of the elements F_{rs} of $\mathbf{F}$, and the two-electron repulsion integrals $(rs|tu)$, $(ru|ts)$ (Eq. (5.82)), as well as the overlap integrals, which are needed to calculate the overlap matrix $\mathbf{S}$ and thus the orthogonalizing matrix $\mathbf{S}^{-1/2}$ (Eq. (5.67)).

Efficient methods have been developed for calculating these integrals [26] and their values will simply be given here. For our calculation the elements F_{rs} of the Fock matrix (Eq. (5.82)) are conveniently written as:

$$
F_{rs} = H_{rs}^{\text{core}}(1) + \sum_{t=1}^{m}\sum_{u=1}^{m} P_{tu}\left[(rs|tu) - \frac{1}{2}(ru|ts)\right]
$$

$$
= T_{rs} + V_{rs}(\text{H}) + V_{rs}(\text{He}) + G_{rs} \tag{5.100}
$$

Here $H^{\text{core}}(1)$ has been dissected into a kinetic energy integral T and two potential energy integrals, $V(\text{H})$ and $V(\text{He})$. From the definition of the operator $\hat{H}^{\text{core}}$ (Eq. (5.64)) and the Roothaan–Hall expression for the integral H^{core} (Eq. (5.79)) we

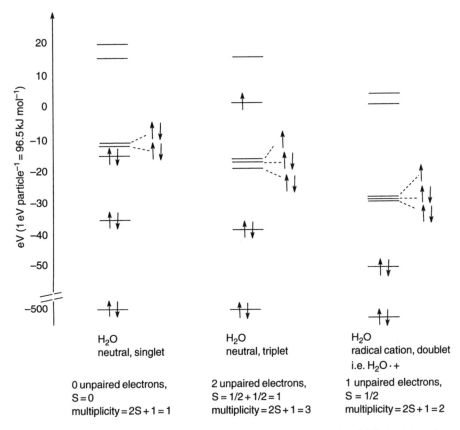

Figure 5.9. Some examples of the results of specification of charge and multiplicity. The calculations used the STO-3G basis set (section 5.3) which has seven basis functions, and so creates seven MOs. All calculations were at the HF/STO-3G geometry of the neutral singlet.

see that (the (1) emphasizes that these integrals involve the coordinates of only one electron):

$$T_{rs}(1) = \int \phi_r \left(-\frac{1}{2}\nabla_1^2\right)\phi_s \, dv = \int \phi_r \left[-\frac{1}{2}\left(\frac{\partial^2}{\partial x^2} + \frac{\partial^2}{\partial y^2} + \frac{\partial^2}{\partial z^2}\right)\right]\phi_s \, dv \tag{5.101}$$

$$V_{rs}(H, 1) = \int \phi_r \left(\frac{Z_H}{r_{H1}}\right)\phi_s \, dv \tag{5.102}$$

and

$$V_{rs}(He, 1) = \int \phi_r \left(\frac{Z_{He}}{r_{He1}}\right)\phi_s \, dv \tag{5.103}$$

In Eq. (5.102) the variable is the distance of the electron ("electron 1" – see the discussion in connection with Eqs (5.18) and (5.19)) from the hydrogen nucleus, and in Eq. (5.103) the variable is the distance of the electron from the helium nucleus; Z_H and Z_{He} are 1 and 2, respectively.

From Eq. (5.100) the two-electron contribution to the each Fock matrix element is

$$G_{rs} = \sum_{t=1}^{m}\sum_{u=1}^{m} P_{tu}\left[(rs|tu) - \frac{1}{2}(ru|ts)\right] \tag{5.104}$$

Each element G_{rs} is calculated from a density matrix element P_{tu} (Eqs (5.80) and (5.81)) and two two-electron integrals $(rs|tu)$ and $(ru|ts)$ (Eqs (5.73) and (5.77)). The required one-electron integrals for calculating the Fock matrix **F** are

$$T_{11} = 0.6249 \quad T_{12} = T_{21} = 0.2395 \quad T_{22} = 1.1609$$

$$V_{11}(H) = -1.0300 \quad V_{12}(H) = V_{21}(H) = -0.4445 \quad V_{22}(H) = -0.6563 \tag{5.105}$$

$$V_{11}(He) = -1.2555 \quad V_{12}(He) = V_{21}(He) = -1.1110 \quad V_{22}(He) = -2.8076$$

To see which two-electron integrals are needed we evaluate the summation in Eq. (5.104) for each of the matrix elements ($G_{11}, G_{12}, G_{21}, G_{22}$):

$$G_{11} = \sum_{t=1}^{2}\sum_{u=1}^{2} P_{tu}\left[(11|tu) - \frac{1}{2}(1u|t1)\right]$$

i.e. $G_{11} = \sum_{t=1}^{2}\left[P_{t1}\left[(11|t1) - \frac{1}{2}(11|t1)\right] + P_{t2}\left[(11|t2) - \frac{1}{2}(12|t1)\right]\right]$

$$= P_{11}\left[(11|11) - \frac{1}{2}(11|11)\right] + P_{12}\left[(11|12) - \frac{1}{2}(12|11)\right]$$

$$+ P_{21}\left[(11|21) - \frac{1}{2}(11|21)\right] + P_{22}\left[(11|22) - \frac{1}{2}(12|21)\right] \tag{5.106}$$

$$G_{12} = G_{21} = \sum_{t=1}^{2}\sum_{u=1}^{2} P_{tu}\left[(12|tu) - \frac{1}{2}(1u|t2)\right]$$

i.e. $G_{12} = G_{21} = \sum_{t=1}^{2}\left[P_{t1}\left[(12|t1) - \frac{1}{2}(11|t2)\right] + P_{t2}\left[(12|t2) - \frac{1}{2}(12|t2)\right]\right]$

$$= P_{11}\left[(12|11) - \frac{1}{2}(11|12)\right] + P_{12}\left[(12|12) - \frac{1}{2}(12|12)\right]$$

$$+ P_{21}\left[(12|21) - \frac{1}{2}(11|22)\right] + P_{22}\left[(12|22) - \frac{1}{2}(12|22)\right] \tag{5.107}$$

$$G_{22} = \sum_{t=1}^{2}\sum_{u=1}^{2} P_{tu}\left[(22|tu) - \frac{1}{2}(2u|t2)\right]$$

i.e. $\quad G_{22} = \sum_{t=1}^{2} \left[P_{t1} \left[(22|t1) - \frac{1}{2}(21|t2) \right] + P_{t2} \left[(22|t2) - \frac{1}{2}(22|t2) \right] \right]$

$= P_{11} \left[(22|11) - \frac{1}{2}(21|12) \right] + P_{12} \left[(22|12) - \frac{1}{2}(22|12) \right]$

$+ P_{21} \left[(22|21) - \frac{1}{2}(21|22) \right] + P_{22} \left[(22|22) - \frac{1}{2}(22|22) \right] \quad (5.108)$

Each element of the electron repulsion matrix **G** has eight two-electron repulsion integrals, and of these 32 there appear to be 14 different ones:

from G_{11} : $(11|11), (11|12), (12|11), (11|21), (11|22), (12|21)$

new with $G_{12} = G_{21}$: $(12|12), (12|22)$

new with G_{22} : $(22|11), (21|12), (22|12), (22|21), (21|22), (22|22)$

However, examination of Eq. (5.73) shows that many of these are the same. It is easy to see that if the basis functions are real (as is almost always the case) then

$(rs|tu) = (rs|ut) = (sr|tu) = (sr|ut) = (tu|rs) = (tu|sr) = (ut|rs) = (ut|sr)$
(5.109)

Taking this into account, there are only six unique two-electron repulsion integrals, whose values are:

$$(11|11) = 0.7283 \quad (21|21) = 0.2192$$
$$(21|11) = 0.3418 \quad (22|21) = 0.4368 \qquad (5.110)$$
$$(22|11) = 0.5850 \quad (22|22) = 0.9927$$

The integrals $(11|11)$ and $(22|22)$ represent repulsion between two electrons both in the same orbital (ϕ_1 or ϕ_2, respectively), while $(22|11)$ represents repulsion between an electron in ϕ_2 and one in ϕ_1; $(21|11)$ could be regarded as representing the repulsion between an electron associated with ϕ_2 and ϕ_1 and one confined to ϕ_1, and analogously for $(22|21)$, while $(21|21)$ can be thought of as the repulsion between two electrons both of which are associated with ϕ_2 and ϕ_1 (Fig. 5.10). Note that in the T and V terms of the Fock matrix elements, the operator in the integrals is $-(1/2)\nabla^2$ and Z_H/r_{H1} or Z_{He}/r_{He1}, while in the G terms it is $1/r_{12}$ (Eqs (5.101)–(5.103) and (5.73)). The overlap integrals are

$$S_{11} = 1.0000 \quad S_{12} = S_{21} = 0.5017 \quad S_{22} = 1.0000 \qquad (5.111)$$

and the overlap matrix is

$$\mathbf{S} = \begin{pmatrix} 1.0000 & 0.5017 \\ 0.5017 & 1.0000 \end{pmatrix} \qquad (5.112)$$

Step 3 Calculating the orthogonalizing matrix
Calculating the orthogonalizing matrix $\mathbf{S}^{-1/2}$ (see Eqs (5.67)–(5.69) and the discussion referred to in chapter 4):

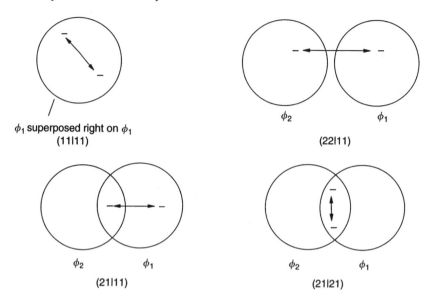

ϕ_1 superposed right on ϕ_1
(11|11)

($22|11$)

($21|11$)

($21|21$)

Figure 5.10. Schematic depictions of the physical meaning of some two-electron repulsion integrals (section 5.2.3.6e). Each basis function ϕ is normally centered on an atomic nucleus. The integrals shown here are one-center and two-center two-electron repulsion integrals – they involve one and two atoms, respectively. In calculations with more than two atoms three-center and four center two-electron integrals can arise.

Diagonalizing **S**:

$$\mathbf{S} = \begin{pmatrix} 0.7071 & 0.7071 \\ 0.7071 & -0.7071 \end{pmatrix} \begin{pmatrix} 1.5017 & 0.0000 \\ 0.0000 & 0.4983 \end{pmatrix} \begin{pmatrix} 0.7071 & 0.7071 \\ 0.7071 & -0.7071 \end{pmatrix} \qquad (5.113)$$

$$\quad\quad\quad\quad \mathbf{P} \quad\quad\quad\quad\quad\quad \mathbf{D} \quad\quad\quad\quad\quad\quad \mathbf{P}^{-1}$$

Calculating $\mathbf{D}^{-1/2}$:

$$\mathbf{D}^{-1/2} = \begin{pmatrix} 1.5017^{-1/2} & 0.0000 \\ 0.0000 & 0.4983^{-1/2} \end{pmatrix} = \begin{pmatrix} 0.8160 & 0.0000 \\ 0.0000 & 1.4166 \end{pmatrix} \qquad (5.114)$$

Calculating $\mathbf{S}^{-1/2}$:

$$\mathbf{S}^{-1/2} = \mathbf{P}\mathbf{D}^{-1/2}\mathbf{P}^{-1} = \begin{pmatrix} 1.1163 & -0.3003 \\ -0.3003 & 1.1163 \end{pmatrix} \qquad (5.115)$$

Step 4   Calculating the Fock matrix
(a) The one-electron matrices
From Eq. (5.100)

$$\mathbf{F} = \mathbf{T} + \mathbf{V}(\mathrm{H}) + \mathbf{V}(\mathrm{He}) + \mathbf{G} = \mathbf{H}^{\mathrm{core}} + \mathbf{G} \qquad (5.116)$$

The one-electron matrices $\mathbf{T}$, $\mathbf{V}(H)$ and $\mathbf{V}(He)$ (i.e. $\mathbf{H}^{core}$) follow immediately from the one-electron integrals. The kinetic energy matrix is

$$\mathbf{T} = \begin{pmatrix} T_{11} & T_{12} \\ T_{21} & T_{22} \end{pmatrix} = \begin{pmatrix} 0.6249 & 0.2395 \\ 0.2395 & 1.1609 \end{pmatrix} \tag{5.117}$$

T_{11} is smaller than T_{22}, as the kinetic energy of an electron in $\phi_1(\phi(H))$ is smaller than that of an electron in $\phi_2(\phi(He))$; this is expected since the larger charge on the helium nucleus results in a larger kinetic energy for an electron in its 1s orbital than for an electron in the hydrogen 1s orbital–classically speaking, the electron must move faster to stay in orbit around the stronger-pulling He nucleus. T_{12} can be regarded as the kinetic energy of an electron in the $H(1s)$–$He(1s)$ overlap region.

The hydrogen potential energy matrix is

$$\mathbf{V}(H) = \begin{pmatrix} V_{11}(H) & V_{12}(H) \\ V_{21}(H) & V_{22}(H) \end{pmatrix} = \begin{pmatrix} -1.0300 & -0.4445 \\ -0.4445 & -0.6563 \end{pmatrix} \tag{5.118}$$

All the $V(H)$ values represent the attraction of an electron to the hydrogen nucleus. $V_{11}(H)$ is the potential energy due to attraction of an electron in ϕ_1 to the hydrogen nucleus, and $V_{22}(H)$ is the potential energy due to attraction of an electron in ϕ_2 to the hydrogen nucleus. As expected, an electron in ϕ_1 ($\phi(H)$) is attracted to the H nucleus more strongly (the potential energy is more negative) than is an electron in ϕ_2 ($\phi(He)$). $V_{12}(H)$ can be regarded as the potential energy of attraction to the hydrogen nucleus of an electron in the $H(1s)$–$He(1s)$ overlap region.

The helium potential energy matrix is

$$\mathbf{V}(He) = \begin{pmatrix} V_{11}(He) & V_{12}(He) \\ V_{21}(He) & V_{22}(He) \end{pmatrix} = \begin{pmatrix} -1.2555 & -1.1110 \\ -1.1110 & -2.8076 \end{pmatrix} \tag{5.119}$$

All the $V(He)$ values represent the attraction of an electron to the helium nucleus. $V_{11}(He)$, the potential energy of attraction of an electron in $\phi(H)$ to the helium nucleus, is of course less negative than the potential energy of attraction of an electron in $\phi(He)$ to this same nucleus. $V_{12}(He)$ can be taken as the potential energy of attraction to the helium nucleus of an electron in the $H(1s)$–$He(1s)$ overlap region. An electron in $\phi(He)$ is attracted to the helium nucleus more strongly than an electron in $\phi(H)$ is attracted to the hydrogen nucleus (-2.8076 in $\mathbf{V}(He)$ cf. -1.0300 in $\mathbf{V}(H)$), due to the greater nuclear charge of helium.

The total one-electron energy matrix, $\mathbf{H}^{core}$, is

$$\mathbf{H}^{core}\mathbf{T} + \mathbf{V}(H) + \mathbf{V}(He) = \begin{pmatrix} -1.6606 & -1.3160 \\ -1.3160 & -2.3030 \end{pmatrix} \tag{5.120}$$

This matrix represents the 1-electron energy (the energy the electron would have if interelectronic repulsion did not exist) of an electron in H–He^+, at the specified geometry, for this STO-1G basis set. The (1,1), (2,2) and (1,2) terms represent, ignoring electron–electron repulsion, the energy of an electron in ϕ_1, ϕ_2, and the $\phi_1 - \phi_2$ overlap region, respectively; the values are the net result of the various kinetic energy and potential energy terms discussed above.

(b) The two-electron matrix
The two-electron matrix $\mathbf{G}$, the electron repulsion matrix (Eq. (5.111)), is calculated from the two-electron integrals and the density matrix elements (Eq. (5.104)). This is intuitively plausible since each two-electron integral describes one interelectronic repulsion in terms of basis functions (Fig. 5.10) while each density matrix element represents (see section 5.4.3) the electron density *on* (the diagonal elements of $\mathbf{P}$ in Eq. (5.80)) or *between* (the off-diagonal elements of $\mathbf{P}$) basis functions. To calculate the matrix elements G_{rs} (Eqs (5.106)–(5.108)) we need the appropriate integrals (Eqs (5.110) and density matrix elements. These latter are calculated from

$$P_{tu} = 2\sum_{j=1}^{n} c_{tj}^* c_{uj} \quad t = 1, 2, \ldots, m \text{ and } u = 1, 2, \ldots, m \qquad (5.121 = 5.81)$$

Each P_{rs} involves the sum over the occupied MO's ($j = 1$ to n; we are dealing with a closed-shell ground-state molecule with $2n$ electrons) of the products of the coefficients of the basis functions ϕ_r and ϕ_s. As pointed out in section 5.2.3.6b the HF procedure is usually started with an "initial guess" at the coefficients. We can use as our guess the extended Hückel coefficients we obtained for HeH^+, with this same geometry (section 4.4.1b); we need the c's only for the *occupied* MO's:

$$c_{11} = 0.249, \quad c_{21} = 0.867 \qquad (5.122)$$

(Usually we need more c's than the small basis set of an extended Hückel or other semiempirical calculation supplies; a *projected* semiempirical wavefunction is then used, with the missing c's extrapolated from the available ones.) Using these c's and Eq. (5.121) we calculate the initial-guess P's for Eqs (5.106)–(5.108); since there is only one occupied MO ($n = 1$ in Eq. 121) the summation has only one term:

$$P_{11} = 2c_{11}c_{11} = 2(0.249)0.249 = 0.1240$$
$$P_{12} = 2c_{11}c_{21} = 2(0.249)0.867 = 0.4318 \qquad (5.123)$$
$$P_{22} = 2c_{21}c_{21} = 2(0.867)0.867 = 1.5034$$

$\mathbf{G}$ may now be calculated. From Eqs (5.106)–(5.108), using the above values of P and the integrals of Eq. (5.110), and recalling that integrals like $(11|12)$ and $(21|11)$ are equal (Eq. (5.109) we get:

$$G_{11} = P_{11}\left[(11|11) - \tfrac{1}{2}(11|11)\right] + P_{12}\left[(11|12) - \tfrac{1}{2}(12|11)\right]$$
$$+ P_{21}\left[(11|21) - \tfrac{1}{2}(11|21)\right] + P_{22}\left[(11|22) - \tfrac{1}{2}(12|21)\right]$$
$$= 0.1240(0.3642) + 0.4318(0.1709)$$
$$+ 0.4318(0.1709) + 1.5034(0.4754) = 0.9075 \qquad (5.124)$$

$$G_{12} = G_{21} = P_{11}\left[(12|11) - \tfrac{1}{2}(11|12)\right] + P_{12}\left[(12|12) - \tfrac{1}{2}(12|12)\right]$$

$$+ P_{21}\left[(12|21) - \tfrac{1}{2}(11|22)\right] + P_{22}\left[(12|22) - \tfrac{1}{2}(12|22)\right]$$

$$= 0.1240(0.1709) + 0.4318(0.1096)$$

$$+ 0.4318(-0.0733) + 1.5034(0.2184) = 0.3652 \tag{5.125}$$

$$G_{22} = P_{11}\left[(22|11) - \tfrac{1}{2}(21|12)\right] + P_{12}\left[(22|12) - \tfrac{1}{2}(22|12)\right]$$

$$+ P_{21}\left[(22|21) - \tfrac{1}{2}(21|22)\right] + P_{22}\left[(22|22) - \tfrac{1}{2}(22|22)\right]$$

$$= 0.1240(0.4754) + 0.4318(0.2184)$$

$$+ 0.4318(0.2184) + 1.5034(0.4964) = 0.9938 \tag{5.126}$$

From the G values based on the initial guess c's the initial-guess electron repulsion matrix is

$$\mathbf{G}_0 = \begin{pmatrix} 0.9075 & 0.3652 \\ 0.3652 & 0.9938 \end{pmatrix} \tag{5.127}$$

The initial-guess Fock matrix is (Eqs (5.116), (5.120) and (5.127))

$$\mathbf{F}_0 = \mathbf{T} + \mathbf{V}(\mathrm{H}) + \mathbf{V}(\mathrm{He}) + \mathbf{G}_0$$

$$= \mathbf{H}^{\mathrm{core}} + \mathbf{G}_0$$

$$= \begin{pmatrix} -1.6606 & -1.3160 \\ -1.3160 & -2.3030 \end{pmatrix} + \begin{pmatrix} 0.9095 & 0.3652 \\ 0.3652 & 0.9938 \end{pmatrix}$$

$$= \begin{pmatrix} -0.7511 & -0.9508 \\ -0.9508 & -1.3092 \end{pmatrix} \tag{5.128}$$

The zero subscripts in Eqs (5.127) and (5.128) emphasize that the initial-guess c's, with no iterative refinement, were used to calculate $\mathbf{G}$; in the subsequent iterations of the SCF procedure $\mathbf{H}^{\mathrm{core}}$ will remain constant while $\mathbf{G}$ will be refined as the c's, and thus the P's, change from SCF cycle to cycle. The change in the electron repulsion matrix $\mathbf{G}$ corresponds to that in the molecular wavefunction as the c's change (recall the LCAO expansion); it is the wavefunction (squared) which represents the time-averaged electron distribution and thus the electron/charge cloud repulsion (sections 5.2.3.2, 5.2.3.5 and 5.2.3.6b).

Step 5   Transforming $\mathbf{F}$ *to* $\mathbf{F}'$, *the Fock matrix that satisfies* $\mathbf{F}' = \mathbf{C}'\varepsilon\mathbf{C}'^{-1}$
As in section 4.4.1b, we use the orthogonalizing matrix $\mathbf{S}^{-1/2}$ (step 3) to transform $\mathbf{F}$ to a matrix $\mathbf{F}'$ which when diagonalized gives the energy levels ε and a coefficient matrix $\mathbf{C}'$ which is subsequently transformed to the matrix $\mathbf{C}$ of the desired c's

(see section 5.2.3.6b):

$$\mathbf{F}_0' = \begin{pmatrix} 1.1163 & -0.3003 \\ -0.3003 & 1.1163 \end{pmatrix} \begin{pmatrix} -0.7511 & -0.9508 \\ -0.9508 & -1.3092 \end{pmatrix} \begin{pmatrix} 1.1163 & -0.3003 \\ -0.3003 & 1.1163 \end{pmatrix}$$

$$\underset{\mathbf{S}^{-1/2}}{} \qquad \underset{\mathbf{F}_0}{} \qquad \underset{\mathbf{S}^{-1/2}}{}$$

$$= \begin{pmatrix} -0.4166 & -0.5799 \\ -0.5799 & -1.0617 \end{pmatrix} \tag{5.129}$$

$$\underset{\mathbf{F}_0'}{}$$

Step 6   Diagonalizing $\mathbf{F}'$ *to obtain the energy level matrix* ε *and a coefficient matrix* $\mathbf{C}'$

$$\mathbf{F}_0' = \begin{pmatrix} 0.5069 & 0.8620 \\ 0.8620 & -0.5069 \end{pmatrix} \begin{pmatrix} -1.4027 & 0.0000 \\ 0.0000 & -0.0756 \end{pmatrix} \begin{pmatrix} 0.5069 & 0.8620 \\ 0.8620 & -0.5069 \end{pmatrix}$$

$$\underset{\mathbf{C}_1'}{} \qquad \underset{\varepsilon_1}{} \qquad \underset{\mathbf{C}_1'^{-1}}{} \tag{5.130}$$

The energy levels (the eigenvalues of $\mathbf{F}_0'$) from this first SCF cycle are -1.4027 and $-0.0756\,h$ (h = hartrees, the unit of energy in atomic units), corresponding to the occupied MO ψ_1 and the unoccupied MO ψ_2. The MO coefficients (the eigenvectors of $\mathbf{F}_0'$) of ψ_1 and ψ_2, *for the transformed, orthonormal basis functions*, are, from $\mathbf{C}_1'$ (actually here $\mathbf{C}_1'$ and its inverse, $\mathbf{C}_1'^{-1}$ are the same):

$$\mathbf{v}_1' = \begin{pmatrix} 0.5069 \\ 0.8620 \end{pmatrix} \quad \text{and} \quad \mathbf{v}_2' = \begin{pmatrix} 0.8620 \\ -0.5069 \end{pmatrix} \tag{5.131}$$

$\mathbf{v}_1'$ is the first column of $\mathbf{C}_1'$ and $\mathbf{v}_2'$ is the second column of $\mathbf{C}_1'$. These coefficients are the weighting factors that with the transformed, orthonormal basis functions give the MO's:

$$\psi_1 = 0.5069\phi_1' + 0.8620\phi_2' \quad \text{and} \quad \psi_2 = 0.8620\phi_1' - 0.5069\phi_2' \tag{5.132}$$

where ϕ_1' and ϕ_2' are linear combinations of our *original* basis functions ϕ_1 and ϕ_2. The original basis functions ϕ were centered on atomic nuclei and were normalized but not orthogonal (section 4.3.3), while the transformed basis functions ϕ' are delocalized over the molecule and are orthonormal (section 4.4.1a)). Note that the sum of the squares of the coefficients of ϕ_1' and ϕ_2' is unity, as must be the case if the basis functions are orthonormal (section 4.3.6). In the next step $\mathbf{C}_1'$ is transformed to obtain the coefficients of the original basis functions ϕ in the MO's. We want the MOs in terms of the original, atom-centered basis functions (roughly, atomic orbitals – section 5.3) because such MOs are easier to interpret.

Step 7   Transforming $\mathbf{C}'$ *to* $\mathbf{C}$, *the coefficient matrix of the original, nonorthogonal basis functions.*

As in section 4.4.1b, we use the orthogonalizing matrix $S^{-1/2}$ to transform C' to C:

$$C_1 = \begin{pmatrix} 1.1163 & -0.3003 \\ -0.3003 & 1.1163 \end{pmatrix} \begin{pmatrix} 0.5069 & 0.8620 \\ 0.8620 & -0.5069 \end{pmatrix} = \begin{pmatrix} 0.3070 & 1.1145 \\ 0.8100 & -0.8247 \end{pmatrix}$$

$$\underset{S^{-1/2}}{} \qquad \underset{C_1'}{} \qquad \underset{C_1}{} \qquad (5.133)$$

This completes the first SCF cycle. We now have the first set of MO energy levels and basis function coefficients:

From Eq. (5.130):

$$\varepsilon_1 = -1.4027 \quad \text{and} \quad \varepsilon_2 = -0.0756 \qquad (5.134)$$

From Eq (5.133) (cf. Eq (5.132)):

$$\psi_1 = 0.3070\phi_1 + 0.8100\phi_2 \quad \text{and} \quad \psi_2 = 1.1145\phi_1 - 0.8247\phi_2 \qquad (5.135)$$

Note that the sum of the squares of the coefficients of ϕ_1 and ϕ_2 is not unity, since these atom-centered functions are not orthogonal (contrast the simple Hückel method, section 4.3.4).

Step 8   Comparing the density matrix from the latest c's with the previous density matrix to see if the SCF procedure has converged
The density matrix elements based on the c's of C_1 (Eq. (5.133)) can be compared with those (Eq. (5.123)) based on the initial guess:

$$P_{11} = 2c_{11}c_{11} = 2(0.3070)0.3070 = 0.1885$$
$$P_{12} = 2c_{11}c_{21} = 2(0.3070)0.8100 = 0.4973 \qquad (5.136)$$
$$P_{22} = 2c_{21}c_{21} = 2(0.8100)0.8100 = 1.3122$$

Suppose our convergence criterion was that the elements of P must agree with those of the previous P matrix to within 1 part in 1000. Comparing Eqs (5.136) with (5.123) we see that this has not been achieved: even the smallest change is $|(1.312 - 1.503)/1.503| = 0.127$, far above the required 0.001. Therefore another SCF cycle is needed.

Step 9   Beginning the second SCF cycle: using the c's of C_1 to calculate a new Fock matrix F_1 (cf. Step 4, (b))
The first Fock matrix F_0 used c's from our initial guess (Step 4, (b)). An improved F may now be calculated using the c's from the first SCF cycle. Calculating G_1 as we did in Step 4, (b) for G_0, but using the new P's:

$$G_{11} = P_{11}\left[(11|11) - \tfrac{1}{2}(11|11)\right] + P_{12}\left[(11|12) - \tfrac{1}{2}(12|11)\right]$$
$$+ P_{21}\left[(11|21) - \tfrac{1}{2}(11|21)\right] + P_{22}\left[(11|22) - \tfrac{1}{2}(12|21)\right]$$
$$= 0.1885(0.3642) + 0.4973(0.1709)$$
$$+ 0.4973(0.1709) + 1.3122(0.4754) = 0.8624 \qquad (5.137)$$

$$G_{12} = G_{21} = P_{11}\left[(12|11) - \tfrac{1}{2}(11|12)\right] + P_{12}\left[(12|12) - \tfrac{1}{2}(12|12)\right]$$
$$+ P_{21}\left[(12|21) - \tfrac{1}{2}(11|22)\right] + P_{22}\left[(12|22) - \tfrac{1}{2}(12|22)\right]$$
$$= 0.1885(0.1709) + 0.4973(0.1096)$$
$$+ 0.4973(-0.0733) + 1.3122(0.2184) = 0.3369 \qquad (5.138)$$

$$G_{22} = P_{11}\left[(22|11) - \tfrac{1}{2}(21|12)\right] + P_{12}\left[(22|12) - \tfrac{1}{2}(22|12)\right]$$
$$+ P_{21}\left[(22|21) - \tfrac{1}{2}(21|22)\right] + P_{22}\left[(22|22) - \tfrac{1}{2}(22|22)\right]$$
$$= 0.1885(0.4754) + 0.4973(0.2184)$$
$$+ 0.4973(0.2184) + 1.3122(0.4964) = 0.9582 \qquad (5.139)$$

From the G values based on the first-cycle c's the electron repulsion matrix is

$$\mathbf{G}_1 = \begin{pmatrix} 0.8624 & 0.3369 \\ 0.3369 & 0.9582 \end{pmatrix} \qquad (5.140)$$

and the Fock matrix from this is

$$\mathbf{F}_1 = \mathbf{H}^{\text{core}} + \mathbf{G}_1$$
$$= \begin{pmatrix} -1.6606 & -1.3160 \\ -1.3160 & -2.3030 \end{pmatrix} + \begin{pmatrix} 0.8624 & 0.3369 \\ 0.3369 & 0.9582 \end{pmatrix}$$
$$= \begin{pmatrix} -0.7982 & -0.9791 \\ -0.9791 & -1.3448 \end{pmatrix} \qquad (5.141)$$

Step 10   Transforming $\mathbf{F}_1$ to $\mathbf{F}'_1$ (cf. Step 5)

$$\mathbf{F}'_1 = \underbrace{\begin{pmatrix} 1.1163 & -0.3003 \\ -0.3003 & 1.1163 \end{pmatrix}}_{\mathbf{S}^{-1/2}} \underbrace{\begin{pmatrix} -0.7982 & -0.9791 \\ -0.9791 & -1.3448 \end{pmatrix}}_{\mathbf{F}_1} \underbrace{\begin{pmatrix} 1.1163 & -0.3003 \\ -0.3003 & 1.1163 \end{pmatrix}}_{\mathbf{S}^{-1/2}}$$

$$= \underbrace{\begin{pmatrix} -0.4595 & -0.5900 \\ -0.5900 & -1.0913 \end{pmatrix}}_{\mathbf{F}'_1} \qquad (5.142)$$

Step 11   Diagonalizing $\mathbf{F}'_1$ to obtain the energy levels ε and a coefficient matrix $\mathbf{C}'$ (cf. Step 6)

$$\mathbf{F}'_1 = \underbrace{\begin{pmatrix} 0.5138 & 0.8579 \\ 0.8579 & -0.5138 \end{pmatrix}}_{\mathbf{C}'_2} \underbrace{\begin{pmatrix} -1.4447 & 0.0000 \\ 0.0000 & -0.1062 \end{pmatrix}}_{\varepsilon_2} \underbrace{\begin{pmatrix} 0.5138 & 0.8579 \\ 0.8579 & -0.5138 \end{pmatrix}}_{\mathbf{C}'^{-1}_2} \qquad (5.143)$$

The energy levels from this second SCF cycle are -1.4447 and $-0.1062\,\text{h}$. To get the MO coefficients corresponding to these MO energy levels in terms of the original basis functions ϕ_1 and ϕ_2 we now transform $\mathbf{C}_2'$ to $\mathbf{C}_2$.

Step 12   Transforming $\mathbf{C}_2'$ to $\mathbf{C}_2$ (cf. Step 7)

$$\mathbf{C}_2 = \begin{pmatrix} 1.1163 & -0.3003 \\ -0.3003 & 1.1163 \end{pmatrix} \begin{pmatrix} 0.5138 & 0.8579 \\ 0.8579 & -0.5138 \end{pmatrix} = \begin{pmatrix} 0.3159 & 1.1120 \\ 0.8034 & -0.8319 \end{pmatrix}$$

$$\mathbf{S}^{-1/2} \qquad\qquad\qquad \mathbf{C}_2' \qquad\qquad\qquad \mathbf{C}_2 \quad (5.144)$$

This completes the second SCF cycle. We now have the MO energy levels and basis function coefficients:

From Eq. (5.143):

$$\varepsilon_1 = -1.4447 \quad \text{and} \quad \varepsilon_2 = -0.1062 \qquad (5.145)$$

From Eq. (5.144):

$$\psi_1 = 0.3159\phi_1 + 0.8034\phi_2 \quad \text{and} \quad \psi_2 = 1.1120\phi_1 - 0.8319\phi_2 \qquad (5.146)$$

Step 13   Comparing the density matrix from the latest c's with the previous density matrix to see if the SCF procedure has converged
The density matrix elements based on the c's of $\mathbf{C}_2$ are

$$P_{11} = 2c_{11}c_{11} = 2(0.3159)0.3159 = 0.1996$$

$$P_{12} = 2c_{11}c_{21} = 2(0.3159)0.8034 = 0.5076$$

$$P_{22} = 2c_{21}c_{21} = 2(0.8034)0.8034 = 1.2909$$

Comparing the above with (5.136) we see that convergence to within our 1-part-in-1000 criterion has not occurred: the largest change in the density matrix is $|(0.1996 - 0.1885)/0.1885| = 0.059$, which is above 0.001, so the SCF procedure is repeated.

Three more SCF cycles were carried out; the results of the "zeroth cycle" (the initial guess) and the five cycles are summarized in Table 5.1. Only with the fifth cycle has convergence been achieved, i.e. have the changes in all the density matrix elements fallen below 1 part in 1000 (the largest change is in P_{11}, $|(0.2020 - 0.2019)/0.2019 = 0.0005 < 0.001$). In actual practice, a convergence criterion of from about 1 part in 10^4 to 1 in 10^8 is used, depending on the program and the particular kind of calculation. The coefficients and the density matrix elements change smoothly, although the energy levels and $E_{\text{HF}}^{\text{total}}$ show some oscillation. To reduce the number of steps needed to achieve convergence, programs sometimes extrapolate the density matrix, i.e. estimate the final P values and use these estimates to initiate the final few SCF cycles.

Often the main result from a HF (i.e. an SCF) calculation is the energy of the molecule (the calculation of energy may be subsumed into a geometry optimization, which is really the task of finding the minimum-energy geometry). The STO-1G energy of HHe$^+$ with an internuclear distance of 0.800 Å may be calculated from our results: the electronic energy is

$$E_{\text{HF}} = \sum_{i=1}^{n} \varepsilon_i + \frac{1}{2}\sum_{r=1}^{m}\sum_{s=1}^{m} P_{rs}H_{rs}^{\text{core}} \qquad (5.147 = 5.90)$$

Table 5.1. Results of initial guess and SCF cycles on HHe^+ at bond length 0.800 Å using the STO-1G basis set. Energies (ε_1, ε_2, and E_{HF}^{total}) are in hartrees

	Initial guess (zeroth cycle)	1st cycle	2nd cycle	3rd cycle	4th cycle	5th cycle
$\varepsilon_1, \varepsilon_2$	—	$-1.4027, -0.0756$	$-1.4447, -0.1062$	$-1.4466, -0.1054$	$-1.4473, -0.1056$	$-1.4470, -0.1051$
c_{11}, c_{21}	$0.249, 0.867$	$0.3070,\ 0.8100$	$0.3159, 0.8034$	$0.3175, 0.8022$	$0.3177, 0.8021$	$0.3178, 0.8020$
c_{12}, c_{22}	—	$1.1145, -0.8247$	$1.1120, -0.8319$	$1.1115, -0.8323$	$1.1115, -0.8325$	$1.1114, -0.8325$
P_{11}	0.1240	0.1885	0.1996	0.2010	0.2019	0.2020
P_{12}	0.4318	0.4973	0.5076	0.5094	0.5097	0.5097
P_{22}	1.5034	1.3122	1.2909	1.2870	1.2867	1.2864
E_{HF}^{total}	—	-2.3992	-2.4419	-2.4428	-2.4443	-2.4438

the internuclear repulsion energy is

$$V_{NN} = \sum_{\text{all } \mu, V} \frac{Z_\mu Z_V}{r_{\mu V}} \qquad (5.148 = 5.92)$$

and the total internal energy of the molecule at 0 K (except for ZPE – section 5.2.3.6d) is

$$E_{HF}^{\text{total}} = E_{HF} + V_{NN} = \sum_{i=1}^{n} \varepsilon_i + \frac{1}{2} \sum_{r=1}^{m} \sum_{s=1}^{m} P_{rs} H_{rs}^{\text{core}} + V_{NN} \qquad (5.149 = 5.93)$$

E_{HF}^{total}, which is what is normally meant by the HF energy, is printed by the program at the end of a single-point calculation or a geometry optimization, or by some programs at the end of each step of a geometry optimization.

Using the energy levels and density matrix elements from the first cycle (Table 5.1), with the H^{core} elements from Eq. (5.120), Eq. (5.147) gives for the electronic energy

$$E_{HF} = \varepsilon_1 + \frac{1}{2} \sum_{r=1}^{2} \sum_{s=1}^{2} P_{rs} H_{rs}^{\text{core}}$$

$$= \varepsilon_1 + \frac{1}{2} \sum_{r=1}^{2} [P_{r1} H_{r1}^{\text{core}} + P_{r2} H_{r2}^{\text{core}}]$$

$$= \varepsilon_1 + \frac{1}{2} [P_{11} H_{11}^{\text{core}} + P_{12} H_{12}^{\text{core}} + P_{21} H_{21}^{\text{core}} + P_{22} H_{22}^{\text{core}}]$$

$$= -1.4027\,h + \frac{1}{2} [0.1885(-1.6606) + 0.4973(-1.3160)$$

$$\quad + 0.4973(-1.3160) + 1.3122(-2.3030)]h$$

$$= -3.7222\,h \qquad (5.150)$$

From Eq. (5.148) the internuclear repulsion energy is

$$V_{NN} = \frac{Z_H Z_{He}}{r_{HHe}} = \frac{1(2)}{1.5117}h = 1.3230\,h \qquad (5.151)$$

and from Eq. (5.149) the total HF energy is

$$E_{HF}^{\text{total}} = E_{HF} + V_{NN} = -3.7222\,h + 1.3230\,h = -2.3992\,h \qquad (5.152)$$

The HF energies for the five SCF cycles are given in Table 5.1.

Instead of starting with eigenvectors from a non-SCF method like the extended Hückel method, as was done in this illustrative procedure, an SCF calculation is occasionally initiated by taking $\mathbf{H}^{\text{core}}$ as the Fock matrix, that is, by initially ignoring electron–electron repulsion, setting equal to zero the second term in Eq. (5.82), or G in Eq. (5.100), whereupon F_{rs} becomes H_{rs}^{core}. This is usually a poor initial guess, but is occasionally useful. You are urged to work your way through several SCF cycles starting with this Fock matrix; this tedious calculation will help you to appreciate the power and utility of modern electronic computers and may enhance your respect for those who pioneered complex numerical calculations when the only arithmetical aids were mathematical tables and mechanical calculators (mechanical calculators were machines with rotating wheels, operated by hand-power or electricity. There were also, in astronomy at least, armies of women arithmeticians called computers – the original meaning of the word).

If we calculate the electronic energy simply as twice the sum of the energies of the occupied MO orbitals, as with the SHM and EHM, we get a much higher value than from the correct procedure (Eq. (5.147)); with a 0.800 Å bond length and the converged results this naive electronic energy is $2(-1.4470) \, h = -2.8940 \, h$, while the correct electronic energy (not given in Table 5.1 – the HF energies there are electronic plus internuclear repulsion) is $-3.7668 \, h$, i.e. 30 percent lower when we correct for the fact that simply double-summing the MO energies counts electron repulsion terms twice (section 5.2.3.6d).

A geometry optimization for HHe$^+$ can be done by calculating the Hartree-Fock energy (electronic plus internuclear) at different bond lengths to get the minimum-energy geometry. The results are shown in Fig. 5.11; the optimized bond length for the STO-1G basis set is ca. 0.86 Å. Note that it is customary to report *ab initio* energies in hartrees to 5 or 6 decimal places (and bond lengths in Å to 3 decimals); the truncated values used here are appropriate for these illustrative calculations.

Summary of the steps in an SCF calculation using the Roothaan–Hall LCAO expansion of the MO's

1. Specify a geometry, basis set, and orbital occupancy (this latter is done by specifying the charge and multiplicity, with an electronic ground state being the default).

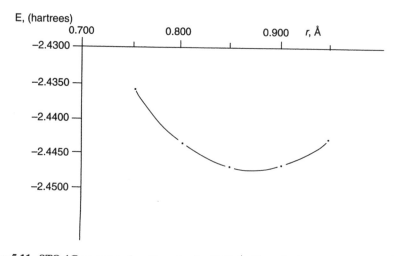

Figure 5.11. STO-1G energy vs. bond length r for H–He$^+$. The calculation for $r = 0.800$ Å was done largely "by hand" (see section 5.2.3.6e); the others were done with the program Gaussian 92 [23].

2. Calculate the integrals: T_{rs}, V_{rs} for each nucleus, and the two-electron integrals $(ru|ts)$ etc. needed for G_{rs}, as well as the overlap integrals S_{rs} for the orthogonalizing matrix derived from **S** (see step 3). Note: in the *direct SCF* method (section 5.3) the two-electron integrals are calculated as needed, rather than all at once.

3. Calculate the orthogonalizing matrix $\mathbf{S}^{-1/2}$
   (a) diagonalize **S** : $\mathbf{S} = \mathbf{PDP}^{-1}$
   (b) Calculate $\mathbf{D}^{-1/2}$ (take the $-\frac{1}{2}$ power of the elements of **D**)
   (c) Calculate $\mathbf{S}^{-1/2} = \mathbf{PD}^{-1/2}\mathbf{P}^{-1}$

4. Calculate the Fock matrix **F**
   (a) Calculate the one-electron matrix $\mathbf{H}^{\text{core}} = \mathbf{T} + \mathbf{V}_1 + \mathbf{V}_2 + \cdots$ using the T and V integrals from step 2.
   (b) The two-electron matrix (the electron repulsion matrix) **G**: Use an initial guess of the coefficients of the occupied MO's to calculate initial-guess density matrix elements:

$$P_{tu} = 2\sum_{j=1}^{n} c_{tj}^* c_{uj} \quad t = 1, 2, \ldots, m \text{ and } u = 1, 2, \ldots, m$$

Use the density matrix elements and the two-electron integrals to calculate **G**:

$$G_{rs} = \sum_{t=1}^{m}\sum_{u=1}^{m} P_{tu}\left[(rs|tu) - \tfrac{1}{2}(ru|ts)\right]$$

The Fock matrix is $\mathbf{F} = \mathbf{H}^{\text{core}} + \mathbf{G}$

5. Transform **F** to $\mathbf{F}'$, the Fock matrix that satisfies $\mathbf{F}' = \mathbf{C}'\boldsymbol{\varepsilon}\mathbf{C}'^{-1}$

$$\mathbf{F}' = \mathbf{S}^{-1/2}\mathbf{F}\mathbf{S}^{-1/2}$$

6. Diagonalize $\mathbf{F}'$ to get energy levels and a $\mathbf{C}'$ matrix

$$\mathbf{F}' = \mathbf{C}'\boldsymbol{\varepsilon}\mathbf{C}'^{-1}$$

7. Transform $\mathbf{C}'$ to **C**, the coefficient matrix of the original basis functions

$$\mathbf{C} = \mathbf{S}^{-1/2}\mathbf{C}'$$

8. Compare the density matrix elements calculated from the **C** of the previous step with those of the step before that one (and/or use other criteria, e.g. the molecular energy); if convergence has not been achieved go back to step 4 and calculate a new Fock matrix using the P's from the latest c's. If convergence has been achieved, stop.

It should be realized modern *ab initio* programs do not rigidly follow the basic SCF procedure described in this section. To speed up calculation they employ a variety of mathematical tricks. Among these are: the use of symmetry to avoid duplicate calculation of identical integrals; testing two-electron integrals quickly to see if they are small enough to be neglected (as is the case for functions on distant nuclei; this decreases the time of a calculation from an n^4 dependence on the number of basis function to about an $n^{2.3}$ dependence); recalculating integrals to avoid the bottleneck of hard-drive access (direct SCF, section 5.3.2); representing the MOs as a set of gridpoints in space (in addition to a basis set expansion), which eliminates the need to explicitly calculate two-electron integrals. This *pseudospectral method* speeds up *ab initio* calculations by a factor of perhaps three or four. Methods of speeding up calculations are explained, with references to the literature, by Levine [27].

The method of calculating wavefunctions and energies that has been described in this chapter applies to *closed-shell*, *ground-state* molecules. The Slater determinant we started with (Eq. (5.12)) applies to molecules in which the electrons are fed pairwise into the MO's, starting with the lowest-energy MO; this is in contrast to free radicals, which have one or more unpaired electrons, or to electronically excited molecules, in which an electron has been promoted to a higher-level MO (e.g. Fig. 5.9, neutral triplet). The HF method outlined here is based on closed-shell Slater determinants and is called the *restricted* HF method or RHF method; "restricted" means that the electrons of α spin are forced to occupy (restricted to) the same spatial orbitals as those of β spin: inspection of Eq. (5.12) shows that we do not have a set of α *spatial* orbitals and a set of β *spatial* orbitals. If unqualified, a HF (i.e. an SCF) calculation means an RHF calculation.

The common way to treat free radicals is with the *unrestricted* HF method or UHF method. In this method, we employ separate spatial orbitals for the α and the β electrons, giving two sets of MOs, one for α and one for β electrons. Less commonly, free radicals are treated by the *restricted open-shell* HF or ROHF method, in which electrons occupy MO's in pairs as in the RHF method, except for the unpaired electron(s). The theoretical treatment of open-shell species is discussed in [1, 10, 1(k, *l*)], in particular, compare the performance of the UHF and ROHF methods.

Excited states, and those unusual molecules with electrons of opposite spin singly occupying different spatial MO's (open-shell singlets) cannot be properly treated with a single-determinant wavefunction. They must be handled with approaches beyond the HF level, such as configuration interaction (section 5.4).

5.3   BASIS SETS

5.3.1   Introduction

We encountered basis sets in sections 4.3.3, and 4.4.1a, and 5.2.3.6a. A basis set is a set of mathematical functions (basis functions), linear combinations of which yield molecular orbitals, as shown in Eqs (5.51) and (5.52). The functions are usually, but not invariably , centered on atomic nuclei (Fig. 5.7). Approximating molecular orbitals as linear combinations of basis functions is usually called the LCAO or linear combination of atomic orbitals approach, although the functions are not necessarily conventional atomic orbitals: they can be any set of mathematical functions that are convenient to manipulate and which in linear combination give useful representations of MOs. With this reservation, LCAO is a useful acronym. Physically, several (usually) basis functions describe the electron distribution around an atom and combining atomic basis functions yields the electron distribution in the molecule as a whole. Basis functions not centered on atoms (occasionally used) can be considered to lie on "ghost atoms"; see basis set superposition error, section 5.4.3.

The simplest basis sets are those used in the SHM and EHM (chapter 4). As applied to conjugated organic compounds (its usual domain), the simple Hückel basis set consists of just *p* atomic orbitals (or "geometrically *p*-type" atomic orbitals, like a lone-pair orbital which can be considered not to interact with the σ framework). The extended Hückel basis set consists of only the atomic *valence* orbitals. In the SHM, we do not worry about the mathematical form of the basis functions, reducing the interactions between them to 0 or -1 in the SHM Fock matrix (e.g. Eqs (4.62) and (4.64)). In the EHM the valence atomic orbitals are represented as Slater functions (section 4.4.1a).

5.3.2 Gaussian functions; basis set preliminaries; direct SCF

The electron distribution around an atom can be represented in several ways. Hydrogen-like functions based on solutions of the Schrödinger equation for the hydrogen atom, polynomial functions with adjustable parameters, Slater functions (Eq. (5.95)), and Gaussian functions (Eq. (5.96)) have all been used [28]. Of these, Slater and Gaussian functions are mathematically the simplest, and it is these that are currently used as the basis functions in molecular calculations. Slater functions are used in semiempirical calculations, like the EHM (section 4.4) and other semiempirical methods (chapter 6). Modern molecular *ab initio* programs employ Gaussian functions.

Slater functions are good approximations to atomic wavefunctions and would be the natural choice for *ab initio* basis functions, were it not for the fact that the evaluation of certain two-electron integrals requires excessive computer time if Slater functions are used. The two-electron integrals (sections 5.2.3.6c, e) of the **G** matrix (Eq. (5.100)) involve four functions, which may be on from one to four centers (normally atomic nuclei). Those two-electron integrals with three or four different functions $((rs|tt), (rs|rt)$ and $(rs|tu))$ and three or four nuclei (three- or four-center integrals) are extremely difficult to calculate with Slater functions, but are readily evaluated with Gaussian basis functions. The reason is that the product of two Gaussians on two centers is a Gaussian on a third center. Consider an *s*-type Gaussian centered on nucleus A and one on nucleus B; we are considering *real* functions, which is what basis functions normally are:

$$g_A = a_A e^{-\alpha_A |\mathbf{r}-\mathbf{r_A}|^2}, \quad g_B = a_B e^{-\alpha_B |\mathbf{r}-\mathbf{r_B}|^2} \tag{5.153}$$

where

$$|\mathbf{r}-\mathbf{r_A}|^2 = (x-x_A)^2 + (y-y_A)^2 + (z-z_A)^2 \quad \text{and}$$
$$|\mathbf{r}-\mathbf{r_B}|^2 = (x-x_B)^2 + (y-y_B)^2 + (z-z_B)^2 \tag{5.154}$$

with the nuclear and electron positions in Cartesian coordinates (if these were not *s*-type functions, the preexponential factor would contain one or more cartesian variables to give the function (the "orbital") nonspherical shape). It is not hard to show that

$$g_A g_B = a_C e^{-\alpha_C |\mathbf{r}-\mathbf{r_C}|^2} = g_C \tag{5.155}$$

The product of g_A and g_B is the Gaussian g_C, centered at $\mathbf{r_C}$. Now consider the general electron-repulsion integral

$$(rs|tu) = \iint \frac{\phi_r^*(1)\phi_s(1)\phi_t^*(2)\phi_u(2)}{r_{12}} \, dv_1 \, dv_2 \qquad (5.156 = 5.73)$$

If each basis function ϕ were a single, real Gaussian, then from Eq. (5.155) this would reduce to

$$(v|w) = \iint \frac{\phi_v(1)\phi_w(2)}{r_{12}} \, dv_1 \, dv_2 \qquad (5.157)$$

i.e. three- and four-center two-electron integrals with four basis functions would immediately simplify to tractable two-center integrals with two functions.

Actually, things are a little more complicated. A single Gaussian is a poor approximation to the nearly ideal description of an atomic wavefunction that a Slater function provides. Figure 5.12 shows that a Gaussian (designated STO-1G) is rounded near $r = 0$ while a Slater function has a cusp there (zero slope vs. a finite slope at $r = 0$); the Gaussian also decays somewhat faster than the Slater function at large r. The solution to the problem of this poor functional behaviour is to use several Gaussians to approximate a Slater function. In Fig. 5.12 a single Gaussian and a linear combination of three Gaussians have been used to approximate the Slater function shown; the nomenclature STO-1G and STO-3G mean "Slater-type orbital (approximated by) one Gaussian" and "Slater-type orbital (approximated by) three Gaussians", respectively. The Slater function shown is one suitable for a hydrogen atom in a molecule ($\zeta = 1.24$ [25]) and the Gaussians are the best fit to this Slater function. STO-1G functions were used in our illustrative HF calculation on HHe$^+$ (section 5.2.3.6e), and the STO-3G function is the smallest basis function used in standard *ab initio* calculations by commercial programs. Three Gaussians are a good speed vs. accuracy compromise between two and four or more [25].

The STO-3G basis function in Fig. 5.12 is a *contracted Gaussian* consisting of three *primitive Gaussians* each of which has a *contraction coefficient* (0.4446, 0.5353 and 0.1543). Typically, an *ab initio* basis function consists of a set of primitive Gaussians bundled together with a set of contraction coefficients. Now consider the two-electron integral $(rs|tu)$ (Eq. (5.156)). Suppose each basis function is an STO-3G contracted Gaussian, i.e.

$$\phi_r = d_{1r}g_{1r} + d_{2r}g_{2r} + d_{3r}g_{3r} \qquad (5.158)$$

and analogously for ϕ_s, ϕ_t, and ϕ_u. Then it is easy to see that

$$(rs|tu) = \iint d_{1r}d_{1s}g_{1r1s}\frac{1}{r_{12}}d_{1t}d_{1u}g_{1t1u} \, dv_1 \, dv_2$$
$$+ \iint d_{1r}d_{1s}g_{1r1s}\frac{1}{r_{12}}d_{1t}d_{2u}g_{1t2u} \, dv_1 \, dv_2 + \cdots$$
$$+ \iint d_{3r}d_{3s}g_{3r3s}\frac{1}{r_{12}}d_{3t}d_{3u}g_{3t3u} \, dv_1 \, dv_2 \qquad (5.159)$$

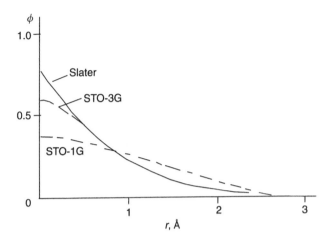

$$\phi(\text{Slater}) = \left(\frac{\zeta^3}{\Pi}\right)^{1/2} e^{-\zeta r} = 0.7790\, e^{-1.24r}$$

$$\phi(\text{STO-1G}) = \left(\frac{2\alpha}{\Pi}\right)^{3/4} e^{-\alpha r^2} = 0.3696\, e^{-0.4166r^2}$$

$$\phi(\text{STO-3G}) = 0.4446 \left(\frac{2\alpha}{\Pi}\right)^{3/4} e^{-\alpha r^2} + 0.5353 \left(\frac{2\alpha}{\Pi}\right)^{3/4} e^{-\alpha r^2} + 0.1543 \left(\frac{2\alpha}{\Pi}\right)^{3/4} e^{-\alpha r^2}$$

$$= 0.0835\, e^{-01689r^2} + 0.2678\, e^{-0.6239r^2} + 0.2769\, e^{-3.4253r^2}$$

Figure 5.12. Comparison of Slater, STO-1G and STO-3G functions for hydrogen. The Slater function shown is the most appropriate one for hydrogen *in a molecular environment*, and the Gaussians are the best 1-G and 3-G fits to this Slater function. Slater and Gaussian functions are usually characterized by parameters designated ζ (zeta) and α, respectively, as shown [25].

where $g_{1r1s} = g_{1r} \times g_{1s}$ and so on. Thus with contracted Gaussians as basis functions, each two-electron integral becomes a sum of easily calculated two-center two-electron integrals. Gaussian integrals can be evaluated so much faster than Slater integrals that the use of contracted Gaussians instead of Slater functions speeds up the calculation of the integrals enormously, despite the larger number of integrals. Discussions of the number of integrals in an *ab initio* calculation usually refer to those at the contracted Gaussian level, rather than the greater number engendered by the use of primitive Gaussians; thus the program Gaussian 92 [23] says that both an STO-1G and an STO-3G calculation on water use the same number (144) of two-electron integrals, although the latter clearly involves more "primitive integrals." The fruitful suggestion to use Gaussians in molecular calculations came from Boys (1950 [29]); it played a major role in making *ab initio* calculations practical, and this is epitomized in the names of the Gaussian series of programs, e.g. Gaussian 92 [23], which are possibly the most widely-used *ab initio* programs.

Fast calculation of integrals is particularly important for the two-electron integrals, as their number increases rapidly with the size of the molecule and the basis set (basis sets are discussed in section 5.3.3). Consider a calculation on water with an STO-1G basis set (and bear in mind that the smallest basis set normally used in *ab initio* calculations is the STO-3G set). In a standard *ab initio* calculation we use at least one basis function for each core orbital and each valence-shell orbital. Thus the oxygen requires five basis functions, for the $1s$, $2s$, $2p_x$, $2p_y$ and $2p_z$ orbitals; we can designate these functions ϕ_1, ϕ_2, ..., ϕ_5, and denote the 1s hydrogen functions, one for each H, ϕ_6 and ϕ_7. In computational chemistry atoms beyond hydrogen and helium in the periodic table are called "heavy atoms", and the computational "first row" is lithium–neon. With experience, the number of heavy atoms in a molecule gives a quick indication of about how many basis functions will be invoked by a specified basis set. Following the procedure for HHe$^+$ in Eq. (5.106):

$$G_{11} = \sum_{t=1}^{7} \sum_{u=1}^{7} P_{tu} \left[(11|tu) - \frac{1}{2}(1u|t1) \right]$$

Now u runs from 1 to 7 and t from 1 to 7, so G_{11} will consist of 49 terms, each containing two two-electron integrals for a G_{11} total of 98 integrals. The Fock matrix for seven basis functions is a 7×7 matrix with 49 elements, $G_{11}, G_{12}, \ldots, G_{17}, \ldots, G_{77}$, so apparently there are $49 \times 98 = 4802$ two-electron integrals. Actually, many of these are duplicates ($G_{ij} = G_{ji}$, so an $n \times n$ Fock matrix has only about $n^2/2$ *different* elements), differ from other integrals only in sign, or are very small, and the number of unique nonvanishing two-electron integrals is 119 (calculated with Gaussian 92 [23]). For an STO-1G calculation on hydrogen peroxide (12 basis functions), there are ca. 700 unique nonvanishing two-electron integrals (cf. a naive theoretical maximum of 41472). The usual formula for estimating the maximum number of unique two-electron integrals for a set of m real basis functions derives from the fact that there are four basis functions in each integral and $(rs|tu)$ is eightfold degenerate (Eq. (5.109)); this approximates the maximum number of these integrals as

$$N_{\max} = m^4/8 \tag{5.160}$$

In the above calculations the symmetry of water (C_{2v}) and hydrogen peroxide (C_{2h}) plays an important role in reducing the number of integrals which must actually be calculated, and modern *ab initio* programs recognize and utilize symmetry where it can be used (most molecules lack symmetry, but the small molecules of particular theoretical interest usually possess it), and are also able to recognize and avoid calculating integrals below a threshold size. Nevertheless the rapid rise in the number of 2-electron integrals with molecular and basis set size portends problems for *ab initio* calculations. An *ab initio* calculation on aspirin, a fairly small ($C_9H_8O_4$, 13 heavy atoms) molecule of practical interest, using the 3-21G basis set (section 5.3.3), which is the smallest that is usually used, requires 133 basis functions, which from Eq. (5.160) could invoke up to 39 million ($133^4/8$) two-electron integrals. Clearly, a modest *ab initio* calculation could require tens of millions of integrals. Information on molecular size, symmetry, basis sets and number of integrals is summarized in Table 5.2 (the 3-21G basis set is

Table 5.2. Molecular size, number of basis functions, and number of two-electron integrals

		Basis functions		Two-electron integrals from			
		STO-3G	3-21G	$m^4/8$	G92*	$m^4/8$	G92
HHe+	$C_{\infty v}$	2	4	2	6	32	55
H_2O	C_{2v}	7	13	300	144	3570	1314
H_2O_2	C_{2h}	12	22	2592	738	29282	7713
H_2O_2	C_1*	12	22	2592	2774	29282	28791
H_2O_3	C_{2v}	17	31	10440	3421	115440	31475
H_2O_3	C_1	17	31	10440	11046	115440	107869

* The coordinates of one of the atoms was altered slightly to get this unnatural symmetry.

explained in section 5.3.3). Note that for those molecules with no symmetry (C_1), the number of two-electron integrals calculated from Eq. (5.160) is about the same as that actually calculated by Gaussian 92.

There are two problems with so many two-electron integrals: the time needed to calculate them, and where to store them. Solutions to the first problem are, as explained, to use Gaussian functions, to utilize symmetry where possible, and to ignore those integrals that a preliminary check reveals are "vanishing". The other problem can be dealt with by storing the integrals in the RAM (the random access memory, i.e. the electronic memory), storing the integrals on the hard drive, or not storing them at all, but rather calculating them as they are required. Calculating all the integrals at the outset and storing them somewhere is called *conventional scf*, being the earlier-used procedure. The latter procedure of calculating only those two-electron integrals needed at the moment, and recalculating them again when necessary, is called *direct scf* (presumably using "direct" in the sense of "just now" or "at the moment"). Calculating all the two-electron the integrals and storing them in the RAM is the fastest approach, since it requires them to be calculated only once, and accessing information from the electronic memory is fast. However, RAM cannot yet store as many integrals as the hard drive. A (currently) respectable memory of 1 GB can store all the integrals generated by perhaps about 1000 basis functions (up to about 50 million); beyond this the computer essentially grinds to a halt. The capacity of the hard drive is typically considerably greater than that of the RAM (say, 50 GB for a respectable hard drive), and storing all the two-electron integrals on the hard drive is often a viable option, but suffers from the disadvantage that the time taken to read data from a mechanical device like the hard drive into the RAM, where it can be used by the cpu, is much greater (perhaps a millisecond compared to a nanosecond) than the time needed to access the information were it stored in a purely electronic device like the RAM (which is the only alternative to direct scf in, e.g. Spartan [30]). For these reasons, *ab initio* calculations with many basis functions (beyond about 120, depending on the size of the RAM) nowadays use direct scf, despite the need to recalculate integrals [31]. These considerations will change with improvements in hardware, and the availability of very large electronic memories may make

Table 5.3. Effect of basis set and symmetry on times for single point, geometry optimization and geometry optimization + frequencies jobs on acetone, $(CH_3)_2CO$

Basis set	Single point Time, s		Geometry optimization Time, s		Geometry optimization + frequencies Time, s	
	C_{2v}	C_1	C_{2v}	C_1	C_{2v}	C_1
STO-3G	0.2	0.2	2	7	13	59
3-21G	0.3	0.5	2	5	20	75
6-31G*	2	3	15	54	172	586

The starting geometry for the *ab initio* jobs was a molecular mechanics (MMFF) one. The C_{2v} geometry is that with two C–H/C=O eclipsed arrangements (the global minimum). The C_1 symmetry starting geometry was obtained by rotating one C–C bond very slightly in the C_{2v} precursor molecular mechanics structure (after MM optimization). These calculations were done with Spartan [30] on a Pentium 4 machine. For times of only a few seconds, the speed advantage of the STO-3G basis over the 3-21G is not always evident.

storage of all the two-electron integrals in RAM the method of choice for *ab initio* calculations.

5.3.3   Types of basis sets and their uses

We have met the STO-1G (section 5.2.3.6e and 5.3.2) and STO-3G (section 5.3.2) basis sets. We saw that a single Gaussian gives a poor representation of a Slater function, but that this approximation can be improved by using a linear combination of Gaussians (Fig. 5.12). In this section the basis sets commonly used in *ab initio* calculations are described and their domains of utility are outlined. Note that the STO-1G basis, although it was useful for our illustrative purposes, is not used in research calculations (Figure 5.12 shows how poorly it approximates a Slater function). We will consider the STO-3G, 3-21G, 6-31G*, and 6-311G* basis sets, which, with variations obtained by adding polarization (*) and diffuse (+) functions, are the most widely-used; other sets will be briefly mentioned. Information on basis sets is summarized in Table 5.3. Good discussions of currently popular basis sets are given in, e.g. [1a,e 1i]; the compilations by Hehre *et al.* [1g,32] are extensive and critically evaluated.

The basis sets described here are those developed by Pople[3] and coworkers [33], which are probably the most popular now, but all general-purpose (those not used just on small molecules or on atoms) basis sets utilize some sort of contracted Gaussian functions to simulate Slater orbitals. A brief discussion of basis sets and references to many, including the widely-used Dunning and Huzinaga sets, is given by Simons and Nichols [34]. There is no one procedure for developing a basis set. One method is

---

[3] John Pople, born in Burnham-on-Sea, Somerset, England, 1925. Ph.D. (Mathematics) Cambridge, 1951. Professor, Carnegie-Mellon University, 1960–1986, Northwestern University (Evanston, Illinois) 1986–present. Nobel Prize in chemistry 1998 (with Walter Kohn, chapter 5, section 7.1).

to optimize *Slater* functions for atoms or small molecules, i.e. to find the values of ζ that give the lowest energy for these, and then to use a least-squares procedure to fit contracted Gaussians to the optimized Slater functions [35].

STO-3G

This is a called a *minimal basis set*, although some atoms actually have more basis functions (which for this basis can be equated with atomic orbitals) than are needed to accommodate all their electrons. For the earlier part of the periodic table (hydrogen to argon) each atom has one basis function corresponding to its usual atomic orbital description, with the proviso that the orbitals used by the later atoms of a row are available to all those of the row. A hydrogen or helium atom has a $1s$ basis function. Each "first-row" atom (lithium to neon) has a $1s$, a $2s$, and a $2p_x$, $2p_y$ and $2p_z$ function, giving 5 basis functions for each of these atoms: although lithium and beryllium are often not thought of as using p orbitals, all the atoms of this row are given the same basis, because this has been found to work better than a literally minimum basis set. Second-row atoms (sodium to argon) have a $1s$ and a $2s$, as well as three $2p$ functions, plus a $3s$ and three $3p$ functions, giving 9 basis functions. In the third row, potassium and calcium, as expected, have the 9 functions of the previous row, plus a $4s$ and three $4p$ functions, for a total of 13 basis functions. Starting with the next element, scandium, five $3d$ orbitals are added, so that scandium to krypton have $13 + 5 = 18$ basis functions. The STO-3G basis is summarized in Fig. 5.13(a).

The STO-3G basis introduces us to the concept of *contraction shells* in constructing contracted Gaussians from primitive Gaussians (section 5.3.2). The Gaussians of a contraction shell share common exponents. Carbon, e.g. has one s shell and one sp shell. This means that the $2s$ and $2p$ Gaussians (belonging to the $2sp$ shell) share common α exponents (which differ from those of the $1s$ function). Consider the contracted Gaussians

$$\phi(2s) = d_{1s}e^{-\alpha_{1s}r} + d_{2s}e^{-\alpha_{2s}r} + d_{3s}e^{-\alpha_{3s}r}$$

$$\phi(2p_x) = d_{1p}xe^{-\alpha_{1p}r} + d_{2p}xe^{-\alpha_{2p}r} + d_{3p}xe^{-\alpha_{3p}r}$$

$$\phi(2p_y) = d_{1p}ye^{-\alpha_{1p}r} + d_{2p}ye^{-\alpha_{2p}r} + d_{3p}ye^{-\alpha_{3p}r}$$

$$\phi(2p_z) = d_{1p}ze^{-\alpha_{1p}r} + d_{2p}ze^{-\alpha_{2p}r} + d_{3p}ze^{-\alpha_{3p}r}$$

The usual practice is to set $\alpha_{1s} = \alpha_{1p}$, $\alpha_{2s} = \alpha_{2p}$, and $\alpha_{3s} = \alpha_{3p}$. Using common α's for the s and p primitives reduces the number of distinct integrals that must be calculated. An STO-3G calculation on CH_4, for example, involves 9 basis functions (5 for C and 1 for each H) in 6 shells: for C one s (i.e. a $1s$) shell, and one sp (i.e. a $2s$ plus $2p$) shell, and for each H one s (i.e. a $1s$) shell. The current view is that the STO-3G basis is not very good, and it would normally be considered unacceptable for research. Nevertheless, one hesitates to endorse Dewar and Storch's assertion that "it must be considered obsolete" [36]. We do not know how many publications report work which began with a preliminary and unreported but valuable investigation using this basis. Its advantages are speed (it is probably the smallest basis set that would even be considered for an *ab initio* calculation) and the ease with which the molecular orbitals can be dissected into atomic orbital contributions. The STO-3G basis is

(a)

$_1$H		$_2$He
1s		1s
1 function		1 function
	$_3$Li–$_{10}$Ne	
	1s	
	2s 2p 2p 2p	
	5 functions	
	$_{11}$Na–$_{18}$Ar	
	1s	
	2s 2p 2p 2p	
	3s 3p 3p 3p	
	9 functions	
$_{19}$K–$_{20}$Ca	$_{21}$Sc–$_{30}$Zn	$_{31}$Ga–$_{36}$Kr
1s	1s	1s
2s 2p 2p 2p	2s 2p 2p 2p	2s 2p 2p 2p
3s 3p 3p 3p	3s 3p 3p 3p	3s 3p 3p 3p
4s 4p 4p 4p	4s 4p 4p 4p	4s 4p 4p 4p
13 functions	3d 3d 3d 3d 3d	3d 3d 3d 3d 3d
	18 functions	18 functions
$_{37}$Rb–$_{38}$Sr	$_{39}$Y–$_{48}$Cd	$_{49}$In–$_{54}$Xe
1s	1s	1s
2s 2p 2p 2p	2s 2p 2p 2p	2s 2p 2p 2p
3s 3p 3p 3p	3s 3p 3p 3p	3s 3p 3p 3p
4s 4p 4p 4p	4s 4p 4p 4p	4s 4p 4p 4p
5s 5p 5p 5p	5s 5p 5p 5p	5s 5p 5p 5p
3d 3d 3d 3d 3d	3d 3d 3d 3d 3d	3d 3d 3d 3d 3d
22 functions	4d 4d 4d 4d 4d	4d 4d 4d 4d 4d
	27 functions	27 functions

Figure 5.13. STO-3G basis set.

(b)

$_1$H		$_2$He
1s'		1s'
1s''		1s''
2 functions		2 functions
	$_3$Li–$_{10}$Ne	
	1s	
	2s' 2p' 2p' 2p'	
	2s'' 2p'' 2p'' 2p''	
	9 functions	
	$_{11}$Na–$_{18}$Ar	
	1s	
	2s 2p 2p 2p	
	3s' 3p' 3p' 3p'	
	3s'' 3p'' 3p″ 3p''	
	13 functions	
$_{19}$K–$_{20}$Ca	$_{21}$Sc–$_{30}$Zn	$_{31}$Ga–$_{36}$Kr
1s	1s	1s
2s 2p 2p 2p	2s 2p 2p 2p	2s 2p 2p 2p
3s 3p 3p 3p	3s 3p 3p 3p	3s 3p 3p 3p
4s' 4p' 4p' 4p'	4s' 4p' 4p' 4p'	4s' 4p' 4p' 4p'
4s'' 4p'' 4p'' 4p''	4s'' 4p'' 4p'' 4p''	4s'' 4p'' 4p'' 4p''
17 functions	3d' 3d' 3d' 3d' 3d' 3d'	3d 3d 3d 3d 3d 3d
	3d''3d'' 3d'' 3d'' 3d'' 3d''	23 functions
	29 functions	
$_{37}$Rb–$_{38}$Sr	$_{39}$Y–$_{48}$Cd	$_{49}$In–$_{54}$Xe
1s	1s	1s
2s 2p 2p 2p	2s 2p 2p 2p	2s 2p 2p 2p
3s 3p 3p 3p	3s 3p 3p 3p	3s 3p 3p 3p
4s 4p 4p 4p	4s 4p 4p 4p	4s 4p 4p 4p
5s' 5p' 5p' 5p'	5s' 5p' 5p' 5p'	5s' 5p' 5p' 5p'
5s'' 5p'' 5p'' 5p''	5s'' 5p'' 5p'' 5p''	5s'' 5p'' 5p'' 5p''
3d 3d 3d 3d 3d 3d	3d 3d 3d 3d 3d 3d	3d 3d 3d 3d 3d 3d
27 functions	4d' 4d' 4d' 4d' 4d' 4d'	4d 4d 4d 4d 4d 4d
	4d'' 4d'' 4d'' 4d'' 4d'' 4d''	33 functionsy
	39 functions	

Figure 5.13. 3-21G basis set.

(c)

$_1$H		$_2$He
1s'		1s'
1s''		1s''
2 functions		2 functions
	$_3$Li–$_{10}$Ne	
	1s	
	2s' 2p' 2p' 2p'	
	2s'' 2p'' 2p'' 2p''	
	9 functions	
	$_{11}$Na–$_{18}$Ar	
	1s	
	2s 2p 2p 2p	
	3s' 3p' 3p' 3p'	
	3s'' 3p'' 3p'' 3p''	
	3d 3d 3d 3d 3d 3d	
	19 functions	
$_{19}$K–$_{20}$Ca	$_{21}$Sc–$_{30}$Zn	$_{31}$Ga–$_{36}$Kr
1s	1s	1s
2s 2p 2p 2p	2s 2p 2p 2p	2s 2p 2p 2p
3s 3p 3p 3p	3s 3p 3p 3p	3s 3p 3p 3p
4s' 4p' 4p' 4p'	4s' 4p' 4p' 4p'	4s' 4p' 4p' 4p'
4s'' 4p'' 4p'' 4p''	4s'' 4p'' 4p'' 4p''	4s'' 4p'' 4p'' 4p''
3d 3d 3d 3d 3d 3d	3d' 3d' 3d' 3d' 3d' 3d'	3d' 3d' 3d' 3d' 3d' 3d'
23 functions	3d'' 3d'' 3d'' 3d'' 3d'' 3d''	3d' 3d' 3d' 3d' 3d' 3d'
	29 functions	29 functions
$_{37}$Rb–$_{38}$Sr	$_{39}$Y–$_{48}$Cd	$_{49}$In–$_{54}$Xe
1s	1s	1s
2s 2p 2p 2p	2s 2p 2p 2p	2s 2p 2p 2p
3s 3p 3p 3p	3s 3p 3p 3p	3s 3p 3p 3p
4s 4p 4p 4p	4s 4p 4p 4p	4s 4p 4p 4p
5s' 5p' 5p' 5p'	5s' 5p' 5p' 5p'	5s' 5p' 5p' 5p'
5s'' 5p'' 5p'' 5p''	5s'' 5p'' 5p'' 5p''	5s'' 5p'' 5p'' 5p''
3d 3d 3d 3d 3d 3d	3d 3d 3d 3d 3d 3d	3d 3d 3d 3d 3d
4d 4d 4d 4d 4d 4d	4d' 4d' 4d' 4d' 4d' 4d'	4d' 4d' 4d' 4d' 4d' 4d'
33 functions	4d'' 4d'' 4d'' 4d'' 4d'' 4d''	4d'' 4d'' 4d'' 4d'' 4d''
	39 functions	4d''
		39 functions

Figure 5.13. 3-21G$^{(*)}$ basis set.

(d)

$_1$H		$_2$He
1s'		1s'
1s''		1s''
2 functions		2 functions
	$_3$Li–$_{10}$Ne	
	1s	
	2s' 2p' 2p' 2p'	
	2s'' 2p'' 2p'' 2p''	
	3d 3d 3d 3d 3d 3d	
	15 functions	
	$_{11}$Na–$_{18}$Ar	
	1s	
	2s 2p 2p 2p	
	3s' 3p' 3p' 3p'	
	3s'' 3p'' 3p'' 3p''	
	3d 3d 3d 3d 3d 3d	
	19 functions	

Figure 5.13. 6-31G* basis set.

roughly twice as fast as the next larger commonly used one, the 3-21G$^{(*)}$ (Table 5.3). Sophisticated semiempirical methods (chapter 6) are perhaps more likely to be used nowadays in preliminary investigations, and to obtain reasonable starting structures for *ab initio* optimizations, but for systems significantly different from those for which the semiempirical methods were parameterized one might prefer to use the STO-3G basis. As for examining atomic contributions to bonding, interpreting bonding in terms of hybrid orbitals and the contribution of particular atoms to MO's is simpler when each atom has just one conventional orbital, rather than split orbitals (as in the basis sets to be discussed). Thus a fairly recent analysis of the electronic structure of three- and four-membered rings used the STO-3G basis explicitly for this reason [37], as did an interpretation of the bonding in the unusual molecule pyramidane [38].

The shortcomings (and virtues) of the STO-3G basis are extensively documented throughout in [1g]. Basically, the drawbacks are that by comparison with the 3-21G$^{(*)}$ basis, which is not excessively more demanding of time, it gives significantly less accurate geometries and energies (this was the reason for the call to abandon this basis [36]). Actually, even for second-row atoms (Na–Ar), where the defects of such a small basis set should be, and are, most apparent, the STO-3G basis supplemented with five

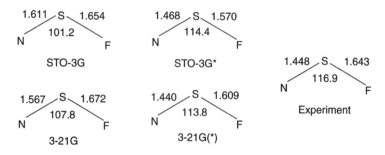

Figure 5.14. Some STO-3G, STO-3G*, 3-21G and 3-21G$^{(*)}$ geometries.

d or *polarization* functions (the STO-3G* basis; polarization functions are discussed below) can give results comparable to those of the 3-21G$^{(*)}$ basis set. Thus for the S–O bond length of Me$_2$SO we get (Å): STO-3G, 1.820; STO-3G*, 1.480; 3-21G, 1.678; 3-21G$^{(*)}$, 1.490; exp., 1.485, and for NSF [39] the geometries shown in Fig. 5.14.

3-21G and 3-21G$^{(*)}$ Split valence and double-zeta basis sets
These bases (the 3-21G$^{(*)}$ basis is described after the "pure" 3-21G) split each valence orbital into two parts, an inner shell and an outer shell. The basis function of the inner shell is represented by two Gaussians, and that of the outer shell by one Gaussian (hence the "21"); the core orbitals are each represented by one basis function, each composed of three Gaussians (hence the "3"). Thus H and He have a 1s orbital (the only valence orbital for these atoms) split into 1s' (1s inner) and 1s'' (1s outer), for a total of 2 basis functions. Carbon has a 1s function represented by three Gaussians, an inner $2s, 2p_x, 2p_y$ and $2p_z$ ($2s', 2p', 2_xp'2p'_y$) function, each composed of two Gaussians, and an outer $2s, 2p_x, 2p_y$ and $2p_z$ ($2s'', 2p''_x, 2p''_y2p''_z$) function, each composed of one Gaussian, making 9 basis functions. The terms inner and outer derive from the fact that the Gaussian of the outer shell has a smaller α than the Gaussians of the inner shell, and so the former function falls off more slowly, i.e. it is more diffuse and effectively spreads out further, into the outer regions of the molecule. The purpose of splitting the valence shell is to give the SCF algorithm more flexibility in adjusting the contributions of the basis functions to the molecular orbitals, thus achieving a more realistic simulated electron distribution. Consider carbene, CH$_2$ (Fig. 5.15). We can denote the basis functions $\phi_1 - \phi_{13}$:

C1s : ϕ_1

C2s', $2p'_x, 2p'_y, 2p'_z$: $\phi_2, \phi_3, \phi_4, \phi_5$ (inner valence shell)

C2s'', $2p''_x, 2p''_y, 2p''_z$: $\phi_6, \phi_7, \phi_8, \phi_9$ (outer valence shell)

H$_1$1s': ϕ_{10} (inner shell)

H$_1$1s'': ϕ_{11} (outer shell)

H$_2$1s': ϕ_{12} (inner shell)

H$_2$1s'': ϕ_{13} (outer shell)

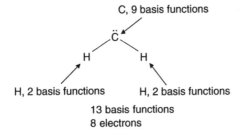

C, 9 basis functions

H, 2 basis functions   H, 2 basis functions

13 basis functions
8 electrons

Figure 5.15. Carbene, with 3-21G basis functions.

Thirteen basis functions ("atomic orbitals") give thirteen LCAO MOs:

$$\psi_1 = c_{11}\phi_1 + c_{21}\phi_2 + \cdots + c_{13,1}\phi_{13}$$
$$\psi_2 = c_{12}\phi_1 + c_{22}\phi_2 + \cdots + c_{13,2}\phi_{13}$$

$$\vdots$$

$$\psi_{13} = c_{1,13}\phi_1 + c_{2,13}\phi_2 + \cdots + c_{13,13}\phi_{13}$$

Note that since there are thirteen MO's but only eight electrons to be accommodated, only the first four MOs ($\psi_1 - \psi_4$) are occupied (recall that we are talking about closed-shell molecules in the ground electronic state). The nine empty MOs are called *unoccupied* or *virtual molecular orbitals*. We shall see that virtual MOs are important in certain kinds of calculations. Now, in the course of the SCF process the coefficients of the various inner- and outer-shell basis functions can be varied independently to find the best wavefunctions ψ (those corresponding to the lowest energy). As the iterations proceed some outer-shell functions, say, could be given greater (or lesser) empha- sis, independently of any inner-shell functions, allowing a finer-tuning of the electron distribution, and a lower energy, than would be possible with unsplit basis functions.

A still more malleable basis set is one with *all* the basis functions, not just those of the valence AO's, split; this is called a *double zeta* (double ζ) basis set (perhaps from the days before Gaussians, with $\exp(-\alpha \mathbf{r}^2)$, had almost completely displaced Slater functions with $\exp(-\zeta \mathbf{r})$ for molecular calculations). Double zeta basis sets are much less widely used than *split valence* sets, since the former are computationally more demanding and for many purposes only the contributions of the "chemically active" valence functions to the MOs need to be fine-tuned, and in fact "double zeta" is sometimes used to refer to split valence basis sets.

Lithium to neon have a $1s$ function and inner and outer $2s$, $2p_x$, $2p_y$ and $2p_z$ ($2s'$, $2s''$, ..., $2p_z''$) functions, for a total of 9 basis functions. These inhabit three contraction shells (see the STO-3G discussion): a $1s$, an sp inner and an sp outer contraction shell. Sodium to argon have a $1s$, a $2s$ and three $2p$ functions, and an inner and outer shell of $3s$ and $3p$ functions, for a total of $1 + 4 + 8 = 13$ basis functions. These are in four shells: a $1s$, an sp ($2s$, $2p$), an sp inner and an sp outer ($3s$ and $3p$ inner, $3s$ and $3p$ outer). Potassium and calcium have a $1s$, a $2s$ and three $2p$, and a $3s$ and three $3p$ functions, plus inner and outer $4s$ and $4p$ functions, for a total of $1+4+4+8 = 17$ basis

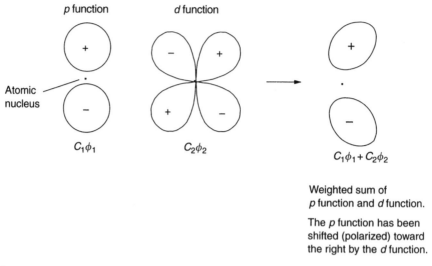

Weighted sum of
p function and d function.

The p function has been
shifted (polarized) toward
the right by the d function.

Figure 5.16. One basis function can be used to shift another in a given direction (to polarize it). In minimizing the energy, the program adjusts the relative contributions of the two functions to shift the electron density where it is needed to get the minimum energy. *p* Functions are also commonly used to polarize the *s* functions on hydrogen atoms, but the main use of polarization functions is the utilization of *d* functions on "heavy" atoms (atoms other than H and He).

functions. For the remaining atoms for which the 3-21G basis is available *d* functions are added. The 3-21G basis set is summarized in Fig. 5.13(b).

For molecules with atoms beyond the first row (beyond neon), this "pure" 3-21G basis tends to give poor geometries. This problem is largely overcome for second-row elements (sodium to argon) by supplementing this basis with *d* functions, called *polarization functions*. The term arises from the fact that *d* functions permit the electron distribution to be polarized (displaced along a particular direction), as shown in Fig. 5.16. Polarization functions enable the SCF process to establish a more anisotropic electron distribution (where this is appropriate) than would otherwise be possible (cf. the use of split valence basis sets to permit more flexibility in adjusting the inner and outer regions of electron density). The 3-21G basis set augmented where appropriate with *d* functions is called the 3-21G$^{(*)}$ basis; the asterisk indicates polarization functions (*d* in this case), and the parentheses mean that the extra (compared to the "pure" 3-21G basis) polarization functions are present only beyond the first row. For H to Ne, the 3-21G and the 3-21G$^{(*)}$ basis sets are identical, and for these first-row atoms the term 3-21G, rather than 3-21G$^{(*)}$, is normally used. The 3-21G basis *without* the supplementary *d* polarization functions is not normally employed for atoms beyond neon: here the usual 3-21G-type basis set is the 3-21G$^{(*)}$. There is also a relatively little-used basis, the 3-21G* (no parentheses) in which a set of six *d* functions is added to the first-row elements, giving carbon, say, 15 rather than nine basis functions. The 3-21G$^{(*)}$ basis (with parentheses) is summarized in Fig. 5.13(c). *p*-Polarization functions can also be added to hydrogens and helium (below).

Examples of geometries calculated with the 3-21G and 3-21G$^{(*)}$ basis sets are shown in Fig. 5.14. The 3-21G$^{(*)}$ gives remarkably good geometries for such a small basis set, and in fact it is used for the geometry optimization step of some high-accuracy energy methods (section 5.5.2). Since it is roughly ten times as fast (Table 5.3) as the next widely-used basis, the 6-31G* (below) and is much less demanding of computer power, the 3-21G$^{(*)}$ basis set is a kind of workhorse for relatively big molecules; see for example a study using it for geometry optimization investigations of pericyclic reactions [40]. The somewhat similar but less popular 4-21G basis was used, with the 3-21G* basis on sulfur, for geometry optimization of the protein crambin, with 46 amino acid residues and 642 atoms. This represented 3597 basis functions, and the job took 260 days [41]. Even where geometry optimizations with larger bases are practical, a survey of the problem with the 3-21G* basis is sometimes useful (it is HF/3-21G$^{(*)}$ *geometries* rather than *relative energies* which are good; consistently getting good relative energies is a more challenging problem – see section 5.5.2).

*6-31G**

This is a split valence basis set with polarization functions (these terms were explained in connection with the 3-21$^{(*)}$ basis set, above). The valence shell of each atom is split into an inner part composed of three Gaussians and an outer part composed of one Gaussian ("31"), while the core orbitals are each represented by one basis function, each composed of six Gaussians ("6"). The polarization functions ($*$) are present on "heavy atoms" – those beyond helium. Thus, H and He have a $1s$ orbital represented by an inner $1s'$ and an outer $1s'$ basis function, making two basis functions. Carbon has a $1s$ function represented by six Gaussians, an inner $2s$, $2p_x$, $2p_y$ and $2p_z$ ($2s'$, $2p_x'$, $2p_y' 2p_z'$) function, each composed of three Gaussians, and an outer $2s$, $2p_x$, $2p_y$ and $2p_z$ ($2s''$, $2p_x''$, $2p_y'' 2p_z''$) function, each composed of one Gaussian, and six (not five) $3d$ functions, making a total of 15 basis functions. A 6-31G* calculation on CH_2 uses $15 + 2 + 2 = 19$ basis functions, and generates 19 MO's. In the closed-shell species the eight electrons occupy four of these MO's, so there are 15 unoccupied or virtual MO's; compare this with a 3-21G calculation on CH_2 (above) where there are a total of 13 MOs and nine virtual MOs. The 6-31G* basis, also often called 6-31G(d), is summarized in Fig. 5.13(d).

The 6-31G* is probably the most popular basis at present. It gives good geometries and, often, reasonable relative energies (section 5.5.2); however, there seems to be little evidence that it is, in general, much better than the 3-21G$^{(*)}$ basis for geometry optimizations. Since it is about 10 times as slow (Table 5.3) as the 3-21G$^{(*)}$ basis, the general preference for the 6-31G* for geometry optimizations may be due to its better relative energies (section 5.5.2). The 3-21G$^{(*)}$ basis *does* have certain geometry deficiencies compared to the 6-31G*, particularly its tendency to overzealously flatten nitrogen atoms (the N of aniline is wrongly predicted to be planar), and this, along with inferior relative energies and less consistency, may be responsible for its being neglected in favor of the 6-31G* basis set [42]. The virtues of the 3-21G$^{(*)}$ and 6-31G* basis sets for geometry optimizations are discussed further in section 5.5.1. Note that the geometries and energies referred to here are those from HF-level calculations. Post-HF (section 5.4) calculations, which can give significantly better geometries and much better relative energies (sections 5.5.1 and 5.5.2), require a basis set of at least the 6-31G* size for meaningful results.

The 6-31G* basis adds polarization functions only to so-called heavy atoms (those beyond helium). Sometimes it is helpful to have polarization functions on the hydrogens as well; a 6-31G* basis with three $2p$ functions on each H and He atom (in addition to their $1s'$ and $1s''$ functions) is called the 6-31G** (or 6-31(d,p)) basis. The 6-31G* and 6-31G** bases are the same except that in the 6-31G** each H and He has five, rather than two, functions. The 6-31G** basis may be preferable to the 6-31G* where the hydrogens are engaged in some special activity like hydrogen bonding or bridging [43]. In high-level calculations on hydrogen bonding or on boron hydrides, for example, polarization functions are placed on hydrogen. For calculations on and references to the hydrogen bonded water dimer, see section 5.4.3.

Diffuse functions
Core electrons or electrons engaged in bonding are relatively tightly bound to the molecular nuclear framework. Lone-pair electrons or electrons in a (previously) virtual orbital, are relatively loosely held, and are on the average at a larger distance from the nuclei than core or bonding electrons. These "expanded" electron clouds are found in molecules with heteroatoms, in anions, and in electronically excited molecules. To simulate well the behaviour of such species *diffuse functions* are used. These are Gaussian functions with small values of α; this causes $\exp(-\alpha r^2)$ to fall off very slowly with the distance $\mathbf{r}$ from the nucleus, so that by giving enough weight to the coefficients of diffuse functions the SCF process can generate significant electron density at relatively large distances from the nucleus. Typically a basis set with diffuse functions has one such function, composed of a single Gaussian, for each valence atomic orbital of the "heavy atoms". The 3-21 + G basis set for carbon (= 3-21 + G$^{(*)}$ for this element) is

$1s$
$2s'2p'2p'2p'$
$2s''2p''2p''2p''$
$2s+, 2p+, 2p+, 2p+$
13 basis functions

and the 6-31 + G* basis for carbon is

$1s$
$2s'2p'2p'2p'$
$2s''2p''2p''2p''$
$3d3d3d3d3d3d$
$2s+, 2p+, 2p+, 2p+$
19 basis functions

Sometimes diffuse functions are added to hydrogen and helium as well as to the heavy atoms; such a basis set is indicated by + +. The 3-21 + +G and 6-31 + +G basis for hydrogen and helium is

   $1s$

   $1s'$

   $1s+$

   3 basis functions

A 3-21++G calculation on CH_2 would use $13 + 3 + 3 = 19$ basis functions, a 6-31 + +G* calculation $19 + 3 + 3 = 25$ basis functions, and a 6-31 + +G** calculation $19 + 6 + 6 = 31$ basis functions.

There is some disagreement over when diffuse functions should be used. Certainly most workers employ them routinely in studying anions and excited states, but not ordinary lone pair molecules (molecules with heteroatoms, like ethers and amines). A reasonable recommendation is to study with and without diffuse functions species representative of the problem at hand, for which experimental results are known, and see if these functions help. A paper by Warner [43] gives useful references and a good account of the efficacy of diffuse functions in treating certain molecules with heteroatoms.

Large basis sets

The 3-21G$^{(*)}$ is a small basis set and the 6-31G* and 6-31G** are moderate-size basis sets. Of those we have discussed, only the 6-31G* and 6-31G** with diffuse functions (6-31+G*, 6-31++G*, 6-31+G** and 6-31++G**) might be considered *fairly* large. A large basis set might have a doubly-split or even triply-split valence shell with d, p and f, and maybe even g, functions on at least the heavy atoms. An example of a large (but not very large) basis set is the 6-311G** (i.e. 6-311(d,p)) set. This is a split valence set with each valence orbital split into three shells, composed of three, one and one Gaussian, while the core orbitals are represented by one basis function composed of six Gaussians; each heavy atom also has five (not six in this case) $3d$ functions and each hydrogen and helium has three $2p$ functions. The 6-311G** basis for carbon is then

   $1s$

   $2s'2p'2p'2p'$

   $2s''2p''2p''2p''$

   $2s'''2p'''2p'''2p'''$

   $3d3d3d3d3d$

   18 basis functions

and for hydrogen

   $1s'$

   $1s''$

   $1s'''$

   $2p2p2p$

   6 basis functions

Unequivocally large basis sets would be triply-split valence shell sets with d and f functions on heavy atoms and p functions on hydrogen. At the smaller end of such sets is the 6-311G(df,p) basis, with five $3d$'s and seven $4f$'s on each heavy atom and three $2p$'s on each hydrogen and helium. For carbon this is

  $1s$
  $2s'2p'2p'2p'$
  $2s''2p''2p''2p''$
  $2s'''2p'''2p'''2p'''$
  $3d3d3d3d3d$
  $4f4f4f4f4f4f4f$
  25 basis functions

and for hydrogen

  $1s'$
  $1s''$
  $1s'''$
  $2p2p2p$
  6 basis functions

A more impressive example of a large basis set would be 6-311G(3df,3pd). This has for each heavy atom three sets of five d functions and one set of seven f functions, and for each hydrogen and helium three sets of three p functions and one set of five d functions, i.e. for carbon

  $1s$
  $2s'2p'2p'2p'$
  $2s''2p''2p''2p''$
  $2s'''2p'''2p'''2p'''$
  $3d3d3d3d3d$
  $3d3d3d3d3d$
  $3d3d3d3d3d$
  $4f4f4f4f4f4f4f$
  35 basis functions

and for hydrogen

  $1s'$
  $1s''$
  $1s'''$
  $2p2p2p$
  $2p2p2p$
  $2p2p2p$
  $3d3d3d3d3d$
  17 basis functions

Note that all these large basis sets can be made still bigger by adding diffuse functions to heavy atoms (+) or to heavy atoms and hydrogen/helium (++). The number of basis functions on CH_2 using some small, medium and large bases is summarized (C+H+H):

STO-3G	$5 + 1 + 1 = 7$ functions
3-21G($=$ 3-21G$^{(*)}$ here)	$9 + 2 + 2 = 13$ functions
6-31G* (6-31G(d))	$15 + 2 + 2 = 19$ functions
6-31G**(6-31G(d,p))	$15 + 5 + 5 = 25$ functions
6-311G**(6-311G(d,p))	$18 + 6 + 6 = 30$ functions
6-311G(df,p)	$25 + 6 + 6 = 37$ functions
6-311G(3df,3pd)	$35 + 17 + 17 = 69$ functions
6-311++G(3df, 3pd)	$39 + 18 + 18 = 75$ functions

Large basis sets are used mainly for post-HF level (section 5.4) calculations, where the use of a basis smaller than the 6-31G* is essentially pointless. At the HF level the largest basis routinely used is the 6-31G* or 6-31G** (augmented if appropriate by diffuse functions), and post- HF geometry optimizations are frequently done using the 6-31G* or 6-31G** basis too. Use of the larger bases (6-311G** and up) tends to be confined to *single-point* calculations on structures optimized with a smaller basis set (section 5.5.2). These are not firm rules: the high-accuracy CBS (complete basis set) methods (section 5.5.2) use as part of their procedure single-point HF (rather than post-HF) level calculations with very large basis sets, and geometry optimizations with large basis sets were performed at both HF and post-HF levels in studies of the theoretically and experimentally challenging oxirene system [44].

Effective core potentials (pseudopotentials)

From about the third row (potassium to krypton) of the periodic table on the large number of electrons (19–36) has a considerable slowing effect on conventional *ab initio* calculations, because of the large number of two-electron repulsion integrals they engender. The usual way of avoiding this problem is to add to the Fock operator a one-electron operator that takes into account the effect of the core electrons on the valence electrons, which latter are still considered explicitly. This "average core effect" operator is called an effective core potential (ECP) or a pseudopotential. With a set of valence orbital basis functions optimized for use with it, it simulates the effect on the valence electrons of the atomic nuclei plus the core electrons. A distinction is sometimes made between an ECP and a pseudopotential, the latter term being used to mean any approach limited to the valence electrons, while ECP is sometimes used to designate a simplified pseudopotential corresponding to a function with fewer orbital nodes than the "correct" functions. However, the terms are often used interchangeably to designate a nuclei-plus-core electrons potential used with a set of valence functions, and that is what is meant here.

So far we have discussed *nonrelativistic ab initio* methods: they ignore those consequences of Einstein's theory of relativity that are relevant to chemistry (section 4.2.3). These consequences arise in the special (rather than the general) theory, from the dependence of mass on velocity [45]. This dependence causes the masses of the inner electrons of heavy atoms to be significantly greater than the electron rest mass; since the Hamiltonian operator in the Schrödinger equation contains the electron mass (Eqs (5.36) and (5.37)), this change of mass should be taken into account. Relativistic effects in heavy-atom molecules affect geometries, energies, and other properties. Relativity is accounted for in the relativistic form of the Schrödinger equation, the Dirac equation (interestingly, Dirac thought his equation would not be relevant to chemistry [46]). This equation is not commonly used explicitly in molecular calculations, but is instead used to develop [47] *relativistic effective core potentials* (relativistic pseudopotentials). Relativistic effects begin to become significant for about third-row elements, i.e. those for which ECPs begin to be useful for speeding up calculations, so it makes sense to take relativistic effects into account in developing these potential operators and their basis functions, and indeed ECPs are generally relativistic. Such ECPs can give accurate results for molecules with third-row-plus atoms by simulating the relativistic mass increase. Comparing such a calculation on silver fluoride using the popular LANL2DZ basis set (a split valence basis) with a 3-21G* calculation, using Gaussian 94 for Windows [48] on a PentiumPro:

LANL2DZ basis, 31 basis functions, 1.9 min; Ag–F = 2.064 Å

3-21G* basis, 48 basis functions, 2.6 min; Ag–F = 2.019 Å

The experimental bond length is 1.983 Å [49].

In this simple case there is no real advantage to the pseudopotential calculation (the 3-21G* geometry is actually better!), but for more challenging calculations on "very-heavy-atom" molecules, particularly transition metal molecules, *ab initio* calculations use pseudopotentials almost exclusively. A concise description of pseudopotential theory and specific relativistic effects on molecules, with several references, is given by Levine [50]. Reviews oriented toward transition metal molecules [51a,b] and the lanthanides [51c] have appeared, as well as detailed reviews of the more "technical" aspects of the theory [52].

Which basis set should I use?

There are books of practical advice [1e,k,53] which help to provide a feel for the appropriateness of various basis sets. By reading the research literature one learns what approaches, including which basis sets, are being applied to a various problems, especially those related to the one's research. This said, one should avoid simply assuming that the basis used in published work was the most appropriate one: in the absence of evidence to the contrary, one may suspect that it was either too small or unnecessarily big. Hehre [32] has shown that in many cases the use of very large bases is pointless; on the other hand some problems yield, if at all, only to very large basis sets (see below). A Goldilocks-like basis can rarely (except for calculations of a cursory or routine nature) be simply picked; rather, one homes in on an it, by experimenting. A rational approach in many cases might be to survey the territory first with a semiempirical method (chapter 6) or with the STO-3G basis and to use one of these to create input structures and input hessians (section 2.4) for higher-level calculations) then to move on to the 3-21G$^{(*)}$ basis or possibly the 6-31G* for a reasonable exploration of the problem. For a novel system for which there is no previous work to serve as a guide one should move up to larger basis sets and to post-HF methods (section 5.4), climbing the ladder of sophistication until reasonable convergence of at least *qualitative* results has been obtained. Janoschek has given an excellent survey indicating the reliability of *ab initio* calculations and the level at which one might need to work to obtain trustworthy results by [54].

Oxirene (oxacyclopropene) provides a canonical example of a molecule which even at the highest current levels of theory has declined to reveal its basic secret: can it exist ("Oxirene: to Be or Not to Be?" [44b]). Very large basis sets and advanced post-HF methods suggest it is a true minimum on the potential energy surface, but its disconcerting tendency to display an imaginary (section 2.5) calculated ring-opening vibrational mode at some of the highest levels used leaves the judicious chemist with no choice but to reserve judgement on its being. The nature of a series of *substituted* oxirenes, studied likewise at high levels, appears to be clearer [44a].

Another system that has yielded results which are dependent on the level of theory used, but which unlike the oxirene problem provides a textbook example of a smooth gradation in the nature of the answers obtained, is the ethyl cation (Fig. 5.17). At the HF STO-3G and 3-21G levels the classical structure is a minimum and the bridged nonclassical structure is a transition state, but with the 6-31G* basis the bridged ion has become a minimum and the classical one, although the global minimum, is not securely ensconced as such, being only 3.4 kJ mol^{-1} lower than the bridged ion. At the post-HF (section 5.4) MP2 level with the 6-31G* basis the bridged ion is a minimum and the classical one has lost the dignity of being even a stationary point. The ethyl cation and several other systems have been reviewed [54].

In summary, in many cases [32] the 3-21G (i.e. 3-21G$^{(*)}$ or 6-31G* basis sets, or even the much faster molecular mechanics (chapter 3) or semiempirical (chapter 6) methods, are entirely satisfactory, but there are problems that require quite high levels of attack. In this connection, whether one chooses to regard the wide variety of basis sets at our disposal as representing a "chaotic proliferation" [55] or rather valuable components of our armamentarium is perhaps a matter of viewpoint.

5.4   POST-HF CALCULATIONS: ELECTRON CORRELATION

5.4.1   Electron correlation

Electron correlation is the phenomenon of the motion of pairs of electrons in atoms or molecules being connected ("correlated") [56]. The purpose of *post-HF calculations* is to treat such correlated motion better than does the HF method. In the HF treatment, electron–electron repulsion is handled by having each electron move in a smeared-out, average electrostatic field due to all the other electrons (sections 5.2.3.2 and 5.2.3.6b), and the probability that an electron will have a particular set of spatial coordinates at some moment is independent of the coordinates of the other electrons at that moment. In reality, however, each electron at any moment moves under the influence of the repulsion, not of an average electron cloud, but rather of *individual* electrons (in fact current physics regards electrons as point particles – with wave properties of course). The consequence of this is that the motion of an electron in a real atom or molecule is more complicated than that for an electron moving in a smeared-out field [57] and the electrons are thus better able to avoid one another. Because of this enhanced (compared to the HF treatment) standoffishness, electron–electron repulsion is really smaller than

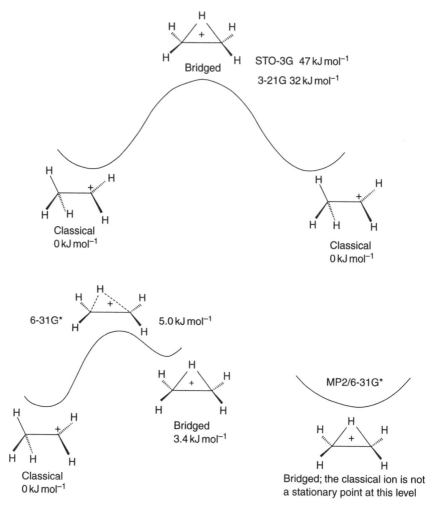

Figure 5.17. The ethyl cation problem at various levels. At the three HF levels the classical cation is a minimum, but at the post-HF (MP2/6-31G*) level only the symmetrical bridged ion is a minimum. The HF/6-31G* results are calculations by the author (ZPE ignored), the other three levels are taken from [54].

predicted by a HF calculation, i.e. the electronic energy is in reality lower (more negative). If you walk through a crowd, regarding it as a smeared-out collection of people, you will experience collisions that could be avoided by looking at individual motions and correlating yours accordingly. The HF method overestimates electron–electron repulsion and so gives higher electronic energies than the correct ones, even with the biggest basis sets, because it does not treat electron correlation properly.

Hartree-Fock calculations are sometimes said to ignore, or at least neglect, electron correlation. Actually, the HF method allows for *some* electron correlation: two electrons of the same spin can't be in the same place at the same time because their spatial and spin coordinates would then be the same and the Slater determinant (section 5.2.3.1) representing the total molecular wavefunction would vanish, since a determinant is zero if two rows or columns are the same (section 4.3.3). This is just a consequence of the antisymmetry of the wavefunction: switching rows or columns of a determinant changes its sign; for two rows/columns the same $D_1 = D_2$ and $D_1 = -D_2$, so $D_1 = D_2 = 0$. If the wavefunction were to vanish so would the electron density, which can be calculated from the wavefunction. This is one way of looking at the Pauli exclusion principle. Now, since the probability is zero that at any moment two electrons of like spin are at the *same* point in space, and since the wavefunction is continuous, the probability of finding them at a given separation should decrease smoothly with that separation. This means that *even if electrons were uncharged*, with no electrostatic repulsion between them, around each electron there would still be a region increasingly (the closer we approach the electron) unfriendly to other electrons of the same spin. This quantum mechanically engendered "Pauli exclusion zone" around an electron is called a *Fermi hole* (after Enrico Fermi; it applies to fermions (section 5.2.2) in general). It can be shown that the HF method overestimates the size of the Fermi hole. Besides the quantum mechanical Fermi hole, each electron *in a real molecule*, not in a "HF molecule", is surrounded by a region unfriendly to all other electrons, regardless of spin, because of the electrostatic (Coulomb) repulsion between point particles (= electrons). This electrostatic exclusion zone is called a Coulomb hole. Since the HF method does not treat the electrons as discrete point particles it essentially ignores the existence of the Coulomb hole, allowing electrons to get too close on the average. This is the main source of the overestimation of electron–electron repulsion in the HF method. Post-HF calculations attempt to allow electrons, even of different spin, to avoid one another better than in the HF approximation.

Hartree-Fock calculations give an electronic energy (and thus a total internal energy, section 5.2.3.6d) that is too high (the variation theorem, section 5.2.3.3, assures us that the HF energy will never be too *low*). This is partly because of the overestimation of electronic repulsion and partly because of the fact that in any real calculation the basis set is not perfect. For sensibly-developed basis sets, as the basis set size increases the HF energy gets smaller, i.e. more negative. The limiting energy that would be given by an infinitely large basis set is called the *HF limit* (i.e. the energy in the HF limit). Table 5.4 and Fig. 5.18 show the results of some HF and post-HF calculations on the hydrogen molecule; the limiting energies are close to the accepted ones [58]. Errors in energy, or in any other molecular feature, that can be ascribed to using a finite basis set are said to be caused by *basis set truncation*. Basis set truncation does not always cause serious errors; for example, the small HF/3-21G basis often gives good geometries (section 5.3.3). Where necessary, the truncation problem can be minimized by using a large (provided the size of the molecule makes this practical), appropriate basis set.

A measure of the extent to which any particular *ab initio* calculation does not deal perfectly with electron correlation is the *correlation energy*. In a canonical exposition [59] Löwdin defined correlation energy thus: "The correlation energy for a certain state with respect to a specified Hamiltonian is the difference between the exact eigenvalue of the Hamiltonian and its expectation value in the HF approximation for the state under consideration." In other words, the correlation energy for a calculation on some molecule or atom is the energy calculated by some perfect quantum mechanical procedure,

Table 5.4. (cf. Fig. 5.18). Dependence of the calculated energy of H_2 on basis set and on correlation level

Basis	No. basis functions	HF energy	Correlated energy	
			Method	Energy
3-21G	4	−1.12292	—	—
6-31G*	10	−1.13127	MP2	−1.15761
6-311++G**	14	−1.13248	MP2	−1.16029
6-311++G(3df,3pd)	36	−1.13303	MP2	−1.16493
6-311++G(3df,3p2d)	46	−1.13307	MP2	−1.16543
6-311++G(3df,3p2d)	46	−1.13307	MP4	−1.17226
6-311++G(3df,3p2d)	46	−1.13307	full CI	−1.17288

All calculations are single-point, without ZPE correction, on H_2 at the experimental bond length of 0.742 Å, using G94W [48]; energies are in hartrees. The accepted HF (E_{HF}^{total}; Eq. (5.149)) and correlated limiting energies are about −1.1336 and −1.1744 h, respectively [58], cf. −1.13307 and −1.17288 h here).

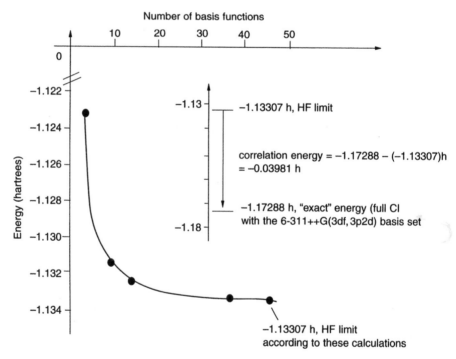

Figure 5.18. (based on Table 5.4). The HF limit and correlation energy for H_2. From the values calculated here, the HF limit, the exact energy (see text) and the correlation energy are −1.13307, −1.17288 and −0.03981 h (see inset); the accepted values [58] are about −1.1336, −1.17439 and −0.04079.

minus the energy calculated by the HF method with a huge ("infinite") basis set, using the same Hamiltonian:

$$E_{correl} = E(exact) - E(HF\ limit) \quad \text{using the same Hamiltonian for both terms}$$

From this definition the correlation energy is negative, since $E(exact)$ is more negative than $E(HF\ limit)$. Usually $E(exact)$ and $E(HF\ limit)$ are taken as the energy from a Hamiltonian that excludes relativistic effects (like that in section 5.2.2, Eqs (5.4), (5.5), (5.6) and associated discussion), which are significant only for heavy atoms, so unless qualified the term correlation energy means nonrelativistic correlation energy. The correlation energy is essentially the energy that the HF procedure fails to account for. *If* relativistic effects (and other , usually small, effects like spin–orbit coupling) are negligible then E_{correl} is the difference between the experimental value (of the energy required to dissociate the molecule or atom into infinitely separated nuclei and electrons) and the limiting HF energy.

A distinction is sometimes made between dynamic and nondynamic or static correlation energy. Dynamic correlation energy is the energy a HF calculation does not account for because it fails to keep the electrons sufficiently far apart; this is the usual "correlation energy". Nondynamic correlation energy is the energy a calculation (HF or otherwise) may not account for because it uses a single determinant, or starts from a single determinant (is based on a single-determinant reference – section 5.4.3); this problem arises with singlet diradicals, e.g. where a closed-shell description of the electronic structure is qualitatively wrong. Dynamic correlation energy can be calculated ("recovered") by Møller-Plesset or configuration interaction methods (sections 5.4.2 and 5.4.3) and static correlation energy can be recovered by basing the wavefunction on more than one determinant, as in the multireference configuration interaction method (section 5.4.3).

Although HF calculations are satisfactory for many purposes (sections 5.5 and 5.6) there are cases where a better treatment of electron correlation is needed. This is particularly true for the calculation of relative energies, although geometries and some other properties are also improved by post-HF calculations, section 5.4). As an illustration of a shortcoming of HF calculations consider an attempt to find the C/C single bond dissociation energy of ethane by comparing the energy of ethane with that of two methyl radicals:

$$H_3C–CH_3 + E_{diss} \longrightarrow H_3C \cdot CH_3$$

Let us simply subtract the energy of two methyl radicals from that of an ethane molecule, and compare with experiment the results of HF calculations and (anticipating section 5.4.2) the post-HF (i.e. correlated) MP2 method. In Table 5.5 the energies shown for $CH_3\cdot$ and CH_3CH_3 are the "uncorrected" *ab initio* energies (the energy displayed at the end of any calculation; this is the electronic energy + the internuclear repulsion), the ZPE, and the "corrected" energy (uncorrected energy + ZPE); see section 5.2.3.6d. The ZPEs used here are from HF/6-31G* optimization/frequency jobs; these are fairly fast and give reasonable ZPEs. The ZPEs were all calculated by multiplying by an empirical correction factor of 0.9135 (this brings them into better agreement with experiment [60]). Although *frequencies* must be calculated with the same method (HF, MP2, etc.)

Table 5.5. The C–C bond energy of ethane calculated by the HF and MP2 methods

Method/basis	Energy		
	CH$_3$·	CH$_3$CH$_3$	$E(2CH_3 \cdot -CH_3CH_3)$
HF/6-31G*	−39.55899	−79.22876	0.09451
	0.02829	0.07285	248
	−39.53070	−79.15591	
HF/6-311++G(3df,3p2d)	−39.57712	−79.25882	0.08831
	0.02829	0.07285	232
	−39.54883	−79.18597	
MP2/6-31G*	−39.66875	−79.49474	0.14097
	0.02829	0.07285	370
	−39.64046	−79.42189	
MP2/6-311++G**	−39.70866	−79.57167	0.13808
	0.02829	0.07285	363
	−39.68037	−79.49882	

The radical CH$_3$· and the closed-shell CH$_3$CH$_3$ were calculated by unrestricted and restricted methods, respectively: UHF and UMP2, vs. RHF and RMP2 – see concluding part of section 5.2.3.6e; the HF method largely ignores electron correlation, while MP2 recovers about 85% of the electron correlation. The set of three numbers for each species are respectively, in hartrees, the uncorrected *ab initio* energy, the corrected (0.9135 factor, see text) HF/6-31G* ZPE, and the corrected *ab initio* energy (uncorrected energy + ZPE). Calculated (by subtraction) bond energies are in hartrees and kJ mol^{-1} (2626 × hartrees). The experimental C–C energy of ethane has been reported at from 368 to 377 kJ mol^{-1} [61]. Each species was optimized at the level shown (i.e. none of these are single-point calculations).

and basis set as were used for the geometry optimization, *ZPEs* from a particular method/basis may be used to correct energies obtained with another method/basis. The only calculations that give reasonable agreement with the experimental ethane C–C dissociation energy (reported at from 368 to 377 kJ mol^{-1} [61]) are the correlated (MP2) ones, 370 and 363 kJ mol^{-1}; because of error in the experimental value the two MP2 results may be equally good. The HF values (248 and 232 kJ mol^{-1}) are very poor, even (especially!) when the very large 6-311++G(3df,3p2d) basis is used.

This inability of HF calculations to model correctly homolytic bond dissociation is commonly illustrated by curves of the change in energy as a bond is stretched, e.g. Fig. 5.19. The phenomenon is discussed in detail in numerous expositions of electron correlation [62]. Suffice it to say here that representing the wavefunction as one determinant (or a few), as is done in Hartree-Fock theory, does not permit correct homolytic dissociation to two radicals because while the reactant (e.g. H$_2$) is a closed-shell species that can (usually) be represented well by one determinant made up of paired electrons in the occupied MOs, the products are two radicals, each with an unpaired electron. Ways of obtaining satisfactory energies, with and without the use of electron correlation methods, for processes involving homolytic cleavage, are discussed further in section 5.5.2.

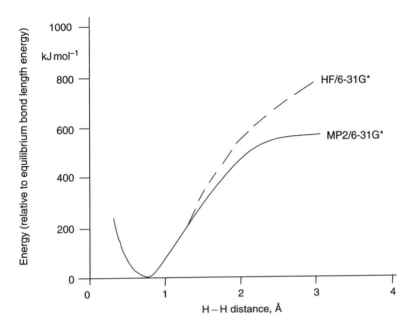

Figure 5.19. Dissociation curves (change in energy as the bond is stretched) for H_2, from HF/6-31G* and MP2/6-31G* calculations. The equilibrium bond lengths are reasonable (HF/6-31G*, 0.730; MP2/6-31G*, 0.737 (cf. experimental, 0.742), but only the MP2 curve approximates the actual dissociation behavior of the molecule.

There are basically three approaches to dealing with electron correlation: explicit use of the interelectronic distances as variables in the Schrödinger equation, treatment of the real molecule as a perturbed HF system, and explicit inclusion in the wavefunction of electronic configurations other than the ground-state one. Using interelectronic distances explicitly quickly seems to become mathematically intractable and is currently limited to atoms and molecules that are very small [63]. The other two methods are general and very important: the perturbation approach is used in the very popular Møller–Plesset[4] methods, and the use of higher electronic configurations in the wavefunction forms the basis of configuration interaction, which in various forms is employed in some of the most advanced *ab initio* methods currently used for dealing with electron correlation. A powerful method that is becoming increasingly popular and incorporates mathematical features of the perturbation and higher-electronic-state methods, the coupled-cluster approach, is also described.

5.4.2   The Møller-Plesset approach to electron correlation

The Møller-Plesset (MP) treatment of electron correlation [64] is based on perturbation theory, a very general approach used in physics to treat complex systems [65]; this particular approach was described by Møller and Plesset in 1934 [66] and developed

---

[4]Møller–Plesset: the Norwegian letter ø is pronounced like French *eu* or German *ö*.

into a practical molecular computational method by Binkley and Pople [67] in 1975. The basic idea behind perturbation theory is that if we know how to treat a simple (often idealized) system then a more complex (and often more realistic) version of this system, if it is not too different, can be treated mathematically as an altered (perturbed) version of the simple one. Møller-Plesset calculations are denoted as MP, MPPT (MP perturbation theory) or MBPT (many-body perturbation theory) calculations. The derivation of the MP method [68] is complicated, and only the flavor of the approach will be given here. There is a hierarchy of MP energy levels: MP0, MP1 (these first two designations are not actually used), MP2, etc...., which successively account more thoroughly for interelectronic repulsion.

"MP0" would use the electronic energy obtained by simply summing the HF one-electron energies (section 5.2.3.6d, Eq. (5.84)). This ignores interelectronic repulsion except for refusing to allow more than two electrons in the same spatial MO. "MP1" corresponds to MP0 corrected with the Coulomb and exchange integrals J and K (Eqs (5.85) and (5.90)), i.e. MP1 is just the Hartree-Fock energy. As we have seen (sections 5.2.3.2 and 5.2.3.6b), this handles interelectronic repulsion in an average way. We could write $E_{MP1} = E_{HF}^{total} = E_{MP0} + E^{(1)}$, where E_{MP0} is the sum of one-electron energies and internuclear repulsions and $E^{(1)}$ is the J, K correction (corresponding respectively to the two terms in Eqs (5.85) and (5.90)), regarding the second term as a kind of perturbational correction to the sum of one-electron energies.

MP2 is the first MP level to go beyond the HF treatment (it is the first "real" MP level). The MP2 energy is the HF energy plus a correction term (a perturbational adjustment) that represents a lowering of energy brought about by allowing the electrons to avoid one another better than in the HF treatment:

$$E_{MP2} = E_{HF}^{total} + E^{(2)} \tag{5.161}$$

The HF term includes internuclear repulsions, and the perturbation correction $E^{(2)}$ is a purely electronic term. $E^{(2)}$ is a sum of terms each of which models the promotion of pairs of electrons (so-called *double excitations* are required by Brillouin's theorem [69]) from occupied to unoccupied MOs (virtual MOs).

Let's do an MP2 energy calculation on HHe^+, the molecule for which a HF (i.e. an SCF) calculation was shown in detail in section 5.2.3.6e. As for the HF calculation, we will take the internuclear distance as 0.800 Å and use the STO-1G basis set; we can then use for our MP2 calculation these HF results that we obtained in section 5.2.3.6e:
The MO coefficients:

For the occupied MO ψ_1, $c_{11} = 0.3178$, $c_{21} = 0.8020$ (recall that these are respectively the coefficient of basis function 1, ϕ_1, in MO1 and the coefficient of basis function 2, ϕ_2, in MO1. In this simple case there is one function on each atom: ϕ_1 and ϕ_2 on atoms 1 and 2 (H and He).

For the unoccupied (virtual) MO ψ_2, $c_{12} = 1.1114$, $c_{22} = -0.8325$:

The two-electron repulsion integrals:

$$(11|11) = 0.7283 \quad (21|21) = 0.2192$$
$$(21|11) = 0.3418 \quad (22|21) = 0.4368$$
$$(22|11) = 0.5850 \quad (22|22) = 0.9927$$

The energy levels: occupied MO, $\varepsilon_1 = -1.4470$, virtual MO, $\varepsilon_2 = -0.1051$

The HF energy: $E_{HF}^{total} = -2.4438$

The MP2 energy correction for a closed-shell two-electron/two-MO system is [70]:

$$E^{(2)} = \frac{\left[\iint \psi_1(1)\psi_1(2)(1/r_{12})\psi_2(1)\psi_2(2)\,dv_1\,dv_2\right]^2}{2(\varepsilon_1 - \varepsilon_2)} \tag{5.162}$$

Applying this formula is straightforward; although the arithmetic is tedious, it is worth doing (as was true for the *HF* calculation in section 5.2.3.6e) in order to appreciate how much work is involved in even this simplest molecular MP2 job. Consider the integral in the numerator of Eq. (5.162); substituting for ψ_1 and ψ_2:

$$\iint \psi_1(1)\psi_1(2)\left(\frac{1}{r_{12}}\right)\psi_2(1)\psi_2(2)\,dv_1\,dv_2$$
$$= \iint [(c_{11}\phi_1(1) + c_{21}\phi_2(1))(c_{11}\phi_1(2) + c_{21}\phi_2(2))\left(\frac{1}{r_{12}}\right)(c_{12}\phi_1(1)$$
$$+ c_{22}\phi_2(1))(c_{12}\phi_1(2)$$

Multiplying out the integrand gives a total of 16 terms (from 4 terms to the left of $1/r_{12}$ and 4 terms to the right), and leads to a sum of 16 integrals:

$$\iint \psi_1(1)\psi_1(2)\left(\frac{1}{r_{12}}\right)\psi_2(1)\psi_2(2)\,dv_1\,dv_2$$
$$= c_{11}^2 c_{12}^2 \int \phi_1(1)\phi_1(2)\left(\frac{1}{r_{12}}\right)\phi_1(1)\phi_1(2)\,dv_1\,dv_2 + \cdots + c_{21}^2 c_{22}^2 \int \phi_2(1)\phi_2(2)\left(\frac{1}{r_{12}}\right)$$
$$= c_{11}{}^2 c_{12}{}^2 (11|11) + \cdots + c_{21}{}^2 c_{22}{}^2 (22|22),$$

recalling the notational degeneracy in the two-electron integrals (section 5.2.3.6e, "Step 2 Calculating the integrals"). Substituting the values of the coefficients and the two-electron integrals:

$$\iint \psi_1(1)\psi_1(2)\left(\frac{1}{r_{12}}\right)\psi_2(1)\psi_2(2)\,dv_1\,dv_2$$
$$= 0.12475(0.7283) + \cdots + 0.44577(0.9927)\,h = 0.12932\,h$$

So from Eq. (5.162)

$$E^{(2)} = \frac{0.12932^2}{2(\varepsilon_1 - \varepsilon_2)}h = \frac{0.12932^2}{2(-1.4470 + 0.1051)}h = -0.00623\,h$$

The MP2 energy is the HF energy plus the MP2 correction (Eq. (5.162)):

$$E_{MP2} = E_{HF}^{total} + E^{(2)} = -2.4438\,h - 0.00623\,h = -2.4500\,h$$

This energy, which includes internuclear repulsion, since E_{HF}^{total} includes this (Eq. (5.93)), is the MP2 energy normally printed out at the end of the calculation. To get an intuitive feel for the physical significance of the calculation just performed look again at Eq. (5.162), which applies to any two-electron/two-basis function species.

The equation shows that the absolute value (the correction is negative since ε_1 is smaller than ε_2 – the occupied MO has a lower energy than the virtual one) of the correlation correction increases, i.e. the energy decreases, with the magnitude of the integral (which is positive). This integral represents the decrease in energy arising from allowing an electron pair in the occupied MO (ψ_1) to spill over into the virtual MO (ψ_2):

$\psi_1(1)$ represents electron 1 in ψ_1 and $\psi_1(2)$ represents electron 2 in ψ_1.

$\psi_2(1)$ represents electron 1 in ψ_2 and $\psi_2(2)$ represents electron 2 in ψ_2.

The operator $1/r_{12}$ brings in coulombic interaction: the coulombic repulsion energy between infinitesimal volume elements $\psi_1(1)\psi_1(2)\,dv_1$ and $\psi_2(1)\psi_2(2)\,dv_2$ separated by a distance r_{12} is $(\psi_1(1)\psi_1(2)\,dv_1)(\psi_2(1)\psi_2(2)\,dv_2)/r_{12}$, and the integral is simply the sum over all such volume elements (cf. the discussion in connection with Fig. 5.3 and the average-field integrals J and K in section 5.2.3.2). Physically, the decrease in energy makes sense: allowing the electrons to be partly in the formally unoccupied virtual MO rather than confining them strictly to the formally occupied MO enables them to avoid one another better than in the HF treatment, which is based on a Slater determinant consisting only of occupied MOs (section 5.2.3.1). *The essence of the MP method* (MP2, MP3, etc.) is that the correction term handles electron correlation by promoting electrons from occupied to unoccupied (virtual) MOs, giving electrons, in some sense, more room to move and thus making it easier for them to avoid one another; the decreased interelectronic repulsion results in a lower electronic energy. The contribution of the "ψ_1/ψ_2 interaction" to $E^{(2)}$ decreases as the occupied/virtual MO gap $\varepsilon_1 - \varepsilon_2$, increases, since this is in the denominator. Physically, this makes sense: the bigger the gap between the occupied and higher-energy virtual MO, the "harder" it is to promote electrons from the one into the other, so the less can such promotion contribute to electronic stabilization. So in the expression for $E^{(2)}$ (Eq. (5.162), the numerator represents the promotion of electrons from the occupied to the virtual orbital, and the denominator represents how hard it is to do this.

   As we just saw, MP2 calculations utilize the HF MOs (their coefficients c and energies ε). The HF method gives the best *occupied* MOs obtainable from a given basis set and a one-determinant total wavefunction ψ, but it does not optimize the *virtual* orbitals (after all, in the HF procedure we start with a determinant consisting of only the occupied MOs – sections 5.2.3.1–5.2.3.4). To get a reasonable description of the virtual orbitals and to obtain a reasonable number of them into which to promote electrons, we need a basis set that is not too small. The use of the STO-1G basis in the above example was purely illustrative; the smallest basis set generally considered acceptable for MP calculations is the 6-31G*, and in fact this is perhaps the one most frequently used for MP2 calculations. The 6-311G** basis set is also widely used for MP2 and MP4 calculations. Both bases can of course be augmented (section 5.3.3) with diffuse functions, and the 6-31G* with H polarization functions (6-31G**). MP2 calculations increase rapidly in complexity with the number of electrons and orbitals, involving as they do a *sum of terms* (rather than just one term as in HHe$^+$), each representing the promotion of an electron pair from an occupied to a virtual orbital; thus an MP2 calculation on CH_2 with the 6-31G* basis involves 8 electrons and 19 MOs (4 occupied and 15 virtual MOs).

In MP2 calculations doubly excited states (doubly excited configurations) interact with the ground state (the integral in Eq. (5.162) involves ψ_1 with electrons 1 and 2, and ψ_2 with electrons 1 and 2). In MP3 calculations doubly excited states interact with one another (there are integrals involving two virtual orbitals). In MP4 calculations singly, doubly triply and quadruply excited states are involved. MP5 and higher expressions have been developed, but MP2 and MP4 are by far the most popular MP levels (also called MBPT(2) and MBPT(4) – many-body perturbation theory). MP2 calculations, which are much slower than HF, can be speeded up somewhat by specifying MP2(FC), MP2 frozen-core, in contrast to MP2(FULL); frozen-core means that the core (non-valence electrons) are "frozen," i.e. not promoted into virtual orbitals, in contrast to full MP2 which takes all the electrons into account in summing the contributions of excited states to the lowering of energy. Most programs (e.g. Gaussian, Spartan) perform MP2(FC) by default when MP2 is specified, and "MP2" usually means frozen-core. MP4 calculations are sometimes done omitting the triply excited terms (MP4SDQ) but the most accurate (and slowest) implementation is MP4SDTQ (singles, doubles, triples, quadruples).

Calculated properties like geometries and relative energies tend to be better (to be closer to the true ones) when done with correlated methods (sections 5.5.1–5.5.4). To save time, energies are often calculated with a correlated method on a HF geometry, rather than carrying out the geometry optimization at the correlated level. This is called a *single-point* calculation (it is performed at a single point on the HF potential energy surface, without changing the geometry). A single-point MP2(FC) calculation using the 6-311G** basis, on a structure that was optimized with the HF method and the 6-31G* basis, is designated as MP2(FC)/6-311G**//HF/6-31G*. A HF/6-31G* (say) geometry optimization, without a subsequent single-point calculation, is sometimes designated HF/6-31*//HF/6-31G*, and an MP2 optimization MP2/6-31G*//MP2/6-31G*. The correlation treatment (HF, MP2, MP4, ...) is often called the *method*, and the basis set (STO-3G, 3-21G, 6-31G*, ...) the *level*, but we will often find it convenient to let *level* denote the combined procedure of method and basis set, referring, say, to an MP2/6-31G* calculation as being at a higher level than an HF/6-31* one.

Figure 5.20 shows the rationale behind the use of single-point calculations for obtaining relative energies. In the diagram a single-point MP2 calculation on a stationary point at the HF geometry gives the same energy as would be obtained by optimizing the species at the MP2 level, which is often approximately true (it would be exactly true if the MP2 and HF geometries were identical). For example, the single-point and optimized energies of butanone are −231.68593 and −231.68818 h, a difference of 0.00225 h (2.3 mh) or 6 kJ mol^{-1}, not large bearing in mind that special high-accuracy calculations (section 5.5.2.2) are needed to reliably get relative energies to within, say, 10 kJ mol^{-1}. Single-point calculations would also give *relative* energies similar to those from the use of optimized correlated geometries if the incremental deviations from the optimized-geometry energies were about the same for the two species being compared (e.g. reactant and TS for an activation energy, reactant and product for a reaction energy).

The method can occasionally give not just quantitatively, but qualitatively wrong results. The HF and correlated surfaces may have different curvatures: for example a minimum on one surface may be a transition state or may not exist (may not be a stationary point) on another. Thus fluoro- and difluorodiazomethane are HF minima but

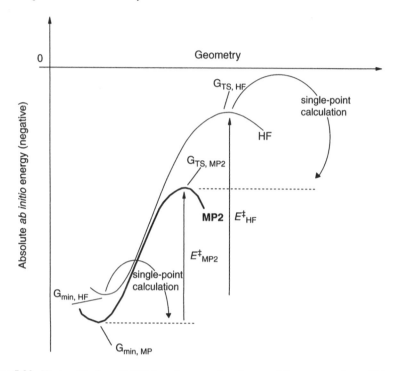

Figure 5.20. Hartree-Fock and MP2 (or other correlated) potential energy surfaces. "Absolute" (as distinct from relative) *ab initio* energies are negative, and correlated energies are lower (more negative) than HF energies. The geometries of the minima and the transition states are designated G_{min} and G_{TS}. Activation energies are denoted by $E^{\ddagger}$. HF activation energies are, as shown, usually bigger than MP2. In this diagram a single-point MP2 calculation on a stationary point at the HF geometry gives the same energy as would be obtained by optimizing the species at the MP2 level; this is often true, but single-point MP2 relative energies would be similar to optimized-MP2 relative energies even if it were not so, provided the incremental energy change were about the same for the two species being compared (e.g. reactant and TS for an activation energy, reactant and product for a reaction energy).

are not MP2/6-31G* stationary points [71]; an attempt to approximate the MP2/6-31G* reaction energy for, say, $CHFN_2 \rightarrow CH_2 + N_2$, using single-point MP2/6-31G* energies on HF geometries, is misguided if $CHFN_2$ does not exist on the MP2 PES. Nevertheless, because HF optimizations followed by single-point correlated energy calculations are much faster ("cheaper") than correlated optimizations, and usually give improved relative energies, the method is widely used. Figure 5.21 compares some MP2 single-point and MP2-optimized energies with experiment or with high-level calculations [72]. Geometries are discussed further in section 5.5.1.

5.4.3   The configuration interaction approach to electron correlation

The configuration interaction (CI) treatment of electron correlation [62,73] is based on the simple idea that one can improve on the HF wavefunction, and hence energy, by

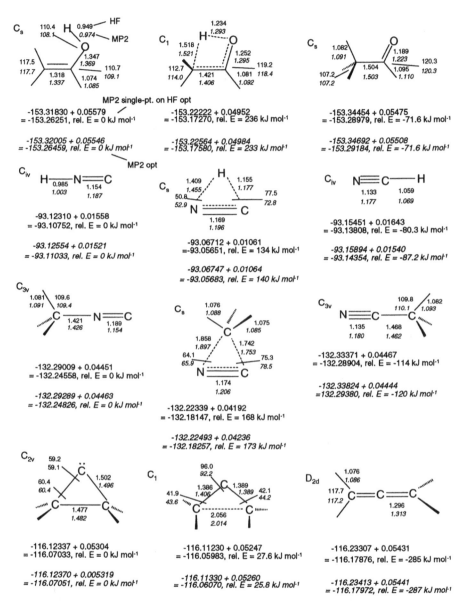

Figure 5.21. Geometries and energies for four reactions (most H's are omitted, for clarity). Geometries are HF/6-31G* and MP2/6-31G*. Energies are MP2/6-31G*//HF/6-31G* (i.e. single-point) with HF ZPE, and MP2/6-31G*//MP2//6-31G* with MP2/6-31G* ZPE. *Ab initio* E (hartrees) + ZPE (hartrees) = corrected *ab initio* E; relative E : E in hartrees × 2626 = kJ mol^{-1}). The ZPEs shown are the *ab initio* ZPEs multiplied by 0.9135 (HF) or 0.967 (MP2) [60]. The correct relative energies (kJ mol^{-1}) are believed to be [72]: $H_2C = CHOH$ reaction, 0, ca. 282, −42; HNC reaction, 0, 129, −58; CH_3NC reaction, 0, 161, −98; cyclopropylidene reaction, 0, 13–20, ca. 293. Calculations are by the author.

adding on to the HF wavefunction terms that represent promotion of electrons from occupied to virtual MOs. The HF term and the additional terms each represent a particular electronic configuration, and the actual wavefunction and electronic structure of the system can be conceptualized as the result of the interaction of these configurations. This electron promotion, which makes it easier for electrons to avoid one another, is as we saw (section 5.4.2) also the physical idea behind the MP method; the MP and CI methods differ in their mathematical approaches.

HF theory (sections 5.2.3.1–5.2.3.6) starts with a total wavefunction or total MO Ψ which is a Slater determinant made of "component" wavefunctions or MOs ψ. In section 5.2.3.1 we approached HF theory by considering the Slater determinant for a four-electron system:

$$\Psi = \frac{1}{\sqrt{4!}} \begin{vmatrix} \psi_1(1)\alpha(1) & \psi_1(1)\beta(1) & \psi_2(1)\alpha(1) & \psi_2(1)\beta(1) \\ \psi_1(2)\alpha(2) & \psi_1(2)\beta(2) & \psi_2(2)\alpha(2) & \psi_2(2)\beta(2) \\ \psi_1(3)\alpha(3) & \psi_1(3)\beta(3) & \psi_2(3)\alpha(3) & \psi_2(3)\beta(3) \\ \psi_1(4)\alpha(4) & \psi_1(4)\beta(4) & \psi_2(4)\alpha(4) & \psi_2(4)\beta(4) \end{vmatrix} \qquad (5.163 = 5.10)$$

To construct the HF determinant we used only occupied MOs: four electrons require only two spatial "component" MOs, ψ_1 and ψ_2, and for each of these there are two spin orbitals, created by multiplying ψ by one of the spin functions α or β; the resulting four spin orbitals ($\psi_1\alpha$, $\psi_1\beta\psi_2\alpha$, $\psi_2\beta$) are used four times, once with each electron. The determinant Ψ, the HF wavefunction, thus consists of the four lowest-energy spin orbitals; it is the simplest representation of the total wavefunction that is antisymmetric and satisfies the Pauli exclusion principle (section 5.2.2), but as we shall see it is not a complete representation of the total wavefunction.

In the Roothaan–Hall implementation of *ab initio* theory each *"component"* ψ is composed of a set of basis functions (Sections 5.2.3.6 and 5.3):

$$\psi_i = \sum_{s=1}^{m} c_{si}\phi_s \quad i = 1, 2, 3, \ldots, m \text{ (component MOs)} \qquad (5.164 = 5.52)$$

Now note that there is no definite limit to how many basis functions ϕ_1, ϕ_2, ... can be used for our four-electron calculation; although only two spatial ψ's, ψ_1 and ψ_2, (i.e. four *spin* orbitals) are *required* to accommodate the four electrons of this ψ, the total number of ψ's can be greater. Thus for the hypothetical H–H–H–H an STO-3G basis gives four ψ's, a 3-21G basis gives 8, and a 6-31G** basis gives 20 (section 5.3.3). The idea behind CI is that a better total wavefunction, and from this a better energy, results if the electrons are confined not just to the four spin orbitals $\psi_1\alpha$, $\psi_1\beta\psi_2\alpha$, $\psi_2\beta$, but are allowed to roam over all, or at least some, of the virtual spin orbitals $\psi_3\alpha$, $\psi_3\beta$, $\psi_4\alpha$, ..., $\psi_m\beta$. To permit this we could write Ψ as a linear combination of determinants

$$\Psi = c_1 D_1 + c_2 D_2 + c_3 D_3 + \cdots + c_i D_i \qquad (5.165)$$

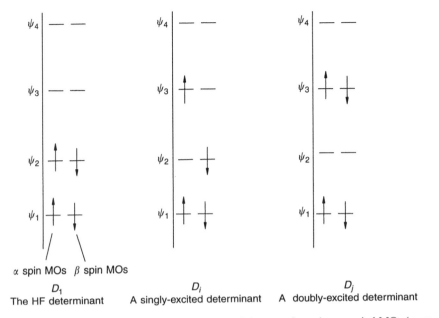

α spin MOs β spin MOs

D_1
The HF determinant

D_i
A singly-excited determinant

D_j
A doubly-excited determinant

Figure 5.22. Configuration interaction: promotion of electrons from the occupied MOs (corresponding to the HF determinant) gives determinants corresponding to excited states. A weighted sum of determinants $D_1, D_2, \ldots, D_i, \ldots$, corresponds to a molecule in which the electrons partly populate virtual MOs and are not strictly confined to the lowest-energy MOs, thus giving them a better chance to avoid one another and decreasing electron–electron repulsion. The method generates a series of wavefunctions and energies; the lowest-energy wavefunction and energy corresponds to the ground electronic state, the others to excited states.

where D_1 is the HF determinant of Eq. (5.163) and D_2, D_3, etc. correspond to the promotion of electrons into virtual orbitals, e.g. we might have

$$D_i = \frac{1}{\sqrt{4!}} \begin{vmatrix} \psi_1(1)\alpha(1) & \psi_1(1)\beta(1) & \psi_3(1)\alpha(1) & \psi_2(1)\beta(1) \\ \psi_1(2)\alpha(2) & \psi_1(2)\beta(2) & \psi_3(2)\alpha(2) & \psi_2(2)\beta(2) \\ \psi_1(3)\alpha(3) & \psi_1(3)\beta(3) & \psi_3(3)\alpha(3) & \psi_2(3)\beta(3) \\ \psi_1(4)\alpha(4) & \psi_1(4)\beta(4) & \psi_3(4)\alpha(4) & \psi_2(4)\beta(4) \end{vmatrix} \qquad (5.166)$$

D_i was obtained from D_1 by promoting an electron from spin orbital $\psi_2\alpha$ to the spin orbital $\psi_3\alpha$. Another possibility is

$$D_j = \frac{1}{\sqrt{4!}} \begin{vmatrix} \psi_1(1)\alpha(1) & \psi_1(1)\beta(1) & \psi_3(1)\alpha(1) & \psi_3(1)\beta(1) \\ \psi_1(2)\alpha(2) & \psi_1(2)\beta(2) & \psi_3(2)\alpha(2) & \psi_3(2)\beta(2) \\ \psi_1(3)\alpha(3) & \psi_1(3)\beta(3) & \psi_3(3)\alpha(3) & \psi_3(3)\beta(3) \\ \psi_1(4)\alpha(4) & \psi_1(4)\beta(4) & \psi_3(4)\alpha(4) & \psi_3(4)\beta(4) \end{vmatrix} \qquad (5.167)$$

Here two electrons have been promoted, from the spin orbitals $\psi_2\alpha$ and $\psi_2\beta$ to $\psi_3\alpha$ and $\psi_3\beta$. D_i and D_j represent promotion into virtual orbitals of one and two electrons, respectively, starting with the HF electronic configuration (Fig. 5.22).

Equation (5.165) is analogous to Eq. (5.164): in (5.164) "component" MOs ψ are expanded in terms of basis functions ϕ, and in (5.165) a total MO Ψ is expanded in terms of determinants, each of which represents a particular electronic configuration. We know that the m basis functions of Eq. (5.164) generate m component MOs ψ (section 5.2.3.6a), so the i determinants of Eq. (5.165) must generate i total wavefunctions Ψ, and Eq. (5.165) should really be written

$$\Psi_1 = c_{11}D_1 + c_{21}D_2 + c_{31}D_3 + \cdots + c_{i1}D_i$$
$$\Psi_2 = c_{12}D_1 + c_{22}D_2 + c_{32}D_3 + \cdots + c_{i2}D_i$$
$$\vdots$$
$$\Psi_i = c_{1i}D_1 + c_{2i}D_2 + c_{3i}D_3 + \cdots + c_{ii}D_i$$

(5.168)

i.e. (cf. Eq. (5.164))

$$\Psi_i = \sum_{s=1}^{i} c_{si}D_s \quad s = 1, 2, 3, \ldots, i \text{ (total MOs)}$$

(5.169)

What is the physical meaning of all these total wavefunctions Ψ? Each determinant D, or a linear combination of a few determinants, represents an idealized (in the sense of contributing to the real electron distribution) configuration, called a *configuration state function* or *configuration function*, CSF (see below). The CI wavefunctions of Eqs (5.168) or Eqs (5.169), then, are linear combinations of CSF's. No single CSF *fully* represents any particular electronic state. Each wavefunction Ψ_i is the total wavefunction of one of the possible electronic states of the molecule, and the weighting factors c in its expansion determine to what extent particular CSF's (idealized electronic states) contribute to any Ψ_i. For the lowest-energy wavefunction Ψ_1, representing the ground electronic state, we expect the HF determinant D_1 to make the largest contribution to the wavefunction. The wavefunctions Ψ_2, Ψ_3 etc. represent excited electronic states. The single-determinant HF wavefunction of Eq. (5.163) (or the general single-determinant wavefunction of Eq. (5.12)) is merely an approximation to the Ψ_1 of Eqs (5.168).

If every possible idealized electronic state of the system, i.e. every possible determinant D, were included in the expansions of Eqs (5.168), then the wavefunctions Ψ would be *full CI* wavefunctions. Full CI calculations are possible only for very small molecules, because the promotion of electrons into virtual orbitals can generate a huge number of states unless we have only a few electrons and orbitals. Consider for example a full CI calculation on a very small system, H–H–H–H with the 6-31G* basis set. We have eight basis functions and four electrons, giving eight spatial MOs and 16 spin MOs, of which the lowest four are occupied. There are two α electrons to be promoted into 6 virtual α spin MOs, i.e. to be distributed among 8 α spin MOs, and likewise for the β electrons and β spin orbitals. This can be done in $[8!/(8-2)!2!]^2 = 784$ ways. The number of configuration state functions is about half this number of determinants; a CSF is a linear combination of determinants for equivalent states, states which differ only by whether an α or a β electron was promoted. CI calculations with more than five billion (*sic*) CSFs have been performed on ethyne, C_2H_2 [74]; rightly called benchmark calculations, such computational tours de force are, although of limited direct application, important for evaluating the efficacy, by comparison, of other methods.

The simplest implementation of CI is analogous to the Roothaan–Hall implementation of the HF method: Eqs (5.168) lead to a CI matrix, as the HF equations (using Eqs (5.164)) lead to a HF matrix (Fock matrix; section 5.2.3.6). We saw that the Fock matrix $\mathbf{F}$ can be calculated from the c's and ϕ's of Eq. (5.164) (starting with a "guess" of the c's), and that $\mathbf{F}$ (after transformation to an orthogonalized matrix $\mathbf{F}'$ and diagonalization) gives eigenvalues ε and eigenvectors c, i.e. $\mathbf{F}$ leads to the energy levels and the wavefunctions ($c\phi$) of the component MOs ψ; all this was shown in detail in section 5.2.3.6e. Similarly, a CI matrix can be calculated in which the determinants D play the role that the basis functions ϕ play in the Fock matrix, since the D's in Eqs (5.168) correspond mathematically to the ϕ's in Eq. (5.164)). The D's are composed of spin orbitals $\psi\alpha$ and $\psi\beta$, and the spin factors can be integrated out, reducing the elements of the CI matrix to expressions involving the basis functions and coefficients of the spatial component MOs ψ. The CI matrix can thus be calculated from the MOs resulting from an HF calculation. Orthogonalization and diagonalization of the CI matrix gives the energies and the wavefunctions of the ground state Ψ_1 and, from i determinants, $i - 1$ excited states. A full CI matrix would give the energies and wavefunctions of the ground state and all the excited states obtainable from the basis set being used. Full CI with an infinitely large basis set would give the exact energies of all the electronic states; more realistically, full CI with a large basis set gives good energies for the ground and many excited states.

Full CI is out of the question for any but small molecules, and the expansion of Eq. (5.169) must usually be limited by including only the most important terms. Which terms can be neglected depends partly on the purpose of the calculation. For example, in calculating the ground state energy quadruply excited states are, unexpectedly, much more important than triply and singly excited ones, but the latter are usually included too because they affect the electron distribution of the ground state, and in calculating excited state energies single excitations are important. A CI calculation in which all the post HF D's involve only single excitations is called CIS (CI singles); such a calculation yields the energies and wavefunctions of excited states and often gives a reasonable account of electronic spectra. Another common kind of CI calculation is CI singles and doubles (CISD, which actually indirectly includes triply and quadruply excited states). Various mathematical devices have been developed to make CI calculations recover a good deal of the correlation energy despite the necessity of (judicious) truncation of the CI expansion. Perhaps the currently most widely-used implementations of CI are *multiconfigurational SCF* (MCSCF) and its variant *complete active space SCF* (CASSCF), and the *coupled-cluster* (CC) and related *quadratic CI* (QCI) methods.

The CI strict analogue of the iterative refinement of the coefficients that we saw in HF calculations (section 5.2.3.6e) would refine just the weighting factors of the determinants (the c's of Eqs (5.168), but in the MCSCF version of CI the spatial MOs *within* the determinants are also optimized (by optimizing the c's of the LCAO expansion, Eq. (5.164)). A widely-used version of the MCSCF method is the CASSCF method, in which one carefully chooses the orbitals to be used in forming the various CI determinants. These *active orbitals*, which constitute the *active space*, are the MOs that one considers to be most important for the process under study. Thus for a Diels–Alder reaction, the two π and two π^* MOs of the diene and the π and π^* MO of the alkene (the dienophile) would be a reasonable minimum [75] as candidates for the active space of the reactants; the six electrons in these MOs would be the *active electrons*, and with the 6-31G* basis this would be a (specifying electrons, MOs) CASSCF (6, 6)/6-31G* calculation. CASSCF calculations are used to study chemical reactions and to calculate electronic spectra. They require judgement in the proper choice of the active space and are not essentially algorithmic like other methods [76]. An extension of the MCSCF method is multireference CI (MRCI), in which the determinants (the CSFs) from an MCSCF calculation are used to generate more determinants, by promoting electrons in them into virtual orbitals (multifererence, since the final wavefunction "refers back" to several, not just one, determinant).

The CC method is actually related to both the perturbation (section 5.4.2) and the CI approaches (section 5.4.3). Like perturbation theory, CC theory is connected to the linked cluster theorem (linked diagram theorem) [77], which proves that MP calculations are size-consistent (see below). Like straightforward CI it expresses the correlated wavefunction as a sum of the HF ground state determinant and determinants representing the promotion of electrons from this into virtual MOs. As with the MP equations, the derivation of the CC equations is complicated. The basic idea is to express the correlated wavefunction Ψ as a sum of determinants by allowing a series of operators $\hat{T}_1, \hat{T}_2, \ldots$ to act on the HF wavefunction:

$$\Psi = \left(1 + \hat{T} + \frac{\hat{T}^2}{2!} + \frac{\hat{T}^3}{3!} + \cdots\right)\Psi_{\mathrm{HF}} = e^{\hat{T}}\Psi_{\mathrm{HF}} \qquad (5.170)$$

where $\hat{T} = \hat{T}_1 + \hat{T}_2 + \cdots$. The operators $\hat{T}_1, \hat{T}_2, \ldots$ are *excitation operators* and have the effect of promoting one, two, etc., respectively, electrons into virtual spin orbitals. Depending on how many terms are actually included in the summation for $\hat{T}$, one obtains the *coupled cluster doubles* (CCD), *coupled cluster singles and doubles* (CCSD) or *coupled cluster singles, doubles and triples* (CCSDT) method:

$$\hat{T}_{\mathrm{CCD}} = e^{\hat{T}_2}\Psi_{\mathrm{HF}}$$

$$\hat{T}_{\mathrm{CCSD}} = e^{(\hat{T}_1 + \hat{T}_2)}\Psi_{\mathrm{HF}}$$

$$\hat{T}_{\mathrm{CCSDT}} = e^{(\hat{T}_1 + \hat{T}_2 + \hat{T}_3)}\Psi_{\mathrm{HF}}$$

Instead of the very demanding CCSDT calculations one often performs CCSD(T) (note the parentheses), in which the contribution of triple excitations is represented in an approximate way (not refined iteratively); this could be called coupled cluster perturbative triples. The *quadratic configuration* method (QCI) is very similar to the CC method. The most accurate implementation of this in common use is QCISD(T) (quadratic CI singles, doubles, triples, with triple excitations treated in an approximate, non-iterative way). The CC method, which is usually only moderately slower than QCI (Table 5.6), is apparently better [78].

Like MP methods, CI methods require reasonably large basis sets for good results. The smallest (and perhaps most popular) basis used with these methods is the 6-31G*

Table 5.6. Energies and times for some calculations involving electron correlation; HF jobs are shown for comparison

Method/basis	Input	Energy	Time, min
HF/6-31G* opt	AM1 geom, hessian	−191.96224	7
HF/6-31G* opt + freq	AM1 geom, hessian	−191.96224	14
MP2/6-31G* sp	HF/6-31G* geom	−192.52160	1
MP2/6-311G** sp	HF/6-31G* geom	−192.64662	7
MP2/6-31G* opt	AM1 geom, hessian	−192.52390	11
MP2/6-31G* opt + freq	AM1 geom, hessian	−192.52390	91
MP4SDTQ/6-31G* sp	MP2/6-31G* geom	−192.57982	33
MP4SDTQ/6-311G** sp	MP2/6-31G* geom	−192.71075	245
QCISD(T)/6-31G* sp	MP2/6-31G* geom	−192.57883	93
QCISD(T)/6-311G** sp	MP2/6-31G* geom	−192.70884	490
CCSD(T)/6-31G* sp	MP2/6-31G* geom	−192.57808	132
CCSD(T)/6-311G** sp	MP2/6-31G* geom	−192.70798	725

The calculations are with Gaussian 94W [48] on C_{2v} acetone, on a 200 MHZ PentiumPro (a relatively slow machine). A lower absolute energy does not guarantee that a method/basis will give a more accurate activation or reaction energy, as these latter two are energy *differences*, not absolute energies. MP2 = MP2(FC), sp = single point. Methods are given in order of the increasing thoroughness with which they usually treat electron correlation; CC is generally superior to QCI [78]. Note that none of the correlation methods is variational: they can give an energy lower than the true energy.

basis, but where practical the 6-311G** basis, developed especially for post-HF calculations, might be preferable (see Table 5.6). Higher-correlated single-point calculations on MP2 geometries tend to give more reliable energies relative energies than do single-point MP2 calculations on HF geometries (section 5.4.2, in connection with Figs 5.20 and 5.21). There is some evidence that when a correlation method is already being used, one tends to get improved geometries by using a bigger basis set rather than by going to a higher correlation level [79]. Figure 5.21 shows the results of HF and MP2 methods applied to chemical reactions. The limitations and advantages of numerous such methods are shown in a practical way in the Gaussian 94 workbook by Foresman and Frisch [1e]. Energies and times for some correlated calculations are given in Table 5.6.

Size-consistency

Two factors that should be mentioned in connection with post-HF calculations are the questions of whether a method is *size-consistent* and whether it is *variational*. A method is size-consistent if it gives the energy of a collection of n widely-separated atoms or molecules as being n times the energy of one of them. For example, the HF method gives the energy of two water molecules 20 Å apart (considered as a single system or "supermolecule") as being twice the energy of one water molecule. The example below gives the result of HF/3-21G geometry optimizations on a water molecule, and on two water molecules at increasing distances (with the two-H_2O supermolecule the O/H internuclear distance r was held constant at 10, 15, ... Å while all the other geometric parameters were optimized):

Energy of H_2O = -75.58596

$2\times$ Energy or H_2O = -151.17192

Energy of $(H_2O)_2$ = -151.17206, at r=10 Å

Energy of $(H_2O)_2$ = -151.17196, at r=15 Å

Energy of $(H_2O)_2$ = -151.17194, at r=20 Å

Energy of $(H_2O)_2$ = -151.17193, at r=25 Å

Energy of $(H_2O)_2$ = -151.17193, at r=30 Å

As the two water molecules are separated a hydrogen bond (equilibrium bond length r = ca. 2.0 Å) is broken and the energy rises, levelling off at 20–25 Å to twice the energy of one water molecule. With the HF method we find that for any number n of molecules M, at large separation the energy of a supermolecule $(M)_n$ equals n times the energy of one M. The HF method is thus size-consistent. A size-consistent method, we see, is one that scales in a way that makes sense.

Now, it is hard to see why, *physically*, the energy of n identical molecules so widely-separated that they cannot affect one another should *not* be n times the energy of one molecule. Any *mathematical* method that does not mimic this physical behaviour would seem to have a conceptual flaw, and in fact lack of size-consistency also places limits on the utility of the program. For instance, in trying to study the hydrogen-bonded water dimer we would not be able to equate the decrease in energy (compared to twice the energy of one molecule) with stabilization due to hydrogen bonding, and it is unclear how we could computationally turn off hydrogen bonding and evaluate the size-consistency error separately (actually, there is a separate problem, basis set superposition error – see below – with species like the water dimer, but this source of error can be dealt with). It might seem that any computational method must be size-consistent (why shouldn't the energy of a large-separation $(M)_n$ come out at n times that of M?). However, it is not hard to show that "straightforward" CI is not size-consistent unless Eqs (5.168) include all possible determinants, i.e. unless it is *full* CI. Consider a CISD calculation with a very large ("infinite") basis set on two helium atoms which are separated by a large ("infinite"; say ca. 20 Å) distance, and are therefore non-interacting. Note that although helium atoms do not form covalent He_2 molecules, at short distances they *do* interact to form van der Waals molecules. The wavefunction for this four-electron system will contain, besides the HF determinant, only determinants with single and double excitations (CISD). Lacking triple and quadruple excitations it will not produce the exact energy of our He–He system, which must be twice that of one helium atom, but instead will yield a higher energy. Now, a CISD calculation with an infinite basis set on a single He atom *will* give the exact wavefunction, and thus the exact energy of the atom (because only single and double promotions are possible for a two-electron system). Thus the energy of the infinitely-separated He–He system is not twice the energy of a single He atom in this calculation.

Variational behavior

The other factor to be discussed in connection with post-HF calculations is whether a particular method is *variational*. A method is variational (see the variation theorem, section 5.2.3.3) if any energy calculated from it is not less than the true energy of the electronic state and system in question, i.e if the calculated energy is an *upper bound* to the true energy. Using a variational method, as the basis set size is increased we get lower and lower energies, levelling off above the true energy (or at the true energy in the unlikely case that our method treats perfectly electron correlation, relativistic effects, and any other minor effects). Figure 5.18 shows that the calculated energy of H_2 using the HF method approaches a limit ($-1.133\dots$ h) with increasingly large basis sets. The calculated energy can be lowered by using a correlated method and an adequate basis: full CI with the very big 6-311 + +G(3df, 3p2d) basis gives -1.17288 h, only $4.0\,kJ\,mol^{-1}$ (small compared with the H–H bond energy of $435\,kJ\,mol^{-1}$) above the accepted exact energy of -1.17439 h (Fig. 5.18).

If we cannot have both, it is more important for a method to be size-consistent than variational. Recall the methods we have seen in this book:

Hartree-Fock

MP (MP2, MP3, MP4, etc.)

full CI

truncated CI: CIS, CISD, etc.

MCSCF and its CASSCF variant

CC and its QCI variants (QCISD, QCISD(T), QCISDT, etc.)

The HF and full CI methods are both size-consistent and variational. All the other methods we have discussed are size-consistent but not variational. Thus we can use these methods to compare the energies of, say, water and the water dimer, but only with the HF or full CI methods can we be sure that the calculated energy is an upper bound to the exact energy, i.e. that the exact energy is really lower than the calculated (only a very high correlation level and basis set are likely to give essentially the *exact* energy; see section 5.5.2).

Basis set superposition error, BSSE

This is not associated with a particular method, like HF or CI, but rather is a basis set problem. Consider what happens when we compare the energy of the hydrogen-bonded water dimer with that of two noninteracting water molecules. Here is the result of an MP2(FC)/6-31G* calculation; both structures were geometry-optimized, and the energies are corrected for ZPE:

$$\text{Energy of } H_2O = -76.27547\,h$$
$$2 \times \text{ Energy of } H_2O = -152.55094\,h$$
$$\text{Energy of } H_2O \text{ dimer} = -152.55658\,h$$
$$(2 \times \text{ Energy of } H_2O) - (\text{Energy of } H_2O \text{ dimer}$$
$$= -152.55094 - (-152.55658)\,h = 0.00564\,h = 14.8\,kJ\,mol^{-1}$$

The straightforward conclusion is that at the MP2(FC)6-31G* level the dimer is stabler than two noninteracting water molecules by $14.8\,kJ\,mol^{-1}$. If there are no other significant intermolecular forces, then we might say the H-bond energy in the water dimer [80] is $14.8\,kJ\,mol^{-1}$ (that it takes this energy to break the bond – to separate the dimer into noninteracting water molecules). Unfortunately there is a problem with this simple subtraction approach to comparing the energy of a weak molecular association AB with the energy of A plus the energy of B. If we do this we are assuming that if there were no interactions at all between A and B at the geometry of the AB species, then the AB energy would be that of isolated A plus that of isolated B. The problem is that when we do a calculation on the AB species (say the dimer $HOH \cdots OH_2$), in this "supermolecule" the basis functions ("atomic orbitals") of B are available to A so A in AB has a bigger basis set than does isolated A; likewise B has a bigger basis than isolated B. When in AB each of the two components can borrow basis functions from the other. The error arises from "imposing" B's basis set on A and vice versa, hence the name basis set superposition error. Because of BSSE A and B are not being fairly compared with AB, and we should use for the energies of separated A and of B lower values than we get in the absence of the borrowed functions. A little thought shows that accounting for BSSE will give a smaller value for the hydrogen bond energy (or van der Waals' energy, or dipole–dipole attraction energy, or whatever weak interaction is being studied) than if it were ignored.

There are two ways to deal with BSSE. One is to say, as we implied above, that we should really compare the energy of AB with that of A with the extra basis functions provided by B, plus the energy of B with the extra basis functions provided by A. This method of correcting the energies of A and B with extra functions is called the counterpoise method [81], presumably because it balances (counterpoises) functions in A and B against functions in AB. In the counterpoise method the calculations on the components A and B of AB are done with *ghost orbitals*, which are basis functions ("atomic orbitals") not accompanied by atoms (spirits without bodies, one might say): one specifies for A, at the positions that *would* be occupied by the various atoms of B in AB, atoms of zero atomic number bearing the same basis functions as the real atoms of B. This way there is no effect of atomic nuclei or extra electrons on A, just the availability of B's basis functions. Likewise one uses ghost orbitals of A on B. A detailed description of the use of ghost orbitals in Gaussian 82 has been given by Clark [81a]. The counterpoise correction gives only an approximate value of the BSSE, and it is rarely applied to anything other than *weakly-bound* dimers, like hydrogen-bonded and van der Waals species: strangely, the correction *worsens* calculated atomization energies (e.g. covalent $AB \rightarrow A + B$), and it is not uniquely defined for species of more than two components [81b].

The second way to handle BSSE is to swamp it with basis functions. If each fragment A and B is endowed with a really big basis set, then extra functions from the other fragment won't alter the energy much – the energy will already be near the asymptotic limit. So if one simply (!) carries out a calculation on A, B and AB with a sufficiently big basis, the straightforward procedure of subtracting the energy of AB from that of A + B should give a stabilization energy essentially free of BSSE. For good results one needs good geometries and adequate accounting for correlation effects. The use of large basis sets and high correlation levels to get high-quality atomization energies (which are of course of the *covalent* AB → A + B type) is explained in the book by Foresman and Frisch [1e]. Energy calculations are discussed further in section 5.5.2.

5.5 APPLICATIONS OF THE *AB INITIO* METHOD

An extremely useful book by Hehre [32] discusses critically the merits of various computational levels (*ab initio* and others) for calculating molecular properties, and contains a wealth of information, admonitory and tabular, on this general subject.

5.5.1 Geometries

It is probably the case that the two parameters most frequently sought from *ab initio* calculations (and most semiempirical and DFT calculations too) are geometries and (section 5.5.2) energies, although this is not to say that other quantities, like vibrational frequencies (section 5.5.3) and parameters arising from electron distribution (section 5.5.4) are unimportant. Molecular geometries are important: they can reveal subtle effects of theoretical importance, and in designing new drugs or materials [82] the shapes of the candidates for particular roles must be known with reasonable accuracy – e.g. docking a putative drug into the active site of an enzyme requires that we know the shape of the drug and the active site. While the creation of new pharmaceuticals or materials can be realized with the aid of molecular mechanics (chapter 3) or semiempirical methods (chapter 6), the increasingly facile application of *ab initio* techniques to large molecules makes it likely that this method will play a more important role in such utilitarian pursuits. Novel molecules of theoretical interest can be studied reliably only by *ab initio* methods, or possibly by density functional theory (chapter 7), which is closer in theoretical tenor to the *ab initio*, rather than semiempirical, approach. The theory behind geometry optimizations was outlined in section 2.4, and some results of optimizations with different basis sets and electron correlation methods have been given (sections 5.3.3 and 5.4). Extensive discussions of the virtues and shortcomings of various *ab initio* levels for calculating geometries can be found in [1e,g,38].

Molecular geometries or structures refer to the bond lengths, bond angles, and dihedral angles that are defined by two, three and four, respectively, *atomic nuclei*. In speaking of the distance, say, between two "atoms" we really mean the *internuclear* distance, unless we are considering nonbonded interactions, when we might also wish to examine the separation of the van der Waals surfaces. In comparing calculated and experimental structures we must remember that calculated geometries correspond to a fictional frozen-nuclei molecule, one with no ZPE (section 5.2.3.6d), while experimental geometries are averaged over the amplitudes of the various vibrations [83]. Furthermore, different methods measure somewhat different things. The most widely-used experimental methods for finding geometric parameters are X-ray diffraction, electron diffraction and microwave spectroscopy. X-ray diffraction determines geometries in a crystal lattice, where they may be somewhat different than in the gas phase to which *ab initio* reactions usually apply (although structures and energies can be calculated taking solvent effects into account, as explained in section 5.5.2). X-ray diffraction depends on the scattering of photons by the electrons around nuclei, while electron diffraction depends on the scattering of electrons by the nuclei, and microwave spectroscopy measures rotational energy levels, which depend on nuclear positions. Neutron diffraction, which is less used than these three methods, depends on scattering by atomic nuclei.

The main differences are between X-ray diffraction (which probes nuclear positions via electron location) on the one hand and electron diffraction, microwave spectroscopy and neutron diffraction (which probe nuclear positions more directly), on the other hand. The differences result from (1) the fact that X-ray diffraction measures distances between *mean* nuclear positions, while the other methods measure essentially *average* distances, and (2) from errors in internuclear distances caused by the nonisotropic (uneven) electron distribution around atoms. The mean vs. average distinction is illustrated here:

Suppose that nucleus A is fixed and nucleus B is vibrating in an arc as indicated. The distance between the mean positions is r (shown), but on the average B is further away than r.

Differences resulting from nonisotropic electron distribution are significant only for H–X bond lengths: X-rays see electrons rather than nuclei, and the simplest inference of a nuclear position is to place it at the center of a sphere whose surface is defined by the electron density around it. However, since a hydrogen atom has only one electron, for a bonded hydrogen there is relatively little electron density left over from covalent sharing to blanket the nucleus, and so the proton, unlike other nuclei, is *not* essentially at the center of an approximate sphere defined by its surrounding electron density:

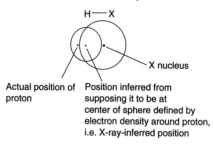

Actual position of proton   Position inferred from supposing it to be at center of sphere defined by electron density around proton, i.e. X-ray-inferred position

Clearly, the X-ray-inferred H–X distance will be less than the internuclear distance measured by electron diffraction, neutron diffraction, or microwave spectroscopy, methods which see nuclei rather than electrons. These and other sources of error that can arise in experimental bond length measurements (like bond length, bond angles and dihedral angles will obviously also depend on nuclear positions) are detailed by Burkert and Allinger [84], who mention nine (!) kinds of internuclear distance r, and a comprehensive reference to the techniques of structure determination may be found in the book edited by Domenicano and Hargittai [85a]. Despite all these problems with defining and measuring molecular geometry (see e.g. [85b], we will adopt the position that it is meaningful to speak of experimental geometries to within 0.01 Å for bond lengths, and to within 0.5° for bond angles and dihedrals [86].

Let us briefly compare HF/3-21G, HF/6-31G* and MP2/6-31G* geometries. Figure 5.23 gives bond lengths and angles calculated at these three levels and experimental bond lengths and angles, for 20 molecules. The geometries shown in Fig. 5.23 are analyzed in Table 5.7, and Table 5.8 provides information on dihedral angles in eight molecules. There should be little difference between MP2(full) geometries and the MP2(FC) geometries used here. This (admittedly limited) survey suggests that:

HF/3-21G$^{(*)}$ geometries are almost as good as HF/6-31G* geometries.

MP2/6-31G* geometries are on the whole slightly but significantly better than HF/6-31G* geometries, although individual MP2 parameters are *sometimes* a bit worse.

HF/3-21G$^{(*)}$ and HF/6-31G* C–H bond lengths are consistently slightly (ca. 0.01–0.03 and ca. 0.01 Å, respectively) shorter than experimental, while MP2/6-31G* C–H bond lengths are not systematically over- or underestimated.

HF/6-31G* O–H bonds are consistently slightly (ca. 0.01 Å) shorter than experimental, while MP2/6-31G* O–H bond lengths are consistently slightly (ca. 0.01 Å) longer. HF/3-21G$^{(*)}$ O–H bond lengths are not consistently over- or underestimated.

None of the three levels consistently over- or underestimates C–C bond lengths.

HF/6-31G* C–X (X = O, N, Cl, S) bond lengths tend to be underestimated slightly (ca. 0.015 Å) while MP2/6-31G* C–X bond lengths may tend to be slightly (ca. 0.01 Å) overestimated. HF/3-21G$^{(*)}$ C–X bond lengths are not consistently over- or underestimated.

HF/6-31G* bond angles may tend to be slightly larger (ca. 1°) than experimental, while MP2/6-31G* angles may tend to be slightly (0.7°) smaller.

HF/3-21G$^{(*)}$ bond angles are not consistently over- or underestimated. Dihedrals do not seem to be consistently over-or underestimated by any of the three levels. The HF/3-21G$^{(*)}$ level breaks down completely for HOOH, where a dihedral angle of 180°, far from the experimental 119.1°, is calculated; omitting this error of 61° and the $ClCH_2CH_2OH$ HOCC dihedral error of 7.6° lowers the HF/3-21G$^{(*)}$ error from 8.8 to 2.5°. The experimental value of 58.4° for the $ClCH_2CH_2OH$ HOCC dihedral is suspect because of its anomalously large deviation from all three calculated results, and because it is among those dihedrals which are said to be suspect or having a large or unknown error (designated X in Harmony *et al.* – see reference in Table 5.8). The error for the HOOH dihedral represents a clear failure of the HF/3-21G$^{(*)}$ level and is an example of a case which provides an argument for using the 6-31G* rather than the 3-21G basis, although the latter is much faster and often of comparable accuracy (of course with correlated methods like MP2 a smaller basis than 6-31G* should not be used, as pointed out in section 5.4). The errors in calculated dihedral angles are ca. 2–3° for HF/6-31G*, and ca. 2° for MP2/6-31G*: omitting the $ClCH_2CH_2OH$ HOCC dihedral errors of 8.6° and 5.9° from the sample lowers the HF error from 2.9° to 2.3° and the MP2 error from 2.3 to 1.9.

The accuracy of *ab initio* geometries is astonishing, in view of the approximations present: the 3-21G$^{(*)}$ basis set is small and the 6-31G* is only moderately large, and so these probably cannot approximate closely the true wavefunction; the HF method does not account properly for electron correlation, and the MP2 method is only the simplest

Table 5.7. Errors in HF/3-21G$^{(*)}$, HF/6-31G*, and MP2(FC)/6-31G* bond lengths and angles, from Fig. 5.23

C–H	O–H, N–H, S–H	C–C	C–O, N, F, Cl, S	Angles
MeOH −0.015/−0.013/−0.004 −0.009/−0.006/0.003	H$_2$O 0.009/−0.010/0.011	Me$_2$CO 0.008/0.007/0.006	MeOH 0.020/−0.021/0.004	H$_2$O (HOH) 3.2/1.0/−0.6
HCHO −0.033/−0.024/−0.012	H$_2$O$_2$ 0.005/−0.016/0.011	CH$_3$CH$_3$ 0.011/−0.010/−0.003	HCHO −0.001/−0.024/0.013	H$_2$O$_2$ (HOO) −0.5/2.1/−1.4
MeF −0.021/−0.018/−0.008	MeOH 0.003/−0.017/0.007	CH$_2$CH$_2$ −0.024/−0.022/−0.002	MeF 0.021/−0.018/0.008	MeOH (HCO) −0.9/0.0/−0.9 (COH) 2.3/1.4/−0.6
HCN −0.015/−0.006/0.004	HOF 0.010/−0.014/0.013	HCCH −0.015/−0.017/0.015	HCN −0.016/−0.020/0.024	HCHO (HCH) −1.6/−0.8/−0.9
MeNH$_2$ −0.009/−0.008/0.001 −0.016/−0.015/−0.007	MeNH$_2$ −0.007/−0.008/0.008	CH$_3$CH$_2$CH$_3$ 0.015/0.002/0.000	MeNH$_2$ 0.000/−0.018/−0.006	MeF (HCH) −1.1/−0.7/−0.8
CH$_3$CH$_3$ −0.012/−0.010/−0.003	HOCl −0.002/−0.024/0.004	CH$_2$CHCH$_3$ 0.009/0.001/−0.002 −0.002/0.000/0.020	Me$_2$CO −0.011/−0.030/0.006	HOF (HOF) 2.2/3.0/0.4
CH$_2$CH$_2$ −0.011/−0.009/0.000	H$_2$S −0.009/−0.010/0.004	HCCCH$_3$ 0.007/0.009/0.004 −0.018/−0.019/0.014	MeCl 0.025/0.004/−0.002	MeNH$_2$ (HCN) 0.9/0.9/1.5
CHCH −0.010/−0.004/0.005	MeSH −0.009/−0.009/0.005		MeSH 0.003/−0.001/−0.003	Me$_2$CO (CCC) −2.2/−0.6/−0.7

			Me₂SO	
MeCl			Me_2SO	CH_3CH_3 (HCH)
−0.020/−0.018/−0.007			−0.008/−0.003/0.010	0.3/−0.1/−0.1
MeSH				CH_2CH_2 (HCH)
−0.010/−0.009/0.000				−1.6/−1.4/−1.2
−0.011/−0.010/−0.001				
				$CH_3CH_2CH_3$ (CCC)
				−0.8/0.4/−0.1
				CH_2CHCH_3 (CCC)
				0.4/0.9/0.2
				MeCl (HCH)
				0.8/0.5/0.0
				H_2S (HSH)
				2.1/2.3/1.2
				MeSH (CSH)
				1.1/1.4/0.3
0+, 13−, none 0	4+, 4−, none 0	5+, 4−, none 0	4+, 4−, one 0	10+, 7−, one 0
0+, 13−, none 0	0+, 8−, none 0	4+, 4−, one 0	1+, 8−, none 0	11+, 5−, two 0
4+, 7−, two 0	8+, 0−, none 0	5+, 3−, one 0	6+, 3−, none 0	6+, 11−, one 0
Mean of 13:	Mean of 8:	Mean of 9:	Mean of 9:	Mean of 18:
0.015/0.012/0.004	0.007/0.014/0.008	0.012/0.010/0.007	0.012/0.015/0.009	1.3/1.0/0.7

Errors are given as HF/3-21G*/HF/6-31G*/MP2/6-31G*. In some cases (e.g. MeOH) errors for two bonds are given, on one line and on the line below. A minus sign means that the calculated value is less than the experimental. The numbers of positive, negative, and zero deviations from experiment are summarized at the bottom of each column. The averages at the bottom of each column are arithmetic means of the absolute values of the errors.

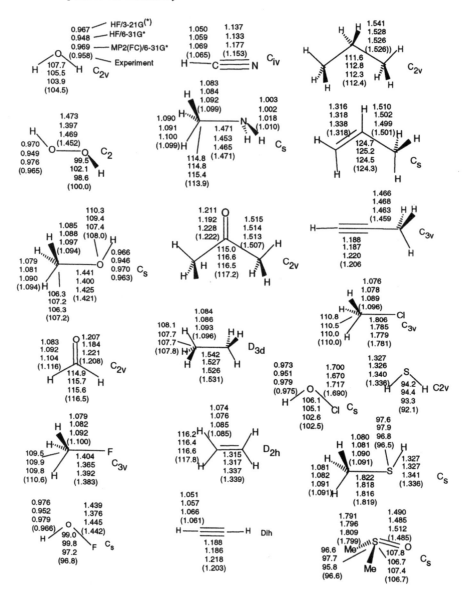

Figure 5.23. A comparison of some HF/3-21G$^{(*)}$, HF/6-31G* and MP2(FC)/6-31G* geometries. Calculations are by the author and experimental geometries are from [1g]. Note that all CH bonds are ca. 1 Å, all other bonds range from ca. 1.2–1.8 Å, and all bond angles (except for linear molecules) are ca. 90–120°.

approach to handling electron correlation; the Hamiltonian in both the HF and the MP2 methods used here neglects relativity and spin-orbit coupling. Yet with all these approximations the largest error (Table 5.7) in bond lengths is only 0.033 Å (HF/3-21$^{(*)}$ level for HCHO) and the largest error in bond angles is only 3.2° (HF/3-21$^{(*)}$ level for

Table 5.8. HF/3-21G$^{(*)}$, HF/6-31G* and MP2(FC)/6-31G* dihedral angles (degrees)

Molecule	Dihedral angles				
	HF/3-21G$^{(*)}$	HF/6-31G*	MP2/6-31G*	Experiment	Errors
HOOH	180.0	116.0	121.3	119.1[a]	61(*sic*)/−3.1/2.2
FOOF	84.1	84.1	85.8	87.5[b]	−3.4/−3.4/−1.7
FCH$_2$CH$_2$F (FCCF)	74.9	69.4	69.0	73[b]	1.9/−4/−4
FCH$_2$CH$_2$OH					
(FCCO)	58.4	61.3	60.1	64.0[c]	−5.6/−2.7/−3.9
(HOCC)	52.7	57.8	54.1	54.6[c]	−1.9/2.5/−0.5
ClCH$_2$CH$_2$OH					
(ClCCO)	65.8	65.7	65.0	63.2[b]	2.6/2.5/1.8
(HOCC)	66.0	67.0	64.3	58.4[b]	7.6/8.6/5.9
ClCH$_2$CH$_2$F (ClCCF)	65.9	67.0	65.9	68[b]	−2.1/−1/−2.1
HSSH	89.8	89.8	90.4	90.6[a]	−0.8/−0.8/−0.2
FSSF	89.4	88.7	88.9	87.9[b]	1.5/0.8/1.0 Deviations: 5+, 5 − /4+, 6 − /4+, 6− mean of 10: 8.8/2.9/ 2.3.*

Omitting the largest error for each of the three methods (61/8.6/8.9 for HF/3-21G$^{(*)}$/HF/6-31G/ MP2(FC)/6-31G*, respectively, the mean of 9 errors for each method is 3.0/2.3/1.9.
[a] Hehre [1g], pp. 151, 152.
[b] M. D. Harmony, V. W. Laurie, R. L. Kuczkowski, R. H. Schwenderman, D. A. Ramsay, F. J. Lovas, W. H. Lafferty, A. G. Makai, "Molecular Structures of Gas-Phase Polyatomic Molecules Determined by Spectroscopic Methods," J. Phys. Chem. Ref. Data, 1979, *8*, 619–721.
[c] J. Huang and K. Hedberg, J. Am. Chem. Soc., 1989, *111*, 6909.
Errors are given in the *Errors* column as HF/3-21G$^{(*)}$/HF/6-31G*/MP2/6-31G*. A minus sign means that the calculated value is less than the experimental. The numbers of positive and negative deviations from experiment and the average errors (arithmetic means of the absolute values of the errors) are summarized at the bottom of the *Errors* column. Calculations are by the author; references to experimental measurements are given for each measurement. Some molecules have calculated minima at other dihedrals in addition to those given here, e.g. FCH$_2$CH$_2$F at 180°. Errors are presented: HF/3-21G$^{(*)}$/HF/6-31G*/MP2/6-31G*.

H$_2$O). The largest error in dihedral angles (Table 5.8), omitting the 3-21G result for H$_2$O$_2$, is 8.6° (HF/6-31G* for ClCH$_2$CH$_2$OH HOCC), but as stated above the reported experimental dihedral of 58.4° is suspect.

From Fig. 5.23 and Table 5.7, the mean error in 39 (13 + 8 + 9 + 9) bond lengths is 0.01–0.015 Å at the HF/3-21$^{(*)}$ and HF/6-31G* levels, and ca. 0.005–0.008 Å at the MP2/6-31G* level. The mean error in 18 bond angles is only 1.3° and 1.0° at the HF/3-21(*) and HF/6-31G* levels, respectively, and 0.7° at the MP2(FC)/6-31G* level. From Table 5.8 the mean dihedral angle error at the HF/3-21(*) level for 9 dihedrals (omitting the questionable ClCH$_2$CH$_2$OH dihedral) is 3.0°; the mean of 8 dihedral errors (omitting the ClCH$_2$CH$_2$OH and the HOOH errors) is 2.5°. For the other two levels the mean of 10 dihedral angles (including the questionable ClCH$_2$CH$_2$OH dihedral) is 2.9° (HF/6-31G*) and 2.3° (MP2/6-31G*). If we agree that errors in calculated bond lengths, angles and dihedrals of up to 0.02 Å, 3° and 4° respectively correspond to fairly good structures, then *all* the HF/3-21(*), HF/6-31G* and MP2/6-31G* geometries, with the exception of the HF/3-21(*) HOOH dihedral, which is simply wrong, and the possible exception of the HOCC dihedral of ClCH$_2$CH$_2$OH, are fairly good. We should, however, bear in mind that, as with the HF/3-21(*) HOOH dihedral, there is the possibility of an occasional nasty surprise. Interestingly, HF/3-21(*) geometries are, for some series of compounds, somewhat better than MP$_2$/6-31G* ones. For example, the RMS errors in geometry for the series H$_2$, CH, NH, OH, HF, CN, N$_2$, H$_2$O, HCN, CH$_3$, and CH$_4$ using UHF/3-21G(*), MP2/6-31G*, and MP2/6-31G† (a modified basis used in CBS calculations – section 5.5.2.2b) are 0.012, 0.016 and 0.015 Å, respectively [86].

The calculations summarized in Tables 7 and 8 are in reasonable accord with conclusions based on information available ca. 1985 and given by Hehre, Radom, Schleyer and Pople [87]: HF/6-31G* parameters for A–H, A/B single and A/B multiple bonds are usually accurate to 0.01, 0.03 and 0.02 Å, respectively, bond angles to ca. 2° and dihedral angles to ca. 3°, with HF/3-21G(*) values being not quite as good. MP2 bond lengths appear to be somewhat better, and bond angles are usually accurate to ca. 1°, and dihedral angles to ca. 2°. These conclusions from Hehre *et al.*, hold for molecules composed of first-row elements (Li to F) and hydrogen; for elements beyond the first row larger errors not uncommon.

The main advantage of MP2/6-31G* optimizations over HF/3-21(*) or HF/6-31G* ones is not that the geometries are *much* better, but rather that for a stationary point, MP2 optimizations followed by frequency calculations are more likely to give the correct curvature of the potential energy surface (chapter 2) for the species than are HF optimizations/frequencies. In other words, the correlated calculation tells us more reliably whether the species is a relative minimum or merely a transition state (or even a higher-order saddle point; see chapter 2). Thus fluorodiazomethane [71] and several oxirenes [44] are (apparently correctly) predicted by MP2 optimizations to be not minima, while HF optimizations indicate them to be minima. The interesting hexaazabenzene ("benzene-N$_6$") is predicted to be a minimum at the HF/6-31G* level, but a hilltop with two imaginary frequencies at the MP2/6-31G* level [88]. For transition states, in contrast to ground states, we don't have experimental geometries, but correlation effects can certainly be important for their *energies* (section 5.5.2.2b), and MP2/6-31G* geometries for transition states are probably significantly better in general than HF/6-31G* ones.

Suppose we want something better than "fairly good" structures? Experienced workers in computational chemistry have said [89]

When we speak of "accurate" geometries, we generally refer to bond lengths that are within about 0.01–0.02 Å of experiment and bond and dihedral angles that are within about 1–2° of the experimentally-measured value (with the lower end of both ranges being more desirable).

Even by these somewhat exacting criteria, MP2/6-31G* and even HF/6-31G* calculations are not, in the cases studied here, far wanting; the worst deviations from experimental values seem to be for dihedral angles, and these may be the least reliable experimentally. However, since some larger deviations from experiment are seen in our sample, it must be conceded that HF/6-31G* and even MP2/6-31G* calculations cannot be *relied* on to provide "accurate" (sometimes called high-quality) geometries. Furthermore, there are some molecules that are particularly recalcitrant to accurate calculation of geometry (and sometimes other characteristics); two notorious examples are FOOF (dioxygen difluoride) and ozone (these have been described as "pathological" [90]). Here are the HF/6-31G*, MP2(FC)/6-31G* and experimental [91] geometries:

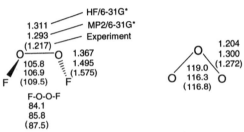

The errors in the calculated geometries are (HF/6-31G*/MP2/6-31G*):

FOOF: FO length, −0.208/−0.080 Å; OO length, 0.094/0.076 Å

FOO angle, −3.7°/−2.6°

FOOF dihedral, −3.4°/−1.7°

O_3:   OO length, −0.068/0.028 Å

OOO angle, 2.2/−0.5

These calculated geometries do not satisfy even our "fairly good" criterion and are well short of being "accurate"; the bond lengths are particularly bad. Using the 6-311++G** basis (for FOOF, 88 vs. 60 basis functions; for O_3, 66 vs. 45 basis functions) we get for calculated geometries (errors) using HF/6-311 + +G**:

FOOF: FO length, 1.353 Å (−0.222); OO length, 1.300 (0.083) Å

FOO angle, 106.5° (−3.0)

FOOF dihedral, 85.3° (−2.2)

O_3:   OO length, 1.194 Å (−0.078)

OOO angle, 119.4° (2.6)

Thus with a much larger basis, but still using the Hartree-Fock method, the FOOF geometry is about the same and the O_3 geometry has become even worse than at the HF/6-31G* level! Better geometries were obtained by going beyond the MP2 correlational level; we get for calculated geometries (errors) using CCSD(T)/6-31G*:

FOOF: FO length, 1.539 Å (−0.036); OO length, 1.276 (0.059) Å

FOO angle, 107.5° (−2.0)

FOOF dihedral, 86.7° (−0.8)

O_3:   OO length, 1.296 Å (0.024)

OOO angle, 116.5° (−0.3)

The O_3 geometry is now on the verge of satisfying ours "accurate" requirement, but the FOOF geometry still has unsatisfactory bond lengths. CCSD(T) with a considerably bigger basis set (similar to 6-311(2d)) has been reported [92] to give for FOOF errors of 0.039, −0.001, −0.6 and 0.3 for the FO and OO lengths and the FOO angle and FOOF dihedral, respectively; even here one of the bond lengths does not meet the "accurate" criterion of being within 0.02 Å of experiment.

The problem with ozone probably arises at least partly from the fact that this molecule has singlet diradical character: it is approximately a species in which two electrons, although having opposite spin, are not paired in the same orbital:

The HF method works best with normal closed-shell molecules, because it uses a single Slater determinant, but ozone has open-shell diradical character: it is, or at least resembles, a species with two half-filled orbitals, one with a single α electron and the other with a single β electron; there are various ways of handling this molecule [91]. The cause of the problems with FOOF are harder to explain, but fluorine is known to be a somewhat troublesome element [92], although not all fluorine-containing species require very large basis sets [93].

5.5.2   Energies

5.5.2.1   Energies: Preliminaries

Along with geometries (section 5.5.1), the molecular features most frequently sought from *ab initio* calculations are probably energies. An *ab initio* calculation gives an energy quantity that represents the energy of the molecule (or atom) relative to its stationary constituent electrons and nuclei at infinite separation, this separated state being taken as the zero of energy. The *ab initio* energy of a species is thus the negative of the energy needed to dissociate it completely (to infinite separation) into the electrons and nuclei, or the negative of the energy given out when the electrons and nuclei "fall together" from infinity to form the species. This was pointed out for HF energies (section 5.2.3.6d, in connection with Eq. (5.93)), and the infinite-separation reference point also holds for correlated *ab initio* energies. By *ab initio* energy, then, we normally mean the electronic energy (whether calculated by the HF or by a correlation method) plus the internuclear repulsion (cf. Eq. (5.93):

ab initio energy (Hartree–Fock):

$$E_{HF}^{total} = E_{HF} + V_{NN} \qquad (5.171)$$

ab initio energy (correlated method):

$$E_{\text{correl}}^{\text{total}} = E_{\text{correl}} + V_{NN} \tag{5.172}$$

If the *ab initio* energy has been corrected by adding the ZPE (cf. Eq. (5.94)), giving the total internal energy at 0 K, this should be pointed out: *ab initio* energy, corrected for ZPE:

$$E_{0K}^{\text{total}} = E^{\text{total}} + \text{ZPE} \tag{5.173}$$

The ZPE-corrected *ab initio* energy is preferred for calculating relative energies (see below). At the end of a calculation E^{total} (HF or correlated) is given; if we wish to include ZPE and get E_{0K}^{total} a frequency calculation is necessary.

What we actually want is usually not these "absolute" *ab initio* energies, because chemistry really deals with *relative* energies (*all* energies are relative to something, but in this context it is useful to restrict the term to the energy difference between reactants and products or transition states, or between two isomers). We are thus interested in the reaction energy (the energy difference between the product and reactant) and what we might call the activation energy (the energy difference between the transition state and reactant; note however – see below – that the well-known Arrhenius activation energy is not simply the difference in calculated energies of transition state and reactants). Calculating the relative stabilities of isomers amounts to calculating the reaction energy of an isomerization reaction.

Figure 5.24 shows what Coulson meant when he said that calculating the relative stabilities of isomers by subtracting absolute energies is like finding the weight of the captain by weighing the ship with and without him [94]. The absolute *ab initio* energies of the two isomers shown are each about 407,000 kJ mol^{-1}, and the difference in their energies is only about 9 kJ mol^{-1}, which is 1 part in 45,000, and these figures are quite typical. If we conservatively assign a captain a weight of 100 kg, the analogy corresponds to a small ship weighing 4,500,000 kg or about 5000 tonnes. Yet the astonishing thing is that modern *ab initio* calculations can, as we shall see, accurately and reliably predict relative energies. A comprehensive account of energy calculations by *ab initio* and other methods is given by Irikura and Frurip [95].

Reaction energies belong to the realm of thermodynamics, and activation energies to that of kinetics: the energy difference between the products and the reactants ("difference" is defined here as product energy minus reactant energy) determines the extent to which a reaction has progressed at equilibrium, i.e the equilibrium constant, and the energy difference between the transition state and the reactants (transition state energy minus reactant energy) determines (partially; see section 5.5.2.2d) the rate of the reaction, i.e the rate constant (Fig. 5.25). The term "energy" in chemistry can mean (to give the three most common entities) potential energy, enthalpy (heat content), H, or Gibbs free energy, G. The potential energy on a computed Born–Oppenheimer surface (the usual "potential energy surface"; section 2.3) represents 0 K enthalpy differences without ZPE. Enthalpy differences, ΔH, and free energy differences, ΔG, are related through the entropy difference:

$$\Delta G = \Delta H - T\Delta S \tag{5.174}$$

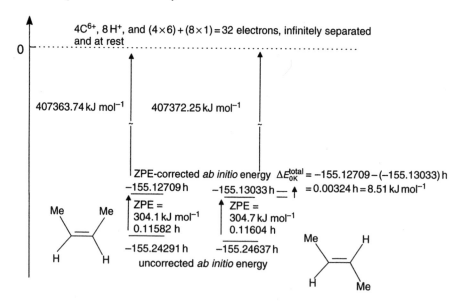

Figure 5.24. Absolute and relative *ab initio* energies, with and without ZPE correction. These are from HF/3-21G calculations. The calculated reaction energy for the (*E*) to (*Z*) (*cis* to *trans*) isomerization is $-8.51\,kJ\,mol^{-1}$.

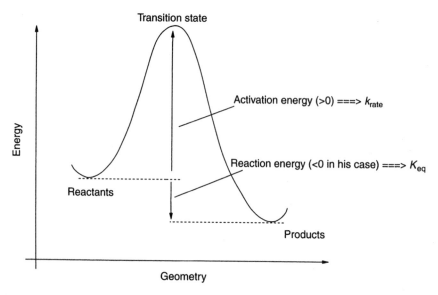

Figure 5.25. The reaction energy, the energy difference of products and reactants, determines the extent of a reaction, i.e its equilibrium constant. The activation energy (the simple *ab initio* energy difference shown here is not exactly the conventional Arrhenius activation energy), the energy difference of transition state and reactants, partially determines the rate of a reaction, i.e. its rate constant. Unfortunately, "energy" is ambiguous, since chemists use the terms potential energy, enthalpy (heat energy), and free energy: see section 5.5.2.1.

More detailed discussion of enthalpy, free energy, and entropy are given in books on thermodynamics, and the relationships between these quantities and processes at the molecular level are explained in books on statistical mechanics [96]; general discussions of these topics are given in physical chemistry texts [97].

To get an intuitive feel for ΔH we can regard it as essentially a measure of the strengths of the bonds in the products or the transition state, compared to the strengths of the bonds in the reactants [98]:

$$\Delta H = H(\text{pdts/TS}) - H(\text{reactants})$$
$$\simeq \sum \text{bond energies (reactants)} - \sum \text{bond energies(pdts/TS)}$$

(5.175)

(pdt or TS depending on whether we are considering reaction enthalpy or activation enthalpy). Thus an exothermic process, which from the definition has $\Delta H < 0$, has stronger bonds in the products than in the reactants; in some sense the bonds lose heat energy, becoming tighter and stabler. The bond energy tables given in most organic chemistry textbooks can be used to calculate rough values of ΔH(reaction), and *accurate* reaction enthalpies can sometimes be obtained from the more sophisticated use of bond energies and similar quantities [99]. To see an application of simple bond energy tables [100], consider the keto/enol reaction:

Using Eq. (5.175):

$$\Delta H \simeq \sum \text{bond energies (reactants)} - \sum \text{bond energies(pdts/TS)}$$
$$= (4\text{C–H} + \text{C–C} + \text{C=O}) - (3\text{C–H} + \text{C=C} + \text{C–O} + \text{O–H})$$
$$= (4 \times 414 + 347 + 749) - (3 \times 414 + 611 + 360 + 464) \text{ kJ mol}^{-1}$$
$$= 2752 - 2677 \text{ kJ mol}^{-1} = 75 \text{ kJ mol}^{-1}$$

The ethanal to ethenol (acetaldehyde to vinyl alcohol) reaction is predicted to be endothermic by 75 kJ mol^{-1}, i.e. neglecting entropy the enol is predicted to lie 75 kJ mol^{-1} above the aldehyde. Because these are only average bond energies, the apparently remarkable agreement with the *ab initio* calculations in Fig. 5.21 (71.6 kJ mol^{-1}; the connection between ΔH from calculations like this and ΔE from *ab initio* calculations is discussed below) must be regarded as a coincidence. In any case, the correct value is about 44 kJ mol^{-1} (section 5.5.2.2e). Crude bond energy calculations like this can be expected to be in error by 50 or more kJ mol^{-1}. Accurate bond energy calculations can however be done [99] using essentially bond energies that refer to quite specific structural environments; for example, a C–H bond on a primary sp^3 carbon that is in turn attached to another sp^3 carbon.

For a reaction taking place at 0 K the enthalpy change is simply the internal energy change at 0 K:

$$\Delta H(0\text{K}) = \Delta E_{0\text{K}}^{\text{total}}$$

(5.176)

Note that although the calculation of E_{0K}^{total} values for ΔE_{0K}^{total} demands frequency jobs, which are relatively time-consuming ("expensive"), accurate relative energy differences require this, and we will regard the ZPE-uncorrected *ab initio* energy E^{total} as only an approximation to E_{0K}^{total} (see 5.2.3.6d and Fig. 2.20). At temperatures other than 0 K, ΔH is ΔE_{0K}^{total} plus the increases in translational, rotational, vibrational and electronic energies on going from 0 K to the higher temperature T, plus the work done by the system in effecting a pressure or volume change:

$$\Delta H(T) = \Delta E_{0K}^{total} + \Delta E_{trans} + \Delta E_{rot} + \Delta E_{vib} + \Delta E_{el} + \Delta(PV) \qquad (5.177)$$

One frequently chooses the standard temperature of 298.15 K, about room temperature. From 0 K to room temperature the increase in electronic energy is negligible and the increase in vibrational energy is small.

The entropy difference ΔS for a process is essentially a measure of the disorder of the products or the transition state, compared to the disorder of the reactants:

$$\Delta S = S(\text{pdts/TS}) - S(\text{reactants})$$

$$= \text{disorder (pdts/TS)} - \text{disorder (reactants)}$$

(pdt or TS depending on whether we are considering reaction entropy or activation entropy)

Entropy is a sophisticated concept, and this is not the place to give a rigorous definition of disorder; suffice it to say that a disordered system is more probable than an ordered one, and the entropy of a system is proportional to the logarithm of its probability [101]. Intuitively, we see that $\Delta S > 0$ for a process in which the product or the transition state is less symmetrical or has more freedom of motion than do the reactants. For example, ring-opening reactions, since they relieve constraints on intramolecular motion, should be accompanied by an increase in entropy. Note that an increase in entropy favors a process: it increases a rate constant (activation entropy) or an equilibrium constant (reaction entropy), while an increase in enthalpy disfavors a process.

More details on the calculation of entropies is given in the book by Hehre, Radom, Schleyer and Pople, who also tabulate the errors in calculated entropy for small molecules composed of elements from H to F [102]. Errors in calculated entropies at 300 K are 1.7, 1.3 and 0.8 J mol^{-1} K^{-1} (0.4, 0.3 and 0.2 cal mol^{-1} K^{-1}) at 300 K, for frequency calculations at the HF/3-21G, 6-31G* and MP2/6-31G* levels, respectively. From Eq. (5.175) this corresponds to an error in free energy at 300 K of $300 \times (0.8 + 0.8)$ J mol^{-1} = 480 J mol^{-1} or 0.5 kJ mol^{-1}, for the MP2/6-31G* calculations. This is much smaller than the enthalpy error of ca. 10 kJ mol^{-1} which can be reliably obtained with high-accuracy methods (see below) and shows that in current *ab initio* work errors in free energies can be expected to come mainly from the enthalpy. Many programs, e.g. Gaussian and Spartan, automatically calculate the correction terms to be added to ΔE_{0K}^{total} in Eq. (5.177) at the end of a frequency calculation, and print out the 298.15 K enthalpy or the correction to the 0 K enthalpy. Reaction entropies are needed to calculate free energies of reaction (from Eq. (5.174)), from which equilibrium constants [103] can be calculated:

$$\Delta G_{react} = -RT \ln K_{eq} \qquad (5.178)$$

where several species are in equilibrium, the ratios are proportional to their Boltzmann exponential factors. For example, if the relative free energies G of A, B and C are 0, 5.0 and 20.0 kJ mol^{-1} (here G for A has been set to zero and B and C lie 5.0 and 20.0 kJ mol^{-1} higher) then

$$[A] : [B] : [C] = \exp(-0/RT) : \exp(-5.0/RT) : \exp(-20.0/RT);$$

at room temperature $RT = 2.48$ kJ mol^{-1} and so at this temperature

$$[A] : [B] : [C] = 1 : 0.133 : 0.000315 = 3175 : 422 : 1$$

Activation entropies are useful because they can give information on the structure of a transition state (as stated above, a more confined transition state is signalled by a negative activation entropy), but the *ab initio* calculation of *rate* constants [104] from activation free energies is not as straightforward as the calculation of *equilibrium* constants from reaction free energies. The crudest way to calculate a rate constant is to use the Arrhenius equation [96c,105]

$$k_r = Ae^{-\Delta E_a/RT} \tag{5.179}$$

and to simply approximate the preexponential factor A by that for a similar reaction (a typical value is 10^{12}–10^{13} [106])and to approximate ΔE_a by ΔE_{0K}^{total} or by the value of $\Delta H^{\ddagger}$ (the activation enthalpy) at the temperature in question. The Arrhenius activation energy and the activation enthalpy are actually related by

$$\Delta H^{\ddagger} = E_a - RT \tag{5.180}$$

for a unimolecular reaction, and by

$$\Delta H^{\ddagger} = E_a - 2RT \tag{5.181}$$

for a bimolecular reaction in the gas phase [107]. The main problem with this is that the preexponential factor A varies by a large factor even for, say, reactions which are formally unimolecular [106]:

$$CH_3NC \rightarrow CH_3CN \qquad 3.98 \times 10^{13}$$
$$\text{cyclopropane} \rightarrow \text{propene} \quad 1.58 \times 10^{15}$$
$$C_2H_6 \rightarrow 2CH_3 \qquad 2.51 \times 10^{17}$$

so that this method of guessing by analogy could give a value for of A that was out by a factor of 10^4 (or more). The exponential factor is prone to smaller errors, since calculating $\Delta H^{\ddagger}$ to within 10 kJ mol^{-1} is now feasible, and an error of this size corresponds to an error factor in $\exp(-\Delta E_a)$ of $\exp(-10/2.48) = 57$ (at $T = 298$ K). This may seem to be itself very big, but a simple method of reliably calculating rate constants to within a factor of 100 would be very useful for estimating the stability of unknown substances. Note that for unimolecular processes like the rearrangement of a molecule to a stabler isomer, the halflife, an intuitively more meaningful quantity than the rate constant, is simply

$$t_{1/2} = \frac{\ln 2}{k_r} = \frac{0.693}{k_r} \tag{5.181}$$

i.e. the halflife of a unimolecular reaction is approximately the reciprocal of its rate constant.

5.5.2.2   Energies: calculating quantities relevant to thermodynamics and to kinetics

5.5.2.2a   Thermodynamics; "direct" methods, isodesmic reactions

Here we are concerned with the relative energies of species other than transition states. Such molecules are sometimes called "stable species," even if they are not at all stable in the usual sense, to distinguish them from transition states, which exist only for an instant on the way from reactants to products. A "stable species," in contrast, sits in a potential energy well and survives at least a few molecular vibrations ($>$ ca. 10^{-13} s). The very useful book by Hehre [32] contains a wealth of information on computational and experimental results concerning thermodynamic quantities.

The *ab initio* reaction energy that is most commonly calculated is simply the difference in ZPE-corrected energies, ΔE_{0K}^{total}, which is the reaction enthalpy change at 0 K (Eq. (5.176)). This provides an easily-obtained indication of whether a reaction is likely to be exothermic or endothermic, or of the relative stabilities of isomers. Table 5.9 illustrates this procedure. The results are only semiquantitatively correct, and the HF/6-31G* method is not necessarily better here than the HF/3-21G. In fact, it has been documented by extensive calculations that such HF/3-21G and HF/6-31G* energy differences generally give only a rough indication of energy changes. Much better results are obtained from MP2/6-31G* calculations on MP2/6-31G*, HF/3-21G* or even semiempirical AM1 geometries, and it is well worth consulting the book by Hehre for details [108].

To get the best results from relatively low-level calculations, one can utilize *isodesmic reactions* (Greek: "same bond," i.e. similar bonding on both sides of the equation). These are reactions in which the number of each kind of bond and each kind of lone pair is conserved. For example,

$$NH_3 + CH_3NH_3^+ \rightarrow NH_4^+ + CH_3NH_2 \tag{5.182}$$

and

$$CH_2F_2 + CH_4 \rightarrow CH_3F + CH_3F \tag{5.183}$$

are isodesmic reactions; the first one has on each side 6 N–H bonds, one C–N bond and one nitrogen lone pair, and the second has on each side 6 C–H and 2 C–F bonds. The reaction

$$H_3C–CH_3 + H_2 \rightarrow 2CH_4 \tag{5.184}$$

is, strictly speaking, not isodesmic, since although it has the same number of bonds, even the same number of single bonds, on both sides, there are 6 C–H, one C–C, and 2 H–H bonds on one side and 8 C–H bonds on the other. Note that an isodesmic reaction does not have to be *experimentally* realizable: it is an artifice to obtain a reasonably accurate energy difference by ensuring that as far as possible errors cancel. This will happen to the extent that particular errors are associated with particular structural features; electron correlation effects are thought to be especially important in calculating energy differences, and such effects tend to cancel when the number of electron pairs of each kind is conserved.

Reactions like (5.182) give fairly good quantitative results for the proton affinities (essentially, the basicity) of nitrogen bases [109], by using NH$_3$ to deprotonate a series

Table 5.9. Reaction energies and relative energies of isomers (HF/3-21G and HF/6-31G*)

Reactants E, h	Products E, h	Reaction energy, or relative energy of isomers	
		Calculated, h/kJ mol^{-1}	Exp, kJ mol^{-1}
$H_2 + Cl_2$	2HCl	$-915.94846 - (-915.86949)$	-192
$-1.11234 + (-914.75715)$	$2(-457.97423) = -915.94846$	$= -0.07897/-207$	
$-1.11625 + (-918.91145) = -920.02770$	$2(-460.05272) = -920.10544$	$-920.10544 - (-920.02770)$	
		$= -0.07774/-204$	
$2H_2 + O_2$	$2H_2O$	$-151.12838 - (-150.99008)$	-523
$2(-1.11234) + (-148.76540) = -150.99008$	$2(-75.56419) = -151.12838$	$= -0.13830/-363$	
$2(-1.11625) + (-149.61336) = -151.84586$	$2(-75.98778) = -151.97556$	$-151.97556 - (-151.84586)$	
		$= -0.12970/-341$	
trans-2-butene	*cis*-2-butene	$-155.12768 - (-155.13032)$	4.60
-155.13032	-155.12768	$= 0.00264/6.93$	
-155.99472	-155.99196	$-155.99196 - (-155.99472)$	
		$= 0.00276/7.2$	
HCN	HNC	$-92.32215 - (-92.33570)$	60.7
-92.33570	-92.32215	$= 0.01355/35.6$	
-92.85721	-92.83828	$-92.83828 - (-92.85721)$	
		$= 0.01893/49.7$	

The energies in hartrees are *ab initio* energies including ZPE. The calculations on O_2 are UHF, on triplet O_2. Calculations are by the author, experimental energies are from [38].

of the conjugate acids. Isodesmic processes like (5.183) (X–CH_2–Y+CH_4 → CH_3X+ CH_3Y) have been used to reveal the nature and amount of interaction between groups X and Y on an sp^3 carbon [110]. If X and Y interact in a stabilizing way, then separating them should be reflected in an endothermic reaction, and conversely mutually destabilizing groups should give an exothermic reaction. For reaction (5.183) at the 3-21G level the product energies minus the reactant energies are 59 kJ mol^{-1}, i.e. the reaction is endothermic by 59 kJ mol^{-1}, while for the corresponding CH_2Cl_2 reaction the reaction energy is -17 kJ mol^{-1}, which indicates a qualitatively different mutual interaction between two fluorines, as compared to two chlorines, on the same carbon (actually, a bigger basis set indicates that the Cl/Cl interaction in CH_2Cl_2 is about zero). Another application of isodesmic reactions has been to estimating aromatic character [111].

There is no *unique* isodesmic reaction for a particular problem. For example, to compare the ring strain in oxacyclopropane (oxirane, ethylene oxide) with that in cyclopropane, one might calculate the reaction energy of the oxygen exchange reaction (remember that isodesmic reactions do not have to be experimentally achievable):

$$(5.185)$$

Here the equilibrium should favor the less-strained ring. Alternatively, one might compare the reaction energies of the two cleavage reactions:

$$(5.186)$$

$$(5.187)$$

Here, the reaction that forms the less-strained ring should be the less endothermic (or the more exothermic, if the reactions turn out to be exothermic).

Let us calculate the reaction energies for the three reactions. For (5.185), the HF/6-31G* (including ZPE) energies are

$$E(\text{pdts}) - E(\text{reactants})$$

$$= (-152.805 - 118.15311) - (-116.971 - 153.97864)\,h$$

$$= -270.95818 - (-270.95023)\,h = -0.00795\,h = -20.9\,\text{kJ mol}^{-1}.$$

The reaction is calculated to be exothermic by 20.9 kJ mol^{-1}, and the simple interpretation is that cyclopropane is more strained than oxacyclopropane by 20.9 kJ mol^{-1}. To compare oxacyclopropane and cyclopropane using (5.186) and (5.187), we do need not calculate the energies of CH_4 and CH_3CH_3, since these will cancel out:

$$E(\text{pdts}) - E(\text{reactants}) = (-152.80507 + 2E(CH_4)) - (-153.97864 + E(CH_3CH_3))$$

$$= 1.17357\,h + 2E(CH_4) - E(CH_3CH_3)$$

and for (187):

$$E(\text{pdts}) - E(\text{reactants}) = (-116.97159 + 2E(CH_4))$$
$$- (-118.15311 + E(CH_3CH_3))\,h$$
$$= 1.18152\,h + 2E(CH_4) - E(CH_3CH_3)$$

Reaction (5.187) is more endothermic than reaction (5.186) by $1.18152 - 1.17357\,h = 0.00795\,h = 20.9\,kJ\,mol^{-1}$, indicating that cyclopropane is more strained than oxa-cyclopropane by this amount [112]. The agreement between the two approaches is not coincidental, since the cancellation of the methane and the ethane energies makes these two conceptually different approaches mathematically identical. Note, however, that for any isodesmic reaction we can always write some *non-equivalent* (unlike the case of (5.185) vs. (5.186)/(5.187)) isodesmic process to get the desired quantity. Also, different isodesmic reactions will give somewhat different results; this is essentially because the "reagents" of one reaction will not be calculated to exactly the same degree of accuracy as those of another reaction.

As we saw in section 5.4.1, calculation of homolytic cleavage energies by simply subtracting HF energies gives very poor results. Let us calculate a homolytic bond dissociation energy by an isodesmic-type approach. The idea is to combine the desired homolytic reaction with one of known (either from experiment or from as high-level calculation) energy in a reaction which, although not strictly isodesmic, conserves the number of single, double, etc. bonds and the number of unpaired electrons. For example, suppose we want $E(C-O)$, the C–O bond dissociation energy in CH_3-OH. We might utilize the scheme

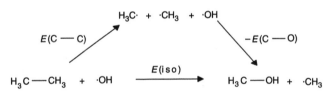

Here $E(C-C)$ and $E(C-O)$ are homolytic dissociation energies(bond energies) and $E(\text{iso})$ is the energy of the isodesmic-type reaction shown. This reaction is not strictly isodesmic, since a C–C bond is replaced by a C–O bond, but it does have the same number of unpaired electrons(one) on each side, so net correlation effects should be much less than for a reaction in which two unpaired electrons are created from a bond. A reaction in which the number of unpaired electrons is conserved is called an *isogyric*(Greek, "same spin") reaction. Since the overall energy change in a process is independent of the path connecting the initial and final states [113] we write

$$E(C-C) - E(C-O) = E(\text{iso})$$

i.e.

$$E(C-O) = E(C-C) - E(\text{iso})$$

$E(C-C)$ must be known, and $E(\text{iso})$ is to be obtained by calculation. Taking an experimental value of $368\,kJ\,mol^{-1}$ for the bond energy of the ethane C–C bond [61], and using HF/6-31G* energies (this level appears to be the lowest for which reasonably accurate results can usually be expected [114]), employing the unrestricted HF(UHF, i.e. UHF/6-31G*; end of section 5.2) method for the two radicals:

$$E(\text{iso}) = (E(CH_3OH) + E(CH_3\cdot)) - (E(CH_3CH_3) + E(OH\cdot))$$

$$= (-114.98009 - 39.52802) - (-79.14900 - 75.37317)\,h$$

$$= -154.50811 - (-154.52217)\,h = 0.01406\,h = 36.9\,kJ\,mol^{-1}.$$

So $E(C-O) = E(C-C) - E(\text{iso}) = 368 - 36.9 = 331\,kJ\,mol^{-1}$.

Now let us repeat this calculation using MP2(FC)/6 − 21G* energies:

$$E(\text{iso}) = (E(CH_3OH) + E(CH_3\cdot)) - (E(CH_3CH_3) + E(OH\cdot))$$

$$= (-115.29353 - 39.63823) - (-79.41756 - 75.51252)\,h$$

$$= -154.93176 - (-154.93008)\,h = -0.00168\,h = -4.41\,kJ\,mol^{-1}.$$

So

$$E(C-O) = E(C-C) - E(\text{iso}) = 368 - (-4.4) = 372\,kJ\,mol^{-1}.$$

In the HF case $E(\text{iso})$ is $+36.9\,kJ\,mol^{-1}$ and in the MP2 case it is $-4.4\,kJ\,mol^{-1}$! Thus $E(C-O)$ comes out in one case significantly weaker than $E(C-C)$, and in the other case slightly stronger. The experimental C−O bond energy for CH_3OH has been reported as $377\,kJ\,mol^{-1}$ [115]. Clearly, at the HF/6-31G* level anyway, this approach to bond energies is unreliable, but may be viable at the MP2 level [116].

Isodesmic reactions help us to make the most of the method/basis set we decide to use(this decision being guided by the size of the molecules and the computational resources available, i.e. by how "expensive" the calculations are). They have on occasion being used with fairly high-level calculations, to obtain high-quality results that would require even higher-level methods were "direct-subtraction" methods to be used [117]. Isodesmic reactions have been discussed and used extensively [109,118].

5.5.2.2b   *Thermodynamics; high-accuracy calculations*

As the previous discussion suggests (section 5.5.2.2a), the calculation of good relative energies is much more challenging than the calculation of good geometries. Nevertheless, it is now possible to reliably calculate energy differences to within about $\pm10\,kJ\,mol^{-1}$. An energy difference with an error of $\pm10\,kJ\,mol^{-1}$ is said to be within *chemical accuracy*. The term was popularized by Pople (biographical footnote section 5.3.3) in connection with the G1 and G2 (see below) methods. An accuracy of about $2\,kcal\,mol^{-1}$ ($8.4\,kJ\,mol^{-1}$, commonly rounded in this context to $10\,kJ\,mol^{-1}$) was, in 1991, considered to be realistic and chemically useful, perhaps because this is small compared to typical bond energies (roughly $400\,kJ\,mol^{-1}$). The *ab initio* energies and methods needed for results of chemical accuracy are called *high-accuracy* energies and methods.

As one might expect, high-accuracy energy methods are based on high-level correlational methods and big basis sets. However, because the straightforward application of such computational levels would require unreasonable times (be very "expensive"), the calculations are broken up into several steps, each of which provides an energy value; summing these gives an energy close to that which would be obtained from the more unwieldy one-step calculation. There are two classes of popular high-accuracy energy methods: the Pople group's G2 and G3 (for Gaussian-2 and Gaussian-3; these have superseded Gaussian-1) methods and their variants, and the Peterson group's CBS (complete basis set) methods. These will now be discussed and briefly compared.

The G2 method and its variants

The G2 and G3 methods [119] are the successors to the first widely-used high-accuracy energy method, the G1 method [120], which they have replaced. Relatively few results are yet available for the G3 method, so only G2 calculations will be discussed. These consist of nine steps:

1. An HF/6-31G* geometry optimization.

2. An HF/6-31G* ZPE calculation.

3. An MP2(full)/6-31G* optimization(at the time G1 and G2 were being developed, analytic frequencies were not available for MP2(FC) (section 5.4.2) optimizations, which are faster and about as good).

4. An MP4/6-311G** single-point calculation.

5. An MP4/6-311 + G** single-point calculation.

6. An MP4/6-311G(2df,p) single-point calculation.

7. A QCISD(T)/6-311G** single-point calculation.

8. An MP2/6-311 + G(3df,2p) single-point calculation.

9. An empirical *higher-level correction* (HLC) to account for any remaining errors due to electron correlation.

The tedious procedure for combining these steps to get the G2 energy is detailed by Foresman and Frisch [121], but with the Gaussian 94 and 98 programs only the keyword "G2" is needed. A G2 calculation is essentially equivalent to a QCISD(T)/6-311 + G(3df, 2p) calculation on an MP2(full)/6-31G* geometry with a HF/6-31G* ZPE correction [122]. Because of the empirical HLC the G2 method is semiempirical, except when this correction cancels out. This happens, for example, in calculating proton affinities as the energy difference of the protonated and unprotonated species [123]. The HLC is based on the number of α- and β-spin electrons, which are the same for both species in a case like this. The G2 method is among the most time-consuming of the high-accuracy energy methods, and several variants have been developed with a view to getting greater speed with little loss of accuracy. This has been largely achieved with the G2(MP2) and G2(MP2,SVP) (SVP = split valence plus polarization basis set) methods. Actually, in the absence of any evidence to the contrary, one may now assume that the G3 method, which seems to be somewhat more accurate and faster than G2, is the Gaussian high-accuracy method of choice, and this, or a faster but nearly as accurate variation, like G3(MP2), is recommended over the G2 family.

CBS methods

Complete basis set methods [86] involve essentially seven or eight steps:

1. A geometry optimization (at the HF/3-21G$^{(*)}$ or MP2/6-31G* level, depending on the particular CBS method).

2. A ZPE calculation at the optimization level.

3. An HF single-point calculation with a very big basis set (6-311 + G(3d2f, 2df, p) or 6-311 + G(3d2f, 2df, 2p), depending on the particular CBS method).

4. An MP2 single-point calculation (basis depending on the particular CBS method).

5. Something called a *pair natural orbital extrapolation* to estimate the error due to using a finite basis set.

6. An MP4 single-point calculation.

7. For some CBS methods, a QCISD(T) single-point calculation.

8. One or more empirical corrections.

There are three basic CBS methods: CBS-4 (for fourth-order extrapolation), CBS-Q (for quadratic CI) and CBS-APNO (or CBS-QCI/APNO, for asymptotic pair natural orbitals), in order of increasing accuracy (and increasing computer time). These methods are available with keywords in the Gaussian 94 and 98 programs.

Comparison of G2-type and CBS methods

The relative merits of the four most popular G2 and CBS methods are apparent from Table 5.10 (taken from the book by Foresman and Frisch [1e]). All four methods are available as a keyword in Gaussian 94 and Gaussian 98. If the test sample is representative, then the CBS-Q method is the most accurate and the G2 the second most accurate (G3 should be somewhat better and faster than G2). As CBS-Q seems to be faster than G2, CBS-Q appears to be the method of choice among these four for most high accuracy energy calculations, unless the system is relatively large, in which case CBS-4, the least accurate of the four, may have to be used. G3 and CBS-Q likely have quite similar errors, and G3 is faster than G2: the times on a Pentium 3

Table 5.10. Comparison of four high accuracy energy methods

Model	MAD	|Max. Error|	Sample relative CPU times		
			PH_3	F_2CO	SiF_4
CBS-4	1.98 (8.28)	7.0 (29.3)	1.0	1.0	1.0
G2(MP2)	1.58 (6.61)	6.3 (26.4)	2.4	10.3	11.5
CBS-Q	1.01 (4.23)	3.8 (15.9)	2.8	8.4	12.7
G2	1.21 (5.06)	4.4 (18.4)	3.2	25.9	59.1

MAD is mean average deviation, the arithmetic mean of absolute deviations from experiment. \|Max. Error\| is the absolute value of the maximum deviation from experiment. The author has placed kJ mol^{-1} in parentheses beside the kcal mol^{-1} values. Taken with permission from J. B. Foresman and Æ. Frisch, Exploring Chemistry with Electronic Structure Methods, 2nd edn, Gaussian Inc., Pittsburgh, PA, 1996. All four methods are available as keywords in Gaussian 94 and Gaussian 98.

for CBS-Q, G2, G3, and G3(MP2) jobs on $H_2C=C(H)OF$ were 59, 206, 136, and 41 minutes, respectively. It thus seems that where size permits and a slight loss of accuracy is tolerable, G3(MP2) is the method of choice. The maximum practical number of heavy atoms for G2, CBS-Q, G2(MP2) and CBS-4 calculations were, at least recently, ca. 5, 7, 7, and 15, respectively [124]. There are more accurate (and more time-consuming!) methods than any of the four in Table 5.10. The CBS-APNO method (available with a keyword in Gaussian 94 and 98), which is limited to about four heavy atoms [124], has a mean absolute deviation of only $2.2 \, \text{kJ mol}^{-1}$ [86], and a method that can give atomization energies accurate to about $0.4 \, \text{kJ mol}^{-1}$ has been reported [125]. Because of the empirical correction terms, the Gaussian and CBS methods are not purely *ab initio*, except where these terms disappear by subtraction [123]. Some applications of high-accuracy energy methods and suggestions about choosing a method (high accuracy or otherwise) will be given in the following sections (5.5.2.2c–e).

5.5.2.3 Thermodynamics; calculating heats of formation

The heat of formation (enthalpy of formation) of a compound is an important thermodynamic quantity, because a table of heats of formation of a limited number of compounds enables one to calculate the heats of reaction (reaction enthalpies) of a great many processes, that is, how exothermic or endothermic these reactions are. The heat of formation of a compound at a specified temperature T is defined [126] as the standard heat of reaction (standard reaction enthalpy) for formation of the compound at T from its elements in their standard states (their reference states). By the standard state of an element we mean the thermodynamically stablest state at $10^5 \, \text{Pa}$ (standard pressure, about normal atmospheric pressure), at the specified temperature (the exception is phosphorus, for which the standard state is white phosphorus; although red phosphorus is stabler under normal conditions, these allotropes are apparently somewhat ill-defined). The specified temperature is usually 298.15 K (about room temperature). The heat of formation of a compound at room temperature is thus the amount of heat energy (enthalpy) that must be put into the reaction to make the compound from its elements in their normal (room temperature and atmospheric pressure) states; it is the "heat content" or enthalpy of the compound compared to that of the elements. For example, at 298 K the heat of formation of CH_4 is $-74.87 \, \text{kJ mol}^{-1}$, and the heat of formation of CF_4 is $-933.20 \, \text{kJ mol}^{-1}$ [127]. To make a mole of CH_4 from solid graphite (carbon in its standard state at 298 K) and hydrogen gas requires $-74.87 \, \text{kJ}$, i.e. 74.87 kJ are given out – the reaction is mildly exothermic. To make a mole of CF_4 from solid graphite and fluorine gas requires $-933.20 \, \text{kJ}$, i.e. 933.20 kJ are given out – the reaction is strongly exothermic. In some sense CF_4 is thermodynamically much stabler with respect to its elements than is CH_4. Note that the standard heat of formation of an *element* is zero, since the reaction in question is the formation of the element from the element, in the same state (no reaction, or a null reaction). Heat of formation is denoted $\Delta H_f^\ominus$ or $\Delta_f H^\ominus$ and heat of formation at, say, 298 K by $\Delta H_{f298}^\ominus$, "delta H sub f standard at 298 K". The delta indicates that this is a difference (enthalpy of the compound minus enthalpy of the elements) and the superscript denotes "standard".

There are extensive tabulations of experimentally-determined heats of formation, mostly at 298 K (one way to determine $\Delta H_{f298}^\ominus$ is from heats of combustion: burning

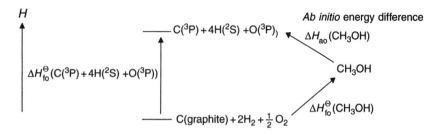

Figure 5.26. The principle behind the *ab initio* calculation of heat of formation (enthalpy of formation) by the atomization method. Methanol is (conceptually) atomized at 0 K into carbon, hydrogen and oxygen atoms; the elements in their standard states are also used to make these atoms, and to make methanol. The heat of formation of methanol at 0 K, $\Delta H_{f0}^{\ominus}(CH_3OH)$, follows from equating the energy needed to generate the atoms via methanol ($\Delta H_{f0}^{\ominus}(CH_3OH) + \Delta H_{a0}^{\ominus}(CH_3OH)$) to that needed to make them directly from the elements in their standard states. The diagram is not meant to imply that methanol *necessarily* lies above its elements in enthalpy.

the compound and the elements and measuring calorimetrically the heat evolved enables one to calculate the heat of formation by subtraction). $\Delta H_{f298}^{\ominus}$ can also be obtained by *ab initio* calculations. This is valuable because (1) it is far easier and cheaper than doing a thermochemical experiment, (2) many compounds have not been subjected to experimental determination of their heats of formation, and (3) highly reactive compounds, or valuable compounds available only in very small quantity cannot be subjected to the required experimental protocol. e.g. combustion. Let us see how $\Delta H_{f298}^{\ominus}$ can be calculated.

Atomization method
Suppose we want to calculate $\Delta H_{f298}^{\ominus}$ for methanol. We will calculate the heat of formation at 0 K ($\Delta H_{f0}^{\ominus}$) and then correct this to 298 K. Figure 5.26 shows the principle behind what has been called the "atomization" method [128]. Methanol is (conceptually) atomized at 0 K into carbon, hydrogen and oxygen atoms(the ground electronic states have been chosen here); it is from this step that the term "atomization" comes. The elements in their normal states are also used to make these atoms, and to make methanol. The heat of formation of methanol at 0 K follows from equating the energy needed to generate the atoms from the elements via methanol ($\Delta H_{f0}^{\ominus}(CH_3OH) + \Delta H_{a0}^{\ominus}(CH_3OH)$) to that needed to make them *directly* from the elements in their normal states:

$$\Delta H_{f0}^{\ominus}(CH_3OH) + \Delta H_{a0}^{\ominus}(CH_3OH) = \Delta H_{f0}^{\ominus}(C(^3P) + 4H(^2S) + O(^3S))$$

i.e.

$$\Delta H_{f0}^{\ominus}(CH_3OH) = \Delta H_{f0}^{\ominus}(C(^3P) + 4H(^2S) + O(^3S)) - \Delta H_{a0}^{\ominus}(CH_3OH) \quad (5.188)$$

$\Delta H_{a0}^{\ominus}(CH_3OH)$ is the *ab initio* atomization energy of methanol, the energy difference between the atoms and methanol. There are a couple points to note about this conceptual scheme. We are converting into carbon atoms graphite, a polymeric material, so strictly speaking Fig. 5.26 should show $nC(graphite) \rightarrow nC(^3P)$, where n is a number large enough to represent the *substance* graphite rather than just some carbon oligomer. All the species in the figure will then be increased in number by a factor of n, but division by this common factor will still give us Eq. (5.188). Another point is that although hydrogen and oxygen are solids at 0 K, we are considering isolated molecules being atomized.

To calculate $\Delta H_{f0}^{\ominus}(CH_3OH)$ we need the 0 K heat of formation of C, H and O atoms, i.e. the atomization energies of graphite, molecular hydrogen, and molecular oxygen, and the 0 K atomization energy of methanol. The atomization energies of hydrogen and oxygen can be calculated *ab initio*, but not that of graphite, which is a very big "molecule". For consistency we will use experimental values of all three elemental atomization energies, as recommended [128]. From Eq. (5.176), the 0 K atomization energy of methanol is simply the *ab initio* energies of its constituent atoms minus the ZPE-corrected *ab initio* of methanol:

$$\Delta H_{a0}^{\ominus}(CH_3OH) = \Delta E_{0K}^{total}(C(^3P) + 4H(^2S) + O(^3S)) - \Delta E_{0K}^{total}(CH_3OH) \quad (5.189)$$

Experimental values of $\Delta H_{f0}^{\ominus}C(^3P)$, $\Delta H_{f0}^{\ominus}H(^2S)$, and $\Delta H_{f0}^{\ominus}O(^3S)$ (as well as $\Delta H_{f0}^{\ominus}$ for other atoms, and references to more extensive tabulations) are given in [113]; in kJ mol^{-1}:

$$C \quad 711.2 \quad H \quad 216.035 \quad O \quad 246.6$$

To calculate $\Delta H_{a0}^{\ominus}(CH_3OH)$ we need (Eq. (5.189)) ΔE_{0K}^{total} for C, H and O atoms (in the states shown) and for methanol. G2 (for comparison with the value in [128]) calculations gave these values (hartrees):

C	−37.78430
H	−0.50000 (there are no correlation effects for the H atom; this is the exact energy)
O	−74.98203
CH$_3$OH	−115.53490

From Eq. (5.189)

$$\Delta H_{a0}^{\ominus}(CH_3OH) = -37.78430 + 4(-0.50000) - 74.98203 - (-115.53490) \, h$$

$$= -114.76633 + 115.53490 \, h = 0.76857 \times 2625.5 \, kJ \, mol^{-1}$$

$$= 2017.88 \, kJ \, mol^{-1}$$

From Eq. (5.188)

$$\Delta H_{f0}^{\ominus}(CH_3OH) = 711.2 + 4(216.035) + 246.6 - 2017.88 \, kJ \, mol^{-1}$$

$$= 1821.94 - 2017.88 \, kJ \, mol^{-1} = -195.9 \, kJ \, mol^{-1}$$

Reference [128] gives the 0 K G2 value by the atomization method as −195.7 kJ mol^{-1} and the experimental value as (two sources) −190.7 or −189.8 kJ mol^{-1}.

To correct the 0 K heat of formation to that at 298.15 K we add the increase in enthalpy of methanol on going from 0 to 298 K and subtract the corresponding increases for the elements in their standard states. The value for methanol is the difference of two quantities provided in the thermochemical summary at the end of the G2 calculation as implemented in Gaussian 94 or Gaussian 98:

$$\Delta\Delta H^{\ominus}(CH_3OH) = \text{G2 Enthalpy} - \text{G2}(0\,K)$$

$$= -115.53061 - (-115.53490)\,h$$

$$= 0.00429 \times 2625.5\,kJ\,mol^{-1}$$

$$= 11.26\,kJ\,mol^{-1}$$

(G2(0 K) is the G2 value for what we have called ΔE_{0K}^{total})

The experimental enthalpy increases for the elements are given in [128]; in kJ mol^{-1}:

$\Delta\Delta H^{\ominus}$(element)	
C(graphite)	1.050
H_2	8.468
O_2	8.680

From these and $\Delta\Delta H_f^{\ominus}(CH_3OH)$:

$$\Delta H_{f298}^{\ominus}(CH_3OH) = \Delta H_{f0}^{\ominus}(CH_3OH) + \Delta\Delta H^{\ominus}(CH_3OH)$$

$$- (\Delta\Delta H^{\ominus}(C) + 2\Delta\Delta H^{\ominus}(H_2) + \tfrac{1}{2}\Delta\Delta H^{\ominus}(O_2)) \quad (5.190)$$

$$= -195.9 + 11.26 - (1.050 + 2(8.468) + \tfrac{1}{2}(8.680))\,kJ\,mol^{-1}$$

$$= -195.9 - 11.07\,kJ\,mol^{-1} = -207.0\,kJ\,mol^{-1}$$

The accepted experimental value [129] is $-201\,kJ\,mol^{-1}$.

Note that if $\Delta H_{f0}^{\ominus}$ is not wanted, $\Delta H_{f298}^{\ominus}$ can be calculated directly, since from Eqs (5.188) and (5.190) the 0 K *ab initio* energy of the compound is subtracted out and it follows that

$$\Delta H_{f298}^{\ominus}(CH_3OH) = \Delta H_{f0}^{\ominus}(C) + 4\Delta H_{f0}^{\ominus}(H) + \Delta H_{f0}^{\ominus}(O) - (\Delta E_{0K}^{total}(C)$$

$$+ 4\Delta E_{0K}^{total}(H) + \Delta E_{0K}^{total}(O)) + \text{G2 Enthalpy} - ((\Delta\Delta H^{\ominus}(C)$$

$$+ 2\Delta\Delta H^{\ominus}(H_2) + \tfrac{1}{2}\Delta\Delta H^{\ominus}(O_2)) \quad (5.191)$$

$$= [(7112 + 4(216.035) + 246.6)\,kJ\,mol^{-1}$$

$$- (-37.78430 + 4(-0.50000) - 74.98203)$$

$$+ (-115.53061)]\,h - \left[(1.050 + 2(8.468) + \tfrac{1}{2}(8.680)\right]\,kJ\,mol^{-1}$$

$$= 1821.94\,kJ\,mol^{-1} - (-114.76633)\,h - 115.53061\,h - 22.33\,kJ\,m\varepsilon$$

$$= 1821.94\,kJ\,mol^{-1} - 0.76428 \times 2625.5 - 22.33\,kJ\,mol^{-1}$$

$$= 1821.94 - 2006.62 - 22.33\,kJ\,mol^{-1} = -207.0\,kJ\,mol^{-1}$$

Formation method

An alternative to the atomization method is what has been called the "formation" method, which is illustrated for methanol in Fig. 5.27. This method utilizes a kind of "pseudo heat of formation"'s, ΔH_{f0}, of the compound from atomic carbon and molecular hydrogen and oxygen (the conventional heat of formation is relative to graphite and molecular hydrogen and oxygen). From Fig. 5.27,

$$\Delta H_{f0}^{\ominus}(CH_3OH) = \Delta H_{f0}^{\ominus}(C(^3P)) + \Delta H_{f0} \qquad (5.192)$$

where the experimental value of $\Delta H_{f0}^{\ominus}(^3C)$ is used, and

$$\Delta H_{f0} = \Delta E_{0\,K}^{total}(CH_3OH) - \Delta E_{0K}^{total}\left(C(^3P) + 2H_2 + \tfrac{1}{2}O_2\right) \qquad (5.193)$$

A calculation using G2 energies gives

$$\Delta H_{f0}^{\ominus}(CH_3OH) = 711.2\,kJ\,mol^{-1} + \Delta H_{f0}$$

$$= 711.2\,kJ\,mol^{-1} + \big[-115.53490 - (-37.78430)$$

$$+ 2(-1.16636) + \tfrac{1}{2}(-150.14821)\big]\,h$$

$$= 711.2\,kJ\,mol^{-1} + [-115.53490 + 115.19113]\,h$$

$$= 711.2\,kJ\,mol^{-1} - 0.34378\,h = 711.2 - 0.34378 \times 2625.5\,kJ\,mol^{-1}$$

$$= 711.1 - 902.59\,kJ\,mol^{-1} = -191.4\,kJ\,mol^{-1}$$

The value calculated by this procedure in [128] is $-191.3\,kJ\,mol^{-1}$. The atomization method usually gives somewhat more accurate heats of formation, at least with the G2-type methods (although for the particular case of methanol this is not so), perhaps

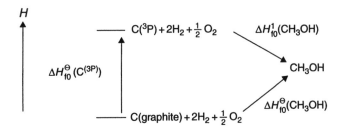

Figure 5.27. The principle behind the *ab initio* calculation of heat of formation (enthalpy of formation) by the formation method. Methanol is (conceptually) formed from atomic carbon and molecular hydrogen and oxygen; the enthalpy input for this resembles that for the heat of formation of methanol (hence the name) except that atomic carbon rather than graphite is used. Graphite is converted to atomic carbon, and the elements in their normal states are also used to make methanol. The heat of formation of methanol at 0 K follows from equating this quantity to the heat of atomization of graphite plus the energy needed to make methanol from atomic carbon and molecular hydrogen and oxygen. The diagram is not meant to imply that methanol *necessarily* lies above its elements in enthalpy.

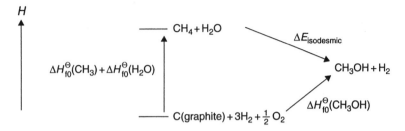

Figure 5.28. The principle behind the *ab initio* calculation of heat of formation (enthalpy of formation) using an isodesmic reaction. Methanol and hydrogen are (conceptually) made from methane and water (other isodesmic reactions could be used); the 0 K enthalpy input for this is the *ab initio* energy difference between the products and reactants. Graphite, hydrogen and oxygen are converted into methane and water and into methanol and hydrogen, with input of the appropriate heats of formation. The heat of formation of methanol at 0 K follows from equating the heat of formation of methanol with the sum of the energy inputs for the other two processes. The diagram is not meant to imply that methanol *necessarily* lies above its elements in enthalpy.

because these methods were optimized (via the semiempirical terms, section 5.5.2.2b) to give accurate atomization energies.

Isodesmic reaction method
Finally, heats of reaction can be calculated by *ab initio* methods with the aid of isodesmic reactions (section 5.5.2.2a), as indicated in Fig. 5.28 (actually, the scheme in Fig. 5.28 is not strictly isodesmic – e.g. only on one side of the "isodesmic" equation is there an H–H bond). From this scheme

$$\Delta H_{f\,0}^{\ominus}(CH_3OH) = \Delta H_{f\,0}^{\ominus}(CH_4) + \Delta H_{f\,0}^{\ominus}(H_2O) + \Delta E_{\text{isodesmic}} \qquad (5.194)$$

where

$$\Delta E_{\text{isodesmic}} = \Delta E_{0\,K}^{\text{total}}(CH_3OH + H_2) - \Delta E_{0\,K}^{\text{total}}(CH_4 + H_2O)$$

Using G2 values:

$$\Delta E_{\text{isodesmic}} = (-115.53490 - 1.16636) - (-40.41090 - 76.33205)\,h$$
$$= -116.70126 + 116.74295\,h = 0.04169\,h$$

With this and the experimental 0 K heats of formation of CH_4 and H_2O [128]:

$$\Delta H_{f\,0}^{\ominus}(CH_3OH) = -66.8 - 238.92 + 0.04169 \times 2625.5\,\text{kJ mol}^{-1}$$
$$= -196.3\,\text{kJ mol}^{-1}.$$

This is very close to our atomization heat of formation value above ($-195.9\,\text{kJ mol}^{-1}$), and a little more negative than the experimental value (-190.7 or $-189.8\,\text{kJ mol}^{-1}$ [128]).
Of the three approaches to calculating heats of formation (atomization, formation and isodesmic), the atomization has been recommended over the formation [128]. The isodesmic (or isodesmic-type, as in Fig. 5.28) should be at least as accurate as the

atomization, because of the ability of isodesmic and related processes to compensate for basis set and correlation deficiencies (section 5.5.2.2a). In concluding our discussion of heats of formation, note that all these calculations of the heat of formation of methanol were not *purely ab initio* (quite apart from the empirical correction term in G2), since they required experimental values of either the heat of atomization of graphite (atomization and formation methods) or the heat of formation of methane (formation method). The inclusion of experimental values makes the calculation of heat of formation with the *aid* of *ab initio* methods a *semiempirical* procedure (do not confuse the term as used here with semiempirical programs like AM1, discussed in chapter 6). Augmentation with experimental data is needed whenever an *ab initio* calculation would involve an extended, solid substance like graphite (see the discussion in connection with the atomization method); other examples are phosphorus and sulfur.

Considerable attention has been given here to heats (enthalpies) of formation, because there are extensive tabulations of these [130] and papers on their calculation appear often in the literature [131]. However, we should remember that equilibria [103] are dependent not just on enthalpy differences, but also on the often-ignored entropy changes, as reflected in free energy differences, and so the calculation of entropies is also important [132].

5.5.2.3a Kinetics; calculating reaction rates

Ab initio kinetics calculations are far more challenging than thermodynamics calculations; in other words, the calculation of rate constants is much more involved than that of equilibrium constants or quantities like reaction enthalpy, reaction free energy, and heat of formation, which are related to equilibrium constants. Why is this so? After all, both rates and equilibria are related to the energy difference between two species: the rate constant to that between the reactant and transition state (TS), and the equilibrium constant to that between the reactant and product (Fig. 5.25). Furthermore, the energies of transition states, like those of reactants and products, can be calculated. The reason for the difference is partly because the energies of transition states are harder to calculate to high accuracy than are those of relative minima ("stable species"). Another problem is that the rate does not depend strictly on the TS/reactant free energy difference (which can, at sufficiently high levels, be accurately calculated).

To understand the problem consider a unimolecular reaction A → B. Figure 5.29 shows the potential energy surface for two reactions of this type, A_1 → B_1 and A_2 → B_2. The reactions have identical calculated free energies of activation. By calculated, we mean here using some computational chemistry method (e.g. *ab initio*) and locating a stationary point with no imaginary frequencies, corresponding to A, and an appropriate stationary point with one imaginary frequency, etc. (section 2.5), corresponding to B. The "traditional" calculated rate constant then follows from a standard expression involving from the energy difference between the TS and reactant (our calculated free energy of activation) and the partition functions of the two species. However, in the TS region the PES for the first process is flatter than for the second process – the saddle-shaped portion of the surface is less steeply-curved for reaction 1 than for reaction 2. If all reacting A molecules followed *exactly* the intrinsic reaction coordinate (IRC; section 2) and passed through the calculated TS species, then we might expect the two

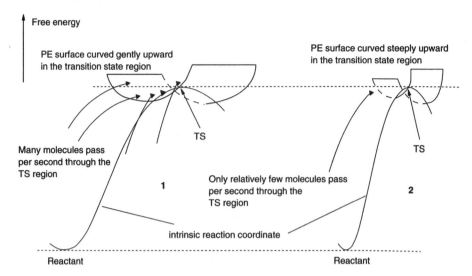

Figure 5.29. Possible potential energy surfaces for two reactions with the same calculated free energy of activation. Reaction 1 is nevertheless faster than reaction 2 because its transition state region is flatter. As a result, in a given time more molecules can stray from the intrinsic reaction coordinate and pass through the transition state region to the product.

reactions to proceed at exactly the same rate, since all A_1 and A_2 molecules would have to surmount identical barriers. However, the IRC is only an idealization [133], and molecules passing through the TS region toward the product frequently stray from this path (dashed lines). Clearly for the reaction $A_1 \rightarrow B_1$ at any finite temperature more molecules (reflected in a Boltzmann distribution) will have the extra energy needed to traverse the higher-energy regions of the saddle, away from the TS point, than in the case of $A_2 \rightarrow B_2$; if the saddle were curved infinitely steeply, *no* molecules could stray outside the reaction path. Thus reaction 1 must be faster than reaction 2, although they have identical computed free energies of activation; the rate constant for reaction 1 must be bigger than that for reaction 2. The difficulty of obtaining good rate constants from accurate calculations on just two PES points, the reactant and the TS, is mitigated by this fact: the vibrational frequencies of the TS sample the curvature of the saddle region both along the reaction path (this curvature is represented by the imaginary frequency) and at "right angles" to the reaction path (represented by the other frequencies). High frequencies correspond to steep curvature. So when we use the TS frequencies in the partition function equation for the rate constant we are, in a sense, exploring regions of the PES saddle other than just the stationary point. The role of the curvature of the PES in affecting reaction rates is nicely alluded to by Cremer, who also shows the place of partition functions in rate equations [134].

Another way to calculate rates is by *molecular dynamics* [135]. Molecular dynamics calculations use the equations of classical physics to simulate the motion of a molecule under the influence of forces; the required force fields can be computed by *ab initio* methods or, for large systems, semiempirical methods (chapter 6) or molecular mechanics (chapter 3). In a molecular dynamics simulation of the reaction $A \rightarrow B$, molecules of A are "shaken" out of their potential well, and some pass through the saddle region at a rate that, for a given temperature, depends on the height of this region and its curvature. The height and curvature can be handled as an analytic function of atomic coordinates $E = f(q_1, q_2, \ldots)$ which has been fitted to a finite number of calculated (e.g. by *ab initio* methods) points. The situation is even more complicated, because the simulation technique just outlined ignores quantum mechanical tunnelling [136], which, particularly where light atoms like hydrogen move, can speed up a reaction by orders of magnitude compared to classical predictions.

For rigorous calculations of rate constants one best utilizes specialized programs [137]. There are many discussions of the theory of reaction rates, in various degrees of detail [138]. In this section we will limit ourselves to gas-phase unimolecular reactions [139] and examine the results of some rough calculations. We will use the equation

$$k_r = \frac{k_B T}{h} e^{-\Delta G^{\ddagger}/RT} \tag{5.195}$$

where k_r = unimolecular rate constant (units = s^{-1}); k_B = Boltzmann constant, 1.381×10^{-23} J K^{-1}; T = temperature, K; h = Planck's constant, 6.626×10^{-34} J s; $\Delta G^{\ddagger}$ = the transition-state-reactant free energy difference (in some calculations we will try the ZPE-corrected 0 K energy difference, ΔE_{0K}^{total}, which is the 0 K enthalpy difference); R = gas constant, 8.314 J K^{-1} mol^{-1}.

This is *not* a rigorous equation for the unimolecular rate constant (which has high- and low-pressure limiting forms anyway [139]). Roughly, Eq. (5.195) results from assuming that the rotational and vibrational partition functions do not change on going from the reactant to the TS in the high-pressure limit [140]; rates at the low-pressure limit appear to be slower than those at the high-pressure limit by a factor of roughly 1000 [141]. In comparing calculated and experimental rate constants, we will consider the conditions to be the high-pressure limit (ca. 100 mmHg or 13000 Pa [141]. In the following calculations we will see if Eq. (5.195) is useful.

Consider the reactions in Fig. 5.30. Reactant and product structures were created with Spartan [30] and transition states were generated with Spartan's transition state routine; semiempirical AM1 geometries from Spartan were used as input to G94W [48] CBS-Q, CBS-4 and G2 (section 5.5.2.2b) calculations. The results of the calculations are summarized in Tables 5.11 and 5.12. Table 5.12 shows that for a given reaction we got the same result, to within a factor of 10, whether we used CBS-Q or G2, $\Delta G^{\ddagger}(298)$ or $\Delta H^{\ddagger}(0)$; the biggest deviation is for CH$_3$NC, a factor of 3.0 cf. 0.3. The CBS-4 method gave rate constants that are smaller than the CBS-Q and G2 rate constants by factors of from about 2 (CH$_2$=CHOH) to 5000 (CH$_3$CN, using $\Delta H^{\ddagger}(0)$). These results may be compared with the experimental facts:

$CH_3NC \rightarrow CH_3CN$.

The experimental rate constant for the isocyanomethane (methyl isocyanide) to propanenitrile (acetonitrile) reaction is 3.6×10^{-15} s^{-1} at 298 K [142]. This compares astonishingly well with the value of 1.4×10^{-15} in Table 5.11, calculated from the G2 value of $\Delta G^{\ddagger}$; the calculated rate constant is only a factor of three too small (using G2 with $\Delta H^{\ddagger}$ gives a rate constant too small by a factor of ten). The CBS-Q

Table 5.11. Calculating rate constants and halflives for the reactions in Fig. 5.30, at 298 K. At 298 K, $k_B T/h = 6.21 \times 10^{12}\,\mathrm{s}^{-1}$ and $RT = 2.478\,\mathrm{kJ\,mol}^{-1}$. $\Delta G^{\ddagger}$ is the 298 K transition state/reactant energy difference and $\Delta H(0K)^{\ddagger}$ is the 0 K ZPE-corrected transition state/reactant energy difference (ΔE_{0K}^{total}). Energies are in $\mathrm{kJ\,mol}^{-1}$. The total time for each calculation with G94W on a 200 MHZ Pentium Pro is shown

rate constant (s^{-1}) or halflife (s), at 298 K:

CBS-Q
CBS-4
G2

II calculation method	III CH$_3$NC etc.	IV CH$_2$ = CHOH etc	V Cyclopropylidene
	3.1 h (CBS-Q)	5.0 h (CBS-Q)	3.6 h (CBS-Q)
	1.0 h (CBS-4)	1.0 h (CBS-4)	1.1 h (CBS-4)
	9.4 h (G2)	17 h (G2)	12 h (G2)
$k_r = \dfrac{k_B T}{h} e^{-\Delta G^{\ddagger}/RT}$	$\Delta G^{\ddagger} = 160.5$	$\Delta G^{\ddagger} = 237.2$	$\Delta G^{\ddagger} = 22.3$
	178.0	238.8	28.2
	157.8	236.9	22.2
	$k_r = 4.6 \times 10^{-16}$	$k_r = 1.7 \times 10^{-29}$	$k_r = 7.7 \times 10^{8}$
	4.0×10^{-19}	8.7×10^{-30}	7.1×10^{7}
	1.4×10^{-15}	1.9×10^{-29}	8.0×10^{8}
$k_r = \dfrac{k_B T}{h} e^{-\Delta H(0K)^{\ddagger}/RT}$	$\Delta H^{\ddagger} = 163.7$	$\Delta H^{\ddagger} = 236.8$	$\Delta H^{\ddagger} = 24.3$
	182.2	238.3	31.3
	161.0	236.5	24.3
	$k_r = 1.3 \times 10^{-16}$	$k_r = 2.0 \times 10^{-29}$	$k_r = 3.4 \times 10^{8}$
	7.3×10^{-20}	1.1×10^{-29}	2.0×10^{7}
	3.8×10^{-16}	1.9×10^{-29}	3.4×10^{8}
$t_{1/2} = \dfrac{\ln 2}{k_r}$, using $\Delta G^{\ddagger}$	1.5×10^{15}	4.1×10^{28}	9.0×10^{-10}
	1.7×10^{18}	8.0×10^{28}	9.8×10^{-9}
	5.0×10^{14}	3.6×10^{28}	8.7×10^{-10}
$t_{1/2} = \dfrac{\ln 2}{k_r}$, using	5.3×10^{15}	3.6×10^{28}	2.0×10^{-9}
	9.5×10^{18}	6.3×10^{28}	3.5×10^{-8}
$\Delta H(0K)^{\ddagger}$	1.8×10^{15}	3.6×10^{28}	2.0×10^{-9}

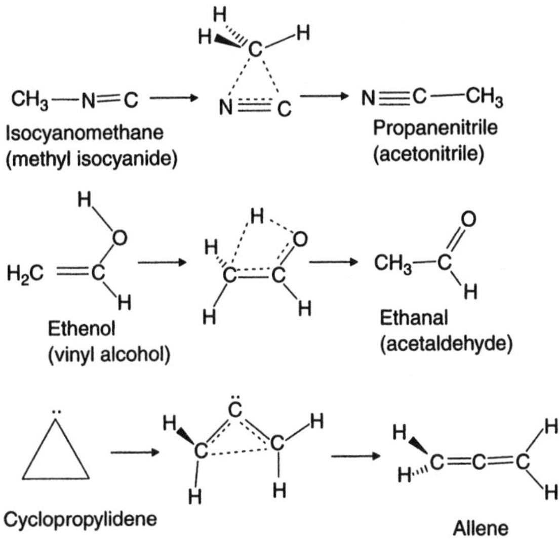

Figure 5.30. Reactions used to illustrate the calculation of rate constants and halflives.

Table 5.12. Relative values of rate constants for the reactions in Fig. 5.30; the data are taken from Table 5.11. For each reaction the rate constant from the CBS-Q value of $\Delta G^{\ddagger}$ has been set equal to unity

	Relative rate constants at 298 K: CBS-Q CBS-4		
	G2		
Calculation method	CH_3NC etc.	$CH_2\!=\!CHOH$ etc.	Cyclopropylidene etc.
$k_{\oplus} = \dfrac{k_o T}{h} e^{-\Delta^{-\ddagger} \equiv \ell\Im}$	1	1	1
	0.001	0.5	0.1
	3.0	1.1	1.0
$k_{\oplus} = \dfrac{k_o T}{h} e^{-\Delta \times \backslash \in \neq \div^{\ddagger} \equiv \ell\Im}$	0.3	1.2	0.4
	0.0002	0.6	0.03
	0.8	1.1	0.4

method gives a rate constant too small by factors of eight (using $\Delta G^{\ddagger}$) and 28 (using $\Delta H^{\ddagger}$). CBS-4 rate constants are too small by factors of from about 10^3 (using $\Delta G^{\ddagger}$) to 5×10^4 (using $\Delta H^{\ddagger}$).

$CH_2\!=\!CHOH \rightarrow CH_3CHO$

The reported halflife of ethenol (vinyl alcohol) in the gas phase at room temperature is ca. 30 min [143], far shorter than our calculated 10^{28}–10^{29} s. However, the 30 min halflife is very likely that for a protonation/deprotonation isomerization catalyzed by the walls of the vessel, rather than for the concerted hydrogen migration (Fig. 5.30) considered here. Indeed, the related *ethynol* has been detected in planetary atmospheres and interstellar space [144], showing that that molecule, in isolation, is long-lived. All three methods predict very long halflives, the same to within a factor of two, regardless of whether one uses $\Delta G^{\ddagger}$ or $\Delta H^{\ddagger}$.

Cyclopropylidene to allene
Cyclopropylidene has never been detected [145], so its halflife must be very short even well below room temperature. Our calculations predict room temperature halflives for cyclopropylidene of 10^{-8}–10^{-9} s. Gaussian 94 and 98 can be instructed to calculate G (and H) at temperatures other than 298.15 K, so that ΔG and $\Delta G^{\ddagger}$, and thus equilibrium and rate constants, could be calculated for other temperatures, but if we just use the 298 K value of $\Delta G^{\ddagger}$, which should not change dramatically with temperature (cf. $\Delta G(298)^{\ddagger}$ and $\Delta G^{\ddagger}(0) = \Delta H(0)^{\ddagger}$ in Table 5.11), we can estimate the halflife at 77 K (attempts to generate cyclopropylidene at 77 K gave allene [146]). Equation (5.195) gives a halflife at 77 K of 590 s (10 min). Since this could easily be out by a factor of 10, the calculation accords with experiment. Cyclopropylidene should be easily observable at 10 K, a routine matrix isolation [147] working temperature, where its halflife is calculated to be ca. 10^{106} s.

These experimental facts and the comparison of the three computational methods suggest these tentative generalizations: the use of Eq. (5.195) with CBS-Q or (preferably?) G2 values of $\Delta G^{\ddagger}$ gives for unimolecular isomerizations rate constants that are qualitatively reliable. The CBS-4 rate constants are smaller than the CBS-Q and G2 ones by a factor of from 2 to 1000. This is not bad for CBS-Q and G2, considering that Eq. (5.195) is quite approximate, and that the CBS and G2 methods (section 5.5.2.2b) were developed to provide reliable *thermodynamic* data, not to handle transition states.

From Eq. (5.195) and the fact that for a unimolecular reaction $t_{1/2} = \ln 2/k_r$ it follows that

$$\log t_{1/2} = \log \left[(\ln 2) \frac{h}{k_{\mathrm{B}} T} \right] + \frac{\Delta G^{\ddagger}}{RT} \cdot \log e \qquad (5.196)$$

At 298 K (about room temperature) this becomes

$$\log t_{1/2} = 0.175 \Delta G^{\ddagger} - 13.0 \qquad (5.197)$$

where $\Delta G^{\ddagger}$ is in kJ mol^{-1}.

Equation (5.196) shows that for $\Delta G^{\ddagger} = 0$ kJ mol^{-1}, $t_{1/2}$ is 10^{-12}–10^{-13} s (at 10–298 K); this is about as expected, since the period of a molecular vibration is about 10^{-13}–10^{-14} s and with no barrier a species should survive for only about one vibrational motion (that along the reaction coordinate, corresponding to the imaginary frequency) as it passes through the saddle region (e.g. Fig. 5.29). Figure 5.31, a graph of Eq. (5.197), can be used to estimate halflives at room temperature from the free energy of activation, for unimolecular isomerizations. We see that the threshold value of $\Delta G^{\ddagger}$ for observability at room temperature for a species that decays by a unimolecular process is predicted to be roughly 80–90 kJ mol^{-1} ($t_{1/2} = 10$ s or 9 min), with

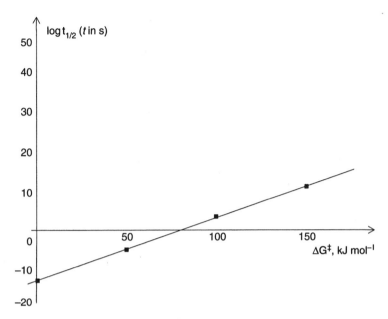

Figure 5.31. Graph of $\log t_{1/2} = 0.175 \Delta G^{\ddagger} - 13.0$. If this equation for the halflife of a unimolecular reaction were strictly true, then the threshold value of $\Delta G^{\ddagger}$ for ready observability at room temperature would be about 80–90 kJ mol^{-1}, corresponding to $t_{1/2} = 10\,\text{s} - 9\,\text{min}$. Actually, a rough rule of thumb is that the threshold barrier for observability at room temperature is about 100 kJ mol^{-1}.

a strong dependence on $\Delta G^{\ddagger}$. Experience shows that in fact the threshold barrier for observing or isolating a compound at room temperature is about 100 kJ mol^{-1}.

5.5.2.3b *Energies: concluding remarks*

Although we have paid considerable attention to high-accuracy energy methods (CBS and G2), many *ab initio* studies are limited to obtaining relative energies at a moderate level like MP2/6-31G*, or even HF/6-31G* or HF/3-21G. In *comparing* the thermodynamic stabilities of isomers (in contrast to calculating an "absolute" thermodynamic parameter like the heat of formation of a compound), simple subtraction of *ab initio* energies calculated at a modest level (e.g. Fig. 5.24) usually gives at least semiquantitatively reliable results. Some examples of energy differences calculated at modest *ab initio* levels are shown in Table 5.13. Here E_{0K}^{total} (section 5.2.3.6d), the ZPE-corrected *ab initio* energy, has been used to calculate energy differences for five pairs of isomers. The *ab initio* energies are from HF/3-21G-optimized geometries, HF/6-31G*-optimized geometries, and single-point HF/6-31G* calculations on HF/3-21G-optimized geometries; the ZPE corrections are all from HF/3-21G* frequencies on HF/3-21G-optimized geometries. Although ZPE corrections have been included here, ignoring them for such modest-level calculations on pairs of isomers evidently makes little difference, as the correction is the same for both isomers, to within a few kJ mol^{-1} (but for comparison of a ground state and a TS the ZPE correction is more likely to be significant, see e.g. Fig. 2.20; and in high-accuracy calculations ZPE's should *always*

Table 5.13. Energy differences ($kJ\,mol^{-1}$) of some isomers, calculated at modest *ab initio* levels

Isomer pair	HF/3-21G* optimization ZPEs	ΔE	HF/6-31G* optimization ΔE	Single point ΔE	Experiment $\Delta H^{\ominus}_{f'298}$
	225.3, 229.2	63.8	37	38.1	32.9 [151]
	304.7, 306.4	9.85	13.1	13.4	10.2 [151]
	304.7, 305.3	6.9	7.2	7.3	3.1 [151]
	157.0, 159.2	37.7	73.4	74.4	43 [72]
$CH_3CN\ CH_3NC$	129.7, 127.8	85.7	85.3	85.4	89.5 [151]

The isomer pairs are propene, cyclopropane; *trans*-2-butene, 1-butene; *trans*-2-butene, *cis*-2-butene; ethanal (acetaldehyde), ethenol (vinyl alcohol); propanenitrile (acetonitrile), methyl isonitrile (methyl isocyanide); for each pair the higher-energy isomer is the second one (e.g. cyclopropane lies above propene). The ZPE's are from frequencies calculated on the HF/3-21G geometry; these were used to correct all sets of *ab initio* energies: those from the 3-21G optimization, the 6-31G* optimization, and the 6-31G* single-point energies on the 3-21G geometries. The experimental energy differences are differences in heats of formation at 298 K, i.e. 298 K enthalpy differences; they are not strictly comparable to the *ab initio* energy differences, which are 0 K enthalpy differences. However, the change in heat of formation from 0 to 298 K should be similar for the two members of an isomer pair, so the table is essentially a comparison of *ab initio* and experimental 0 or 298 K enthalpy differences.

be included). This very small sample does not permit one to draw conclusions about 3-21G vs. 6-31G* energies, but this has been discussed in sections 5.3.3 and 5.5.2.2a. The single-point 6-31G* energies are almost identical to the optimized-geometry ones (correlated single-point energies were discussed in sections 5.4.2 and 5.4.3). Agreement with experiment (the 0 and 298 K enthalpy values are approximately comparable) is good except for ethanal/ethenol; the reported experimental value of $43\,kJ\,mol^{-1}$ for this pair is supported by a high-accuracy energy calculation (CBS-Q, section 5.5.2.2b) that gives a 0 K enthalpy difference of $44.4\,kJ\,mol^{-1}$ ($43.7\,kJ\,mol^{-1}$ at 298 K, bolstering the assertion that ΔH changes little from 0 to 298 K). The two 6-31G* values are evidently in error by about $73 - 43 = 30\,kJ\,mol^{-1}$, while the 3-21G* error is only $37 - 43 = -6\,kJ\,mol^{-1}$. A "higher" level is not guaranteed to give more accurate results, at least not until we reach very high levels. As might be suspected, the relative energies of conformers is also treated quite well by these modest levels [148].

Foresman and Frisch [149], in a chapter with very useful data and recommendations, show large mean absolute deviations (MAD) and enormous maximum errors for HF and even MP2 methods; e.g.

HF/6-31+G**	MAD, $195\,kJ\,mol^{-1}$ $(46.7\,kcal\,mol^{-1})$
	Max. Error, $753\,kJ\,mol^{-1}$ $(179.9\,kcal\,mol^{-1})$
MP2/6-311+G(2d,p)	MAD, $37\,kJ\,mol^{-1}$ $(8.9\,kcal\,mol^{-1})$
	Max. Error, $164\,kJ\,mol^{-1}$ $(39.2\,kcal\,mol^{-1})$

How can this be reconciled with the results shown in this chapter and the modest levels endorsed by Hehre [32]? As hinted ("Don't Panic!"; [1e], p. 146)], the large errors reported are a composite including some "tough cases" [150] like atomization energies (section 5.4.1). If the calculation set had been limited to, say, comparing the energies of isomers containing only carbon, hydrogen and oxygen, the errors would have been much smaller. A good feel for the accuracy of various levels of calculation will emerge from examining the extensive data in Hehre's book [32], while Foresman and Frisch remind us that there *are* cases in which high-accuracy methods are required.

5.5.3 Frequencies

The calculation of normal-mode frequencies (section 2.5) is important because:

1. The number of imaginary frequencies tells us the curvature of the potential energy surface at the point corresponding to a particular stationary point: whether an optimized structure (i.e. a stationary point) is a minimum, a transition state (a first-order saddle point), or a higher-order saddle point. See section 2.5. Routinely checking optimized structures with a frequency calculation is a good idea, if the size of the job does not make this impractical (frequencies take longer than optimizations).
2. The frequencies must be calculated to get the zero point energy of the molecule. This is needed for accurate energy comparisons (section 2.5).
3. The normal-mode vibrational frequencies of a molecule correspond (with qualifications) to the bands seen in the infrared (IR) spectrum of the substance. Thus the IR spectrum of a substance that has never been made can be calculated to serve as a guide for the experimentalist. Unidentified IR bands observed in an experiment can sometimes be assigned to a particular substance on the basis of the calculated spectrum of a suspect; if the spectra of the usual suspects are not available from experiment (they might be extremely reactive, transient species), we can calculate them.

The characterization of stationary points by the number of imaginary frequencies was discussed in chapter 2, and ZPEs in chapter 2 and earlier sections of this chapter. Here we will examine the utility of *ab initio* calculations for the prediction of IR spectra [152]. It is important to remember that frequencies should be calculated at the same level (e.g. HF/3-21G, MP2/6-31G*, ...) as was used for the geometry optimization (section 2.5).

Positions (frequencies) of IR bands
In section 2.5, we saw that diagonalization of the force constant matrix gives an eigenvalue matrix whose elements are the force constants of the normal modes, and an eigenvector matrix whose elements are their "direction vectors". Mass-weighting the force constants gives the wavenumbers ("frequencies") of the normal-mode vibrations, and their motions can be identified by using the direction vectors to animate them. So

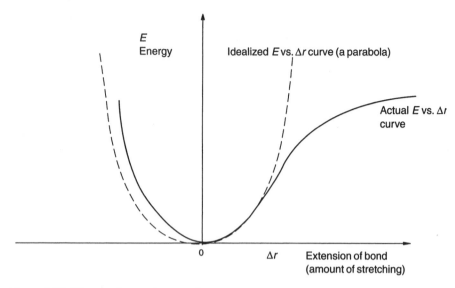

Figure 5.32. The actual curve for potential energy vs. stretch for a bond is not really a parabola, i.e. not really $E = (\Delta r)^2$, but near the equilibrium bond length ($\Delta r = 0$) the parabola fits the actual curve fairly well.

we can calculate the wavenumbers of IR bands and associate each band with some particular vibrational mode. The wavenumbers from *ab initio* calculations are larger than the experimental ones, i.e. the "frequencies" are too high. There are two reasons why this might be so: the principle of equating second derivatives of energy (with respect to geometry changes) with force constants might be at fault, or the basis set and/or correlation level might be deficient.

The principle of equating a second derivative with a stretching or bending force constant is not exactly correct. A second derivative $\partial^2 E/\partial q^2$ is strictly equal to a force constant only for cases where the energy is a quadratic function of the geometry (simple harmonic motion), i.e. where a graph of E vs. q is a parabola, but vibrational curves are not parabolas (Fig. 5.32). For a parabolic E/q relationship, and considering a diatomic molecule for simplicity:

$$E = \frac{k}{2}(q - q_{eq})^2 \tag{5.198}$$

where q_{eq} is the equilibrium geometry. Here k is by definition the force constant, the second derivative of E, and $\partial^2 E/\partial q^2 = k$. For a real molecule, however, the E/q relationship is more complicated, being a power series in q^2, q^3, etc., terms, and there is not just one constant. Equation (5.198) holds for what is called *simple harmonic motion*, and the coefficients of the higher-power terms in the more accurate equation are called *anharmonicity corrections*. Assuming that bond vibrations are simple harmonic is the *harmonic approximation*.

For small molecules it is possible to calculate from the experimental IR spectrum the simple harmonic force constant k and the anharmonicity corrections. Using k, *harmonic frequencies* can be calculated [153]. These correspond to a parabolic E/q relationship (Fig. 5.32), i.e. to a steeper curve than the real one, and thus to stiffer bonds. Stiffer bonds need more energy to stretch them (or bend them, for bending force constants), and thus absorb higher-frequency infrared light. Harmonic frequencies derived from experimental IR spectra are higher than the *observed* (the "raw") experimental frequencies, and are closer to *ab initio* frequencies than are the observed frequencies [154]. Since both theoretically calculated (e.g. by *ab initio* methods) frequencies and experimentally-derived harmonic frequencies are based on a parabolic E/q relationship, it is sometimes considered better to compare calculated frequencies with harmonic frequencies rather than observed experimental frequencies [155].

Because both *ab initio* and experimentally-derived harmonic frequencies rest on second derivatives, we might expect *ab initio* frequencies to converge not toward the observed experimental, but rather toward the experimentally-derived harmonic frequencies, as correlation level/basis set are increased. This is indeed the case, as has been shown by calculations on water with high correlation levels (CCSD(T); section 5.4.3) and large basis sets (polarization functions and triply- or quadruply-split valence shells (section 5.3.3). The deviations fell from 269, 282, and $127\,cm^{-1}$ at the Hartree-Fock level to values only 9, 13, and $10\,cm^{-1}$ higher than the experimentally-derived harmonic values of 3943, 3832, and $1649\,cm^{-1}$ [156]. The *observed* water frequencies are 3756, 3657 and $1595\,cm^{-1}$; experimentally-derived harmonic frequencies are typically about 5% higher, and *ab initio* frequencies about 5–10% higher, than observed frequencies.

From the foregoing discussion it appears that *ab initio* frequencies are too high because of the harmonic approximation: equating of $\partial^2 E/\partial q^2$ with a force constant. There is no theoretical reason why high-level calculations should converge toward the *observed* frequencies; this statement applies to frequencies calculated, as is almost always the case, by the harmonic approximation (above). However, we wish, ideally, to compute *observed* IR spectra. Fortunately, calculated and observed frequencies differ by a fairly constant factor, and *ab initio* frequencies can be brought into reasonable agreement with experiment by multiplying them by a correction factor. An extensive comparison by Scott and Radom of calculated and experimental frequencies [60] has provided empirical correction factors for frequencies calculated by a variety of methods. A few of the correction factors from this compilation are:

HF/3-21G	0.9085
HF/6-31G*	0.8953
HF/6-311G(df,p)	0.9054
MP2(FC)/6-31G*	0.9434
MP2(FC)/6-311G**	0.9496

The correction factors at the HF level with the three basis sets are very similar, 0.90–0.91; the factors at the MP2 level are significantly closer to 1, but Scott and Radom say that "MP2/6-31(d) does not appear to offer a significant improvement in performance over HF/6-31G(d) and occasionally shows large errors", and "The most cost-effective procedures found in this study for predicting vibrational frequencies are HF/6-31(d) and [certain density functional methods]". Scott and Radom have also derived separate correction factors for zero-point vibrational energies, although it was at least hitherto common practice to use the same correction factor for frequencies and for ZPE's. Better agreement with experiment can be obtained by using empirical correction factors for specific kinds of vibrations (Scott and Radom give separate factors for low-frequency vibrations, as opposed to the relatively high-frequency ones to which the factors above refer), but this is rarely done.

Intensities of IR bands
The bands in an IR spectrum have not just *positions* ("frequencies", denoted by various wavenumbers), but also *intensities* (not routinely quantified, but commonly described as weak, medium, or strong). To calculate an IR spectrum for comparison with experiment it is desirable to compute both wavenumbers and intensities. The intensity of an IR vibration is determined by the change in dipole moment accompanying the vibration. If a vibrational mode leads to no change in dipole moment, the mode will, theoretically, not result in absorption of an IR photon, because the oscillating electric fields of the radiation and the vibrational mode will be unable to couple. Such a vibrational mode is said to be *IR-inactive*, i.e. it should cause no observable band in the IR spectrum. Stretching vibrations that, because of symmetry, are not accompanied by a change in dipole moment, are expected to be IR-inactive. These occur mainly in homonuclear molecules like O_2 and N_2, and in linear molecules; thus the C/C triple bond stretch in symmetrical akynes, and the symmetric OCO stretch in carbon dioxide, do not engender bands in the IR spectrum. For Raman spectroscopy (in which one measures the scattered rather than the transmitted IR light), the requirement for observing a vibrational mode (i.e. for absorption of a photon) is that the vibration occur with a change in polarizability. Raman spectra are routinely calculable (e.g. by the Gaussian programs [23,48]; the IR and Raman *frequencies* are the same) along with IR spectra. A band which should be IR-inactive or at least very weak can in fact sometimes be seen because of coupling with other vibrational modes; thus the triple-bond stretch of benzyne (C_6H_4, dehydrobenzene) has been observed [157], although it apparently should be accompanied by only a very small change in dipole moment. Bands like this are expected to be, at best, weak.

As might be expected from the foregoing discussion, the intensity of an IR normal mode can be calculated from the change in dipole moment with respect to the change in geometry accompanying the vibration. The intensity is proportional to the square of the change in dipole moment with respect to geometry:

$$I = \text{constant} \times \left(\frac{d\mu}{dq}\right)^2 \qquad (5.199)$$

This can be used to calculate the relative intensities of IR bands (the calculation of dipole moments is discussed in the next section). One way to calculate the derivative is to approximate it as a ratio of finite increments d becomes Δ) and calculate the change in dipole moment with a small change in geometry; there are also analytical methods for calculating the derivative [158].

It has been reported that at the HF level calculated IR-band intensities often differ from experiment by a factor of over 100 percent, but at the MP2 level are typically within 30 percent of experiment [159]. A few calculated (frequencies and intensities) IR

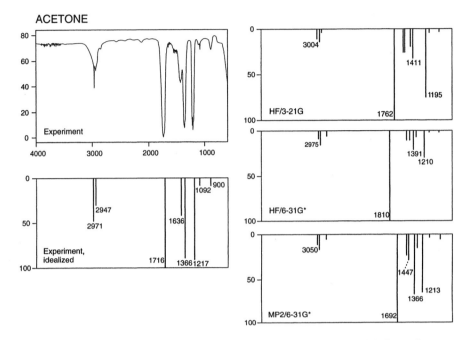

Figure 5.33. Experimental (gas phase) and *ab initio* calculated IR spectra (Table 5.14) of acetone.

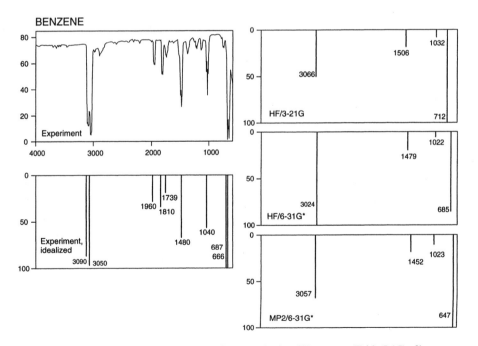

Figure 5.34. Experimental (gas phase) and *ab initio* calculated IR spectra (Table 5.15) of benzene.

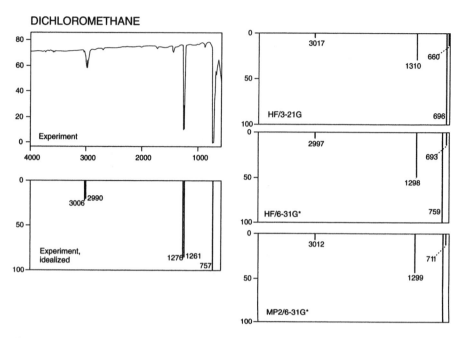

Figure 5.35. Experimental (gas phase) and *ab initio* calculated IR spectra (Table 5.16) of dichloromethane.

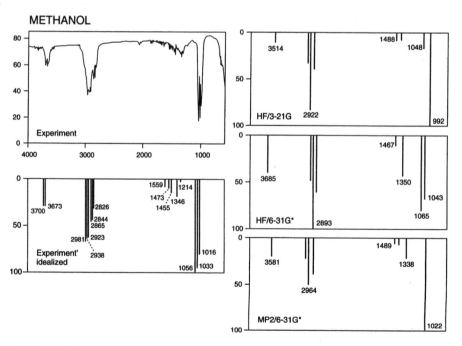

Figure 5.36. Experimental (gas phase) and *ab initio* calculated IR spectra (Table 5.17) of methanol.

Table 5.14. Acetone, calculated IR spectrum after discarding IR-inactive or very weak (intensity less than 2 per cent of the strongest band) frequencies, ignoring frequencies below 600 cm^{-1}, correcting frequencies, combining degenerate frequencies and their intensities, and normalizing intensities. Frequency correction factors [60]: HF/3-21G, 0.9085; HF/6-31G*, 0.8953; MP2(FC)/6-31G*, 0.9434

	HF/3-21G		HF/6-31G*		MP2(FC)/6-31G*	
	cm^{-1}	Intensity	cm^{-1}	Intensity	cm^{-1}	Intensity
1	891	4.9	880	1.6	875	4.1
2	1142	5.6	1109	1.3	1091	3.3
3	1195	77	1210	33	1213	68
4	1411	36	1384	3.5	1361	16
5	1419	17	1391	21	1366	74
6	1482	23	1445	9.2	1447	28
7	1510	23	1461	7.9	1461	24
8	1762	100	1810	100	1692	100
9	2912	4.5	2876	5.2	2931	5.2
10	2959	15	2925	17	3008	17
11	3004	13	2975	13	3050	13

Table 5.15. Benzene, calculated IR spectrum after discarding IR-inactive or very weak (intensity less than 2 percent of the strongest band) frequencies, ignoring frequencies below 600 cm^{-1}, correcting frequencies, combining degenerate frequencies and their intensities, and normalizing intensities. Frequency correction factors [60]: HF/3-21G, 0.9085; HF/6-31G*, 0.8953; MP2(FC)/6-31G*, 0.9434

	HF/3-21G		HF/6-31G*		MP2(FC)/6-31G*	
	cm^{-1}	Intensity	cm^{-1}	Intensity	cm^{-1}	Intensity
1	712	100	685	85	647	100
2	1032	4.9	1022	4.6	1023	7.6
3	1506	21	1479	18	1452	19
4	3066	51	3024	100	3057	68

spectra are shown in Figs 5.33–5.36 (based on experiment, and the data in Tables 5.14–5.17). This sample, although very limited, gives one an idea of the kind of similarity one can expect between experimental and *ab initio* IR spectra. A detailed resemblance cannot be expected, but the general features of a spectrum are reproduced. Probably the main utility of calculated *ab initio* IR spectra is in predicting the IR spectra of unknown molecules, as an aid to their synthesis. It should be possible to increase the accuracy of predicted spectra by performing calculations on a series of known compounds and fitting the experimental to the calculated wavenumbers, and perhaps intensities, to obtain empirical corrections tailored specifically to the functional group of interest.

Table 5.16. Dichloromethane, calculated IR spectrum after discarding IR-inactive or very weak (intensity less than 2 percent of the strongest band) frequencies, ignoring frequencies below $600\,cm^{-1}$, correcting frequencies, combining degenerate frequencies and their intensities, and normalizing intensities. Frequency correction factors [60]: HF/3-21G, 0.9085; HF/6-31G*, 0.8953; MP2(FC)/6-31G*, 0.9434

	HF/3-21G		HF/6-31G*		MP2(FC)/6-31G*	
	cm^{-1}	Intensity	cm^{-1}	Intensity	cm^{-1}	Intensity
1	660	14	693	14	711	11
2	696	100	755	100	770	100
3	1310	30	1298	47	1299	42
4	3017	3.5	2997	7.7	3012	6.7

Table 5.17. Methanol, calculated IR spectrum after discarding IR-inactive or very weak (intensity less than 2 percent of the strongest band) frequencies, ignoring frequencies below $600\,cm^{-1}$, correcting frequencies, combining degenerate frequencies and their intensities, and normalizing intensities. Frequency correction factors [60]: HF/3-21G, 0.9085; HF/6-31G*, 0.8953; MP2(FC)/6-31G*, 0.9434

	HF/3-21G		HF/6-31G*		MP2(FC)/6-31G*	
	cm^{-1}	Intensity	cm^{-1}	Intensity	cm^{-1}	Intensity
1	992	100	1043	69	1022	100
2	1048	19	1065	78	1338	23
3	1345	6.4	1350	44	1452	5.5
4	1488	5.0	1467	11	1489	2.8
5	2886	41	2852	63	2901	41
6	2922	84	2893	100	2964	55
7	2991	34	2959	49	3039	23
8	3514	8.8	3685	41	3581	20

5.5.4 Properties arising from electron distribution: dipole moments, charges, bond orders, electrostatic potentials, atoms-in-molecules

We have seen three applications of *ab initio* calculations: finding the shapes (geometries), the relative energies, and the frequencies of stationary points (usually minima and transition states) on a potential energy surface. The *shape* of a molecular species is one of its fundamental characteristics. It can, for example, provide clues to the existence of theoretical principles (why is it that benzene has six equal-length CC bonds, but cyclobutadiene has two "short" and two "long" bonds [160]?), or act as a guide to designing useful molecules (docking a candidate drug into the active site of an enzyme requires a knowledge of the shapes of the drug and of the active site [82]). The relative *energies* of molecular species is fundamental to a knowledge of their kinetic and thermodynamic behaviour, and this can be important in attempts to synthesize them. The *vibrational frequencies* of a molecule, provide information about the electronic nature of its bonds, and prediction of the spectra represented by these frequencies may be useful to experimentalists.

A fourth important characteristic of a molecule is the *distribution of electron density* in it. Calculation of the electron density distribution enables one to predict the dipole moment, the charge distribution, the bond orders, and the shapes of various molecular orbitals.

Dipole moments

The dipole moment [161] of a system of two charges Q and $-Q$ separated by a distance r is, by definition, the vector $\mathbf{Qr}$; the direction of the vector is officially from $-Q$ toward $+Q$, but chemists usually assign a molecular or bond dipole (see below) the direction from the positive end of the bond or molecule to the negative (Fig. 5.37(a)). The dipole moment of a collection of charges $Q_1, Q_2, \ldots, Q_n$, with corresponding position vectors $\mathbf{r}_1, \mathbf{r}_2, \ldots, \mathbf{r}_n$ (Fig. 5.37(b)) is

$$\mu = \sum_{1}^{n} Q_i \mathbf{r_i} \tag{5.200}$$

and the so the dipole moment of a molecule is seen to arise from the charges and positions of its component electrons and nuclei. The dipole moment of a molecule is an experimental observable [162], with which we can compare calculated moments. It is often convenient to think of the molecular dipole moment as the vector sum of bond moments (Fig. 5.37(c)). Two points should be noted: we are discussing an *average* dipole moment, because electron and nuclear motions will cause the dipole moment to fluctuate, so that even a spherical atom can have temporary nonzero dipole moments. Another point is that we usually consider the dipole moments of *neutral* molecules only, not of ions; this is because the dipole moment of a charged species is not unique, but depends on the choice of the point in the coordinate system from which the position vectors are measured.

Let us look at the calculation of the dipole moment within the HF approximation. The quantum mechanical analogue of Eq. (5.200) for the electrons in a molecule is

$$\mu = \left\langle \Psi \left| \sum_{j=1}^{2n} e r_j \right| \Psi \right\rangle \tag{5.201}$$

Here the summation of charges times position vectors is replaced by the integral over the total wavefunction Ψ (the square of the wavefunction is a measure of charge) of the dipole moment operator (the summation over all electrons of the product of an electronic charge and the position vectors of the electrons). To perform an *ab initio* calculation of the dipole moment of a molecule we want an expression for the moment in terms of the basis functions ϕ, their coefficients c, and the geometry (for a molecule of specified charge and multiplicity these are the only "variables" in an *ab initio* calculation). The HF

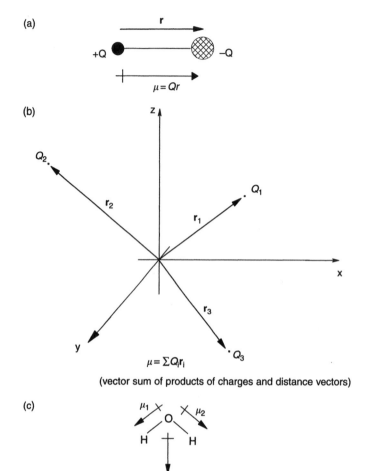

Figure 5.37. (a) Chemists usually consider the dipole moment of a diatomic molecule, the vector $Q\mathbf{r}$, to be directed from the positive to the negative atom. (b) The dipole moment of a collection of charges, such as a molecule, arises from the magnitudes of the charges, and their locations (i.e. distances and directions from the origin). (c) The dipole moment of a molecule can be thought of as the vector sum of bond moments.

total wavefunction Ψ is composed of those component orbitals ψ which are occupied, assembled into a Slater determinant (section 5.2.3.1), and the ψ's are composed of basis functions and their coefficients (sections 5.3). Eq. (5.201), with the inclusion of the contribution of the nuclei to the dipole moment, leads to the dipole moment in Debyes as ([1g], p. 41)

$$\mu = -2.5416 \left[\sum_{A}^{N} Z_A \mathbf{R}_A - \sum_{r}^{m} \sum_{s}^{m} P_{rs} \langle \phi_r | \mathbf{r} | \phi_s \rangle \right] \tag{5.202}$$

Here the first term refers to the nuclear charges and position vectors and the second term (the double summation) refers to the electrons. P_{rs} = the density matrix elements (sections 5.2.3.6d and 5.2.3.6e), cf.:

$$P_{tu} = 2 \sum_{j=1}^{n} c_{tj}^* c_{uj} \qquad (5.203 = 5.81)$$

The P summation is over the occupied orbitals ($j = 1, 2, \ldots, n$; we are considering closed-shell systems, so there are $2n$ electrons) and the double summation is over the m basis functions. The operator $\mathbf{r}$ is the electronic position vector.

How good are *ab initio* dipole moments? Hehre's extensive survey of practical *ab initio* methods [32] indicates that fairly good results are given by HF/6-31G*//HF/6-31G* (dipole moment from a HF/6-31G* calculation on a HF/6-31G* geometry) calculations, and that MP2/6-31G*//MP2/6-31G* calculations are usually not much better. Some calculated and experimental dipole moments are compared in Table 5.18. These results, which are quite typical, indicate that calculated values tend to be about 0.0–0.5 D higher than experimental, with a mean deviation of about 0.3 D; negative deviations are rare. HF/3-21G//HF/3-21G (the lowest *ab initio* level likely to be used) calculations may show the largest deviations. Single-point HF/6-31G* calculations on HF/3-21G geometries appear to give results about as good as (or better than? Note CH_3NH_2, and [32], pp. 76, 77) those from MP2(FC)/6-31G*//MP2(FC)/6-31G* calculations. As is the case for other properties, 3-21G calculations of dipole moments on molecules with atoms beyond neon require

Table 5.18. Some calculated dipole moments compared to experimental ones

Compound	HF/3-21G*// HF/3-21G*	HF/6-31G* //HF/3-21G*	HF/6-31G* //HF/6-31G*	MP2(FC)/6-31G* //MP2(FC)/6-31G*	exp.
CH_3NH_2	1.44	1.30	1.53	1.60	1.31
H_2O	2.39	2.18	2.20	2.24	1.85
HCN	3.04	3.20	3.21	3.26	2.99
CH_3OH	2.12	1.95	1.87	1.95	1.70
Me_2O	1.85	1.64	1.48	1.60	1.30
H_2CO	2.66	2.79	2.67	2.84	2.34
CH_3F	2.34	2.18	1.99	2.11	1.85
CH_3Cl	2.31	2.32	2.25	2.21	1.87
Me_2SO	4.27	4.55	4.50	4.63	3.96
CH_3CCH	0.71	0.64	0.64	0.66	0.75
Deviations	9+, 1−	8+, 2−	9+, 1−	9+, 1−	
Mean	0.33	0.31	0.26	0.34	

Dipole moments are in Debyes; the computational levels are arranged, from left to right, in what is conventionally considered lowest to highest. Calculations are by the author; experimental values are taken from [1g, 38]. For each level is given the number of positive and negative deviations and the arithmetic mean of the absolute values of the deviations.

polarization functions for reasonable results (the 3-21G$^{(*)}$ basis; [32], pp. 23–30). The 3-21G* calculations in Table 5.18 show a mean deviation 0.33; the HF/6-31G* calculations are only slightly better (mean deviation 0.26) and the MP2/6-31G* calculations appear to be, if anything, slightly worse (mean error 0.34). If high-accuracy calculated dipole moments (0.1 D or better) are needed, high-level correlation and large basis sets must be used; such calculations may be needed to reproduce the magnitude and even the direction of small dipole moments, e.g. in carbon monoxide [163].

Charges and bond orders
Chemists make extensive use of the idea that the atoms in a molecule can be assigned electrical *charges*. Thus in a water molecule each hydrogen atom is considered to have an equal, positive, charge, and the oxygen atom to have a negative charge (equal in magnitude to the sum of the hydrogen charges). This concept is clearly related to the dipole moment: in a diatomic (for simplicity) molecule one expects the negative end of the dipole vector to point toward the atom assigned the negative charge. However, there are two problems with the concept: first, the charge on an atom in a molecule, unlike the dipole moment of a molecule, cannot be measured (it is not an observable). Second, there is no unique, correct theoretical method for calculating the charge on an atom in a molecule.

Both the measurement and calculational problems arise from the difficulty of defining what we mean by "an atom in a molecule". Consider the hydrogen chloride molecule. As we move from the hydrogen *nucleus* to the chlorine *nucleus*, where does the hydrogen atom end and the chlorine atom begin? If we had a scheme for partitioning the molecule into atoms (Fig. 5.38(a)), the charge on each atom could be defined as the net electric charge within the space of the atom, i.e. the algebraic sum of the electronic and the nuclear charges. The electronic charge in the defined space could be found by integrating the electron density (essentially the square of the wavefunction; only the wavefunction composed of the occupied orbitals need be considered) over that region of space.

Bond order is a term with conceptual difficulties related to those associated with atom charges. The simplest electronic interpretation of a bond is that it is a pair of electrons shared between two nuclei, somehow [164] holding them together. From this criterion and Lewis structures the C/C bond order in ethane is 1, in ethene 2, and in ethyne 3, in accordance with the classical assignment of a single, a double, and a triple bond, respectively. However, if a bond is a manifestation of the electron density between two nuclei, then the bond order need not be an integer; thus the C=C bond in $H_2C=CH-CHO$ might be expected to have a lower bond order than the C=C in $H_2C=CH-CH_3$, because the C=O group might drain electron density away toward the electronegative oxygen. However, an attempt to calculate bond order from electron density (the square of the wavefunction) runs into the problem that in a polyatomic molecule, at any rate, it is not clear how to define precisely the region "between" two atomic nuclei (Fig. 5.38(b)).

Assigning atom charges and bond orders involves calculating the number of electrons "belonging to" an atom or shared "between" two atoms, i.e. the "population" of electrons on or between atoms; hence such calculations are said to involve *population analysis*. Earlier schemes for population analysis bypassed the problem of defining the space occupied by atoms in molecules, and the space occupied by bonding electrons, by partitioning electron density in a somewhat arbitrary way. The earliest such schemes

(a) Where does atom A end and B begin?

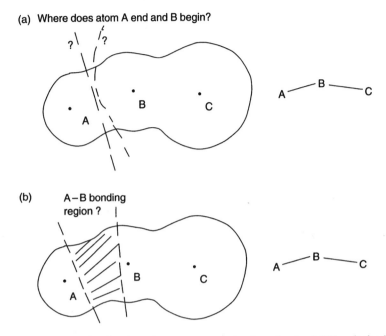

(b) A–B bonding
    region ?

Figure 5.38. (a) In a molecule where does one atom end and another begin? How is the dividing surface to be drawn? (b) How is the bonding region between two nuclei to be defined?

were utilized in the simple Hückel or similar methods [165], and related these quantities to the basis functions (which in these methods are essentially valence, or even just p, atomic orbitals; see section 4.3.4). The simplest scheme used in *ab initio* calculations is *Mulliken population analysis* [166].

Mulliken population analysis is in the general spirit of the scheme used in the simple Hückel method, but allows for several basis functions on an atom and does not require the overlap matrix to be a unit matrix. In *ab initio* theory each molecular orbital has a wavefunction ψ (section 5.2.3.6a):

$$\psi_1 = c_{11}\phi_1 + c_{21}\phi_2 + c_{31}\phi_3 + \cdots + c_{m1}\phi_m$$
$$\psi_2 = c_{12}\phi_1 + c_{22}\phi_2 + c_{32}\phi_3 + \cdots + c_{m2}\phi_m$$
$$\psi_3 = c_{13}\phi_1 + c_{23}\phi_2 + c_{33}\phi_3 + \cdots + c_{m3}\phi_m \qquad (5.204 = 5.51)$$
$$\vdots$$
$$\psi_m = c_{1m}\phi_1 + c_{2m}\phi_2 + c_{3m}\phi_3 + \cdots + c_{mm}\phi_m$$

Here the basis set $\{\phi_1, \phi_2, \ldots, \phi_m\}$ engenders MOs $\psi_1, \psi_2, \ldots, \psi_m$. Several basis functions can reside on each atom, so c_{si} is the coefficient of basis function s (not, as in simple Hückel theory, atom s) in MO i. For any MO ψ_i, squaring and integrating over all space gives

$$\int |\psi_i|^2 \, dv = 1$$

$$= c_{1i}c_{1i}S_{11} + c_{2i}c_{2i}S_{22} + \cdots$$

$$+ 2c_{1i}c_{2i}S_{12} + 2c_{1i}c_{3i}S_{13} + 2c_{2i}c_{3i}S_{23} + \cdots \qquad (5.205)$$

The integral equals one because the probability that the electron is *somewhere* in the MO (which, strictly, extends over all space) is one; the S_{ii} (both ϕ's the same) overlap integrals are also unity, since the basis functions are normalized (cf. section 4.4.1b).

In the Mulliken scheme each electron in ψ_i is taken to contribute a "fraction of an electron" $c_{1i}c_{1i}S_{11} = c_{1i}^2$ to basis function ϕ_1, and a fraction of an electron $2c_{1i}c_{2i}S_{12}$ (see Eq. (5.205)) to the ϕ_1/ϕ_2 overlap region, and in general to contribute a fraction of an electron c_{ri}^2 to the basis function ("orbital") ϕ_r, and a fraction of an electron $2c_{ri}c_{si}S_{rs}$ to the ϕ_r/ϕ_s overlap space; see Fig. 5.39(a). This seems reasonable since (1), the terms sum to one (the "fractions" of the electron must add to one), and (2) it seems reasonable to partition the contribution of electrons to basis functions and overlap regions according to the "electron density sum" in Eq. (5.205). Now if there are n_i electrons in MO ψ_i, then the contributions of ψ_i to the electron population of basis function ϕ_r and of the overlap region between ϕ_r and ϕ_s are

$$n_{r,i} = n_i c_{ri}^2 \qquad (5.206)$$

and

$$n_{r/s,i} = n_i(2c_{ri}c_{si}S_{rs}) \qquad (5.207)$$

The total contributions from all the MOs to the electron population in ϕ_r and in the overlap region between ϕ_r and ϕ_s are

$$n_r = \sum_i n_{r,i} = \sum_i n_i c_{ri}^2 \qquad (5.208)$$

and

$$n_{r/s} = \sum_i n_{r/s,i} = \sum_i n_i(2c_{ri}c_{si}S_{rs}) \qquad (5.209)$$

The sums are over the occupied MOs, since $n_i = 0$ for the virtual MOs. The number n_r is the *Mulliken net population* in the basis function ϕ_r, and the number $n_{r/s}$ is the *Mulliken overlap population* for the pair of basis functions ϕ_r and ϕ_s. The net population summed over all r plus the overlap population summed over all r/s pairs equals the total number of electrons in the molecule.

The quantities n_r and n_s are used to calculate atom charges and bond orders. The *Mulliken gross population in the basis function ϕ_r* is defined as the Mulliken net population n_r (Eq. (5.206)) plus one half of all those Mulliken overlap populations $n_{r/s}$ (Eq. (5.207)) which involve ϕ_r (of course for some ϕ_s, $n_{r/s}$ may be negligible; e.g. for well-separated atoms S_{rs} is very small):

$$N_r = n_r + \frac{1}{2} \sum_{s \neq r} n_{r/s} \qquad (5.210)$$

(a)

MO ψ_i, formed from basis functionsand ϕ_1 and ϕ_2

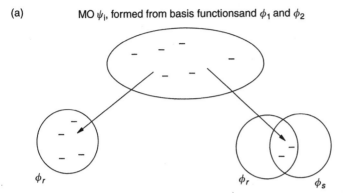

ϕ_r ϕ_r ϕ_s

In the Mulliken scheme (Eqs (5.206) and (5.207)):
Each electron in MO ψ_i contributes a fraction of an electron c_{ri}^2 to ϕ_r and $2c_{ri}c_{si}S_{rs}$ to the ϕ_r/ϕ_s overlap region. If there are n_i electrons in ψ_i, then the MO contributes to the electron population of basis function ϕ_r $n_{r1}=n_ic_{ri}^2$ electrons, and to the electron population of the ϕ_r/ϕ_s overlap region $n_{r/s,i}= n_i(2c_{ri}c_{si}S_{rs})$ electrons.

(b)

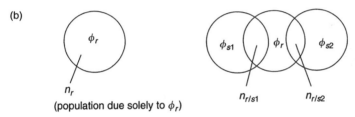

ϕ_r ϕ_{s1} ϕ_r ϕ_{s2}

n_r $n_{r/s1}$ $n_{r/s2}$

(population due solely to ϕ_r)

In the Mulliken scheme (Eqs (5.210)):
The gross electron population in ϕ_r is n_r (due solely to ϕ_r), plus half the sum of all overlap populations: $Nr= n_r+1/2n_{r/s1}+1/2n_{r/s2}$.

Figure 5.39. The Mulliken scheme for partitioning electron density.

The gross population N_r is an attempt to represent the *total* electron population in the basis function ϕ_r; this is considered here to be the net population n_r, the population that all the occupied MOs contribute to ϕ_r through the representation of ϕ_r in each ψ_i by its coefficient c_{ri} (Eq. (5.208), plus one-half of the all the populations in the overlap regions involving ϕ_r (Fig. 5.39(b)). Assigning to ϕ_r one-half, rather than some other fraction, of the electron population in an overlap region with ϕ_s is said to be arbitrary. Of course it is not arbitrary, in the sense that Mulliken thought about it carefully and decided that one-half was at least as good as any other fraction. One might imagine a more elaborate partitioning in which the fraction depends on the electronegativity difference between the atoms on which ϕ_r and ϕ_s reside, with the more electronegative atom getting the larger share of the electron population. To get the charge on an atom A we calculate the *gross atomic population for A*:

$$N_A = \sum_{r \in A} N_r \qquad (5.211)$$

This is the sum over all the basis functions ϕ_r on atom A ($r \in$ A qualifying the summation means "r belonging to A") of the gross populations in each ϕ_r (Eq. (5.210); it involves all the basis functions on A and all the overlap regions these functions have with other basis functions ϕ_s. We can regard N_A as the total electron population on atom A (within the limits of the Mulliken treatment). The *Mulliken charge* on atom A, the *net charge* on A, is then simply the algebraic sum of the charges due to the electrons and the nucleus:

$$q_A = Z_A - N_A \tag{5.212}$$

The *Mulliken bond order* for the bond between atoms A and B is the total population for the A/B overlap region:

$$b_{AB} = \sum_{r,s \in A,B} n_{r/s} \tag{5.213}$$

The overlap population for basis functions ϕ_r and ϕ_s (Eq. (5.209)) is summed over all the overlaps between basis functions on atoms A and B.

Since the formulas for calculating Mulliken charges and bond orders (Eqs (5.206)–(5.213)) involve summing basis function coefficients and overlap integrals, it is not too surprising that they can be expressed neatly in terms of the density matrix (section 5.2.3.6d) **P** and the overlap matrix **S** (section 4.3.3). The elements of the density matrix **P** are

$$P_{rs} = 2 \sum_{i=1}^{n} c_{ri} c_{si} \tag{5.214}$$

The matrix element P_{rs} is summed over all filled MOs (from ψ_1 to ψ_n for the ground electronic state of a $2n$-electron closed-shell molecule); an example of the calculation of **P** was given in section 5.2.3.6e. The elements of the overlap matrix **S** are simply the overlap integrals:

$$S_{rs} = \int \phi_r \phi_s \, dv \tag{5.215}$$

From Eq. (5.214) it follows that the matrix (**PS**) obtained by multiplying corresponding elements of **P** and **S**,

$$(\mathbf{PS}) = \begin{pmatrix} (PS)_{11} & (PS)_{12} & (PS)_{13} & \cdots & (PS)_{1m} \\ (PS)_{21} & (PS)_{22} & (PS)_{23} & \cdots & (PS)_{2m} \\ \vdots & \vdots & \cdots & \vdots \\ (PS)_{m1} & (PS)_{m2} & (PS)_{m3} & \cdots & (PS)_{mm} \end{pmatrix} \tag{5.216}$$

has elements

$$(PS)_{rs} = P_{rs} S_{rs} = 2 \sum_{i=1}^{n} c_{ri} c_{si} S_{rs} \tag{5.217}$$

Note that (**PS**) is *not* the matrix **PS** obtained by matrix multiplication of **P** and **S**; each element of *that* matrix would result from series multiplication: a row of **P** times a column of **S** (section 4.3.3).

The diagonal elements of (**PS**) are

$$(PS)_{rr} = P_{rr}S_{rr} = 2\sum_{i=1}^{n} c_{ri}^2 \tag{5.218}$$

Compare this with Eq. (5.208): for a ground-state closed-shell molecule there are 2 electrons in each occupied MO and Eq. (5.208) can be written as:

$$n_r = 2\sum_{i=1}^{n} c_{ri}^2 \tag{5.219}$$

i.e.

$$n_r = (PS)_{rr} \tag{5.220}$$

The off-diagonal elements of (**PS**) are given by Eq. (5.217), $r \neq s$. Compare this with Eq. (5.209): for a ground-state closed-shell molecule there are 2 electrons in each occupied MO and Eq. (5.209) can be written as:

$$n_{r/s} = 2\sum_{i=1}^{n}(2c_{ri}c_{si}S_{rs}) \tag{5.221}$$

i.e.

$$n_{r/s} = 2(PS)_{rs} \tag{5.222}$$

Thus the matrix (**PS**) can be written as:

$$(\textbf{PS}) = \begin{pmatrix} n_1 & \frac{1}{2}n_{1/2} & \frac{1}{2}n_{1/3} & \cdots & \frac{1}{2}n_{1/m} \\ \frac{1}{2}n_{2/1} & n_2 & \frac{1}{2}n_{2/3} & \cdots & \frac{1}{2}n_{2/m} \\ \vdots & \vdots & & \cdots & \vdots \\ \frac{1}{2}n_{m/1} & \frac{1}{2}n_{m/2} & \frac{1}{2}n_{m/3} & \cdots & n_m \end{pmatrix} \tag{5.223}$$

The matrix (**PS**) (or sometimes 2(**PS**)) is called a *population matrix*.

An example of population analysis: $H-He+$
As a simple illustration of the calculation of atom charges and bond orders, consider $H-He^+$. From our HF calculations on this molecule (section 5.2.3.6e) we have

$$\textbf{P} = \begin{pmatrix} 0.2020 & 0.5097 \\ 0.5097 & 1.2864 \end{pmatrix} \quad \text{and} \quad \textbf{S} = \begin{pmatrix} 1.0000 & 0.5017 \\ 0.5017 & 1.0000 \end{pmatrix} \tag{5.224}$$

Therefore,

$$(\textbf{PS}) = \begin{pmatrix} 0.2020 & 0.2557 \\ 0.2557 & 1.2864 \end{pmatrix} \tag{5.225}$$

From Eq. (5.223), (**PS**) gives us

$$n_1 = 0.2020$$

$$n_2 = 1.2864$$

$$n_{1/2} = n_{2/1} = 2(0.2557) = 0.5114$$

Charge on H, q_H For this we need N_H, the sum of all the N_r on H (Eqs (5.211) and (5.210)). There is only one basis function on H, ϕ_1, so there is only one relevant N_r for H, and for ϕ_1 there is only one overlap, with ϕ_2, so the summation involves only one term, $n_{1/2}$. Using Eq. (5.210):

$$N_r = N_1 = n_r + \frac{1}{2}\sum_{s \neq r} n_{r/s} = n_1 + \frac{1}{2}(n_{1/2}) = 0.2020 + \frac{1}{2}(0.5114) = 0.4577$$

The sum of all the N_r on H has only one term, N_1, since there is only one basis function on H. Using Eq. (5.211):

$$N_A = N_H = \sum_{r \in H} N_r = N_1 = 0.4577$$

The charge on H, q_H, is the algebraic sum of the gross electronic population and the nuclear charge (Eq. (5.212)):

$$q_A = q_H = Z_H - N_H = 1 - 0.4577 = 0.5423$$

Charge on He, q_{He} For this we need N_{He}, the sum of all the N_r on He (Eq. (5.211). There is only one basis function on He, ϕ_2, so there is only one relevant N_r for He, and for ϕ_2 there is only one overlap, with ϕ_1, so the summation involves only one term, $n_{2/1}(= n_{2/1})$:

$$N_r = N_2 = n_r + \frac{1}{2}\sum_{s \neq r} n_{r/s} = n_2 + \frac{1}{2}(n_{2/1}) = 1.2864 + \frac{1}{2}(0.5114) = 1.5421$$

The sum of all the N_r on He has only one term, N_2, since there is only one basis function on He:

$$N_A = N_{He} = \sum_{r \in He} N_r = N_2 = 1.5421$$

The charge on He, q_{He}, is the algebraic sum of the gross electronic population and the nuclear charge:

$$q_A = q_{He} = Z_{He} - N_{He} = 2 - 1.5421 = 0.4579$$

The charges sum to $0.5423 + 0.4579 = 1.000$, the total charge on the molecule. The less positive charge on helium is in accord with the fact that electronegativity increases from left to right along a row of the periodic table. *H−He bond order* For this we use Eq. (5.213); $n_{r/s}$ is summed for all overlaps between basis functions on atoms A and B. There is only one such overlap, that between ϕ_1 and ϕ_2, so

$$b_{AB} = b_{HHe} = \sum_{r,s \in A,B} n_{r/s} = n_{1/2} = 2(0.2557) = 0.5114$$

Note that the elements of the population matrix (**PS**) sum to the number of electrons in the molecule: $0.2020 + 1.2864 + 0.2557 + 0.2557 = 2.000$. This is expected, since the diagonal elements are the number of electrons in the "atomic space" of the basis functions, and the off-diagonal elements are the number of electrons in the overlap space of the basis functions.

The Mulliken approach to population analysis has certain problems; for example, it sometimes assigns more than two electrons, and sometimes a negative number of electrons, to an orbital. It is also fairly basis-set dependent (Hehre, Radom, Schleyer and Pople compare Mulliken charges for a variety of molecules using the STO-3G, 3-21G$^{(*)}$ and 6-31G* basis sets: [1g], pp. 337–339). Other approaches to partitioning electrons among orbitals and thus calculating charges and bond orders are the Löwdin method [167] and the natural atomic orbitals (NAO) population analysis of Weinhold [168].

Electrostatic potential

The *electrostatic potential* (ESP) is a measure of charge distribution that also provides other useful information [169]. The electrostatic potential at a point P in a molecule is defined as the amount of energy (work) needed to bring a unit point positive "probe charge" (e.g. a proton) from infinity to P. The electrostatic potential can be thought of as a measure of how positive or negative the molecule is at P: a positive value at the point means that the net effect experienced by the probe charge as it was brought from infinity was repulsion, while a negative value means that the probe charge was attracted to P, i.e. energy was *released* as it fell from infinity to P. The ESP at a point is the net result of the effect of the positive nuclei and the negative electrons. The calculation of the effect of the nuclei is trivial, following directly from the fact that the potential due to a point charge Z at a distance r away from the unit charge is

$$V(P) = \int_r^\infty \frac{Z \times 1}{r^2}\, dr = \frac{Z}{r} \tag{5.226}$$

Thus the ESP created by the nuclei is

$$V(P)_{nuc} = \sum_A \frac{Z_A}{|r_P - r_A|} \tag{5.227}$$

where $|r_p - r_A|$ is the distance from nucleus A to the point P, i.e. the absolute value of the difference of two vectors. To obtain the expression for the ESP due to the electrons, we can modify Eq. (5.227) by replacing the summation over the nuclei by an integral over infinitesimal volume elements of the electron density or charge density $\rho(r)$ (see *Atoms-in-molecules*). The total ESP is:

$$V(P)_{tot} = V(P)_{nuc} + V(P)_{el} = \sum_A \frac{Z_A}{|r_P - r_A|} + \int \frac{\rho(r)}{|r_P - r|}\, dr \tag{5.228}$$

The ESP at many points on the surface of the molecule can be calculated (section 5.5.6) and a set of atom charges then calculated to fit (by a least-squares procedure) the ESP values, and also to sum to the net charge on the molecule (the use of visualization of the ESP is discussed in section 5.5.6).

Table 5.19. Comparing Mulliken, electrostatic potential and natural charges, and Mulliken and Lowdin bond orders, at various levels, for hydrogen fluoride. The geometry used in each case corresponds to the method/basis set for that charge or bond order, but any reasonable geometry should give essentially the same results

Level	Charge on H (= −charge on F)			Bond order	
	Mulliken	Electrostatic	Natural	Mulliken	Löwdin
HF/STO-3G	0.19	0.28	0.23	0.96	0.98
HF/3-21G	0.45	0.49	0.50	0.78	0.93
HF/6-31G*	0.52	0.45	0.56	0.72	0.82
HF/6-31G**	0.39	0.45	0.56	0.86	1.07
HF/6-311G**	0.32	0.46	0.54	0.95	1.32
6-31+G*	0.57	0.48	0.58	0.64	0.75
6-311++G**	0.30	0.47	0.55	0.98	1.27
MP2/6-31G*	0.52	0.45	0.56	0.72	0.81

Values of Mulliken and Löwdin bond orders, and Mulliken, natural and ESP atom charges, are compared in Table 5.19, for hydrogen fluoride. We see that the Mulliken charges vary considerably, but apart from the STO-3G values, the electrostatic charges vary very little, and the natural charges little, with the level of calculation. Bond orders, however, are quite sensitive to the level of calculation. The utility of charges and bond orders lies not in their absolute values, but rather in the fact that a comparison of, say, Löwdin charges or bond orders, calculated at the same level for a *series* of molecules, can provide insights into a trend. For example, one might argue that the electron-withdrawing power of a series of groups A, B, etc. could be compared by comparing the C/C bond orders in $A-CH=CH_2$, $B-CH=CH_2$, etc. Bond orders have been used to judge whether a species is free or really covalently bonded, and have been proposed as an index of progress along a reaction coordinate [170].

Atoms-in-molecules
A method of calculating charges and bond orders that may be less arbitrary than any of those mentioned so far is based on the theory of atoms in molecules (AIM), developed by Bader and coworkers [171]. The AIM approach rests on analyzing the variation from place to place in a molecule of the electron density function (electron probability function, charge density function, charge density), ρ. This is a function $\rho(x, y, z)$ which gives the variation of the total electron density from point to point in the molecule; $\rho(x, y, z)\, dx\, dy\, dz = \rho(x, y, z)\, dv$ is the probability of finding an electron in an infinitesimal volume dv centered on the point with coordinates (x, y, z) (the probability of finding *more than one* electron in dv is infinitesimal). The electron probability density can be calculated from the wavefunction. It is not, as one might have thought, simply $|\Psi|^2$, where Ψ is the multielectron wavefunction of space and spin coordinates (section 5.2.3.1). *This* latter is the function for the variation from point to point of the probability of finding electron 1 with a specified spin, electron 2 with a specified spin, etc., at points (x, y, z). If we think of the electrons being smeared out in a fog around

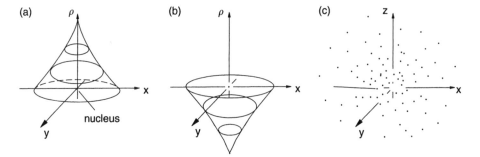

Figure 5.40. The distribution of electron density (charge density) ρ for an atom; the nucleus is at the origin of the coordinate system. (a) Variation of ρ with distance from the nucleus. Moving away from the nucleus ρ decreases from its maximum value and fades asymptotically toward zero. (b) Variation of $-\rho$ with distance from the nucleus; $-\rho$ becomes less negative and approaches zero. The $-\rho$ picture is useful for molecules (Fig. 5.41) because it makes clearer analogies with a potential energy surface. (c) A "4D" picture (ρ vs. x, y, z) of the variation of ρ in an atom: the density of the dots (number of dots per unit volume) indicates qualitatively electron density ρ in various regions.

the molecule, then the variation of ρ from point to point corresponds to the varying density of the fog, and $\rho(x, y, z)$ centered on a point $P(x, y, z)$ corresponds to the amount of fog in the volume element dxdydz. Alternatively, in a scatterplot of electron density in a molecule, the variation of ρ with position can be indicated by varying the volume density of the points. For more on the electron density function see sections 7.1 and 7.2.1.

Consider first the electron density ρ around an atom. As we approach the nucleus this rises toward a maximum, or the *negative* of the electron density, $-\rho$, falls toward a minimum (Fig. 5.40). Viewing the electron distribution in terms of $-\rho$ rather than ρ is useful because it more easily enables us to discern analogies between the variation of ρ in a molecule (in a ρ vs. location-in-molecule graph), and a potential energy surface, which is the variation of energy with geometry (an E vs. geometry graph). Examine the distribution of $-\rho$ in a homonuclear diatomic molecule X_2 (Fig. 5.41). This shows a plot of $-\rho$ vs. two of the three Cartesian coordinates needed to assign positions to all the points in the molecule. The graph retains the internuclear axis (by convention the z-axis) and one other axis, say y; the molecule is symmetrical with respect to reflection in the yz plane. The negative of the electron density, $-\rho$, dips toward a minimum at the atomic nuclei (ρ goes toward a maximum). The nuclei correspond to stationary points, where the surface has zero slope (if we ignore the fact that, *strictly speaking*, the nonrelativistic density forms a cusp at a nucleus, i.e. the derivative of $\rho(x, y, z)$ becomes discontinuous) and from whence it goes upward in all directions. Thus the first derivative of $-\rho$ with respect to all spatial coordinates is zero, and the second derivative is positive:

$$\frac{\partial(-\rho)}{\partial z} = 0, \quad \frac{\partial(-\rho)}{\partial y} = 0, \quad \frac{\partial(-\rho)}{\partial x} = 0 \qquad (5.229)$$

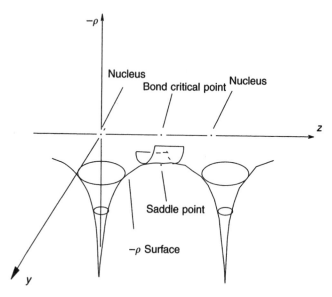

Figure 5.41. The distribution of the electron density ρ for a homonuclear diatomic molecule X_2. One nucleus lies at the origin, the other along the positive z-axis (the z-axis is commonly used as the molecular axis). The xz plane represents a slice through the molecule along the z-axis. The vertical axis is the $-\rho$ axis (negative of the electron density). The $-\rho = f(x, z)$ surface is analogous to a potential energy surface $E = f(\text{nuclear coordinates})$, and has minima at the nuclei (maximum value of ρ) and a saddle point, corresponding to a bond critical point, along the z-axis (midway between the two nuclei since the molecule is homonuclear).

and

$$\frac{\partial^2(-\rho)}{\partial z^2} > 0, \quad \frac{\partial^2(-\rho)}{\partial y^2} > 0, \quad \frac{\partial^2(-\rho)}{\partial x^2} > 0 \tag{5.230}$$

Moving along the internuclear line we find a point in a saddle-shaped region, analogous to a transition state, where the surface again has zero slope (all first derivatives zero), and is negatively curved along the z-axis but positively curved in all other directions (Fig. 5.41), i.e.

$$\frac{\partial^2(-\rho)}{\partial z^2} < 0, \quad \frac{\partial^2(-\rho)}{\partial y^2} > 0, \quad \frac{\partial^2(-\rho)}{\partial x^2} > 0 \tag{5.231}$$

This transition-state-like point is called a *bond critical point*. All points at which the first derivatives are zero are critical points, so the nuclei are also critical points. Analogously to the energy/geometry Hessian of a potential energy surface, an electron density function critical point (stationary point; relative maximum or minimum or saddle point) can be characterized in terms of its second derivatives by diagonalizing the ρ/q Hessian ($q = x$, y, or z) to get the number of positive and negative eigenvalues:

$$\rho/q \text{ Hessian} = \begin{pmatrix} \partial^2\rho/\partial x^2 & \partial^2\rho/\partial xy & \partial^2\rho/\partial xz \\ \partial^2\rho/\partial yx & \partial^2\rho/\partial y^2 & \partial^2\rho/\partial yz \\ \partial^2\rho/\partial zx & \partial^2\rho/\partial zy & \partial^2\rho/\partial z^2 \end{pmatrix} \tag{5.232}$$

For the $-\rho/q$ surface of Fig. 5.41 the number of positive and negative eigenvalues for a nuclear critical point are 3 and 0, and for a bond critical point, 2 and 1. Thus for the ρ/q surface to which the Hessian of Eq. (5.232) refers (the mirror image of the $-\rho/q$ surface), the number of positive and negative eigenvalues is, respectively, 0 and 3 (for a nucleus), and 1 and 2 (for a bond critical point).

The minimum-$(-\rho)$ path (maximum-ρ path) from one X nucleus to the other is the *bond path*; with certain qualifications this can be regarded as a bond. It is analogous to the minimum-energy path connecting a reactant and its products, i.e. to the intrinsic reaction coordinate. Such a bond is not necessarily a straight line: in strained molecules it may be curved. The bond passes through the bond critical point, (which for a homonuclear diatomic molecule X_2 is the midpoint between of the internuclear line). Now consider Fig. 5.42, which shows the X_2 molecule viewed along the ρ-axis. The contour lines represent electron density, which rises as we approach a nucleus and falls off as we go to and beyond the van der Waals surface. If it is true that the molecule can be divided into atoms, then for X_2 the dividing surface S (represented as a vertical line in Fig. 5.42) must lie midway between the nuclei, with the internuclear line being normal to S and meeting S at the bond critical point. The electron density defines a *gradient vector field*, the totality of the *trajectories* each of which results from starting at infinity and moving along the path of steepest increase in ρ. Figure 5.42 shows that only two of the trajectories (of those in the plane of the paper) that originate at infinity do not end at the nuclei; these end at the bond critical point. These two trajectories define the intersection of S with the plane of the paper. None of the trajectories cross S, which is

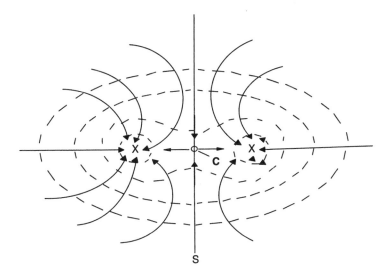

Figure 5.42. Contour lines for ρ, the electron density distribution, in a homonuclear diatomic molecule X_2. The lines originating at infinity and terminating at the nuclei and at the bond critical point **C** are trajectories of the gradient vector field (the lines of steepest increase in ρ; two trajectories also *originate* at **C**). The line S represents the dividing surface between the two atoms (the line is where the plane of the paper cuts this surface). S passes through the bond critical point and is not crossed by any trajectories.

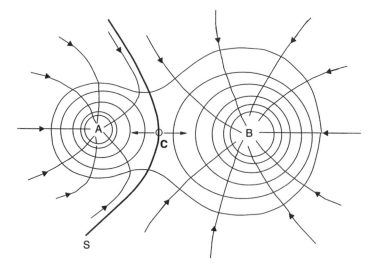

Figure 5.43. Heteronuclear (as well as homonuclear; cf. Fig. 5.42) molecules can be partitioned into atoms. S represents a slice through the zero-flux surface that defines the atoms A and B in a molecule AB. The lines with arrows are the trajectories of the gradient vector field. S passes through the bond critical point C and is not crossed by any trajectory lines.

thus called a *zero-flux* surface (the gradient vector field is analogous to an electric field whose "flux lines" point along the direction of attraction of a positive charge toward a central negative charge). The space within a molecule bounded by one (for a diatomic molecule) or more zero-flux surfaces is an *atomic basin* (away from the nuclei the basin extends outward to infinity, becoming shallower as the electron density fades toward zero). The nucleus and the electron density in an atomic basin constitute an atom in a molecule. Even for molecules other than homonuclear diatomics, atoms are still defined by atomic basins partitioned off by unique zero-flux surfaces, as illustrated in Fig. 5.43.

In the AIM (atoms-in-molecules method), the charge on an atom is calculated by integrating the electron density function over the volume of its atomic basin; the charge is the algebraic sum of the electronic charge and the nuclear charge (the atomic number of the nucleus minus the number of electrons in the basin). An AIM bond order ρ_b can be defined in terms of the electron density [172], and the bond order b_{AB} for two particular atoms A and B is then defined by an empirical equation obtained by fitting ρ_b to a few accepted A–B bond orders. For example, for nitrogen/nitrogen bonds a linear equation $b_{AB} = a_{NN}\rho_b + b_{NN}$ correlates b_{AB} and ρ_b for, say, $H_2N–NH_2$, HN=NH and N≡N; from this equation bond orders can be assigned to other nitrogen/nitrogen bonds from their ρ_b values.

5.5.5   Miscellaneous properties – UV and NMR spectra, ionization energies, and electron affinities

A few other properties that can be calculated by *ab initio* methods are briefly treated here.

Table 5.20. Calculated and experimental UV spectra of methylenecyclopropene, using the RCIS/6-3+G* method on the B3LYP/6-31G* (a density functional method, chapter 7) geometry. The procedure and the experimental values are given in [173]

Calculated		Experimental	
Wavelength (nm)	Relative intensity	Wavelength (nm)	Relative intensity
224	15	309	13
209	6	242	0.6
196	0	206	100
194	8		
193	100		

UV spectra

Ultraviolet spectra result from the promotion of an electron in an occupied MO of a ground electronic state molecule into a virtual MO, thus forming an electronically excited state [152] (excited state-to-excited state spectra are not normally studied by experimentalists). Calculation of UV spectra with reasonable accuracy requires some method of dealing with excited states. Simply equating energy differences between HOMO and LUMO with $h\upsilon$ does not give satisfactory results for the absorption frequency/wavelength, because the energy of a virtual orbital, unlike that of an occupied one, is not a good measure of its energy (of the energy needed to remove an electron from it; this is dealt with in connection with ionization potentials and electron affinities) and because this method ignores the energy difference between a singlet and a triplet state.

Electronic spectra of moderate accuracy can be calculated by the configuration interaction CIS method (section 5.4.3) [173]. Compare, for example, the UV spectra of methylenecyclopropene calculated by the CIS/6-31+G* method (diffuse functions appear to be desirable in treating excited states, as the electron cloud is relatively extended) with the experimental spectrum [173], in Table 5.20. The agreement in wavelength is not particularly good for the longest-wavelength band, although this result can be made more palatable by noting that both calculation and experiment agree reasonably well on relative intensities (the two bands that were not observed are calculated to be relatively weak and to lie very near the strongest band). The CIS approach to excited states has been said [173] to be analogous to the HF approach to ground states in that both give at least qualitatively useful results.

NMR spectra

NMR spectra result from the transition of an atomic nucleus in a magnetic field from a low-energy to a high-energy state [152]. Quantum-mechanical calculations of NMR spectra focus on predicting the chemical shift (magnetic field strength needed for the transition relative to that needed for some reference) of a nucleus. This requires calculation of the magnetic shielding of the nuclei of the molecule of interest, and of the reference nuclei, usually those of tetramethylsilane, TMS. The chemical shift of the

^{13}C or ^{1}H nucleus is then its (absolute) shielding value minus that of the TMS ^{13}C or ^{1}H nucleus. The theory of magnetic shielding of nuclei involves a treatment of how the energy of a nucleus varies with a magnetic field and with nuclear magnetic moments [174]. NMR spectra can be calculated with remarkable accuracy even at the HF level [175] (although there is some evidence that improved results are obtained using the

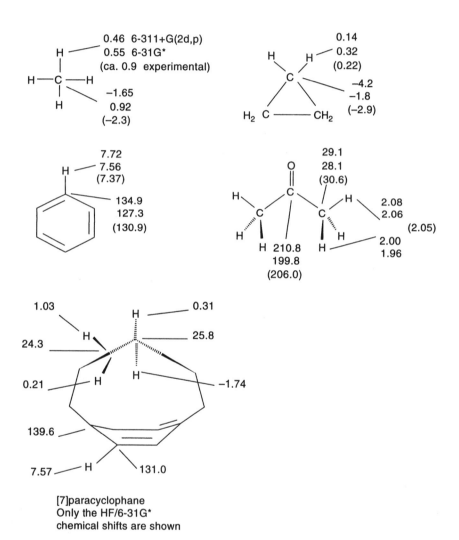

[7]paracyclophane
Only the HF/6-31G*
chemical shifts are shown

Figure 5.44. Calculated and experimental ^{1}H and ^{13}C NMR spectra: chemical shifts relative to TMS H and C, respectively. The calculations were done on the B3LYP/6-31G* geometry (B3LYP is a density functional method; chapter 7) at the HF/6-311 + G(2d,p) and HF/6-31G* levels using the default NMR method (GIAO) implemented in Gaussian 94W [48]. The experimental values are from [151], except for the values for [7]paracyclophane [177]. The larger basis set *may* be somewhat more accurate, but takes 10–20 times as long.

MP2 method [175b,176]), as is clearly shown by the results in Fig. 5.44. The remarkable shielding effect of a benzene ring in [7]paracyclophane [177] is nicely reproduced. The calculation of NMR spectra has become an important tool in probing the electronic structure of theoretically interesting molecules [178].

Ionization energies and electron affinities
Ionization energies (also called ionization potentials) and electron affinities are related in that both involve transfer of an electron between a molecular orbital and infinity: in one case (IE) we have removal of an electron from an occupied orbital and in the other (EA) addition of an electron to a virtual (or a half-occupied) orbital. The IE for an orbital is defined as the energy needed to remove an electron from the orbital (to infinite separation), while the EA of an orbital is the energy released when the orbital accepts an electron from infinity [179]. The term IE when applied to a molecule normally means the *minimum* energy needed to remove an electron to infinity, i.e. to form the radical (for an originally closed-shell molecule) cation, and the term electron affinity normally means the *maximum* energy released when the molecule accepts an electron to form the radical anion (for an originally closed-shell molecule). The IE of a "stable" species, i.e. any molecule or atom that can exist (a relative minimum on the potential energy surface), is always positive. The EA of a molecule is positive if the accepted electron is bound, i.e. if it is not spontaneously ejected; if the new electron is ejected in microseconds or less (is unbound), the molecule has a negative EA (is a "resonance state" – this has nothing to do with the term resonance as in resonance hybrid).

IEs and EAs may be *vertical* or *adiabatic*: the energy difference between the precursor molecule M_1 and the species M_2 formed by removing or adding an electron gives the vertical value if M_2 is at the same geometry as M_1, while the adiabatic value is obtained if M_2 has its own actual, equilibrium geometry. Since the equilibrium geometry of M_2 is clearly of lower energy than the unrelaxed geometry corresponding to M_1, vertical IEs are larger than adiabatic IEs, and vertical EAs are smaller than adiabatic EAs. *Experimental* IEs and EAs may be vertical or adiabatic. The adiabatic values appear to be of more interest to chemists, since it is these that represent the energy difference between two "stable" molecules (the neutral and the charged; at least in those cases where the charged species does not instantly decompose), but compilations of IEs and EAs often do not state explicitly whether their listed values are adiabatic or vertical; a welcome exception is the book by Levin and Lias [180]. A good brief discussion of IEs and EAs, including various measurement techniques, is to be found in the compilation by Lias et al. [130b]. Many IEs and EAs are available on the worldwide web [130a].

Ionization energies and electron affinities can be calculated simply as the energy difference between the neutral and the ion. Approximate IEs can be obtained by applying Koopmans' (not Koopman's) theorem [181], which says that the energy required to remove an electron from an orbital is the negative of the orbital energy. Thus the IE of a molecule is approximately the negative of the energy of its HOMO (the principle does not work as well for ionization of electrons more tightly bound than those in the HOMO). This makes it simple to obtain approximate IEs for comparison with photoelectron spectroscopy [182] results. Unfortunately, the principle does not work well for EAs: the EA of a molecule is not reasonably well approximated as the negative of the LUMO energy. In fact, *ab initio* calculations normally give virtual MOs (vacant MOs) positive

Table 5.21. Some ionization energies (eV). The basis set is 6-31G*; the calculations are based on the data in Table 5.22. The experimental values are from ref. [180], except for CH_3SH [185]

	IE from ΔE		IE from Koopmans' theorem		
	HF	MP2(FC)	HF	MP2(FC)	Exp.
CH_3OH adiabatic	9.38	10.57	—	—	10.9
CH_3OH vertical	9.66	10.82	12.06	12.12	10.95
CH_3SH adiabatic	8.34	8.97	—	—	9.44
CH_3SH vertical	8.38	9.03	9.69	9.69 (*sic*)	—
CH_3COCH_3 adiabatic	8.19	9.63	—	—	9.71, 9.74
CH_3COCH_3 vertical	8.37	9.78	11.07	11.19	9.5, 9.72

energies, implying that molecules will not accept electrons to form anions (i.e. that they have negative EAs), which is often false. Koopmans' theorem works because of a cancellation of errors in the IE case (which actually leads to modest overestimation of the IE) but not for EAs. Errors arise from approximate treatment of electron correlation, and from the fact that when an electron is removed from or added to a molecule electronic relaxation (not to be confused with geometry relaxation) occurs. A further problem for EAs is that the procedure for minimizing the energies of MOs (section 5.2.3.4) gives, within the limits of the HF procedure, the best *occupied*, but not virtual, MOs.

Some calculated and experimental IEs are given in Table 5.21, based on the raw data in Table 5.22. The calculations (experimental data are sparse) indicate vertical IEs to be indeed slightly (about 0.2 eV) higher than adiabatic. The HF/6-31G* ΔE values underestimate the IE by about 1–1.5 eV while MP2(FC)/6-31G* ΔE values underestimate it by only about 0.1–0.4 eV (others have reported them to be generally too low by 0.3–0.7 eV [183]). The Koopmans' theorem (-HOMO) energies for both the HF and MP2 level calculations are about 1–1.5 eV too high. Electron affinities (which seem to be of less interest than ionization energies) can be calculated as the energy difference between the neutral molecule and its anion. High-accuracy IEs and EAs can be calculated by the G2 (or a variation like G2(MP2)) method or one of the CBS methods [86,184,185] (section 5.5.2.2b), although the convenient procedures implemented for these methods in the Gaussian 94 and 98 programs do not allow calculation of vertical energies since the geometry of the ion will be automatically optimized.

5.5.6   Visualization

Modern computer graphics have given visualization, the pictorial presentation of the results of calculations, a very important place in science. Not only in chemistry, but in physics, aerodynamics, meteorology, and even mathematics, the remarkable ability of the human mind to process visual information is being utilized [186]. Gone are the days when it was *de rigeur* to pore over tables of numbers to comprehend the factors at work in a system, whether it be a galaxy, a supersonic airliner, a thunderstorm, or a novel mathematical entity. We will briefly examine the role of computer graphics

Table 5.22. The raw data for Table 5.21: energies, ZPEs and HOMO values, for calculating ionization energies

	HF/6-31G*	MP2(FC)/6-31G*
CH_3OH	−115.03542	−115.34514
	0.05055	0.05086
	−114.98487	−115.29528
$CH_3OH^{\cdot +}$ cation geom.	−114.68722	−114.95358
	0.04723	0.04665
	−114.63999	−114.90693
$CH_3OH^{\cdot +}$ neutral geom.	−114.68040	−114.94849
	0.05055	0.05086
	−114.62985	−114.89763
CH_3OH, HOMO	−0.44328	−0.44526
CH_3SH	−437.70032	−437.95267
	0.04534	0.04621
	−437.65498	−437.90646
$CH_3SH^{\cdot +}$ cation geom.	−437.39316	−437.62211
	0.04468	0.04526
	−437.34848	−437.57685
$CH_3SH^{\cdot +}$ neutral geom.	−437.39227	−437.62089
	0.04534	0.04621
	−437.34693	−437.57468
CH_3SH, HOMO	−0.35596	−0.35627
CH_3COCH_3	−191.96224	−192.52391
	0.08214	0.08309
	−191.88010	−192.44082
$CH_3COCH_3^{\cdot +}$ cation geom.	−191.65994	−192.16837
	0.08071	0.08128
	−191.57923	−192.08709
$CH_3COCH_3^{\cdot +}$ neutral geom.	−191.65451	−192.16448
	0.08214	0.08309
	−191.57237	−192.08139
CH_3COCH_3, HOMO	−0.40692	−0.41119

The numbers are hartrees, and represent (other than the HOMO energies): uncorrected *ab initio* energy, ZPE, corrected *ab initio* energy. For the cations at the geometry of the neutrals, the ZPE of the corresponding neutral was used. The ZPEs shown have been multiplied [60] by 0.9135 (HF) or 0.9670 (MP2(FC)). Hartrees were converted to eV in Table 5.21 by multiplying by 27.2116.

in computational chemistry, limiting ourselves to molecular vibrations, van der Waals surfaces, charge distribution, and molecular orbitals.

Molecular vibrations
Animation of normal-mode frequencies usually readily enables one to ascribe a band in the calculated vibrational (i.e. IR) spectrum to a particular molecular motion (a stretching, bending, or torsional mode, involving particular atoms). It sometimes requires a little ingenuity to *describe* clearly the motion involved, but animation is far superior

to trying to discern the motion from the printed direction vectors (section 2.5; these show the extent of motion in the x, y, and z directions), which are supplied by some programs. Useful, however, are the visualized direction vectors that some programs, e.g. GaussView [187], can attach to a picture of the molecule, catching the vibration in the act so to speak.

Animating vibrations is useful not only for predicting or interpreting an IR spectrum, but can sometimes also help in geometry optimizations. Suppose we wish to locate computationally the intermediate through which the chair conformers of cyclohexane interconvert $1 \leftrightharpoons 1'$, Fig. 5.45). This reaction, although degenerate, can be studied by NMR spectroscopy [188]. One might surmise that the intermediate is the boat conformation **2**, but a geometry optimization and frequencies calculation on this C_{2v} structure (note that in a quantum mechanical calculation, whether *ab initio* or otherwise, the input symmetry is normally preserved) followed by animation of the vibrations, shows otherwise. There is one imaginary vibration (section 2.5), and the transition state wants to escape from its saddle point by twisting to a D_2 structure **3**, called the twist or twist-boat, which latter is the true intermediate. The enantiomeric twist structures **3** and **3'** go to **1** and **1'**, respectively, over a high-energy form **4** (or **4'**) called the half-chair. A geometry optimization starting with a D_2 structure leads to the desired relative minimum. Similarly, if one obtains a second-order saddle point (one kind of hilltop), animation of the two imaginary frequencies often indicates what the species seeks to do to escape from the hilltop to a become a first-order saddle point (a transition state) or

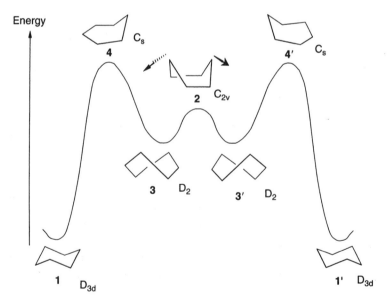

Figure 5.45. One might have guessed that the chair cyclohexane conformations **1** and **1'** are connected by a boat-shaped intermediate **2**. However, this C_{2v} structure shows an imaginary frequency: it is a transition state which wants to twist toward **3** (arrows) or **3'** (arrows in opposite directions, not shown), which are the actual intermediates (no imaginary frequencies) between **1** and **1'**. The chair conformation reaches the twist via a half-chair **4**.

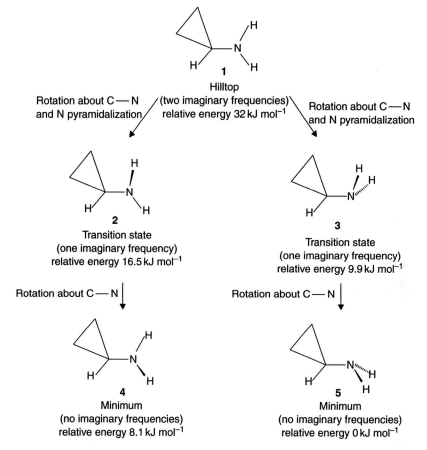

Figure 5.46. Cyclopropylamine conformations at the B3LYP/6-31G* (chapter 7) level. The structure **1** is a hilltop whose two imaginary frequencies indicate that it wants to undergo nitrogen pyramidalization and rotation about the C–N bond to form, eventually, the other four conformations.

a minimum, and it often possible to obtain the desired transition state or minimum by altering the shape of the input structure so that it has the symmetry and approximates the shape of the desired structure. An example is provided by cyclopropylamine (Fig. 5.46) [189]; the structure **1** is a hilltop at the B3LYP/6-31G* level (chapter 7), whose two imaginary frequencies indicate that it wants to undergo nitrogen pyramidalization and rotation about the C–N bond to form the other four conformations shown.

Electrostatic potential
Electrostatic potential, the net electrostatic potential energy (roughly, the charge) due to nuclei and electrons was mentioned in section 5.5.4 in connection with calculation of atom charges. The ESP can be displayed (visualized) by color-coding it onto the van der Waals surface, by displaying it as a surface itself, or by showing it with contour lines on a slice through the molecule; the three possibilities are shown for the water molecule

(a)   (b)   (c)

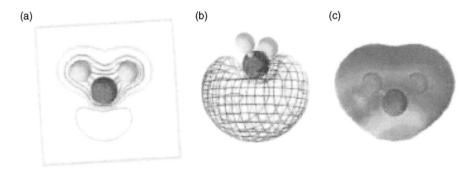

Figure 5.47. Distribution of net charge in the water molecule (electrostatic charge calculated with AM1 – chapter 6). Negative to positive: red to blue (R O Y G B). (a) Slice through the plane of the molecule; the contour lines show the decrease in net negative charge. (b) Charge in space; this corresponds essentially to the lone pairs. (c) Charge on the van der Waals surface.

in Fig. 5.47. Color-coding (mapping) the ESP onto the surface of the molecule, (c), enables one to see how an approaching reagent would perceive the charge distribution. Showing the ESP as a surface residing in the region of space where the net charge is negative, (b), gives a very useful picture of those parts of a molecule where the electrostatic effect of the electrons wins out over that of the nuclei; this is a particularly good way of seeing the presence of lone pairs, as Fig. 5.48, also, makes clear. Note that in Fig. 5.47(c) and (b) (mapping of the ESP on the van der Walls surface and depicting the ESP itself as a surface) the lone pairs do not stick out like rabbit ears [190]. This is because as electron density which can be ascribed to one orbital falls off, that due to another increases: there is no "electron hole" between the two lone pairs (for the same reason the electron density cross section through a $\sigma-\pi$ double bond is elliptical and through a $\sigma-\pi-\pi$ triple bond circular; see section 4.3.2). Showing the ESP as a surface made clear that the remarkable cycloalkane pyramidane [191] has a lone pair, like the carbene CH_2 (Fig. 5.48). Depicting the ESP by contour lines on a slice through the molecule reveals its internal structure, but this is probably not as relevant to reactivity as the picture seen by mapping it onto the van der Waals surface, which is the picture presented to the outside molecular world. Examining the ESP interactions between a molecule and the active site of an enzyme can be important in drug design [82]. Various applications of the ESP are discussed by Politzer and Murray [192] and Brinck [169a].

Molecular orbitals
Visualization of molecular orbitals shows the location of those regions where the highest-energy electrons are concentrated (the highest occupied MO, the HOMO), and those regions which offer the lowest-energy accommodation to any donated electrons (the lowest unoccupied MO, the LUMO). Electrophiles should bond to the atom where the HOMO is "strongest" (where the electron density due to the highest-energy electron pair is greatest) and nucleophiles to the atom where the LUMO is strongest, at least as seen on the van der Waals surface by an approaching reagent. The information pro- vided by inspection of the HOMO and LUMO (the frontier orbitals) is thus somewhat akin to that given by visualizing the ESP (electrophiles should tend to go to regions of

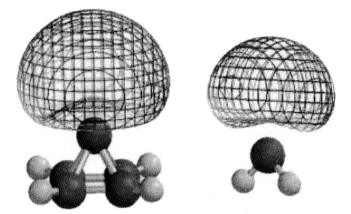

Figure 5.48. The hydrocarbon pyramidane, C_5H_4, evidently (pyramidane has not been synthesized) has a lone pair of electrons on its pyramidal carbon atom, like carbene (methylene), CH_2. While the lone pair on CH_2 is no surprise (draw the Lewis structure for the singlet), a cycloalkane with an unshared electron pair is remarkable.

negative ESP, nucleophiles to regions of positive ESP). Figure 5.49 shows the LUMOs of the ketones norcamphor and camphor, mapped onto their van der Waals surfaces. For norcamphor (Fig. 5.49(a)), the prominence of its LUMO at the carbonyl carbon as seen from the "top" or *exo* face (the face with the CH_2 bridge) rather than the bottom (*endo*) face, suggests that nucleophiles should attack from the *exo* direction. In accord with this, hydride donors, for example, approach from the *exo* face to give mainly the *endo* alcohol. For camphor, where the bridge is $C(CH_3)_2$ instead of CH_2, the *exo* face is shielded by a CH_3 group which sterically thwarts the electronically preferred attack from this direction, and so nucleophiles tend to approach rather the *endo* face, a fact nicely rationalized by visualizing simultaneously the LUMO and the van der Waals surface (Fig. 5.49(b)) [193].

Figure 5.50 shows the LUMOs of three bicyclo[2.2.1]heptane derivatives (camphor and norcamphor, above, also have this carbon skeleton). The LUMOs are shown here as 3D regions of space, rather than mapping them onto a surface as was done in Fig. 5.49. Comparing the "composite" molecule (Fig. 5.50(c)) with the cation and the alkene clearly shows electronic interaction between the p orbital of the cationic carbon and the antibonding MO of the double bond.

Visualization–closing remarks
Other molecular properties and phenomena that can benefit from the aid of visualization are the distribution of unpaired electron spin in radicals and the changes in orbitals and charge distribution as a reaction progresses. These and many other visualization exercises are described in publications (e.g. [53c]) by Wavefunction, Inc. and in their visualization CD [194].

(a)   (b)

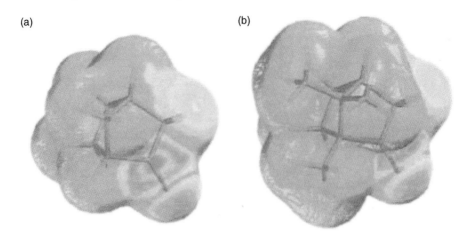

Figure 5.49. (a) Norcamphor, with the LUMO mapped onto the van der Waals surface. The LUMO as seen on the surface is most prominent at the carbonyl carbon, on the "top" of the molecule (the *exo* face), as shown by the blue area. Viewed from the bottom of the molecule (not shown here), the LUMO still lies at the C=O carbon, but is less prominent (the blue is less intense). We can thus predict that nucleophiles will attack the C=O carbon, from the *exo* direction. (b) Camphor (norcamphor with three methyl groups): the carbonyl carbon is shielded from *exo* attack by a methyl group, so for steric reasons nucleophiles tend to attack this carbon from the *endo* direction, despite *exo* attack being electronically favored.

5.6   STRENGTHS AND WEAKNESSES OF *AB INITIO* CALCULATIONS

Strengths
Ab initio calculations are based on a fundamental physical equation, the Schrödinger equation, without empirical adjustments. This makes them esthetically satisfying, and ensures (if the Schrödinger equation is true) that they will give correct answers *provided the approximations needed to obtain numerical results (to solve the Schrödinger equation) are not too severe for the problem at hand*. The level of theory needed for a reliable answer to a particular problem must be found by experience – comparison with experiment for related cases – so in this sense current *ab initio* calculations are not fully *a priori* [36,195]). A few "*ab initio* methods" do not even fully eschew empirical factors: the G2 and G3 and the CBS series of methods have empirical factors which, unless they cancel(as in proton affinity calculations, section 5.2.2.2b) make these methods, strictly speaking, semiempirical. A consequence of the (usual) absence of empirical parameters is that *ab initio* calculations can be performed for any kind of molecular species (including transition states and even non-stationary points), rather than only species for which empirical parameters are available (see chapter 6). These characteristics of reliability (with the caveats alluded to) and generality are the strengths of *ab initio* calculations.

Weaknesses
Compared to other methods (molecular mechanics, semiempirical calculations, density

(a) (b)

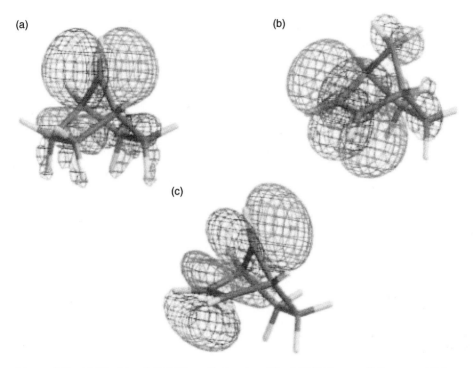

(c)

Figure 5.50. (a) The bicyclo[2.2.1]hept-7-yl cation. The LUMO is essentially a *p* orbital on the cationic carbon. (b) Bicyclo[2.2.1]hept-2-ene. The LUMO is essentially the Π^* MO of this alkene. (c) The bicyclo[2.2.1]hept-2-ene-7-yl cation, which combines the cationic and alkene features of (a) and (b). The LUMO is clearly an orbital resulting from extensive electronic interaction of the cationic center and the double bond.

functional calculations – chapters 3, 6 and 7, respectively) *ab initio* calculations are slow, and they are relatively demanding of computer resources (memory and disk space, depending on the program and the particular calculation). These disadvantages, which increase with the level of the calculation, have been to a large extent overcome by the tremendous increase in computer power, accompanied by decreases in price, that have taken place since the invention of electronic computers. In 1959 Coulson doubted the possibility (he also questioned the desirability, but in this regard visualization has been of enormous help) of calculations on molecules with more than 20 electrons, but 30 years later computer speed had increased by a factor of 100,000 [196], and *ab initio* calculations on molecules with 100 electrons (about 15 heavy atoms) were common.

5.7   SUMMARY OF CHAPTER 5

Ab initio calculations rest on solving the Schrödinger equation; the nature of the necessary approximations determine the level of the calculation. In the simplest approach, the HF method, the total molecular wavefunction Ψ is approximated as a Slater determinant composed of occupied spin orbitals (each spin orbital is a product of a conventional spatial orbital ψ and a spin function). Writing the molecular energy as the expectation value of the wavefunction ($E = \langle \Psi | \hat{H} | \Psi \rangle$), i.e. invoking the Schrödinger equation, then differentiating E with respect to the spin orbitals that compose the wavefunction (= the Slater determinant), we get the HF equations. To use these in practical calculations the spatial orbitals are approximated as a linear combination (a weighted sum) of basis functions. These are usually identified with atomic orbitals, but can really be any mathematical functions that give a reasonable wavefunction, i.e. a wavefunction which gives reasonable answers when we do the calculations. The main defect of the HF method is that it does not treat electron correlation properly: each electron is considered to move in an electrostatic field represented by the average positions of the other electrons, whereas in fact electrons avoid each other better than this model predicts, since any electron A sees any other electron B as a moving particle and the two mutually adjust (correlate) their motions to minimize their interaction energy. Electron correlation is treated better in post-HF methods, such as the MP, CI and CC methods. These methods lower electron-electron interaction energy by allowing the electrons to reside not just in conventionally occupied MOs (the n lowest MOs for a $2n$-electron species), but also in formally unoccupied MOs (virtual MOs).

The main uses of the *ab initio* method are calculating molecular geometries, energies, vibrational frequencies, spectra (IR, UV, NMR), ionization potentials and electron affinities, and properties like dipole moments which are directly connected with electron distribution. These calculations find theoretical and practical applications, since, for example, enzyme-substrate interactions depend on shapes and charge distributions, reaction equilibria and rates depend on energy differences, and spectroscopy plays an important role in identifying and understanding novel molecules. The visualization of calculated phenomena, such as molecular vibrations, charge distributions, and molecular orbitals, can be very important in interpreting the results of calculations.

REFERENCES

[1] General discussions of and references to *ab initio* calculations are found in: (a) I. N. Levine, "Quantum Chemistry," 4th edn, Prentice Hall, Engelwood Cliffs, New Jersey, 2000. (b) J. P. Lowe, "Quantum Chemistry," 2nd edn, Academic Press, New York, 1993. (c) F. L. Pilar, Elementary Quantum Chemistry," 2nd edn, McGraw-Hill, New York, 1990. (d) An advanced book: A. Szabo and N. S. Ostlund, "Modern Quantum Chemistry," McGraw-Hill, New York, 1989. (e) J. B. Foresman and Æ. Frisch, "Exploring Chemistry with Electronic Structure Methods," Gaussian Inc., Pittsburgh, PA, 1996. (f) A. R. Leach, "Molecular Modelling," Longman, Essex, England, 1996. (g) An important reference is still: W. J. Hehre, L. Radom, P. v. R. Schleyer, and J. A. Pople, "*Ab Initio* Molecular Orbital Theory," Wiley, New York, 1986. (h) A recent evaluation of the state and future of quantum chemical calculations, with the emphasis on *ab initio* methods: M. Head-Gordon, J. Phys. Chem., 1996, *100*, 13213. (i) F. Jensen, "Introduction to Computational Chemistry," Wiley, New York, 1999. (j) M. J. S. Dewar, "The Molecular Orbital Theory of Organic Chemistry," McGraw-Hill, New York, 1969. This book contains many trenchant comments by one of the major contributors to computational chemistry; begins with basic quantum mechanics and *ab initio* theory, although it later stresses semiempirical

theory. (k) D. Young, "Computational Chemistry. A Practical Guide for Applying Techniques to Real World Problems," Wiley, New York, 2001. (l) C. J. Cramer, "Essentials of Computational Chemistry," Wiley, New York, 2002.

[2] D. R. Hartree, Proc. Cambridge Phil. Soc., 1928, *24*, 89.

[3] (a) The relativistic Schrödinger equation is called the Dirac equation; see [1a], pp. 602–604. (b) For a brief discussion of spin–orbit interaction see [1a], *loc. cit.*

[4] For treatments of the hydrogen and helium atoms see the appropriate sections of references 1.

[5] See [1b], pp. 127–132.

[6] (a) J. C. Slater, Phys. Rev., 1930, *35*, 210. (b) V. Fock, Z. Physik, 1930, *61*, 126.

[7] Reference [1a], pp. 187–189 and 282–285.

[8] Although it is sometimes convenient to speak of electrons as belonging to a particular atomic or molecular orbital, and although they sometimes behave as if they were localized, no electron is really confined to a single orbital, and in a sense all the electrons in a molecule are delocalized; see [1j], pp. 139–143.

[9] See, e.g. [1c], p. 200.

[10] J. A. Pople and D. L. Beveridge, "Approximate Molecular Orbital Theory," McGraw-Hill, New York, 1970, chapters 1 and 2.

[11] Reference [1b], Appendix 7.

[12] Reference [1a], p. 284.

[13] Reference [1j], chapter 2.

[14] Reference [1a], p. 474.

[15] Reference [1a], chapter 8.

[16] See, e.g. C. L. Perrin, "Mathematics for Chemists," Wiley-Interscience, New York, 1970, pp. 39–41.

[17] Reference [1b], pp. 354–355.

[18] How do we know that iterations *improve* psi and epsilon? This is not always the case, see e.g. [1j], p. 35, but in practice "initial guess" solutions to the Hartree–Fock equations usually converge fairly smoothly to give the best wavefunction and orbital energies (and thus total energy) that can be obtained by the HF method from the particular kind of guess wavefunction (e.g. basis set; section 5.2.3.6e).

[19] Reference [1a], pp. 305–315.

[20] C. C. J. Roothaan, Rev. Mod. Phys., 1951, *23*, 69; G. G. Hall, Proc. Roy. Soc. (London), 1951, *A205*, 541.

[21] Reference [1c], pp. 288–299.

[22] Frequencies and zero-point energies are discussed in [1g], section 6.3.

[23] GAUSSIAN 92, Revision F.4: M. J. Frisch, G. W. Trucks, M. Head-Gordon, P. M. W. Gill, M. W. Wong, J. B. Foresman, B. G. Johnson, H. B. Schlegel, M. A. Robb, E. S. Repogle, R. Gomperts, J. L. Andres, K. Raghavachari, J. S. Binkley, C. Gonzales, R. L. Martin, D. J. Fox, D. J. Defrees, J. Baker, J. J. P. Stewart, and J. A. Pople; Gaussian, Inc., Pittsburgh, PA, 1992.

[24] See, e.g. "Interactive Linear Algebra: A Laboratory Course Using Mathcad," G. J. Porter and D. R. Hill, Springer Verlag, New York, 1996.

[25] Cf. [1d], pp. 152–171.

[26] Reference [1d], Appendix A.

[27] Reference [1a], pp. 494–498.

[28] See [1a–i].

[29] S. F. Boys, Proc. Roy. Soc. (London), 1950, *A200*, 542.

[30] Spartan is an integrated molecular mechanics, *ab initio* and semiempirical program with an outstanding input/output graphical interface that is available in UNIX workstation and PC versions: Wavefunction Inc., 18401 Von Karman, Suite 370, Irvine CA 92715; http://www.wavefun.com.

[31] Reference [1e], pp. 32–33.

[32] W. J. Hehre, "Practical Strategies for Electronic Structure Calculations," Wavefunction, Inc., Irvine, CA, 1995.

[33] Reference [1g], pp. 65–88.

[34] J. Simons and J. Nichols, "Quantum Mechanics in Chemistry," Oxford University Press, New York, 1997, pp. 412–417.

[35] Reference [1a], pp. 490–493.

[36] M. J. S. Dewar and D. M. Storch, J. Am. Chem. Soc., 1985, *107*, 3898.

[37] S. Inagaki, Y. Ishitani, and T. Kakefu, J. Am. Chem. Soc., 1994, *116*, 5954.

[38] E. Lewars, J. Mol. Struct. (Theochem), 1998, *423*, 173.

[39] The experimental geometries of Me_2SO and NSF are taken from [1g], Table 6.14.

[40] O. Wiest, D. C. Montiel, and K. N. Houk, J. Phys. Chem. A, 1997, *101*, 8378, and references therein.

[41] C. Van Alsenoy, C.-H. Yu, A. Peeters, J. M. L. Martin, and L. Schäfer, J. Phys. Chem. A, 1998, *102*, 2246.

[42] Basis sets without polarization functions evidently make lone-pair atoms like tricoordinate N and tricoordinate O^+ too flat: C. C. Pye, J. D. Xidos, R. A. Poirer, and D. J. Burnell, J. Phys. Chem. A, 1997, *101*, 3371. Other problems with the 3-21G basis are that cation-metal distances tend to be too short (e.g. W. Rudolph, M. H. Brooker, and C. C. Pye, J. Phys. Chem., 1995, *99*, 3793) and that adsorption energies of organics on aluminosilicates are overestimated, and charge separation is exaggerated (private communication from G. Sastre, Instituto de Technologica Quimica, Universidad Polytechnica de Valencia). Nevertheless, the 3-21G basis apparently usually gives good geometries (section 5.5.1).

[43] P. M. Warner, J. Org. Chem., 1996, *61*, 7192.

[44] (a) J. E. Fowler, J. M. Galbraith, G. Vacek, and H. F. Schaefer, J. Am. Chem. Soc., 1994, *116*, 9311. (b) G. Vacek, J. M. Galbraith, Y. Yamaguchi, H. F. Schaefer, R. H. Nobes, A. P. Scott, and L. Radom, J. Phys. Chem., 1994, *98*, 8660.

[45] The special theory of relativity (the one relevant to chemistry) and its chemical consequences are nicely reviewed in K. Balasubramanian, "Relativistic Effects in Chemistry," Parts A and B, Wiley, New York, 1997.

[46] P. A. M. Dirac, Proc. R. Soc. 1929, *A123*, 714: "[relativity is] . . . of no importance in the consideration of atomic and molecular structure, and ordinary chemical reactions . . ."

[47] M. Krauss and W. J. Stevens, Annu. Rev. Phys. Chem., 1984, *35*, 357; L. Szasz, "Pseudopotential Theory of Atoms and Molecules," Wiley, New York, 1985. (b) Relativistic Dirac–Fock calculations on closed-shell molecules: L. Pisani and E. Clementi, J. Comput. Chem., 1994, *15*, 466.

[48] Gaussian 94 for Windows (G94W): Gaussian 94, Revision E.1, M. J. Frisch, G. W. Trucks, H. B. Schlegel, P. M. W. Gill, B. G. Johnson, M. A. Robb, J. R. Cheeseman, T. Keith, G. A. Petersson, J. A. Montgomery, K. Raghavachari, M. A. Al-Laham, V. G. Zakrzewski, J. V. Ortiz, J. B. Foresman, J. Cioslowski, B. B. Stefanov, A. Nanayakkara, M. Challacombe, C. Y. Peng, P. Y. Ayala, W. Chen, M. W. Wong, J. L. Andres, E. S. Replogle, R. Gomperts, R. L. Martin, D. J. Fox, J. S. Binkley, D. J. Defrees, J. Baker, J. P. Stewart, M. Head-Gordon, C. Gonzalez, and J. A. Pople, Gaussian, Inc., Pittsburgh PA, 1995. G94 and G98 are available for both UNIX workstations and PCs.

[49] Reference [1g], p. 191.

[50] Reference [1a], pp. 444, 494, 602–604.

[51] (a) A detailed review: G. Frenking, I. Antes, M. Böhme, S. Dapprich, A. W. Ehlers, V. Jonas, A. Neuhaus, M. Otto, R. Stegmann, A. Veldkamp, and S. Vyboishchikov, chapter 2 in Reviews in Computational Chemistry, Volume 8, K. B. Lipkowitz and D. B. Boyd, Eds., VCH, New York, 1996. (b) The main points of [51a] are presented in G. Frenking and U. Pidun, J. Chem. Soc., Dalton Trans., 1997, 1653. (c) T. R. Cundari, S. O. Sommerer, L. Tippett, J. Chem. Phys., 1995, *103*, 7058.

[52] J. Comp. Chem., 2002, *23*, issue no. 8.

[53] (a) W. J. Hehre, W. W. Huang, P. E. Klunzinger, B. J. Deppmeier, and A. J. Driessen, "A Spartan Tutorial," Wavefunction Inc., Irvine, CA, 1997. (b) W. J. Hehre, J. Yu, and P. E. Klunzinger, "A Guide to Molecular Mechanics and Molecular Orbital Calculations in Spartan," Wavefunction Inc., Irvine, CA, 1997. (c) "A Laboratory Book of Computational Organic Chemistry," W. J. Hehre, A. J. Shusterman, and W. W. Huang, Wavefunction Inc., Irvine, CA, 1996.

[54] R. Janoschek, Chemie in unserer Zeit, 1995, *29*, 122.

[55] M. J. S. Dewar, "A Semiempirical Life," Profiles, Pathways and Dreams series, J. I. Seeman, Ed., American Chemical Society, Washington, D.C., 1992, p. 185.

[56] (a) K. Raghavachari and J. B. Anderson, J. Phys. Chem., 1996, *100*, 12960. (b) A historical review: P.-O. Löwdin, Int. J. Quantum Chem., 1995, *55*, 77. (c) Fermi and Coulomb holes and correlation: [1c], pp. 296–297.

[57] Reference [1c], p. 286.

[58] For example, A. C. Hurley, "Introduction to the Electron Theory of Small Molecules," Academic Press, New York, 1976, pp. 286–288, or W. C. Ermler and C. W. Kern, J. Chem. Phys., 1974, *61*, 3860.

[59] P.-O. Löwdin, Advan. Chem. Phys., 1959, *2*, 207.

[60] A. P. Scott and L. Radom, J. Phys. Chem., 1996, *100*, 16502.

[61] $368 \, kJ \, mol^{-1}$: R. T. Morrison and R. N. Boyd, "Organic Chemistry," 6th edn., Prentice Hall, Englewood Cliffs, New Jersey, 1992, p. 21; $377 \, kJ \, mol^{-1}$: K. P. C. Vollhardt and N. E. Schore, "Organic Chemistry," 2nd edn., Freeman, New York, 1987, p. 75.

[62] For example, R. J. Bartlett and J. F. Stanton, chapter 2 in "Reviews in Computational Chemistry," Vol. 5, K. B. Lipkowitz and D. B. Boyd, Eds., VCH, New York, 1994.

[63] For example, the helium atom: [1a], pp. 256–259.

[64] Brief introductions to the MP treatment of atoms and molecules: [1a], pp. 563–568; [1b], pp. 369–370; [1f], pp. 83–85.

[65] Reference [1a], chapter 9.

[66] C. Møller and M. S. Plesset, Phys. Rev., 1934, *46*, 618.

[67] J. S. Binkley and J. A. Pople, Int. J. Quantum Chem., 1975, *9*, 229.

[68] Reference [1d], chapter 6.

[69] For example, ref. [1b], pp. 367–368.

[70] For example, ref. [1d], p. 353, [1f], p. 85.

[71] A. Boldyrev, P. v. R. Schleyer, D. Higgins, C. Thomson, and S. S. Kramarenko, J. Comput. Chem., 1992, *9*, 1066. Fluoro- and difluorodiazomethanes are minima by HF calculations, but are not viable minima by the MP2 method.

[72] $H_2C = CHOH$ *reaction*: E. Lewars and I. Bonnycastle, J. Mol. Struct. (Theochem), 1997, *418*, 17 and references therein. *HNC reaction* V. S. Rao, A. Vijay, A. K. Chandra, Can. J. Chem., 1996, *74*, 1072. CH_3NC *reaction*: The reported experimental activation energy is $161 \, kJ \, mol^{-1}$: F. W. Schneider and B. S. Rabinovitch, J. Am. Chem. Soc., 1962, *84*, 4215; B. S. Rabinovitch and P. W. Gilderson, J. Am. Chem. Soc., 1965, *87*, 158. The energy of CH_3CN relative to CH_3NC by a high-level (G2) calculation is $-98.3 \, kJ \, mol^{-1}$ (E. Lewars). An early *ab initio* study of the reaction: D. H. Liskow, C. F. Bender, and H. F. Schaefer, J. Am. Chem. Soc., 1972, *95*, 5178. A comparison of CH_3CN, CH_3NC, other isomers and radicals, cations and anions: P. M. Mayer, M. S. Taylor, M. Wong, and L. Radom, J. Phys. Chem. A, 1998, *102*, 7074. *Cyclopropylidene reaction*: H. F. Bettinger, P. R. Schreiner, P. v. R. Schleyer, and H. F. Schaefer, J. Phys. Chem., 1996, *100*, 16147.

[73] (a) A superb brief introduction to CI is given in [1a], pp. 444–451, 557–562, and 568–573. (b) A comprehensive review of the development of CI: I. Shavitt, Mol. Phys., 1998, *94*, 3. (c) See also [1b], pp. 363–369; [1c], pp. 388–393; [1d], chapter 4; [1g], pp. 29–38.

[74] N. Ben-Amor, S. Evangelisti, D. Maynau, and E. P. S. Rossi, Chem. Phys. Lett., 1998, *288*, 348.

[75] R. B. Woodward and R. Hoffmann, "The Conservation of Orbital Symmetry," Academic Press, New York, 1970, chapter 6.

[76] Reference [1e], pp. 228–236 shows how to do CASSCF calculations. For CASSCF calculations on the Diels-Alder reaction, see Y. Li and K. N. Houk, J. Am. Chem. Soc., 1993, *115*, 7478.

[77] Reference [1d], chapter 6.

[78] A paper boldly titled "Quadratic CI versus Coupled-Cluster theory…": J. Hrusak, S. Ten-no, and S. Iwata, J. Chem. Phys., 1997, *106*, 7185.

[79] I. L. Alberts and N. C. Handy, J. Chem. Phys., 1988, *89*, 2107.

[80] The water dimer has been extensively studied, theoretically and experimentally: (a) M. Schuetz, S. Brdarski, P.-O. Widmark, R. Lindh, and G. Karlström, J. Chem. Phys., 1997, *107*, 4597; these workers report an interaction energy of $20.7 \, kJ \, mol^{-1}$ ($4.94 \, kcal \, mol^{-1}$). (b) M. W. Feyereisen, D. Feller, and D. A. Dixon, J. Phys. Chem., 1996, *100*, 2993; these workers report an interaction energy of $20.9 \, kJ \, mol^{-1}$ ($5.0 \, kcal \, mol^{-1}$). (c) A. Halkier, H. Koch, P. Jorgensen, O. Christiansen, M. B. Nielsen, and T. Halgaker, Theor. Chem. Acc., 1997, *97*, 150. (d) M. S. Gordon and J. H. Jensen, Acc. Chem. Res., 1996, *29*, 536.

[81] (a) For discussions of BSSE and the counterpoise method see: T. Clark, "A Handbook of Computational Chemistry," Wiley, New York, 1985, pp. 289–301. (b) J. M. Martin in "Computational Thermochemistry," K. K. Irikura and D. J. Frurip, Eds., American Chemical Society, Washington, D.C., 1998, p. 223. (c) References [80] give leading references to BSSE and [80(a)] describes a method for bringing the counterpoise correction closer to the basis set limit. (d) Methods designed to be free of BSSE: G. J. Halasz, A. Vibok, I. Mayer, J. Comput. Chem., 1999, *20*, 274.

[82] (a) L. M. Balbes, S. W. Mascarella, D. B. Boyd, chapter 7 in Reviews in Computational Chemistry, Volume 5, K. B. Lipkowitz and D. B. Boyd, Eds., VCH, New York, 1994. (b) A. Tropsha and J. P. Bowen, chapter 17 in "Using Computers in Chemistry and Chemical Education," T. J. Zielinski and M. L. Swift, Eds., American Chemical Society, Washington D.C., 1997. (c) H.-D. Höltje and G. Folkers, "Molecular Modelling," VCH, New York, 1997. (d) C. E. Bugg, W. M. Carson, J. A. Montgomery, Scientific American, 1993, December, 92. (e) J. L. Vinter and M. Gardner, "Molecular Modelling and Drug Design," Macmillan, London, 1994. (f) (e) P. M. Dean, "Molecular Foundations of Drug–Receptor Interactions," Cambridge University Press, Cambridge, 1987.

[83] Reference [62], p. 106.

[84] U. Burkert and N. L. Allinger, "Molecular Mechanics," ACS Monograph 177, American Chemical Society, Washington, D.C., 1982; pp. 6–10. See also B. Ma, J.-H. Lii, H. F. Schaefer, N. L. Allinger, J. Phys. Chem., 1996, *100*, 8763; M. Ma, J.-H. Lii, K. Chen, N. L. Allinger, J. Am. Chem. Soc., 1997, *119*, 2570.

[85] (a) A. Domenicano and I. Hargittai, Eds., "Accurate Molecular Structures," Oxford University Press, New York, 1992. (b) A " wake-up call": V. G. S. Box, Chem. & Eng. News, 2002, February 18, 6.

[86] G. A. Peterson in chapter 13, "Computational Thermochemistry," K. K. Irikura and D. J. Frurip, Eds., American Chemical Society, Washington, D.C., 1998.

[87] Reference [1g], pp. 133–226; note the summary on p. 226.

[88] Observations by the author. Others have noted that planar hexaazabenzene is a relative minimum with Hartree–Fock calculations, but a hilltop at correlated levels, e.g. R. Engelke, J. Phys. Chem., 1992, *96*, 10789 (HF/4-31G, HF/4-31G*, MP2/6-31G*).

[89] Reference [1e], p. 118.

[90] Reference [1e], pp. 118 (ozone) and 128 (FOOF).

[91] Reference [1e], p. 36; other calculations on ozone are on pp. 118, 137 and 159.

[92] Other examples of problems with fluorine and/or other halogens: (a) the G2 method (section 5.5.2) is relatively inaccurate (errors of up to $33 \, kJ \, mol^{-1}$ for molecules with multiple halogens, cf. the average error of $6.6 \, kJ \, mol^{-1}$ for a set of 148 molecules: L. A. Curtiss, K. Raghavachari, P. C. Redfern, J. A. Pople, J. Chem. Phys., 1997, *106*, 1063. (b) A (very?) good *ab initio* geometry for H_2CCH_2 required a very large basis (HF/6-311 + +G(3df,2pd)): D. Diederdorf, Applied Research Associates Inc., personal communication).

[93] For example, Fluoroethanes: R. D. Parra and X. C. Zeng, J. Phys. Chem. A, 1998, *102*, 654. Fluoroethers: D.A. Good and J. S. Francisco, J. Phys. Chem. A, 1998, *102*, 1854.

[94] C. A. Coulson, "Valence," 2nd edn, Oxford University Press, London, 1961, p. 91.

[95] "Computational Thermochemistry," K. K. Irikura and D. J. Frurip, Eds., American Chemical Society, Washington, D.C., 1998.

[96] (a) M. L. McGlashan, "Chemical Thermodynamics," Academic Press, London, 1979. (b) L. K. Nash, "Elements of Statistical Thermodynamics," Addison-Wesley, Reading, MA, 1968. (c) A good, brief introduction to statistical thermodynamics is given by K. K. Irikura in [95], Appendix B.

[97] For example, P. W. Atkins, "Physical Chemistry," 6th edn, Freeman, New York, 1998.

[98] R. S. Treptow, J. Chem. Educ., 1995, *72*, 497.

[99] See K. K. Irikura and D. J. Frurip, chapter 1, S. W. Benson and N. Cohen, chapter 2, and M. R. Zachariah and C. F. Melius, chapter 9, in [95].

[100] The bond energies were taken from M. A. Fox and J. K. Whitesell, "Organic Chemistry," Jones and Bartlett, Boston, 1994, p. 72.

[101] For good accounts of the history and meaning of the concept of entropy, see: (a) H. C. von Baeyer, "Maxwell's Demon. Why warmth disperses and time passes," Random House, New York, 1998. (b) G. Greenstein, "Portraits of Discovery. Profiles in Scientific Genius," chapter 2 ("Ludwig Boltzmann and the second law of thermodynamics"), Wiley, New York, 1998.

[102] Reference [1g], section 6.3.9.

[103] A sophisticated study of the calculation of gas-phase equilibrium constants: F. Bohr and E. Henon, J. Phys. Chem. A, 1998, *102*, 4857.

[104] A very comprehensive treatment of rate constants, from theoretical and experimental viewpoints, is given in J. I. Steinfeld, J. S. Francisco, and W. L. Hase, "Chemical Kinetics and Dynamics," Prentice Hall, New Jersey, 1999.

[105] For the Arrhenius equation and problems associated with calculations involving rate constants and transition states see J. L. Durant in [95], chapter 14.

[106] Reference [97], p. 949.

[107] P. W. Atkins, "Physical Chemistry," 4th edn, Freeman, New York, 1990, p. 859.

[108] Reference [32], chapter 2.

[109] Reference [32], section 4.2.

[110] Reference [1g], section 7.2.2.

[111] (a) Reference [1g], section 7.2.4. (b) D. B. Chestnut, J. Comput. Chem., 1995, *16*, 1227; D. B. Chestnut, J. Comput. Chem., 1997, *18*, 584 (c) P. K. Freeman, J. Am. Chem. Soc, 1998, *120*, 1619. (d) P. Politzer, M. E. Grice, J. S. Murray, and J. M. Seminario, Can J. Chem., 1993, *71*, 1123. (e) For calculation of aromatic character by nonisodesmic methods (G2 calculations on cyclopropenone) see D. W. Rogers, F. L. McLafferty, and A. W. Podosenin, J. Org. Chem, 1998, *63*, 7319.

[112] By another approach (diagonal strain energy, relating three-membered rings to essentially strainless six-membered rings), the strain energies of cyclopropane and oxirane have been calculated to be 117 and 105 kJ mol^{-1}, respectively: A. Skancke, D. Van Vechten, J. F. Liebman, and P. N. Skancke, J. Mol. Struct., 1996, *376*, 461.

[113] This is one way to state *Hess's law*. It follows from the law of conservation of energy: if it were not true then we could go from A to B by one path, putting in a certain amount of energy, then go from B to A by another path and get back more energy than we put in, creating energy.

[114] For example, [32], section 4.2. One problem with the 3-21G basis is that it tends to flatten tricoordinate nitrogen too much [42].

[115] Reference [105], p. 940.

[116] Errors of about 30–60 kJ mol^{-1} have been noted for isogyric reactions: [102], p. 13.

[117] For example K. B. Wiberg and J. W. Ochterski, J. Comput. Chem., 1997, *18*, 108.

[118] General discussions: (a) D. A. Ponomarev and V. V. Takhistov, J. Chem. Educ., 1997, *74*, 201. (b) Ref. [1e], pp. 181–184 and 204–207. (c) Ref. [1g], pp. 271, 293, 298, 356. (d) Ref. [1a], pp. 595–597.

[119] (a) G2 method: L. A. Curtiss, K. Raghavachari, G. W. Trucks, and J. A. Pople, J. Chem. Phys., 1991, *94*, 7221. There have been many reviews of the G2 method, e.g. L. A. Curtiss and K. Raghavachari, in [102], chapter 10. (b) G3 method: L. A. Curtiss, K. Raghavachari, P. C. Redfern, V. Rassolov, and J. A. Pople, J. Chem. Phys., 1998, *109*, 7764. Variations on G3: G3(MP2), L. A. Curtiss, P. C. Redfern, K. Raghavachari, V. Rassolov, and J. A. Pople, J. Chem. Phys., 1999, *110*, 4703; G3(B3) and G3(MP2B3), A. G. Baboul, L. A. Curtiss, P. C. Redfern, and K. Raghavachari, J. Chem. Phys., 1999, *110*, 7650.

[120] J. A. Pople, M. Head-Gordon, D. J. Fox, K. Raghavachari, and L. A. Curtiss, J. Chem. Phys., 1989, *90*, 5622.

[121] Reference [1e], chapter 7.

[122] L. A. Curtiss, J. E. Carpenter, K. Raghavachari, and J. A. Pople, J. Chem. Phys., 1992, *96*, 9030.

[123] B. J. Smith and L. Radom, J. Am. Chem. Soc., 1993, *115*, 4885.

[124] Reference [86], p. 244.

[125] J. M. L. Martin in chapter 12, [95].

[126] Reference [97], section 2.8.

[127] M. W. Chase, Jr, J. Phys. Chem. Ref. Data, Monograph 9, 1998, 1-1951. NIST-JANAF Thermochemical Tables, 4th edn.

[128] A. Nicolaides, A. Rauk, M. N. Glukhovtsev, and L. Radom, J. Phys. Chem., 1996, *100*, 17460.

[129] J. Hine and K. Arata, Bull. Soc. Chem. Jpn., 1976, *49*, 3089: −201 kJ mol^{-1}; J. H. S. Gree, Chem. Ind. (London), 1960, 1215: −201 ± 0.2 kJ mol^{-1}.

[130] (a) Very convenient and useful: the NIST (National Institute of Standards and Technology, Gaithersburg, MD, USA) website: http://webbook.nist.gov/chemistry/ (b) S. G. Lias, J. E. Bartmess, J. F. L. Holmes, R. D. Levin, W. G. Mallard, J. Phys. Chem. Ref. Data, 1988, *17*, Suppl. 1., American Chemical Society and American Institute of Physics, 1988. (c) J. B. Pedley, "Thermochemical Data and Structures of Organic Compounds," Thermodynamics Research Center, College Station, Texas, 1994.

[131] (a) Worked examples, with various fine points: K. K. Irikura and D. J. Frurip, in [102], Appendix C. (b) Heats of formation of neutral and cationic chloromethanes: C. F. Rodrigues, D. K. Bohme, A. C. Hopkinson, J. Phys. Chem., 1996, *100*, 2942. (c) Heats of formation, entropies and enthalpies of neutral and cationic enols: F. Turecek and C. J. Cramer, J. Am. Chem. Soc., 1995, *117*, 12243. of neutral and cationic enols: (d) Heats of formation by *ab initio* and molecular mechanics: D. F. DeTar, J. Org. Chem., 1995, *60*, 7125. (e) Heats of formation and antiaromaticity in strained molecules: M. N. Glukhovtsev, S. Laiter, A. Pross, J. Phys. Chem., 1995, *99*, 6828. (f) Heats of formation of organic molecules with the aid of *ab initio* and group equivalent methods: L. R. Schmitz and Y. R. Chen, J. Comput. Chem., 1994, *15*, 1437. (f) Isodesmic reactions in *ab initio* calculation of heat of formation of cyclic C_6 hydrocarbons and benzene isomers: Z. Li, D.

W. Rogers, F. J. McLafferty, M. Mandziuk, and A. V. Podosenin, J. Phys. Chem. A, 1999, *103*, 426. (g) Isodesmic reactions in *ab initio* calculation of heat of formation of benzene isomers: Y.-S. Cheung, C.-K. Wong, and W.-K. Li, Mol. Struct. (Theochem), 1998, *454*, 17.

[132] DeL. F. DeTar, J. Org. Chem., 1998, *102*, 5128.

[133] S. S. Shaik, H. B. Schlegel, S. Wolfe, "Theoretical Aspects of Physical Organic Chemistry. The SN2 mechanism," Wiley, New York, 1992; pp. 50–51.

[134] Reference [11], pp. 474–489. Discussion of computation of activation energies: pp. 495–497.

[135] (a) J. M. Haile, "Molecular Dynamics Simulation. Elementary Methods," Wiley, New York, 1992. (b) A tutorial in MD is available from the website of S. Deiana and B. Manunza, University of Sassani, Italy: http://antas.agraria.uniss.it (c) An application of MD to chemical reactions: B. K. Carpenter, Ang. Chem. Int. Ed. Engl., 1998, *37*, 3341.

[136] (a) Ref. [1a], section 2.5. (b) Ref. [104], section 12.3. (c) R. P. Bell, "The Tunnel Effect in Chemistry," Chapman and Hall, London, 1980.

[137] Programs for calculating rate constants are available from groups at the University of Minnesota: http://comp.chem.umn.edu/. Of particular relevance are: POLYRATE, probably the most widely-used program for calculating rate constants; GAUSSRATE, which interfaces between POLYRATE and Gaussian; ABCRATE, which interfaces with various programs; AMSOLRATE, which interfaces between POLYRATE and the semiempirical program AMSOL, for calculating rates in solution and in the gas phase; MORATE, which interfaces between POLYRATE and a modified version of the semiempirical program MOPAC (MOPAC 5.07 mn), for the calculation of gas-phase reaction rates.

[138] (a) S. S. Shaik, H. B. Schlegel, S. Wolfe, "Theoretical Aspects of Physical Organic Chemistry. The SN2 mechanism," Wiley, New York, 1992; pp. 84–88. (b) Ref. [104], chapters 25, 26, 27. (c) Kinetics of halocarbons reactions: R. J. Berry, M. Schwartz, P. Marshall in [95], chapter 18. (d) Ref. [104], particularly chapters 7, 8, 10, 11,12. (e) The *ab initio* calculation of rate constants is given in some detail in these two references: D. M. Smith, A. Nicolaides, B. T. Golding, and L. Radom, J. Am. Chem. Soc., 1998, *120*, 10223; J. P. A. Heuts, R. G Gilbert, L. Radom, Macromolecules, 1995, *28*, 8771; [11], pp. 471–489, 492–497.

[139] Reference [104], chapter 11.

[140] Reference [104], p. 350.

[141] Reference [104], p. 352, Fig. 11-11.

[142] Calculated from the Arrhenius parameters in [97], p. 949, Table 25.4.

[143] J. F. McGarrity, A. Cretton, A. A. Pinkerton, D. Schwarzenbach, and H. D. Flack, Angew. Chem. int. Ed. Engl., 1983, *22*, 405; C. E. Blom and A. Bauder, J. Am. Chem. Soc., 1984, *106*, 4029; B. Capon, B. Guo, F. C. Kwok, A. K. Siddhanta, C. Zucco, Acc. Chem. Res., 1988, *21*, 135.

[144] D. J. DeFrees and A. D. McLean, J. Chem Phys., 1982, *86*, 2835.

[145] (a) W. Runge in "The Chemistry of Ketenes, Allenes and Related Compounds," S. Patai, Ed., Wiley, New York, 1980, p. 45. (b) E. Hirota and C. Matsumura, J. Chem Phys., 1973, *59*, 3038.

[146] O. L. Chapman, Pure Appl. Chem., 1974, *40*, 511.

[147] I. R. Dunkin, "Matrix Isolation Techniques: A Practical Approach. The Practical Approach in Chemistry Series," Oxford University Press, New York, 1998.

[148] Reference [32], chapter 6.

[149] Reference [1e], pp. 146–148, 157–158.

[150] J. B. Foresman, personal communication, October 1998.

[151] Reference [32], pp. 58–62.

[152] For introductions to the theory and interpretation of IR, UV and NMR spectra, see R. M. Silverstein and F. X. Webster, "Spectrometric Identification of Organic Compounds," 6th edn, Wiley, New York, 1997.

[153] K. P. Huber and G. Herzberg, "Molecular Spectra and Molecular Structure. IV. Constants of Diatomic Molecules," Van Nostrand Reinhold, New York, 1979.

[154] Reference [1g], pp. 234, 235.

[155] For example, "... it is unfair to compare frequencies calculated within the harmonic approximation with experimentally observed frequencies...": A. St-Amant, chapter 2, p. 235 in Reviews in Computational Chemistry, Volume 7, K. B. Lipkowitz and D. B. Boyd, Eds., VCH, New York, 1996.

[156] Reference [1i], pp. 271–274.

[157] (a) J. G. Radziszewski, B. A. Hess, Jr., and R. Zahradnik, J. Am. Chem. Soc., 1992, *114*, 52. (b) C. Wentrup, R. Blanch, H. Briel, G. Gross, J. Am. Chem. Soc., 1988, *110*, 1874. (c) O. L. Chapman, C. C. Chang, J. Kolc, N. R. Rosenquist, H. Tomioka, J. Am. Chem. Soc., 1975, *97*, 6586. (d) O. L. Chapman, K. Mattes, C. L. McIntosh, J. Pacansky, G. V. Calder, and G. Orr, J. Am. Chem. Soc., 1973, *95*, 6134.

[158] (a) A. Komornicki and R. L. Jaffe, J. Chem. Phys., 1979, *71*, 2150. (b) Y. Yamaguchi, M. Frisch, J. Gaw, H. F. Schaefer, and J. S. Binkley, J. Chem. Phys., 1986, *84*, 2262. (c) M. Frisch, Y. Yamaguchi, H. F. Schaefer, and J. S. Binkley, J. Chem. Phys., 1986, *84*, 531. (d) R. D. Amos, Chem. Phys. Lett, 1984, *108*, 185. (e) J. E. Gready, G. B. Bacskay, and N. S. Hush, J. Chem. Phys., 1978, *90*, 467.

[159] (a) H. Lampert, W. Mikenda, and A. Karpfen, J. Phys. Chem. A, 1997, *101*, 2254. This paper shows actual pictures of experimental and calculated (HF, MP2, DFT) spectra for phenol, benzaldehyde and salicylaldehyde. (b) M. D. Halls and H. B. Schlegel, J. Chem. Phys., 1998, *109*, 10587. For a series of small molecules absolute intensities compared with QCISD as benchmark. (c) Ref. [69], pp. 118–121. (d) D. H. Magers, E. A. Salter, R. J. Bartlett, C. Salter, B. A. Hess, Jr., and L. J. Schaad, J. Am. Chem. Soc., 1988, *110*, 3435 (comments on intensities on p. 3439).

[160] For a good review of the cyclobutadiene problem, see B. K. Carpenter in "Advances in Molecular Modelling," D. Liotta, Ed., JAI Press Inc., Greenwich, Connecticut, 1988.

[161] (a) Theoretical calculation of dipole moments: [1a], pp. 399–402. (b) Measurement and applications of dipole moments: O. Exner, "Dipole Moments in Organic Chemistry," Georg Thieme Publishers, Stuttgart, 1975.

[162] Tables: A. L. McClellan, "Tables of Experimental Dipole Moments," vol. 1, W. H. Freeman, San Francisco, CA, 1963; vol. 2, Rahara Enterprises, El Cerrita, CA, 1974.

[163] For example, [62], p. 152.

[164] The reason why an electron pair forms a covalent bond has apparently not been settled. See (a) Ref. [1a], pp. 362–363. (b) G. B. Backsay, and J. R. Reimers, S. Nordholm, J. Chem. Ed., 1997, *74*, 1494.

[165] For example, (a) Electron density on an atom: G. W. Wheland and L. Pauling, J. Am. Chem. Soc., 1935, *57*, 2086. (b) Pi-bond order: C. A. Coulson, Proc. Roy. Soc., 1939, *A169*, 413.

[166] (a) R. S. Mulliken, J. Chem Phys., 1955, *23*, 1833. (b) R. S. Mulliken, J. Chem Phys., 1962, *36*, 3428. (c) Ref. [1a], pp. 475–478.

[167] P.-O. Löwdin, Advances in Quantum Chemistry, 1970, *5*, 185.

[168] A. E. Reed, L. A. Curtiss, and F. Weinhold, Chem. Rev., 1988, *88*, 899.

[169] (a) T. Brinck, "Theoretical Organic Chemistry," C. Parkanyi, Ed., Elsevier, New York, 1998. (b) D. S. Marynick, J. Comp. Chem., 1997, *18*, 955.

[170] (a) Use of bond orders in deciding if a covalent bond is present: P. v. R. Schleyer, P. Buzek, T. Müller, Y. Apelloig, and H.-U. Siehl, Angew. Chem. Int. Ed. Engl., 1993, *32*, 1471. (b) Use of bond order in estimating progress along a reaction coordinate: G. Lendvay, J. Phys. Chem., 1994, *98*, 6098.

[171] (a) R. F. W. Bader, "Atoms in Molecules," Oxford University Press, Oxford, 1990. (b) R. W. F. Bader, P. L. A. Popelier, T. A. Keith, Angew. Chem. Int. Ed. Engl., 1994, *33*, 620. Applications of AIM theory: (c) I. Rozas, I. Alkorta, J. Elguero, J. Phys. Chem., 1997, *101*, 9457. (d) Can. J. Chem., 1996, *74*, issue dedicated to R. F. W. Bader. (e) R. J. Gillespie, and E. A. Robinson, Angew. Chem. int. Ed. Engl., 1996, *35*, 495. (f) S. Grimme, J. Am. Chem. Soc., 1996, *118*, 1529. (g) O. Mo, M. Yanez, M. Eckert-Maksoc, Z. B. Maksic, J. Org. Chem., 1995, *60*, 1638. (h) K. M. Gough and J. Millington, Can. J. Chem., 1995, *73*, 1287. (i) R. Glaser and G. S.-C. Choy, J. Am. Chem. Soc., 1993, *115*, 2340. (j) J. Cioslowski and T. Mixon, J. Am. Chem. Soc., 1993, *115*, 1084.

[172] Reference [170a], p. 175.

[173] Reference [1e], chapter 9.

[174] T. Helgaker, M. Jaszuński, and K. Ruud, Chem. Rev., 1999, *99*, 294.

[175] (a) Ref. [1e], pp. 53–54 and 104–105. (b) J. R. Cheeseman, G. W. Trucks, T. A. Keith. and M. J. Frisch, J. Chem. Phys., 1996, *104*, 5497.

[176] (a) Review: J. Gauss, Ber. Bunsenges. Phys. Chem., 1995, *99*, 1001. (b) J. Gauss, J. Chem. Phys., 1993, *99*, 3629. (c) P. v. R. Schleyer, J. Gauss, M. Bühl, R. Greatrex, and M. A. Fox, Chem. Commun., 1993, 1766. (d) Electron correlation and coupling constants: S. A. Perera, M. Nooijen, and R. J. Bartlett, J. Chem. Phys., 1996, *104*, 3290.

[177] A. D. Wolf, V. V. Kane, R. H. Levin, and M. Jones, J. Am. Chem. Soc., 1973, *95*, 1680.

[178] (a) V. I. Minkin, M. N. Glukhovtsev, and B. Ya. Simkin, "Aromaticity and Antiaromaticity: Electronic and Structural Aspects," Wiley, New York, 1994; chapter 2. (b) A simple magnetic criterion for aromaticity (NICS): P. v. R. Schleyer, C. Maerker, A. Dransfeld, H. Jiao, and N. v. E. Hommes, J. Am. Chem. Soc., 1996, *118*, 6317. (c) NICS(2.0): T. Veszprémi, M. Tahahashi, B. Hajgató, J. Ogasawara, K. Skamoto, and M. Kira, J. Phys. Chem., 1998, *102*, 10530–10535; R. West, J. J. Buffy, M. Haaf, T. Müller, B. Gehrhus, M. F. Lappert, and Y. Apeloig, J. Am. Chem. Soc., 1998, *120*, 1639. (d) Analysis of NICS: I. Morao and F. P. Cossío, J. Org. Chem., 1999, *64*, 1868.

[179] Reference [1b], pp. 276–277, 288, 372–373.

[180] R. D. Levin and S. G. Lias, "Ionization Potential and Appearance Potential Measurements, 1971–1981," National Bureau of Standards, Washington, DC, 1982.

[181] (a) See e.g. [1b], pp. 361–363; [1c], pp. 278–280; [1d], pp. 127–128; [1g], pp. 24, 116. A novel look at Koopmans' theorem: C. Angeli, J. Chem. Ed., 1998, 75, 1494. (b) T. Koopmans, Physica, 1934, 1, 104.

[182] M. B. Smith and J. March, "March's Advanced Organic Chemistry," 5th edn, Wiley, New York, 2001, pp. 10–12.

[183] J. E. Lyons, D. R. Rasmussen, M. P. McGrath, R. H. Nobes, and L. Radom, Ang. Chem. Int. Ed. Engl., 1994, 33, 1667.

[184] L. A. Curtiss and K. Raghavachari, chapter 10 in [102].

[185] L. A. Curtiss, R. H. Nobes, J. A. Pople, and L. Radom, J. Chem Phys., 1992, 97, 6766.

[186] (a) "Data Visualization in Molecular Science: Tools for Insight and Innovation," J. E. Bower, Ed., Addison-Wesley, Reading, MA, 1995. (b) "Frontiers of Scientific Visualization," C. Pickover and S. Tewksbury, Wiley, 1994. (c) http://www.tc.cornell.edu/~richard/.

[187] GaussView: Gaussian Inc., Carnegie Office Park, Bldg. 6, Pittsburgh, PA 15106, USA.

[188] E. L. Eliel and S, H. Wilen, "Stereochemistry of Carbon Compounds," Wiley, New York, 1994, pp. 502–507 and 686–690.

[189] E. Lewars, unpublished.

[190] The term is not just whimsy on the author's part: certain stereoelectronic phenomena arising from the presence of lone pairs on heteroatoms in a 1,3-relationship were once called the "rabbit-ear effect," and a photograph of the eponymous creature even appeared on the cover of the Swedish journal *Kemisk Tidskrift*. History of the term, photograph: E. L. Eliel, "From Cologne to Chapel Hill," American Chemical Society, Washington, DC, 1990, pp. 62–64.

[191] E. Lewars, J. Mol. Struct. (Theochem), 2000, 507, 165; E. Lewars, J. Mol. Struct. (Theochem), 1998, 423, 173.

[192] P. Politzer and J. S. Murray, chapter 7 in Reviews in Computational Chemistry, Volume 2, K. B. Lipkowitz and D. B. Boyd, Eds., VCH, New York, 1996.

[193] Reference [53c], pp. 141–142.

[194] W. J. Hehre, A. J. Shusterman, and J. E. Nelson, "The Molecular Modelling Workbook of Organic Chemistry," Wavefunction Inc., Irvine, CA, 1998 (book and CD).

[195] M. J. S. Dewar, "A Semiempirical Life," Profiles, Pathways and Dreams series, J. I. Seeman, Ed., American Chemical Society, Washington, D.C., 1992, p. 129.

[196] (a) Coulson's remarks: J. D. Bolcer and R. B. Hermann, chapter 1 in Reviews in Computational Chemistry, Volume 5, K. B. Lipkowitz and D. B. Boyd, Eds., VCH, New York, 1996, p. 12. (b) The increase in computer speed is also dramatically shown in data provided in *Gaussian News*, 1993, 4, 1. The approximate times for a single-point HF/6-31G** calculation on 1,3,5-triamino-2,4,6-trinitrobenzene (300 basis functions) are reported as: ca. 1967, on a CDC 1604, 200 years (estimated); ca. 1992, on a 486 DX personal computer, 20 hours. This is a speed factor of 90,000 in 25 years. The price factor for the machines may not be as dramatic, but suffice it to say that the CDC 1604 was not considered a personal computer.

EASIER QUESTIONS

1. In the term *Hartree-Fock*, what, essentially, were the contributions of each of these two people?

2. What is a spin orbital? A spatial orbital?

3. At which step in the derivation of the HF energy does the assumption that each electron sees an "average electron cloud" appear?

4. For a closed-shell molecule the number of occupied molecular orbitals is half the number of electrons, but there is no limit to the number of virtual orbitals. Explain.

5. In the simple Hückel method, c_{si} denotes the basis function coefficient for the contribution of atom number s (in whatever numbering scheme we choose) to MO number i. In the *ab initio* method, c_{si} still refers to MO number i, but the s does not necessarily denote atom number s. Explain.

6. The derivation of the Roothaan–Hall equations involves some key concepts: Slater determinant, Schrödinger equation, explicit Hamiltonian operator, energy minimization, and LCAO. Using these, summarize the steps leading to the Roothaan–Hall equations $\mathbf{FC} = \mathbf{SC\varepsilon}$.

7. What are the similarities and the differences between the basis set of the extended Hückel method and the *ab initio* STO-3G basis set?

8. In the simple and extended Hückel methods, the molecular orbitals are calculated and then filled from the bottom up with the available electrons. However, in *ab initio* calculations the occupancy of the orbitals is taken into account as they are being calculated. Explain. (Hint: look at the expression for the Fock matrix elements in terms of the density matrix.)

9. Isodesmic reactions have been used to investigate aromatic stabilization, but there is not a unique isodesmic reaction for each problem. Write two isodesmic reactions for the ring-opening of benzene, both of which have on each side of the equation the same number of each kind of bond. Have you any reason to prefer one of the equations to the other?

10. List the strengths and weaknesses of *ab initio* calculations compared to molecular mechanics and extended Hückel calculations. State the molecular features that can be calculated by each method.

HARDER QUESTIONS

1. Does the term *ab initio* imply that such calculations are "exact"? In what sense might *ab initio* calculations be said to be semiempirical – or at least not *a priori*?

2. Can the Schrödinger equation be solved *exactly* for a species with two protons and one electron? Why or why not?

3. The input for an *ab initio* calculation (or a semiempirical calculation of the type discussed in chapter 6, or a DFT calculation – chapter 7) on a molecule is usually just the cartesian coordinates of the atoms (plus the charge and multiplicity). So how

does the program know where the bonds are, i.e. what the structural formula of the molecule is?

4. Why is it that (in the usual treatment) the calculation of the internuclear repulsion energy term is easy, in contrast to the electronic energy term?

5. In an *ab initio* calculation on H_2 or HHe^+, one kind of interelectronic interaction does not arise; what is it, and why?

6. Why are basis functions not necessarily the same as atomic orbitals?

7. One desirable feature of a basis set is that it should be "balanced." How might a basis set be unbalanced?

8. In a HF calculation, you can always get a lower energy (a "better" energy, in the sense that it is closer to the true energy) for a molecule by using a bigger basis set, as long as the HF limit has not been reached. Yet a bigger basis set does not *necessarily* give better geometries and better relative (i.e. activation and reaction) energies. Why is this so?

9. Why is size-consistency in an *ab initio* method considered more important than variational behavior (MP2 is size-consistent but not variational)?

10. A common alternative to writing a HF wavefunction as an explicit Slater determinant is to express it using a *permutation operator* $\hat{P}$ which permutes (switches) electrons around in MOs. Examine the Slater determinant for a two-electron closed-shell molecule, then try to rewrite the wavefunction using $\hat{P}$.

Chapter 6

Semiempirical Calculations

> *Current "ab initio" methods were limited to very inaccurate*
> *calculations for very small molecules.*
> M J. S. Dewar, *A Semiempirical Life, 1992.*

6.1 PERSPECTIVE

We have already seen examples of semiempirical (SE) methods, in chapter 4: the simple Hückel method (SHM, Erich Hückel, ca. 1931) and the extended Hückel method (EHM, Roald Hoffman, 1963). These are semiempirical (semiexperimental) because they combine physical theory with experiment. Both methods start with the Schrödinger equation (theory) and derive from this a set of secular equations which may be solved for energy levels and molecular orbital coefficients (most efficiently by diagonalizing a Fock matrix; see chapter 4). However, the SHM gives energy levels in units of a parameter (β) that can be translated into actual quantities only by comparing SHM results with experiment, and the EHM uses experimental ionization energies to translate the Fock matrix elements into actual energy quantities. SE calculations stand in contrast to empirical methods, like molecular mechanics (MM, chapter 3), and theoretical methods, like *ab initio* calculations (chapter 5). MM starts with a model of a molecule as balls and springs, a model that works but whose *theoretical justification* lies outside MM. The *ab initio* method, like the Hückel methods, starts with the Schrödinger equation but does not appeal to experiment (beyond invoking, when actual quantities are needed, experimental values for Planck's constant, the charge on the electron and proton, and the masses of the electron and atomic nuclei – fundamental physical constants which could be calculated only by some deep theory of the origin and nature of our universe [1].

The Hückel methods were discussed in chapter 4 rather than here because extensive application of these methods came before widespread use of *ab initio* methods, and because the simple Hückel, extended Hückel and *ab initio* methods form a conceptual progression in which the first two methods aid understanding of the next one in this hierarchy of complexity. The SE methods treated in this chapter are logically regarded as simplifications of the *ab initio* method, since they use the SCF procedure (chapter 5) to refine the Fock matrix, but do not evaluate these matrix elements *ab initio*. The SHM was developed (1931) outside the realm of SCF theory (which was invented for atoms: Hartree, 1928 [2]), as the first application of the Schrödinger equation to molecules of reasonable size, and the EHM is a straightforward extension of this. In contrast, the methods of this chapter were created in a conscious attempt to provide practical alternatives to the *ab initio* approach, the application of which to molecules of reasonable size understandably seemed hopeless in the infancy of electronic computers (the PPP method, one of the first SCF SE methods, was published in 1953, just when the first electronic computers began to be available to chemists [3]). SE calculations are much less demanding than *ab initio* ones, because parameterization and approximations drastically reduce the number of integrals which must be calculated. The pessimism with which the *ab initio* approach was viewed is clear in the words of several pioneers of the application of quantum mechanics to chemistry:

C. A. Coulson, 1959: "I see little chance – and even less desirability – of dealing in this accurate manner with systems containing more than 20 electrons..." [4]

M. J. S. Dewar[1], 1969: "How then shall we proceed? The answer lies in abandoning attempts to carry out rigorous a priori calculations." [5].

Neither Coulson nor Dewar could have foreseen the enormous increase in computer power that was to come over the next few decades. What Coulson meant by "even less desirability" was perhaps that the computed results would be too complex to interpret; one factor which has obviated this problem is the visual display of information (sections 5.5.6, 6.3.6). The development of improved algorithms and far faster computers has altered the situation almost out of recognition: computers in 2000 were about one million times faster than in 1959 (computers were said [4] to be 100 000 times faster in 1989 than in 1959, the date of Coulson's remarks; it seems safe to say that they increased in speed by a factor of 10 in the subsequent decade). A calculation that in 1967 would have taken 200 years can now be run on a cheap computer in less than an hour [6]. Why, then, are SE calculations still used? Because they are still about 100–1000 times faster than *ab initio* (chapter 5) or density functional (chapter 7) methods. The increase in computer speed means that we can now routinely examine by *ab initio* methods moderately large molecules – up to, say, steroids, with about 30 heavy atoms (nonhydrogen atoms), and by semiempirical methods (and faster with MM, chapter 3) huge molecules, like proteins and nucleic acids.

6.2   THE BASIC PRINCIPLES OF SCF SE METHODS

6.2.1   Preliminaries

The SE methods we saw in chapter 4 simply construct a Fock matrix and diagonalize it to get molecular orbital (MO) energy levels and MOs (i.e. the coefficients of the basis

---

[1]Michael J. S. Dewar, born Ahmednagar, India, 1918. Ph.D. Oxford, 1942. Professor of chemistry at Universities of London, Chicago, Texas at Austin, and University of Florida. Died Florida, 1997.

functions that make up the MOs). The simple Hückel Fock matrix elements were simply relative energies 0 and -1 (0 and $-1|\beta|$ units, relative to the nonbonding level α), or in the EHM the Fock matrix elements were calculated from ionization energies. A single matrix diagonalization gave the energy levels and MO coefficients. This chapter is concerned with SE methods that are closer to the *ab initio* method in that the SCF procedure (sections 5.2.3.6d and 5.2.3.6e) is used to refine the energy levels and MO coefficients. As in *ab initio* calculations each Fock matrix element is calculated from a core integral H_{rs}^{core}, density matrix elements P_{tu}, and electron repulsion integrals $(rs|tu)$, $(ru|ts)$:

$$F_{rs} = H_{rs}^{\text{core}}(1) + \sum_{t=1}^{m}\sum_{u=1}^{m} P_{tu}\left[(rs|tu) - \frac{1}{2}(ru|ts)\right] \qquad (6.1 = 5.82)$$

As stated above, the following discussion applies to SE methods that use the SCF procedure and so pay some service to Eq. (6.1). As with an *ab initio* calculation, to initiate the process we need an *initial guess* of the coefficients, to calculate the density matrix values P_{tu}; the guess can come from a simple Hückel calculation (for a π electron theory like the PPP method) or from an extended Hückel calculation (for an all-valence-electron theory, like CNDO and its descendants). The Fock matrix of F_{rs} elements is diagonalized repeatedly to refine energy levels and coefficients.

The divergence from the *ab initio* method lies in (1) treating only valence or π electrons, i.e. in the meaning of the "core," (2) the mathematical functions used to expand the MOs (the basis set functions), (3) how the core and two-electron repulsion integrals are evaluated, and (4) the treatment of the overlap matrix. These approximations are discussed in detail by Dewar [7]. An excellent yet compact survey of the principles behind all the major SE methods is given by Levine [8], and SE methods have also been reviewed by Thiel [9]; a detailed exposition of the basic (pre-1970) theory behind these methods can be found in the book by Pople and Beveridge [10]. Expanding on points (1)–(4):

(1) *Treating only valence or π electrons, i.e. the meaning of the "core".* In an *ab initio* calculation H_{rs}^{core} is the kinetic energy of an electron moving in the force-field of the atomic nuclei, plus the potential energy of attraction of the electron to these atomic nuclei: the electron is moving under the influence of a positive core composed of atomic nuclei. SE calculations handle at most valence electrons (the PPP method handles only π electrons), so each element of the core becomes an atomic nucleus *plus its core electrons* (for the PPP method, a nucleus with the core electrons plus all σ valence electrons). Instead of a cloud of all the electrons moving in a framework of nuclei, we have a cloud *valence* electrons (for the PPP method, π electrons) moving in a framework of atomic cores (atomic core = nucleus + valence electrons, or for PPP, nucleus + all electrons that don't contribute to the π system). The SCF SE energy is calculated in a manner analogous to that of an *ab initio* calculation of the Hartree–Fock energy (cf. Eq. (5.149)), but n in Eq. (6.2) is not half the total number of electrons, but rather half the number of valence electrons (half the number of π electrons for a PPP calculation), i.e. n is the number of MOs formed from the those electrons being included in the basis set. E_{SE} is the valence electronic (π electronic for the PPP method) energy, rather than the total electronic energy, and V_{CC} is the core–core repulsion, rather than

the nucleus–nucleus repulsion:

$$E_{SE}^{total} = E_{SE} + V_{CC} = \sum_{i=1}^{n} \epsilon_i + \frac{1}{2} \sum_{r=1}^{m} \sum_{s=1}^{m} P_{rs} H_{rs}^{core} + V_{CC} \tag{6.2}$$

Treating the core electrons in effect as part of the atomic nuclei means that we need basis functions only for the valence electrons. With a minimal basis set (section 5.3.3) an *ab initio* calculation on ethene, C_2H_4, needs five basis functions ($1s$, $2s$, $2p_x$, $2p_y$, $2p_z$) for each carbon and one basis function ($1s$) for each hydrogen, a total of 14 basis functions, while a SE calculation needs four functions for each carbon and one for each hydrogen, for a total of 12; for cholesterol, $C_{27}H_{46}$, the numbers of basis functions are 181 and 154 for *ab initio* and SE, respectively. In both cases the SE calculation needs only about 85% as many basis functions as an *ab initio* calculation; the SE basis set advantage is small compared to minimal basis set *ab initio* calculations, but large compared to *ab initio* calculations with split valence and split valence plus polarization (section 5.3.3) basis sets. For ethene, comparing a 6-31G* *ab initio* calculation with a minimal basis SE calculation, the numbers of basis functions are 38 and 12, a ratio of 32%; for cholesterol, 497 and 154, a ratio of 31%. SE calculations use only a minimal basis set and hope to compensate for this by parameterization of the two-electron integrals (below).

(2) *The basis set functions.* In SE methods the basis functions correspond to atomic orbitals (valence AOs or *p*-π AOs), while in *ab initio* calculations this is strictly true only for a minimal basis set, since an *ab initio* calculation can use many more basis functions than there are conventional AOs. The SCF-type SE methods we are considering in this chapter use Slater functions, rather than approximating Slater functions as sums of Gaussian functions (section 5.3.2). Recall that the only reason *ab initio* calculations use Gaussian, rather than the more accurate Slater, functions, is because calculation of the electron–electron repulsion two-electron integrals is far faster with Gaussian functions (section 5.3.2). In SE calculations these integrals have been parameterized into the calculation (see below). Mathematical forms of the basis functions ϕ are still needed, to calculate overlap integrals $\langle \phi_r | \phi_s \rangle$, for although these methods treat the overlap matrix as a unit matrix, some overlap integrals are evaluated (approximate MO theory has some apparent logical contradictions [7]) and used to help calculate core integrals and electron-repulsion integrals. As in *ab initio* calculations linear combinations of the basis functions are used to construct MOs, which in turn are multiplied by spin functions and used to represent the total molecular wavefunction as a Slater determinant (section 5.2.3.1).

(3) *The integrals.* The core integrals and the two-electron repulsion integrals (electron-repulsion integrals), Eq. (6.1), are not calculated from first principles (i.e. not from an explicit Hamiltonian and basis functions, as illustrated in section 5.2.3.6e), but rather many integrals are taken as zero, and those that are used are evaluated in an empirical way from the kinds of atoms involved and their distances apart. Recall that calculation of the two-electron integrals, particularly the three- and four-center ones (those involving three or four different atoms) takes up most of the time in an *ab initio* calculation. The integrals to be ignored (set equal to zero) are determined from the extent to which *differential overlap* is neglected. Differential overlap *dS* is the

differential of the overlap integral (e.g. section 4.3.3) S:

$$S = \int \phi_r(1)\phi_s(1)\, dv_1 \qquad (6.3)$$

$$dS = \phi_r(1)\phi_s(1)\, dv_1 \qquad (6.4)$$

SE methods differ amongst themselves in (amongst other ways) the criteria for setting $dS = 0$, i.e. for applying *zero differential overlap* (ZDO).

(4) *The overlap matrix.* SCF-type SE methods take the overlap matrix as a unit matrix, $\mathbf{S} = \mathbf{1}$, so $\mathbf{S}$ vanishes from the Roothaan–Hall equations $\mathbf{FC} = \mathbf{SC}\epsilon$ without the necessity of using an orthogonalizing matrix to transform these equations into standard eigenvalue form $\mathbf{FC} = \mathbf{C}\epsilon$ (which latter enables the Fock matrix to be diagonalized to give the MO coefficients and energy levels; sections 4.4.3, 4.4.1, and 5.2.3.6b).

6.2.2 The Pariser-Parr-Pople (PPP) method

The first SE SCF-type method to gain widespread use was the PPP method (1953) [11,12]. Like the SHM, PPP calculations are limited to π electrons, with the other electrons forming a σ framework to hold the atomic p orbitals in place. The Fock matrix elements are calculated from Eq. (6.1); for a PPP calculation H_{rs}^{core} represents the nuclei plus all non-π-system electrons, P_{tu} is calculated from the coefficients of those p AOs contributing to the π system, and the two-electron repulsion integrals refer to electrons in the π system. The one-center core integrals H_{rr}^{core} are estimated empirically from the ionization energy of a $2p$ AO and (see below) the two-electron integral $(rr|ss)$. The two-center core integrals H_{rs}^{core} are calculated from

$$H_{rs}^{core} = k\langle \phi_r(1)|\phi_s(1)\rangle, \qquad r \neq s \qquad (6.5 = 5.82)$$

where k is an empirical parameter chosen to give the best agreement with experiment of the wavelength of UV absorption bands, and the overlap integral $\langle \phi_r|\phi_s\rangle$ is calculated from the basis functions, with the proviso that if ϕ_r and ϕ_s are on atoms that are not connected then the integral is taken as zero.

The two-electron integrals are evaluated by applying the ZDO approximation (above) to all different orbitals r and s:

$$dS = \phi_r(1)\phi_s(1)\, dv_1 = 0, \qquad \text{for } r \neq s \qquad (6.6)$$

From Eq. (6.6) and the definition of the two-electron integral

$$(rs|tu) = \iint \frac{\phi_r^*(1)\phi_s(1)\phi_t^*(2)\phi_u(2)}{r_{12}}\, dv_1\, dv_2 \qquad (6.7 = 5.73)$$

it follows that (1) for $r \neq s$, $(rs|tu) = 0$, and (2) for $r = s$ and $t = u$, $(rs|tu) = (rr|tt)$. Both cases are taken into account by writing

$$(rs|tu) = \delta_{rs}\delta_{tu}(rr|tt) \qquad (6.8)$$

where the δ's are Kronecker deltas ($= 1$ if the subscripts are the same, zero otherwise). Thus the four-center (i.e. $(rs|tu)$) and three-center (i.e. $(rr|tu)$) two-electron integrals are ignored, but not the two-center (i.e. $(rr|tt)$) and one-center (i.e. $(rr|rr)$) two-electron integrals. The one-center integrals $(rr|rr)$ are taken as the difference between the valence-state ionization energy and the electron affinity of the atom bearing ϕ_r (these valence-state parameters refer to a hypothetical isolated atom in the same hybridization state as in the molecule, and can be found spectroscopically). The two-center integrals $(rr|tt)$ are estimated from $(rr|rr)$ and $(tt|tt)$ and the distance between the ϕ_r and ϕ_t atoms.

Although the overlap integrals $\langle f_r | f_s \rangle$ are actually calculated for the evaluation of H_{rs}^{core} (Eq. (6.5)), the overlap matrix is taken as a unit matrix as far as the matrix Roothaan–Hall equations $\mathbf{FC} = \mathbf{SC\epsilon}$ go; thus $\mathbf{FC} = \mathbf{C\epsilon}$ or $\mathbf{F} = \mathbf{C\epsilon C^{-1}}$ and the Fock matrix is diagonalized to give the MO coefficients and energy levels without transforming it with an orthogonalizing matrix. That the overlap matrix is a unit matrix is a corollary of the ZDO approximation of Eq. (6.6), from which it follows that the off-diagonal matrix elements are zero; the diagonal elements are of course unity if normalized AO basis functions are used. PPP energies are π-electron electronic energies E_{SE}, or electronic energies plus core–core repulsions, E_{SE}^{total}, if V_{CC} is added (Eq. (6.2)).

The PPP method has been used to calculate the UV spectra of conjugated compounds, especially dyes [13], a task it performs fairly well. The accuracy of these calculations can be improved by incorporating electron correlation (section 5.4), using the configuration interaction (CI) method. The calculations were usually done at a fixed geometry, although an empirical bond length–bond order relation permits optimization of bond length. The classical PPP method is not much used, having evolved into other neglect of differential overlap (NDO) methods, especially those parameterized for spectra, like INDO/S and ZINDO/S (below).

6.2.3　The complete neglect of differential overlap (CNDO) method

The first SE SCF-type method to go beyond just π electrons was the complete neglect of differential overlap method (1965) [14]. This was a general-geometry method, since it is not limited to planar π systems (molecules with conjugated π electron systems, like benzene, are usually planar). Like the other early general-geometry method, the EHM, which appeared in 1963 (section 4.4), CNDO calculations use a minimal valence basis set of Slater-type orbitals, in which each atom has the usual number of valence AOs. The Fock matrix elements are calculated from Eq. (6.1); for a CNDO calculation H_{rs}^{core} represents the nuclei plus all core electrons, P_{tu} is calculated from the coefficients of the valence AOs, and the two-electron repulsion integrals refer to valence electrons.

There are two versions of CNDO, CNDO/1 and an improved version, CNDO/2. First look at CNDO/1. Consider the core integrals $H_{r_A r_A}^{core}$, where both orbitals are the same (i.e. the same orbital occurs twice in the integral $\langle \phi_r(1) | \hat{H}_{rr}^{core} | \phi_r(1) \rangle$) and are on the same atom A. Recall the example of an *ab initio* calculation on HHe$^+$ (section 5.2.36e).

Consider, say, element (1,1) of that $\mathbf{H}^{core}$ matrix. From Eq. (5.116):

$$H_{11}^{core} = \langle \phi_1(1)|\hat{T}|\phi_1(1)\rangle + \langle \phi_1(1)|\hat{V}_H|\phi_1(1)\rangle + \langle \phi_1(1)|\hat{V}_{He}|\phi_1(1)\rangle$$

$$= \langle \phi_1(1)|\hat{T} + \hat{V}_H|\phi_1(1)\rangle + \langle \phi_1(1)|\hat{V}_{He}|\phi_1(1)\rangle \tag{6.9}$$

Eq. (6.9) can be generalized to a matrix element (r,r) and a molecule with atoms A, B, C, ..., giving

$$H_{rArA}^{core} = \langle \phi_{rA}(1)|\hat{T} + \hat{V}_A|\phi_{rA}(1)\rangle + \langle \phi_{rA}(1)|\hat{V}_B|\phi_{rA}(1)\rangle$$

$$+ \langle \phi_{rA}(1)|\hat{V}_C|\phi_{rA}(1)\rangle + \cdots$$

$$= U_{rr} + \sum_{B \neq A} \langle \phi_{rA}(1)|\hat{V}_B|\phi_{rA}(1)\rangle = U_{rr} + V_{AB} \tag{6.10}$$

where ϕ_{rA} is a basis function on atom A. The U_{rr} term in Eq. (7.0) is regarded as the energy of an electron in the AO on A corresponding to the function ϕ_{rA}, and is taken as the negative of the valence-state ionization energy of such an electron. The integrals in the V_{AB} term are simply calculated as the potential energy of a valence s orbital in the electrostatic field of the core of atom A, B, etc., e.g.

$$\langle \phi_{rA}(1)|\hat{V}_B|\phi_{rA}(1)\rangle = \left\langle s_A(1) \left| \frac{C_B}{r_{1B}} \right| s_A(1) \right\rangle \tag{6.11}$$

where C_B is the charge on the core of atom B, i.e. the atomic number minus the number of core (non-valence) electrons, and the variable r_{1B} is the distance of the $2s$ electron from the center of the core (from the atomic nucleus). The core integrals with different orbitals ϕ_r and ϕ_s, on the same atom (A = B; one-center integrals) or on different atoms are taken as being proportional to the overlap integral of the relevant orbitals:

$$H_{rAsB}^{core} = \beta_{AB}\langle \phi_r(1)|\phi_s(1)\rangle, \quad r \neq s \tag{6.12}$$

The overlap integral here is calculated from the basis functions, although (as for the PPP method, above) the overlap matrix is taken as a unit matrix as far as the matrix Roothaan–Hall equations are concerned. The proportionality constant β_{AB} is taken as the arithmetic mean of parameters for atoms A and B, these parameters being those that give the best fit of CNDO MO coefficients to those of minimal-basis-set *ab initio* calculations. Since different AOs on the same atom are orthogonal, when A=B these integrals are zero. Note that calculating β_{AB} from a best-fit to minimal-basis-set *ab initio* calculations means that CNDO parameterization is not purely empirical, but rather, to some extent attempts to match (low-level) *ab initio* results. This is a weakness of CNDO and a potential weakness of its successors INDO and NDDO (below). As repeatedly emphasized by Dewar, this deficiency was avoided in his methods (section 6.2.5.1) by consistently parameterizing to match *experiment*.

As with the PPP method, the two-electron repulsion integrals are evaluated by applying the ZDO approximation to all different orbitals r and s (Eq. (6.6)). Thus the two-electron integrals reduce to $(rs|tu) = \delta_{rs}\delta_{tu}(rr|tt)$ (Eq. (6.8)), i.e. only one- and two-center two-electron integrals are considered. All one-center integrals on the same atom A are given the same value, γ_{AA}, and all two-center integrals between atoms A and B are given the same value, γ_{AB}. These integrals are calculated from valence *s* Slater functions on A and B.

CNDO/2 differs from CNDO/1 in two modifications to the $H^{core}_{r_A r_A}$ matrix elements (Eq. (7.0)): (1) to account better for both ionization energy and electron affinity, U_{rr} is evaluated not just from ionization energy but as a kind of average of ionization energy and electron affinity, and (2) the integrals in the V_{AB} term are calculated from the two-electron integrals γ_{AB}, as $V_{AB} = -C_B \gamma_{AB}$. This latter evaluation amounts to neglecting so-called *penetration integrals*; these integrals make nonbonded atoms attract one another, and causes bond lengths to be too short and bond energies to be too large [14–17]. CNDO energies are valence electron electronic energies E_{SE}, or electronic energies plus core–core repulsions, E^{total}_{SE}, if V_{CC} is added (Eq. (6.2)). CNDO is not much used nowadays, having evolved into the less approximate semiempirical methods INDO and NDDO (below).

6.2.4 The intermediate neglect of differential overlap (INDO) method

INDO [18] goes beyond CNDO by curtailing the application of the ZDO approximation. Instead of applying it to all different ($r \neq s$) atomic orbitals in the two-electron integrals (Eq. (6.6)), as in the PPP and CNDO methods, in INDO ZDO is not applied to those one-center two-electron integrals, $(rs|tu)$, with ϕ_r, ϕ_s, ϕ_t, and ϕ_u all on the same atom; obviously, these repulsion integrals should be the most important. Although more accurate than CNDO, INDO is nowadays used mostly only for calculating UV spectra, in specially parameterized versions called INDO/S and ZINDO/S [19].

6.2.5 The neglect of diatomic differential overlap (NDDO) method

NDDO [20] goes beyond INDO in that the ZDO approximation (section 6.2.1, point (3)) is not applied to orbitals on the same atom, i.e. ZDO is used only for atomic orbitals on different atoms. NDDO is the basis of the currently popular semiempirical methods developed by Dewar and coworkers: modified NDDO (MNDO), Austin method 1 (AM1) and parametric method (PM3).

6.2.5.1 NDDO-based methods from the Dewar group: MNDO, AM1, PM3 and SAM1 – preliminaries

SCF-type (see section 6.1) SE theories are based to a large extent on the approximate MO theory (see the book of this title [10]) developed by Pople and coworkers. The Pople school, however, went on to concentrate on the development of *ab initio* methods, and indeed it is for his contributions to these, which are largely encapsulated in the Gaussian series of programs [21], that Pople was awarded the 1998 Nobel Prize in chemistry [22] (shared with Walter Kohn, a pioneer in density functional theory; see chapter 7). In contrast, Dewar pursued the SE approach almost exclusively [23], and continued till the end of his career to stoutly maintain that at least as far as molecules of real chemical interest go his SE methods were superior to *ab initio* ones. ("There is clearly little point in using a procedure that requires thousands of times more computing time than ours do if it is no better than ours, let alone one that is inferior.") [24]. The rivalry between the Dewar school and the adherents of the *ab initio* approach began relatively early in the development of Dewar methods (see, e.g. [25–27]), intensified to actual polemic [28], and is passionately described from an unabashedly partisan viewpoint in Dewar's autobiography [23]. The *ab initio* vs. Dewar SE controversy was largely rooted in a difference of viewpoints and in a focus by Dewar on the inability of *ab initio* calculations to give reasonably accurate absolute molecular energies (an absolute molecular energy is the energy needed to dissociate a molecule into its nuclei and electrons, infinitely separated and at rest). In the absence of error cancellation, errors in absolute energies lead to errors in activation and reaction energies, and the errors in absolute energies were, ca. 1970, commonly in the region of a thousand kilojoules per mole. Cancellation (actually not as untrustworthy as Dewar thought – section 5.5.2.1) could not, he held, be relied on to provide chemically useful relative energies (reaction and activation energies), say up to some tens of kilojoules per mole. The exchange with Halgren, Kleir and Lipscomb nicely illustrates the viewpoint difference [27]: one side held that even when inaccurate, *ab initio* calculations can teach us something fundamental, while SE calculations, no matter how good, do not contribute to fundamental theory. Dewar focussed on the study of reactions of "real" chemical interest. Interestingly, the high-accuracy *"ab initio"* methods that in recent years have achieved chemical accuracy (section 5.5.2.2b), now considered to be about $10 \, kJ \, mol^{-1}$, employ some empirical parameters, a fact that would have amused Dewar (section 6.1, footnote 1).

In contrast to the viewpoint of the *ab initio* school, Dewar regarded the SE method not merely as an approximation to *ab initio* calculations, but rather as an approach that, carefully parameterized, could give results far superior to those from *ab initio* calculations, at least for the foreseeable future: "The situation [ca. 1992] could be changed only by a huge increase in the speed of computers, larger than anything likely to be attained before the end of the century, or by the development of some fundamentally better *ab initio* approach." [29]. The conscious decision to achieve experimental accuracy rather than merely to approximate *ab initio* results (note the remarks in connection with Eq. (6.12)) was clearly stated several times [26,28,30] in the course of the development of these SE methods: "We set out to parametrize [semiempirical methods] in an entirely different manner, to reproduce the results of experiment rather than those of dubious *ab initio* calculations." [30].

The first (1967) of the Dewar-type methods was PNDDO [31] (partial NDDO). Because further development of the NDDO approach turned out to be "unexpectedly formidable" [30], Dewar's group temporarily turned to INDO, creating MINDO/1 [32] (modified INDO, version 1). The third version of this method, MINDO/3, was said [30] "[to have] so far survived every test without serious failure," and it became the first widely-used Dewar-type method, but keeping their promise to return to NDDO the group soon came up with MNDO. MINDO/3 was made essentially obsolete by MNDO, except perhaps for the study of carbocations (Clark has summarized the strengths and weaknesses of MINDO/3, and the early work on MNDO [33]). MNDO and its descendants AM1 and PM3 are discussed below. A modification of AM1, SAM1, is briefly mentioned; it has not (yet?) gained much popularity.

6.2.5.2 Heats of formation from SE electronic energies

Of the several experimental parameters that the Dewar methods are designed to reproduce, probably the two most important are geometry and heat of formation. For heat of formation the procedure encoded in the methods is the following [34]. As with *ab initio* calculations, SCF-type SE calculations initially give electronic energies E_{SE}; these are calculated using Eq. (6.2). Inclusion of the core–core repulsion V_{CC}, which is necessary for geometry optimization, gives the total semiempirical energy E_{SE}^{total}, normally expressed in atomic units (hartrees), as in an *ab initio* calculation (e.g. section 5.2.3.6e). This energy E_{SE}^{total}, the total internal energy of the molecule except for zero point vibrational energy, is used to calculate the heat of formation (enthalpy of formation) of the molecule. Figure 6.1 will help to make it clear how this is done. The quantities in Fig. 6.1 are

(1) $\Delta H_{f298}^{\ominus}(M)$, the experimental 298 K heat of formation of the molecule M, i.e. the heat energy needed to make M from its elements. This is the quantity we want.

(2) The atomization energy of M, which is the energy of the atoms minus the energy of M. The energy of the atoms is $F\Sigma E_{SE}(A_i)$; the conversion factor F converts $E_{SE}(A_i)$, the energy per atom in hartrees, into the same units, kJ mol^{-1} or kcal mol^{-1}, as is used for the experimental heats of formation of the atoms; F is 2625.5 kJ mol^{-1} per hartree atom^{-1} (or molecule^{-1}). The energy of the molecule M is $FE_{SE}^{total}(M)$, the optimized geometry being used. The same SE method is used to calculate atomic and molecular energies, both of which are negative quantities, the energy of the species relative to electrons and one or more atomic cores infinitely separated. $E_{SE}(A_i)$ is

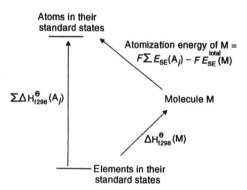

Atoms in their
standard states

Atomization energy of M =
$F\Sigma E_{SE}(A_i) - F E_{SE}^{total}(M)$

$\Sigma \Delta H_{f298}^{\ominus}(A_i)$

Molecule M

$\Delta H_{f298}^{\ominus}(M)$

Elements in their
standard states

Figure 6.1. The principle behind the SE calculation of heat of formation (enthalpy of formation). The molecule is (conceptually) atomized at 298 K; the elements in their standard states are also used to make these atoms, and to make the molecule M. The heat of formation of M at 298 K follows (with some approximations) from equating the energy needed to generate the atoms via M to that needed to make them directly from the elements.

purely electronic, since an atom has no core–core repulsion (i.e. it has no atoms to separate), while the molecular energy $E_{SE}^{total}(M)$ includes core–core repulsion.

(3) $\sum \Delta H_{f'298}^{\ominus}(A_i)$, the sum, over all the atoms A of M, of the experimental 298 K heats of formation of these atoms.

Equating the two paths from the elements in their standard states at 298 K to atoms we get

$$\Delta H_{f298}^{\ominus}(M) = \sum \Delta H_{f298}^{\ominus}(A_i) - F \sum E_{SE}(A_i) + F E_{SE}^{total}(M) \qquad (6.13)$$

Thus the desired quantity, the heat of formation of the molecule, can be calculated from the experimental heats of formation of the atoms and the semiempirical energies of the atoms and the molecule. The calculation of Eq. (6.13) is automatically done by the program using stored values for atomic heats of formation and semiempirical atomic energies, and the "freshly calculated" calculated molecular energy, and one normally never sees $E_{SE}^{total}(M)$. These calculations are for the gas phase, and if one wants the heat of formation of a liquid or a solid, then the experimental heat of vaporization or sublimation must be taken into account. Note that this procedure is conceptually almost the same as the atomization method for *ab initio* calculation of heats of formation (section 5.5.2.2c). However, the purpose here is to obtain the heat of formation at room temperature (298 K) from the molecular "total semiempirical energy," the electronic energy plus core–core repulsion; in the *ab initio* atomization method the 0 K heat of formation is calculated with the aid of the molecular energy including ZPE (the 0 K heat of formation can be corrected to 298 K – see section 5.5.2.2c). The SE procedure involves some approximations. The ZPE of the molecule is not used, and the increase in thermal energy from 0 to 298 K is not calculated. Thus if $E_{SE}^{total}(M)$ were fully analogous to the *ab initio* 0 K electronic energy plus internuclear repulsion then the calculated atomization energy would be at 0 K, not 298 K, and furthermore would employ a frozen-nucleus approximation to the true 0 K energy. The good news is that $E_{SE}^{total}(M)$ is parameterized (below) to reproduce $\Delta H_{f'298}^{\ominus}(M)$; to the extent that this parameterization succeeds the neglect of ZPE and of the 0–298 K increase in thermal energy are overcome (and electron correlation is also implicitly taken into account). The key to obtaining reasonably accurate heats of formation from these methods is thus their parameterization to give the values of $E_{SE}(A_i)$ and $E_{SE}^{total}(M)$ used in Eq. (6.13). This parameterization, which is designed to also give reasonable geometries and dipole moments, is discussed below.

6.2.5.3 MNDO

MNDO [33], a modified NDDO (section 6.2.5) method, was reported in 1977 [35]. MNDO is conveniently explained by reference to CNDO (section 6.2.3). MNDO is a general geometry method with a minimal valence basis set of Slater-type orbitals, the one-center core integrals. The Fock matrix elements are calculated using Eq. (6.1). We discuss the core and two-electron integrals in the same order as for CNDO.

The core integrals H_{rArA}^{core}, with the same orbital ϕ_r twice on the same atom A are calculated using Eq. (6.10). Unlike the case in CNDO, where U_{rr} is found from ionization energies (CNDO/1) or ionization energies and electron affinities (CNDO/2), in

MNDO U_{rr} is one of the parameters to be adjusted. The integrals in the summation term V_{AB} are evaluated similarly to the CNDO/2 method from a two-electron integral (see below) involving ϕ_{rA} and the valence s orbital on atom B:

$$\langle \phi_{rA}(1)|\hat{V}_B|\phi_{rA}(1)\rangle = -C_B(\phi_r\phi_r|s_Bs_B) \tag{6.14}$$

The core integrals H^{core}_{rAsA} with different orbitals ϕ_r and ϕ_s, on the same atom A are not simply taken as being proportional to the overlap integral, as in CNDO (Eq. (6.12)), but rather are also (like the case of both orbitals on the same atom) evaluated from Eq. (6.10), which in this case becomes

$$H^{core}_{rAsA} = \langle \phi_{rA}(1)|\hat{T} + \hat{V}_A|\phi_{sA}(1)\rangle + \langle \phi_{rA}(1)|\hat{V}_B|\phi_{sA}(1)\rangle$$
$$+ \langle \phi_{rA}(1)|\hat{V}_C|\phi_{sA}(1)\rangle + \cdots$$
$$= U_{rs} + \sum_{B \neq A} \langle \phi_{rA}(1)|\hat{V}_B|\phi_{sA}(1)\rangle \tag{6.15}$$

The first term is zero from symmetry [36]. Each integral of the summation term is again evaluated, as in CNDO/2, from a two-electron integral:

$$\langle \phi_{rA}(1)|\hat{V}_B|\phi_{sA}(1)\rangle = -C_B(\phi_{rA}\phi_{sA}|s_Bs_B) \tag{6.16}$$

The core integrals H^{core}_{rAsB} with different orbitals ϕ_r and ϕ_s, on different atoms A and B are taken, as in CNDO (cf. Eq. (6.12)), to be proportional to the overlap integral between ϕ_r and ϕ_s, where again the proportionality constant is the arithmetic mean of parameters for atoms A and B:

$$H^{core}_{rAsB} = \tfrac{1}{2}(\beta_{rA} + \beta_{sB})\langle \phi_r(1)|\phi_s(1)\rangle, \quad r \neq s \tag{6.17}$$

The overlap integral is calculated from the basis functions although the overlap matrix is taken as a unit matrix as far as the Roothaan–Hall equations go (section 6.2.2). These core integrals are sometimes called core resonance integrals.

The two-electron integrals are evaluated applying ZDO (section 6.2.1) within the framework of the NDDO approximation (section 6.2.5). As with the PPP (section 6.2.2) and CNDO (section 6.2.3) methods, this makes all two-electron integrals become $(rs|tu) = \delta_{rs}\delta_{tu}(rr|tt)$, i.e only one- and two-center two-electron integrals are nonzero. The one-center integrals are evaluated from valence-state ionization energies. The two-center integrals are evaluated from the one-center integrals and the separation of the nuclei by an involved procedure in which the integrals are expanded as sums of multipole–multipole interactions [35a,37] that make the two-center integrals show correct limiting behavior at zero and infinite separation.

As in CNDO the penetration integrals are neglected (section 6.2.3, CNDO/2). A consequence of this is that the core–core repulsions (V_{cc} in Eq. (6.2)) cannot be realistically calculated simply as the sum of pairs of classical electrostatic interactions between point charges centered on the nuclei. Instead, Dewar and coworkers chose [35a] the expression

$$V_{CC} = \sum_{B>A}\sum_{A}[C_AC_B(s_As_B|s_Bs_B) + f(R_{AB})] \tag{6.18}$$

where C_A and C_B are the core charges of atoms A and B and s_A and s_B are the valence s orbitals on A and B (the two-electron integral in Eq. (6.18) is actually approximately proportional to $1/R_{AB}$, so there is some connection with the simple electrostatic model). The $f(R_{AB})$ term is a correction increment to make the result come out better; it depends on the core charges and the valence s functions on A and B, their separation R, and empirical parameters α_A and α_B:

$$f(R_{AB}) = C_A C_B (s_A s_A | s_B s_B)(e^{-\alpha_A R_{AB}} + e^{-\alpha_B R_{AB}}) \qquad (6.19)$$

The above mathematical treatment constitutes the creation of the *form* of the semiempirical equations; to actually use these equations, they must be parameterized using experimental data. This is analogous to the situation in molecular mechanics (chapter 3), where a force field, defined by the form of the functions used (e.g. a quadratic function of the amount by which a bond is stretched, for the bond-stretch energy term) is constructed, and must then be parameterized by inserting specific quantities for the parameters (e.g. values for the stretching force constants of various bonds). For each kind of atom A (a maximum of) six parameters is needed:

(1) The kinetic-energy-containing term U_{rr} of Eq. (6.10) (as explained above, this CNDO equation is also used in MNDO to evaluate H^{core}_{rArA}) where ϕ_{rA} is a valence s AO.
(2) The term U_{rr} of Eq (6.10) where ϕ_{rA} is a valence p AO.
(3) The parameter ζ in the exponent of the Slater function (e.g. section 5.3.2, Fig. 5.12) for the various valence AOs (MNDO uses the same ζ for the s and p AOs).
(4) The parameter β (Eq. (6.17)) for a valence s AO.
(5) The parameter β for a valence p AO.
(6) the parameter α in the correction increment ($f(R_{AB})$, (Eq. (6.19)) to the core–core repulsion (Eq. (6.18)).

Some atoms have five parameters because for them MNDO takes β to be the same for s and p orbitals, and hydrogen has four parameters because MNDO does not assign it p orbitals.

We want the parameters that will give the best results, for a wide range of molecules. What we mean by "results" depends on the molecular characteristics of most interest to us. MNDO (and its siblings AM1 and PM3, below) was parameterized [35a] to reproduce heat of formation, geometry, dipole moment, and the first vertical ionization energy (from Koopmans' theorem; section 5.5.5). To parameterize MNDO a training set of molecules (a "molecular basis set" is Dewar's term – no connection with a basis set of functions used to construct MOs) of small, common molecules (e.g. methane, benzene, dinitrogen, water, methanol; 34 molecules were used for the C, H, O, N set) was chosen and the six parameters above (U_{rr} etc.) were adjusted in an attempt to give the best values of the four molecular characteristics (heat of formation, geometry, dipole moment, and ionization energy). Specifically, the objective was to minimize Y, the sum of the weighted squares of the deviations from experiment of the four molecular characteristics:

$$Y = \sum_{i=1}^{N} W_i [Y_i(calc) - Y_i(\exp)]^2 \tag{6.20}$$

where N is the number of molecules in the training set, and W_i is a weighting factor chosen to determine the relative importance of each characteristic Y_i. The actual process of assigning values to the parameters is formally analogous to the problem of geometry optimization (section 2.4). In geometry optimization we want the set of atomic coordinates that correspond to a minimum (sometimes to a transition state) on a potential energy hypersurface. In parameterizing a SE method we want the set of parameters that correspond to the minimum overall calculated deviation of the chosen characteristics from their experimental values – the parameters that give the minimum Y, above. Details of the parameterization process for MNDO have been given by Dewar [35a] and by Stewart [38].

The results of MNDO calculations on 138 compounds limited to the elements C, H, O, N were reported by Dewar and Thiel [35b]. The absolute mean errors were: in heat of formation, 26 kJ mol^{-1} for all 138 compounds; in geometry, 0.014 Å for bond lengths for 228 bonds, $2°$ for angles at C for acyclics (less for cyclic molecules); in dipole moment, 0.30 D for 57 compounds; in ionization energy, 0.48 eV for 51 compounds. To put the errors in perspective, typical values of these quantities are, respectively, roughly $-600 - 600 \text{ kJ mol}^{-1}$, $1.0–1.5 \text{ Å}$, $0–3 \text{ D}$, and $10–15 \text{ eV}$. Although MNDO can reproduce these and other properties of a wide variety of molecules [33,39], it is little used nowadays, having been largely superseded by AM1 and, to a somewhat lesser extent, PM3 (below). More results from MNDO calculations are given in section 6.3.

6.2.5.4   AM1

Austin method 1 (AM1, developed at the University of Texas at Austin [40]) was introduced by Dewar, Zoebisch, Healy, and Stewart in 1985 [41]. AM1 is an improved version of MNDO in which the main change is that the core–core repulsions (Eq. (6.18)) were modified to overcome the tendency of MNDO to overestimate repulsions between atoms separated by about their van der Waals distances (the other change is that the parameter ζ in the exponent of the Slater function – see parameter 3 in the listing of the six parameters above – need not be the same for s and p AOs on the same atom). The core–core repulsions were modified by introducing attractive and repulsive Gaussian functions centered at internuclear points [42], and the method was then reparameterized. The great difficulties experienced in the parameterization of AM1 and its predecessors are emphasized by Dewar and coworkers in many places, e.g.: "All our work has therefore been based on a very laborious purely empirical technique..." for the MINDO methods [30]; parameterization is a "purely empirical affair" and "needs infinite patience and enormous amounts of computer time," for AM1 [41]. In his autobiography Dewar says [43] "This success [of these methods] is no accident and it has not been obtained easily," and summarizes the problems with parameterizing these methods: (1) the parametric functions are of unknown form, (2) the choice of molecules for the training set affects the parameters to some extent, (3) the parameters are not unique, there is no way to tell if the set of values found is the best one, and there is no systematic way to find alternative sets, (4) deciding if a set of parameters is acceptable is a matter of judgment. Dewar *et al.* chose to call their modified MNDO method AM1, rather than MNDO/2, because they felt that their methods were being confused (presumably because of the "INDO" and "NDO" components of the appellations) with "grossly inaccurate" [41] ZDO SCF SE methods like CNDO and INDO.

Dewar, Zoebisch, Healy, and Stewart reported [41] that AM1 calculations on compounds containing nitrogen and/or oxygen gave an absolute mean error in heat of formation of $25 \, \text{kJ mol}^{-1}$ for 80 compounds, "generally satisfactory" agreement with experiment for the geometries of 138 molecules, absolute mean error in dipole moment of 0.26 D for 46 compounds, and absolute mean error in ionization energy of 0.40 eV for 29 compounds. These results are slightly better than those for MNDO, but the real advantages of AM1 over MNDO were said [41] to lie in its better treatment of hydrogen bonding, crowded molecules, four-membered rings, and activation energies. AM1 is the most widely-used SE method nowadays. MNDO and AM1 (and PM3, below) are compared further in section 6.3.

6.2.5.5 PM3

PM3 is a variation of AM1, differing mainly in how the parameterization is done. When PM3 was first published [38], two parameterizations of MNDO-type methods, MNDO and AM1, had been carried out, and PM3 was at first called MNDO-PM3, meaning MNDO parametric method 3. Three papers [38,44,45] define the PM3 method. The Dewar school's approach to parameterization was a painstaking one (section 6.2.5.4), making liberal use of chemical intuition. The developer of PM3, Stewart, employed a faster, more algorithmic approach, "several orders of magnitude faster than those previously employed" [38]. Although it is based on AM1, PM3 did not enjoy Dewar's blessing. The reasons for this appear to be at least twofold: (1) Dewar evidently felt (on the basis of very early results [46]) that PM3 represented at best an only marginal improvement over AM1, and that a new SE method should make previous ones essentially obsolete, as MNDO made MINDO/3 obsolete, and AM1 largely replaced MNDO. Stewart defended his approach [47] with the rejoinder, *inter alia*, that if PM3 was only a only marginal improvement over AM1, then AM1 was only a marginal improvement over MNDO. (2) Dewar objected strongly to any proliferation of computational chemistry methods, whether it be in the realm of *ab initio* basis sets [48] or of SE methods [46,48].

For compounds containing H, C, N, O, F, Cl, Br, and I, Holder *et al.* reported [49] that PM3 calculations gave an absolute mean error in heat of formation of $22 \, \text{kJ mol}^{-1}$ for 408 compounds (cf. $27 \, \text{kJ mol}^{-1}$ for AM1), and Dewar *et al.* reported an absolute mean error in bond lengths of 0.022 Å for 344 bonds (cf. 0.027 for AM1), 2.8° for 146 angles (cf. 2.3° for AM1) [50], and 0.40 D for 196 compounds (cf. 0.35 D for AM1) [50]. PM3 is the second most widely-used semiempirical method nowadays (after AM1). MNDO, AM1, and PM3 are compared further in section 6.3.

6.2.5.6   SAM1

Semi *ab initio* method number 1 (SAM1) was the last SE method to be reported (1993, [50]) by Dewar's group. SAM1 is essentially a modification of AM1 in which the two-electron integrals are calculated *ab initio* using contracted Gaussians (an STO-3G basis set) as in standard *ab initio* calculations (section 5.3.2). This is in contrast to AM1, where the two-center two-electron integrals are calculated from the one-center two-electron integrals, which are estimated spectroscopically. As Holder and Evleth point out in a brief but lucid outline of the basis of AM1 and SAM1 [51], a key distinguishing feature of each SE method is how it calculates the two-electron repulsion integrals. Since the NDDO approximation discards all the three- and four-center two-electron integrals, the number of two-electron integrals to be calculated is greatly reduced. This, and the limitation to valence electrons, makes SAM1 only about twice as slow as AM1 [51].

One of the main reasons for developing SAM1 was to improve the treatment of hydrogen bonding (this was also a primary reason for developing AM1 from MNDO; evidently success there was only limited). SAM1 is indeed an improvement over AM1 in this respect, and "appears to be the first semiempirical parameterization to handle a wide variety of [hydrogen bonded] systems correctly"; in fact, it was said that "the results from SAM1 for virtually every system has improved over AM1 and PM3, fulfilling the criteria for SAM1 to be a reasonable successor to AM1 and PM3 for general purpose semiempirical calculations" [51]. An extensive list of experimental heats of formation compared with those calculated by SAM1, AM1, and PM3 has been published [49]. Actually, despite its apparent generally significant superiority over AM1, there have been relatively few publications using SAM1. This is probably because the program at present is available only in the commercial SE package AMPAC [52], which seems to be used mainly by chemists working in industry who cannot always publish freely, and because the parameterization of SAM1 has not yet been fully disclosed in the open literature (researchers are perhaps uncomfortable about using a black box, or even a gray one).

6.2.5.7   Inclusion of d orbitals: MNDO/d and PM3t; explicit electron correlation: MNDOC

The original (and most widely-used) versions of MNDO, AM1, and PM3 do not use *d* orbitals. Hence they might be expected to show reduced accuracy for elements in the "second-row" (computational chemists' lingo) and beyond like, P, S, Cl, Br, and I, and cannot be used for transition metals. Actually, because of appropriate parameterization AM1 and PM3 are able to treat monovalent Cl, Br and I as standard elements (C, H, O, N, F), and they handle divalent S reasonably well. To make them able to work better with elements in the second row and beyond, and/or to handle transition metals (note that in Zn, Cd, and Hg the *d* electrons are not normally involved in bonding), *d* orbitals have been incorporated into some SE methods. MNDO/d [53] uses *d* orbitals for some post-first row nonmetals and has been parameterized for several transition metals. Some versions of SPARTAN [54] have PM3 (tm), PM3 with *d* orbitals for many transition metals. PM3 (tm) geometries have been compared with experimental and *ab initio* ones;

the results were said to range from excellent for dihydrogen complexes to very poor for H-BR$_2$ complexes [55].

The parameterization of SE methods is supposed to simulate, amongst other effects, electron correlation, so it might seem pointless to introduce electron correlation explicitly, by the Møller-Plesset method or by configuration interaction (section 5.4). However, the parameterization of these SE methods is done using ordinary stable molecules. Surprisingly, MNDO, AM1, PM3 (and presumably SAM1) also reproduce reasonably well the energies and geometries of reactive intermediates like carbocations, carbanions and carbenes. However, the parameterization is unlikely to be as reliable for transition states as for ground states, so activation energies are expected to be less accurate than reaction energies. A method that explicitly calculates electron correlation might improve calculated activation energies. In the SE field the standard program for this is MNDOC (MNDO with correlation), developed by Thiel and coworkers [56]. MNDOC is said to perform better than MNDO (and presumably better than the other MNDO-related methods) in calculating activation energies and electronic excitation energies [39]. For more on the accuracy of MNDO and MNDOC see sections 6.3.1 (geometries) and 6.3.2.2 (activation energies).

6.3 APPLICATIONS OF SE METHODS

A good, brief overview of the performance of MNDO, AM1, and PM3 is given by Levine [57]. Hehre has compiled an extremely useful book comparing AM1 with MM (chapter 3), *ab initio* (chapter 5) and DFT (chapter 7) methods for calculating geometries and other properties [58], and an extensive collection of AM1 and PM3 geometries is to be found in Stewart's second PM3 paper [44].

6.3.1 Geometries

Many of the general remarks on molecular geometries in section 5.5.1, preceding the discussion of results of specifically *ab initio* calculations, apply also to SE calculations. Geometry optimizations of large biomolecules like proteins and nucleic acids, which a few years ago were limited to MM, can now be done routinely [59] with SE methods on inexpensive personal computers with the program MOZYME [60], which uses localized orbitals to solve the SCF equations [61].

Let us compare AM1, PM3, MP2(FC)/6-31G* (chapter 5) and experimental geometries; the MP2(FC)/6-31G* method is the highest-level *ab initio* method routinely used. Figure 6.2 gives bond lengths and angles calculated by these three methods and experimental bond lengths and angles, for the same 20 molecules as in Fig. 5.23. The geometries shown in Fig. 6.2 are analyzed in Table 6.1, and Table 6.2 provides information on dihedral angles for the same eight molecules as in Table 5.8. Fig. 6.2 corresponds exactly to Fig. 5.23, Table 6.1 to Table 5.7, and Table 6.2 to Table 5.8.

This survey suggests that: AM1 and PM3 give quite good geometries (although dihedral angles, below, show quite significant errors): bond lengths are mostly within 0.02 Å of experimental (although the AM1 C-S bonds are about 0.06 Å too short), and angles are usually within 3° of experimental (the worst case is the AM1 HOF angle, which is 7.1° too big).

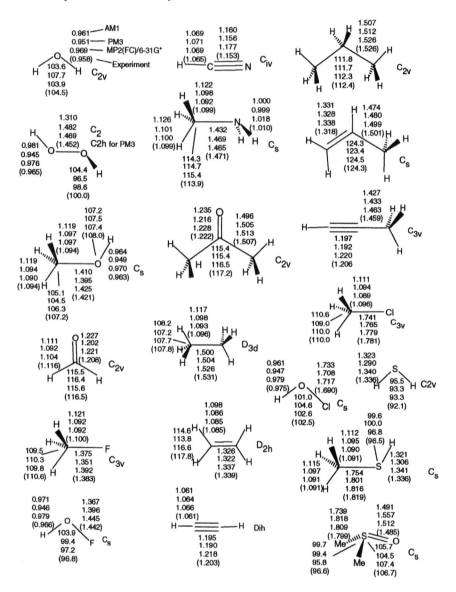

Figure 6.2. A comparison of some AM1, PM3, MP2(FC)/6-31G* and experimental geometries. Calculations are by the author and experimental geometries are from Ref. [63]. Note that all CH bonds are ca. 1 Å, all other bonds range from ca. 1.2–1.8 Å, and all bond angles (except for linear molecules) are ca. 90°–120°.

Of AM1 and PM3, neither has a clear advantage over the other in predicting geometry, although PM3 C–H and C–X (X=O, N, F, Cl, S) bond lengths appear to be more accurate than AM1. MP2 geometries are considerably better than AM1 and PM3, but HF/3-21G and HF/6-31G* geometries (Fig. 5.23 and Table 5.7) are only moderately better.

Table 6.1. Errors in AM1, PM3, and MP2(FC)/6-31G* bond lengths and angles, from Fig. 6.2

Bond length errors, $r-r_{exp}$ (Å)				Bond angle errors, $a - a_{exp}$ Angles
C–H	O–H, N–H, S–H	C–C	C–O, N, F, Cl, S	
MeOH 0.025/0.000/−0.004 0.025/0.003/0.003	H$_2$O 0.003/−0.007/0.011	Me$_2$CO −0.011/−0.002/0.006	MeOH 0.001/−0.014/0.007	H$_2$O(HOH) −0.9/3.2/−0.6
HCHO −0.005/−0.024/−0.012	H$_2$O$_2$ 0.016/−0.020/0.011	CH$_3$CH$_3$ −0.031/−0.027/−0.005	HCHO 0.019/−0.006/0.013	H$_2$O$_2$ (HOO) 4.4/−3.5/−1.4
MeF 0.021/−0.008/−0.008	MeOH 0.001/−0.014/0.007	CH$_2$CH$_2$ −0.013/−0.017/−0.002	MeF −0.008/−0.032/0.009	MeOH (HCO) −2.2/−2.7/−0.9 (COH) −0.2/−0.5/−0.6
HCN 0.004/0.006/0.004	HOF 0.005/−0.020/0.013	HCCH −0.008/−0.013/0.015	HCN 0.007/0.003/0.024	HCHO(HCH) −1.0/−0.1/−0.9
MeNH$_2$ 0.021/0.002/0.001 0.023/−0.001/−0.007	MeNH$_2$ −0.010/−0.011/0.008	CH$_3$CH$_2$CH$_3$ −0.019/−0.014/0.000	MeNH$_2$ −0.039/−0.002/−0.006	MeF (HCH) −1.1/−0.3/−0.8
CH$_3$CH$_3$ 0.021/0.002/−0.003	HOCl −0.014/−0.028/0.004	CH$_2$CHCH$_3$ −0.027/−0.021/−0.002 0.013/0.010/0.020	Me$_2$CO 0.013/−0.006/0.006	HOF (HOF) 7.1/2.6/0.4
CH$_2$CH$_2$ 0.013/0.001/0.000	H$_2$S −0.013/−0.046/0.004	HCCCH$_3$ −0.032/−0.026/0.004 −0.009/−0.014/0.014	MeCl −0.040/−0.016/−0.002	MeNH$_2$ (HCN) 0.4/0.8/1.5

Table 6.1. Continued

	Bond length errors, $r - r_{exp}$ (Å)				Bond angle errors, $a - a_{exp}$ Angles
C–H	O–H, N–H, S–H	C–C	C–O, N, F, Cl, S		
CHCH 0.000/0.003/0.005	MeSH −0.015/−0.030/0.005		MeSH −0.065/−0.018/−0.003		Me₂CO (CCC) −1.8/−1.8/−0.8
MeCl 0.015/−0.002/−0.007			Me₂SO −0.060/0.019/0.010		CH₃CH₃ (HCH) 0.4/−0.6/−0.1
MeSH 0.024/0.006/0.000					CH₂CH₂ (HCH) −3.2/−4.0/−1.2
0.021/0.004/−0.001					
					CH₃CH₂CH₃(CCC) −0.6/−0.7/−0.1
					CH₂CHCH₃ (CCC) 0.0/−0.9/0.2
					MeCl (HCH) 0.6/−1.0/0.0
					H₂S (HSH) 3.4/1.2/1.2
					MeSH (CSH) 3.1/3.5/0.3
12+, 1−, none 0	4+, 4−, none 0	1 + /8−, none 0	4+, 5−, none 0		8+, 9−, one 0
8+, 4−, one 0	0+, 8−, none 0	1+, 8−, none 0	2+, 7−, none 0		6+, 12−, one 0
4+, 7−, two 0	8+, 0−, none 0	5+, 3−, one 0	6+, 3−, none 0		6+, 11−, none 0
Mean of 13:	Mean of 8:	Mean of 9:	Mean of 9:		Mean of 18:
0.017/0.005/0.004	0.010/0.022/0.008	0.018/0.016/0.008	0.028/0.013/0.009		1.9/1.8/0.7

Errors are given as AM1/PM3/MP2. In some cases (e.g. MeOH) errors for two bonds are given, on one line and on the line below. A minus sign means that the calculated value is less than the experimental. The numbers of positive, negative, and zero deviations from experiment are summarized at the bottom of each column. The averages at the bottom of each column are arithmetic means of the absolute values of the errors.

Table 6.2. AM1, PM3, MP2(FC)/6-31G* and experimental dihedral angles (degrees)

| Molecule | Dihedral angles | | | | Errors |
	AM1	PM3	MP2/6-31G*	Exp.	
HOOH	128	180	121.3	119.1[a]	9/61(*sic*)/2.2
FOOF	89	90	85.8	87.5[b]	1.5/2.5/−1.7
FCH$_2$CH$_2$F	81	57	69.0	73[b]	8/−16/−4
(FCCF)					
FCH$_2$CH$_2$OH					
(FCCO)	65	66	60.1	64.0[c]	1/2/−3.9
(HOCC)	58	62	54.1	54.6[c]	3/7/−0.5
ClCH$_2$CH$_2$OH					
(ClCCO)	74	65	65.0	63.2[b]	11/2/1.8
(HOCC)	62	59	64.3	58.4[b]	4/1/5.9
ClCH$_2$CH$_2$F					
(ClCCF)	79	61	65.9	68[b]	11/−7/−2.1
HSSH	99	93	90.4	90.6[a]	8/2/−0.2
FSSF	89	87	88.9	87.9[b]	1/−1/1.0
					Deviations:
					10+, 0 − /7+, 3 − /4+, 6−
					Mean of 10:
					6/10/2.3;
					Mean of 9,
					omitting 9/61/2.2
					Errors: 5/4.5/1.9

Errors are given in the *Errors* column as AM1/PM3/MP2/6-31G*. A minus sign means that the calculated value is less than the experimental. The numbers of positive and negative deviations from experiment and the average errors (arithmetic means of the absolute values of the errors) are summarized at the bottom of the *Errors* column. Calculations are by the author; references to experimental measurements are given for each measurement. The AM1 and PM3 dihedrals vary by a fraction of a degree depending on the input dihedral. Some molecules have calculated minima at other dihedrals in addition to those given here, e.g. FCH$_2$CH$_2$F at FCCF 180°.

[a]Hehre [63], pp. 151, 152.
[b]M. D. Harmony, V. W. Laurie, R. L. Kuczkowski, R. H. Schwenderman, D. A. Ramsay, F. J. Lovas, W. H. Lafferty, A. G. Makai, "Molecular Structures of Gas-Phase Polyatomic Molecules Determined by Spectroscopic Methods," J. Physical and Chemical Reference Data, 1979, *8*, 619–721.
[c]J. Huang and K. Hedberg, J. Am. Chem. Soc., 1989, *111*, 6909.

AM1 and PM3 C–H bond lengths are almost always (AM1) or tend to be (PM3) longer than experimental, by ca. 0.004–0.025 (AM1) or ca. 0.002 Å (PM3). AM1 O–H bonds tend to be slightly longer (up to 0.016 Å) and PM3 O–H bonds to be somewhat shorter (up to 0.028 Å) than experimental. Both AM1 and PM3 consistently underestimate C–C bond lengths (by about 0.02 Å).

C–X (X=O, N, F, Cl, S) bond lengths appear to be consistently neither over- nor underestimated by AM1, while PM3 tends to underestimate them; as stated above, the PM3 lengths seem to be the more accurate (mean errors 0.013 vs. 0.028 Å for AM1).

Both AM1 and PM3 give quite good bond angles (largest error ca. 4°, except for HOF for which the AM1 error is 7.1°).

AM1 tends to overestimate dihedrals (10+, 0−), while PM3 may do so to a lesser extent (7+, 3−). PM3 breaks down for HOOH (calculated 180°, experimental 119.1°, and does poorly for FCH_2CH_2F (calculated 57°, experimental 73°). Omitting the case of HOOH, the mean dihedral angle errors for AM1 and PM3 are 5° and 4.5°; however, the variation here is from 1° to 11° for AM1 and from −1° to −16° for PM3 (although not wildly out of line with the AM1, PM3 or MP2 calculations, the reported experimental $ClCH_2CH_2OH$ HOCC dihedral of 58.4° is suspect; see section 5.5.1).

The accuracy of AM1 and PM3 then is quite good for bond lengths and angles, but fairly approximate for dihedrals. The largest error (Table 6.1) in bond lengths is 0.065 Å (AM1 for MeSH) and in bond angles 7.1° (AM1 for HOF). The largest error in dihedrals (Table 6.2), omitting the PM3 result for HOOH, is 16° (PM3 for FCH_2CH_2F).

From Fig. 6.2 and Table 6.1, the mean error in 39(13 + 8 + 9 + 9) bond lengths is ca. 0.01–0.03 Å for the AM1 and PM3 methods, with PM3 being somewhat better except for O–H and O–S. The mean error in 18 bond angles is ca. 2° for both AM1 and PM3. From Table 6.2, the mean dihedral angle error for 9 dihedrals for AM1 and PM3 (omitting the case of HOOH, where PM3 simply fails) is ca. 5°; if we include HOOH, the mean errors for AM1 and PM3 are 6° and 10°, respectively.

Schröder and Thiel have compared MNDO (section 6.2.5.3) and MNDOC (section 6.2.5.7) with *ab initio* calculations for the study of the geometries and energies of 47 transition states [62]. AM1 and PM3 calculations should give somewhat better results than MNDO for these systems, since these two methods are essentially improved versions of MNDO. The general impression is that the SE and *ab initio* transition states are qualitatively similar in most cases, with MNDOC geometries being sometimes a bit better. The SE and *ab initio* geometries were in most cases fairly similar, so that as far as geometry goes one would draw the same qualitative conclusions.

Semiempirical and *ab initio* geometries are compared further in Fig. 6.3, which presents results for four reactions, the same as for the *ab initio* calculations summarized in Fig. 5.21. As expected from the results of Fig. 6.2, the SE geometries of the reactants and products (the energy minima) are quite good (taking the MP2/6-31G* results as our standard). The SE transition state geometries, however, are also surprisingly good: with only small differences between the AM1 and PM3 results, in all four cases the SE transition states resemble the *ab initio* ones so closely that qualitative conclusions based on geometry would be the same whether drawn from the AM1 or PM3, or from the MP2/6-31G* calculations. The largest bond length error (if we accept the MP2 geometries as accurate) is about 0.09 Å (for the CH_3NC transition state, 1.897–1.803), and the largest angle error is 9° (for the HNC transition state, 72.8°–63.9°; most of the angle errors are less than 3°).

These results, together with those of Schröder and Thiel [62] indicate that SE geometries are usually quite good, even for transition states. Exceptions might be expected for hypervalent compounds, and for unusual structures like the C_2H_5 cation; for the latter AM1 and PM3 predict the classical CH_3CH_2 structure, but MP2/6-31G* calculations predict this species to have a hydrogen-bridged structure (Fig. 5.17). SE energies are considered in section 6.3.2.

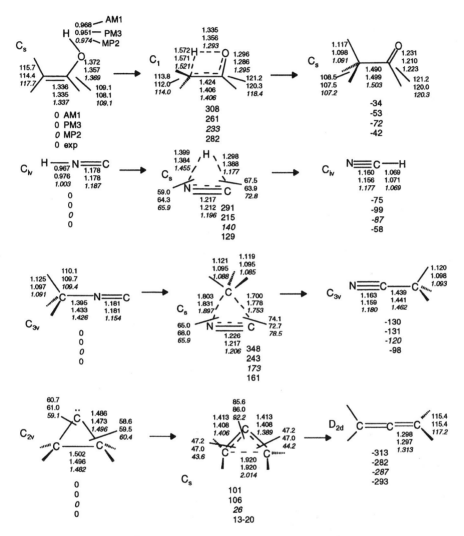

Figure 6.3. Geometries (Å, degrees) and relative energies (kJ mol⁻¹) for four reactions, the same as for the *ab initio* calculations of Fig. 5.21; most Hs are omitted, for clarity (in Fig. 5.21, for the MP2(FC)/6-31G* calculations the raw energies in hartrees, and the ZPEs, are given). The experimental relative energies [67] are somewhat approximate for the ethenol (vinyl alcohol) and cyclopropylidene reactions.

6.3.2 Energies

6.3.2.1 Energies: preliminaries

As with *ab initio* (chapter 5) and MM (chapter 4) calculations, the molecular parameters usually sought from SE calculations are geometries (preceding section) and energies. As explained (section 6.2.5.2), the most widely-used SE methods, AM1 and PM3, give (room temperature, 298 K) heats of formation. This is in distinct contrast to *ab initio* calculations, which give (the negative of) the energy for total dissociation of the molecule into nuclei and electrons, starting from a hypothetical zero-vibrational energy state or from the 0 K state with ZPE included (section 5.5.2.1). *Ab initio* methods can be made to provide heats of formation, by slightly roundabout methods (section 5.5.2.2c). The errors in SE heats of formation might at first strike one as being very large; thus for the common diatomic molecules, which by definition have standard heats of formation of zero, AM1/PM3 give these heats of formation (kJ mol^{-1}): H_2, $-21.7/-56.0$; N_2, $+46.7/+73.5$; O_2 (triplet), $-116/-17.5$; F_2, $-94.0/-90.8$/Cl_2, $-59.2/-48.5$; Br_2, $-22.0/+20.6$. An extensive compilation of AM1 and PM3 heats of formation (which corrects errors in earlier values) [44] gave for 657 normal-valent compounds these average errors for the absolute deviations (AM1/PM3, kJ mol^{-1}): 53/33; for 106 hypervalent compounds 348 (sic)/57. These results are not as bad as they may at first seem if we note that (1) the heats of formation of organic compounds are commonly in the region of $\pm400-800$ kJ mol^{-1}, (2) often we are interested in trends, which are more likely to be qualitatively right than actual numbers are to be quantitatively accurate, and (3) usually chemists are concerned with energy *differences*, i.e. *relative* energies (below). AM1 heats of formation for hypervalent compounds (above and Ref. [47]) appear to be distinctly inferior to those from PM3. Thiel has compared MNDO, AM1, PM3, and MNDO/d heats of formation with those from some *ab initio* methods [64].

The discussion of enthalpy, free energy, and reaction and activation energies in section 5.5.2.1 applies to SE calculations too. Now let's retrace some of the calculations of chapter 5, using AM1 and PM3 rather than *ab initio* methods.

6.3.2.2 Energies: calculating quantities relevant to thermodynamics and kinetics

We are usually interested in *relative* energies. An *ab initio* energy difference (for isomers, or isomeric systems like reactants cf. products) represents a 0 K energy difference, i.e. a 0 K enthalpy difference, whereas a semiempirical (AM1 or PM3) energy difference represents a room temperature enthalpy difference; thus even if the *ab initio* and SE calculations both had negligible errors, they would not be expected to give exactly the same relative energy, unless the 0–298 K enthalpy change on both sides of the equation cancelled. A typical change in heat of formation is shown by methanol; the (*ab initio* calculated) heats of formation of methanol at 0 and 298 K are -195.9 and -207.0 kJ mol^{-1}, respectively (section 5.5.2.2c). This change of 11 kJ mol^{-1} is small compared to the errors in SE and many *ab initio* calculations, so discrepancies between energy changes calculated by the two approaches must be due to factors other than the 0–298 K enthalpy change. The errors in heats of formation cannot be counted on to consistently cancel when we subtract to obtain relative energies, and because of the quite large errors in the heats of formation (53 and 33 kJ mol^{-1} for AM1 and PM3, for a large sample of "normal compounds"; section 6.6.2.1) errors of about 100–60 kJ mol^{-1} should not be uncommon, although much smaller errors are often obtained. Consider the relative energies of (Z)- and (E)-2-butene (Fig. 5.24). The HF/3-21G energy difference, corrected for ZPE (although in this case the ZPE is practically the same for both isomers) is $(Z) - (E) = -155.12709 - (-155.13033)h = 0.00324\,h = 8.5$ kJ mol^{-1}.

AM1 calculations (ZPE is not considered here, since as explained in section 6.2.5.2, this is taken into account in the parameterization) give $(Z) - (E) = -9.24 - (-14.01) = 4.8\,\mathrm{kJ\,mol^{-1}}$. The experimental heats of formation (298 K, gas phase) are $(Z) = -29.7$, $(E) = -47.7\,\mathrm{kJ\,mol^{-1}}$, i.e. $(Z) - (E) = 18.0\,\mathrm{kJ\,mol^{-1}}$ [65].

The comparison by Schröder and Thiel [62] (section 6.3.1) of SE (MNDO and MNDOC) and *ab initio* geometries and energies concluded that the SE methods usually overestimate activation energies. Of 21 activation energies (Table IV in Ref. [62], entries I, K, W omitted), MNDO overestimated (compared with "best" correlated *ab initio* calculations) 19 and underestimated 2; the overestimates ranged from $8\text{–}201\,\mathrm{kJ\,mol^{-1}}$ and the underestimates were 46 and $13\,\mathrm{kJ\,mol^{-1}}$. MNDOC overestimated 16 and underestimated 5; the overestimates ranged from $2\text{–}109\,\mathrm{kJ\,mol^{-1}}$ and the underestimates $4\text{–}63\,\mathrm{kJ\,mol^{-1}}$. Thus for calculating activation energies MNDOC is significantly better than MNDO, and it is probably better than AM1 for this purpose, since, like MNDO but unlike MNDOC, AM1 does not explicitly take into account electron correlation, which can be important for activation energies. For these 21 reactions, restricted Hartree–Fock calculations overestimated 18 activation energies and underestimated 3; the overestimates of energies ranged from $3\text{–}105\,\mathrm{kJ\,mol^{-1}}$ and the underestimates $13\text{–}28\,\mathrm{kJ\,mol^{-1}}$. The mean absolute deviations from the "best" correlated *ab initio* calculations for the 21 reactions were: MNDO, $92\,\mathrm{kJ\,mol^{-1}}$; MNDOC, $38\,\mathrm{kJ\,mol^{-1}}$; RHF, $50\,\mathrm{kJ\,mol^{-1}}$. Evidently MNDOC is somewhat better than RHF (uncorrelated) calculations for activation energies. Correlated-level *ab initio* calculations, however, appear to be superior to MNDOC; in particular, MNDOC predicts substantial barriers for isomerization of carbenes by hydrogen migration. Other work showed that AM1 greatly overestimates the barrier for decomposition or rearrangement of some highly reactive species [66].

Some SE reaction energies and relative energies of isomers are given in Table 6.3; these are analogous to the *ab initio* results in Table 5.9. These calculations suggest that, like the Hartree–Fock-level calculations of Table 5.9, AM1 and PM3 can give useful, although sometimes only rough, indications of the magnitude of energy differences. Further information on the reliability of these methods is provided by the calculations for the four reactions summarized in Fig. 6.3, which were discussed in section 6.3.1 in connection with geometries. Fig. 6.4, based on the energies in Fig. 6.3, makes these results clear. In all four cases the SE methods give the relative energies of the products semiquantitatively; the worst deviation from experiment is for the PM3 relative energy of HCN, which is $41\,\mathrm{kJ\,mol^{-1}}$ (-99 cf. $-58\,\mathrm{kJ\,mol^{-1}}$) too low. In fact, in two of the four cases ($H_2C{=}CHOH$ and HNC reactions) the AM1 product relative energies are the best (and in the other two cases, the MP2 energies are the best); however, this is likely to be due to an atypical cancellation of errors. The transition state relative energies are best-approximated in one case ($H_2C{=}CHOH$ reaction) by AM1 and PM3, and in the other three cases by MP2; for these three latter reactions the SE relative energies are considerably higher than the experimental and MP values, which accords with other work mentioned above [62,66].

From the available information, then, we can conclude that SE heats of formation and reaction energies (reactant cf. product) are semiquantitatively reliable. Activation energies (reactant cf. transition state) are usually considerably overestimated by AM1 and PM3, but are handled better by MNDOC, which actually gives results somewhat

Table 6.3. Reaction energies (kJ mol^{-1}) of isomers (AM1 and PM3)

Reactants	Products	Reaction energy, or relative energy of isomers, calculated	Exp (kJ mol^{-1})
$H_2 + Cl_2$	2HCl		-192
$-21.7 + (-59.3)$ $= -81.0$	$2(-103.0) = -206.0$	$-206.0 - (-81.0) = -125$	
$-56.0 + (-48.4)$ $= -104.4$	$2(-85.6) = -171.2$	$-171.2 - (-104.4) = -67$	
$2H_2 + O_2$	$2H_2O$		-523
$2(-21.7) + (-116.0)$ $= -159.4$	$2(-247.9) = -495.8$	$-495.8 - (-159.4) = -336$	
$2(-56.0) + (-17.5)$ $= -129.5$	$2(-223.5) = -447.0$	$-447.0 - (-129.5) = -318$	
(E)-2-butene	(Z)-2-butene		4.60
-14.0	-9.2	$-9.2 - (-14.9) = 5.7$	
-15.8	-14.9	$-14.9 - (-15.8) = 0.9$	
HCN	HNC		60.7
129.7	204.3	$204.3 - 129.7 = 74.6$	
137.9	236.8	$236.8 - 137.9 = 98.9$	

The calculations on O_2 are UHF, on triplet O_2. Calculations are by the author, experimental energies are from Ref. [58].

better than those from RHF calculations, at least in many cases. An extensive comparison of AM1 with *ab initio* and density functional methods for calculating geometries and relative energies is given in Hehre's book [58]. Consistently good calculated reaction energies and especially activation energies require correlated *ab initio* methods (sections 5.4.2 and 5.5.2) or DFT methods (chapter 7). It is interesting that AM1 and PM3, which were parameterized mainly to give good energies (heats of formation) actually provide quite good geometries but energies of only modest quality.

6.3.3   Frequencies

The general remarks and the theory concerning frequencies in section 5.5.3, apply to SE frequencies too, but the zero-point energies are usually not needed, since the SE energy is normally not adjusted by adding the ZPE (section 6.2.5.2). As with *ab initio* calculations, SE frequencies are used to characterize a species as a minimum or a transition state (or a higher-order saddle point), and to get an idea of what the IR spectrum looks like. As with *ab initio* frequencies too, in SE methods the wavenumbers (frequencies) of vibrations are calculated from a mass-weighted second-derivative matrix (a hessian) and intensities are calculated from the changes in dipole moment accompanying the vibrations. Like their *ab initio* counterparts, SE frequencies are higher than the experimental ones; presumably this is at least partly due to the harmonic approximation, as was discussed in section 5.5.3.

Correction factors improve the fit between SE calculated and experimentally measured spectra, but the agreement does not become as good as does the fit of corrected

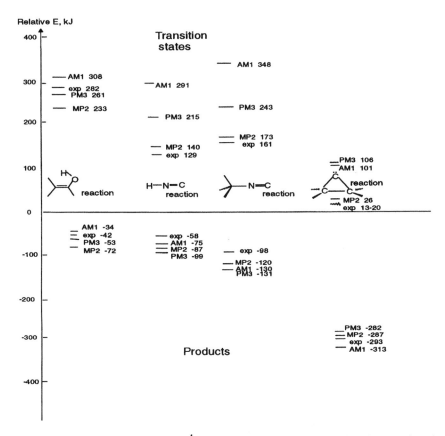

Figure 6.4. Relative energies (kJ mol^{-1}) for the four reactions of Fig. 6.3. Compared to the reactants (the four species shown), the transition state energies are all positive and the product energies all negative.

ab initio to experimental spectra. This is because deviations from experiment are less systematic for SE than for *ab initio* methods (a characteristic that has been noted for errors in SE energies [68]). For AM1 calculations, correction factors of 0.9235 [69] and 0.9532 [70], and for PM3, factors of 0.9451 [69] and 0.9761 [70], have been recommended. A factor of 0.86 has been recommended for SAM1 for non-H stretches [71]. However, the variation of the correction factor with the *kind* of frequency is bigger for SE than for *ab initio* calculations; for example, for correcting carbonyl stretching frequencies, examination of a few molecules indicated (author's work) that (at least for C, H, O compounds) correction factors of 0.83 (AM1) and 0.86 (PM3) give a much better fit to experiment. In Table 6.4 the factors of Ref. [69] were used. The after-correction deviations from experiment are considerably larger for AM1 and PM3 than for even moderate-level *ab initio* calculations: about 35% of the SE frequencies deviated by more than 10% from experiment, compared with 21% for HF/3–21G and 10% for HF/6-31G* [70].

Table 6.4. Acetone, benzene, dichloromethane and methanol, calculated IR spectrum after discarding IR-inactive or very weak (intensity less than 2 per cent of the strongest band) frequencies, ignoring frequencies below $600\,cm^{-1}$, correcting frequencies (uncorrected frequencies are in parentheses), combining degenerate frequencies and their intensities, and normalizing intensities. Frequency correction factors [69]: AM1, 0.9235 , PM3, 0.9451

	AM1		PM3	
	cm^{-1}	Relative intensity	cm^{-1}	Relative intensity
Acetone	(2060) 1902	100	(1979) 1870	100
Benzene	(744) 687	87	(712) 673	100
	(1579) 1458	36	(1548) 1463	47
	(3194) 2950	100	(3072) 2903	83
Dichloromethane	(773) 714	100	(671) 634	100
	(1327) 1225	2	(1353) 1279	8
Methanol	(1131) 1044	4	(991) 937	3
	(1362) 1258	100	(1164) 1100	78
	(1382) 1276	9	(1367) 1292	4
	(1471) 1358	5	(1408) 1331	100
	(1522) 1406	13	(3069) 2901	4

The calculated intensities of SE vibrations are much more approximate than those for *ab initio* vibrations (the latter are typically within 30% of the experimental intensity at the MP2 level), which is somewhat surprising, since SE (AM1 and PM3) dipole moments, from the vibrational changes of which intensities are calculated, are fairly accurate (section 6.3.4). Note that unlike the case with UV spectra, IR intensities are rarely actually measured; rather, one usually simply visually classifies a band as strong, medium, etc., by comparison with the strongest band in the spectrum. An idea of the reliability of SE frequencies and intensities is given by the IR spectra in Figs 6.5–6.8, which compare experimental spectra with semiempirical (AM1 and PM3) and *ab initio* (MP2/6-31G*) spectra, for the same four compounds (acetone, benzene, dichloromethane, methanol) shown in Figs 5.33–5.36. The SE spectra are based on the data in Table 6.4; the experimental and MP2 spectra are the ones used in Figs 5.33–5.36. Recall that the MP2/6-31G* IRs of these four compounds do not differ dramatically from the spectra calculated at the HF/3-21G$^{(*)}$ and HF/6-31G* levels. The indication is that SE IR spectra tend to overemphasize the intensities of the strongest bands at the expense of the weaker, with the result that the spectra (after one discards the weakest bands) show fewer bands than experimental or *ab initio* calculated spectra. Thus if one ignores bands with less than 2% of the intensity of the strongest band (the carbonyl stretch), the AM1 and PM3 spectra of acetone show only carbonyl stretch, while with the 2% cutoff the MP2/6-31G* spectrum has 11 bands and matches the experimental spectrum tolerably well (Fig. 6.5). The semiempirical IRs of benzene (Fig. 6.6) are not bad, but those of dichloromethane (Fig. 6.7) and methanol Fig. 6.8) are also significantly sparser than the experimental and MP2 spectra. Of these four compounds, the SE methods do the best job for benzene, which has only two kind of

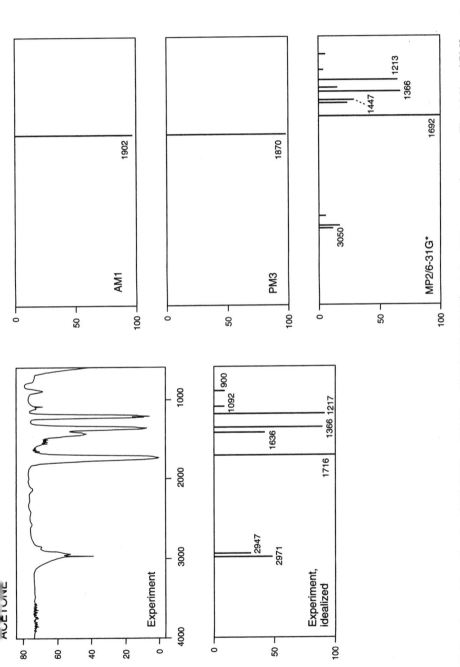

Figure 6.5. Experimental (gas phase) and SE (AM1 and PM3) and MP2(FC)/6-31G* calculated IR spectra of acetone. The AM1 and PM3 spectra are based on the data in Table 6.4; the MP2 spectrum is that shown in Fig. 5.33.

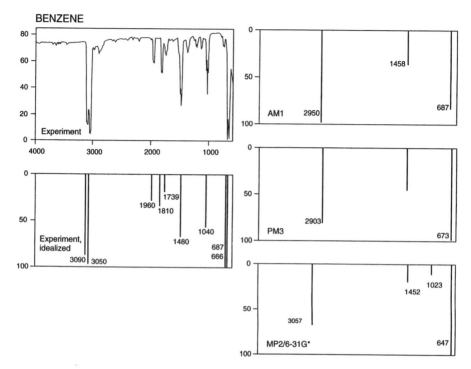

Figure 6.6. Experimental (gas phase) and SE (AM1 and PM3) and MP2(FC)/6-31G* calculated IR spectra of benzene. The AM1 and PM3 spectra are based on the data in Table 6.4; the MP2 spectrum is that shown in Fig. 5.34.

bonds, C/H and C/C, both of which are nonpolar. A problem with AM1 and PM3 IR spectra, then, is that a "missing" band may not be particularly weak, and may represent a prominent structural feature, e.g. the O–H stretch of methanol.

The wavenumbers (frequencies) of SE vibrations are more reliable than the intensities. All the normal modes are actually present in the results of an AM1 or PM3 frequency calculation, and animation of these will usually give, approximately, the frequencies of these vibrations. A very extensive compilation of experimental, MNDO and AM1 frequencies has been given by Healy and Holder, who conclude that the AM1 error of 10% can be reduced to 6% by an empirical correction, and that entropies and heat capacities are accurately calculated from the frequencies [72]. In this regard, Coolidge *et al.* conclude – surprisingly, in view of our results for the four molecules in Figs 6.5–6.8 – from a study of 61 molecules that (apart from problems with ring- and heavy atom-stretch for AM1 and S–H, P–H and O–H stretch for PM3) "both AM1 and PM3 should provide results that are close to experimental gas phase spectra" [73].

6.3.4 Properties arising from electron distribution: dipole moments, charges, bond orders

The discussion in section 5.5.4 on dipole moments, charges and bond orders applies in a general way to the calculation of these quantities by SE methods too. Electrostatic

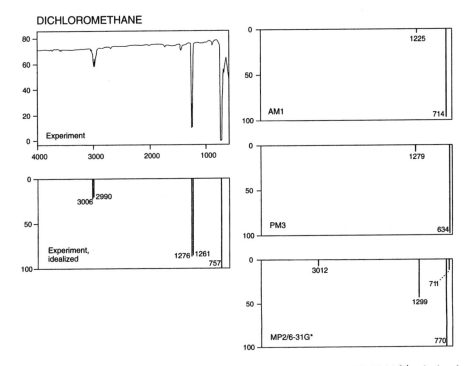

Figure 6.7. Experimental (gas phase) and SE (AM1 and PM3) and MP2(FC)/6-31G* calculated IR spectra of dichloromethane. The AM1 and PM3 spectra are based on the data in Table 6.4; the MP2 spectrum is that shown in Fig. 5.35.

potentials, whether visualized as regions of space or mapped onto van der Waals surfaces, are usually qualitatively the same for AM1 and PM3 as for *ab initio* methods. Atoms-in-molecules calculations are not viable for SE methods, because the core orbitals, lacking in these methods, are important for AIM calculations.

Dipole moments

Hehre's extensive survey of practical computational methods reports the results of *ab initio* and DFT single point dipole moment (μ) calculations on AM1 geometries [74]. There does not appear to be much advantage to calculating HF/6-31G* dipole moments on HF/6-31G* geometries (HF/6-31G*//HF/6-31G* calculations) rather than on the much more quickly- obtained AM1 geometries (HF/6-31G*//AM1 calculations). Indeed, even the relatively time-consuming MP2/6-31G*//MP2/6-31G* calculations seem to offer little advantage over fast HF/6-31G*//AM1 calculations as far as dipole moments are concerned (Tables 2.19 and 2.21 in Ref. [74]). This is consistent with our finding that AM1 geometries are quite good (section 6.3.1). Table 6.5 compares calculated and experimental dipole moments for 10 molecules,

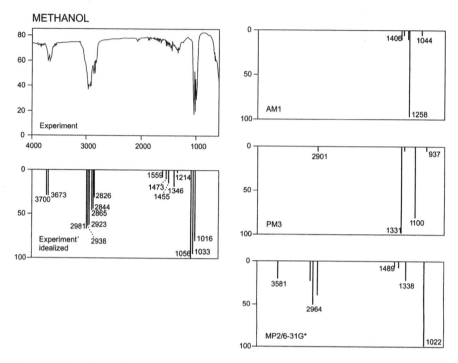

Figure 6.8. Experimental (gas phase) and SE (AM1 and PM3) and MP2(FC)/6-31G* calculated IR spectra of methanol. The AM1 and PM3 spectra are based on the data in Table 6.4; the MP2 spectrum is that shown in Fig. 5.36.

using these methods: AM1 (using the AM1 method to calculate μ for the AM1 geometry, AM1//AM1), HF/6-31*//AM1, PM3 (PM3//PM3), HF/6-31G*//PM3, and MP2/6-31G* (MP2/6-31G*//MP2/6-31G*). For this set of molecules, the smallest deviation from experiment, as judged by the arithmetic mean of the absolute deviations from the experimental values, is shown by the AM1 calculation (0.21 Debyes), and the largest deviation is shown by the "highest" method, MP2/6-31G* (0.34 D). The other three methods give essentially the same errors (0.27–0.29 D). It is of course possible that AM1 gives the best results (for this set on molecules, at least) because errors in geometry and errors in the calculation of the electron distribution cancel. A study of 196 C, H, N, O, F, Cl, Br, I molecules gave these mean absolute errors: AM1, 0.35 D; PM3, 0.40 D; SAM1, 0.32 D [50]. Another study with 125 H, C, N, O, F, Al, Si, P, S, Cl, Br, I molecules gave mean absolute errors of: AM1, 0.35 D and PM3, 0.38 D [44]. So with these larger samples the AM1 errors were somewhat bigger. Nevertheless, all these results taken together do indicate that unless one is prepared to use the slower approach of large basis sets with density functional (chapter 7) methods (errors of ca. 0.1 D [75]; this paper also gives some results for *ab initio* calculations), AM1 dipole moments using AM1 geometries may be as good a way as any to calculate this quantity.

Table 6.5. Some calculated dipole moments compared to experimental ones

	AM1	HF/6-31G* //AM1	PM3	HF/6-31G* //PM3	MP2/6-31G*	Exps
			Computational method			
CH_3NH_2	1.50	1.42	1.40	1.54	1.60	1.31
H_2O	1.86	2.25	1.74	2.16	2.24	1.85
HCN	2.36	3.24	2.70	3.24	3.26	2.99
CH_3OH	1.62	1.90	1.49	1.88	1.95	1.70
Me_2O	1.43	1.54	1.25	1.51	1.60	1.30
H_2CO	2.32	2.87	2.16	2.76	2.84	2.34
CH_3F	1.62	2.00	1.44	1.91	2.11	1.85
CH_3Cl	1.51	2.07	1.38	2.14	2.21	1.87
Me_2SO	3.95	4.56	4.49	4.83	4.63	3.96
CH_3CCH	0.40	0.58	0.36	0.60	0.66	0.75
Deviations	4+, 6−	9+, 1−	2+, 8−	9+, 1−	9+, 1−	
	Mean 0.21	Mean 0.29	Mean 0.27	Mean 0.29	Mean 0.34	

Dipole moments are in Debyes. Calculations are by the author; experimental values are taken from Refs. [58, 63]. For each method is given the number of positive and negative deviations from experiment and the arithmetic mean of the absolute values of the deviations.

This applies, of course, only to conventional molecules; molecules of exotic structure (note the remarks for the geometries of hypervalent molecules and molecules of unusual structure in section 6.3.1) may defy accurate SE predictions.

Charges and bond orders

The conceptual and mathematical bases of these concepts were outlined in chapter 5 (section 5.5 4). We saw that unlike, say, frequencies and dipole moments, charges and bond orders cannot even in principle be measured experimentally; as physicists say, they are not observables. Thus there are no "right" values to calculate, and in fact no single, correct, definitions of these terms, since as with *ab initio* calculations, SE charges and bond orders can be defined in various ways. The concepts are nevertheless useful, and electrostatic potential charges and Löwdin bond orders are preferred nowadays to the Mulliken parameters.

Figure 6.9 shows charges and bond orders calculated for an enolate (the conjugate base of ethenol or vinyl alcohol) and for a protonated enone system (protonated propenal). Consider first Mulliken charges and bond orders of the enolate (Fig. 6.9A). The AM1 and PM3 charges, which are essentially the same, are a bit surprising in that the carbon which shares charge with the oxygen in the alternative resonance structure is given a bigger charge than the oxygen; intuitively, one expects most of the negative charge to be on the more electronegative atom, oxygen (this "defect" of AM1 and PM3 has been noted by Anh *et al.* [76]). The HF/3-21G method gives the oxygen the bigger charge (-0.80 vs. -0.67). The two SE and the HF methods all give C/C and C/O bond orders of about 1.5; this, and the rough equality of O and C charges, suggests approximately equal contributions from the O-anion and C-anion resonance structures.

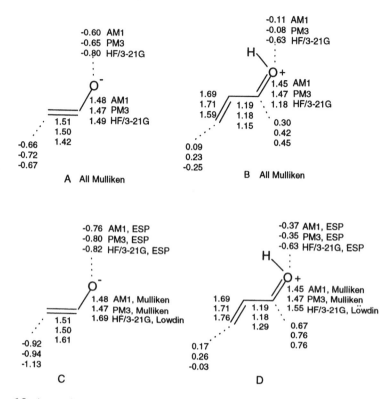

Figure 6.9. Atom charges and bond orders calculated using the AM1, PM3 and HF/3-21G methods. In A and B the charges and bond orders are all from the Mulliken approach. In C and D the charges are all electrostatic potential charges, and the bond orders are Mulliken for AM1 and PM3, and Löwdin for HF/3-21G* (Löwdin bond orders were not available for AM1 and PM3 from the Spartan program used). Note that charges and bond orders involving hydrogens have been omitted.

The Mulliken charges of the protonated enone system (Fig. 6.9B) make the oxygen negative, which may seem surprising. However, this is normal for protonated oxygen and nitrogen (though not protonated sulfur and phosphorus): the hetero atom in H_3O^+ and in NH_4^+ is calculated to be negative (i.e. the positive charge is on the hydrogens) and the hetero atom is also negative in $H_2C=OH^+$ and $H_2C=NH_2^+$. On the oxygen and the carbon furthest from the oxygen (C_3) the HF/3-21G charges differ considerably from the SE ones: the HF calculations make the O much more negative, and make C_3 negative, suggesting that they place more positive charge on the hydrogens than do the semiempirical calculations (in all cases the charge on C_2 is 0.3–0.5). The three methods do not differ as greatly in their bond order results, although HF method makes the formal C/O double bond essentially a single bond (bond order 1.18).

Finally, electrostatic potential (ESP) charges and, for the HF/3-21G calculations, Löwdin bond orders, are shown (Figs 6.9C and D). For the enolate, all three methods make the ESP charge on carbon more negative than that on oxygen, but the bond orders are not greatly altered. For the protonated enone system, AM1 and PM3 suggest more polarization of electrons toward the O in the C/O bond than is shown by the Mulliken charges, but while the HF ESP charge on this carbon is greater than the Mulliken (0.76 vs. 0.45), the charge on oxygen is unchanged. The HF Löwdin bond orders for all three bonds of the CCCO framework (1.55, 1.29, 1.76) are all somewhat bigger than the Mulliken bond orders (1.18, 1.15, 1.59).

These results indicate that charges are more dependent than are bond orders on the method used to calculate them, and that charges are also harder to interpret than are bond orders. As with *ab initio* charges and bond orders, the semiempirically calculated parameters may be useful in revealing trends in a series of compounds or changes as a reaction proceeds. For example, *ab initio* bond order changes along a reaction coordinate have been shown to be useful [77], but presumably semiempirically calculated bond orders would also yield similar information, at least if the species being studied were not too exotic. Clearly, one must use the same semiempirical method (e.g. AM1) and the same procedure (e.g. the Mulliken procedure) in studying a series of species.

6.3.5 Miscellaneous properties – UV spectra, ionization energies, and electron affinities

All the properties that can be calculated by *ab initio* methods can in principle also be calculated semiempirically, bearing in mind that the more the molecule of interest differs from the training set used to parameterize the SE program, the less reliable the results will be. For example, a program parameterized to predict the UV spectra of aromatic hydrocarbons may not give good predictions for the UV spectra of heterocyclic compounds. NMR spectra are usually calculated with *ab initio* (section 5.5.5) or density functional (chapter 7) methods. UV spectra, and ionization energies (ionization potentials) and electron affinities will be discussed here.

UV spectra

As pointed out in section 5.5.5, although ultraviolet spectra result from the promotion of electrons from occupied to unoccupied orbitals, UV spectra cannot be calculated with reasonable accuracy simply from the HOMO/LUMO gap of the ground electronic state, since the UV bands represent energy differences between the ground and *excited* states. Furthermore the HOMO/LUMO gap does not account for the presence of the several bands often found in UV spectra, and gives no indication of the intensity of a band. Accurate prediction of UV spectra requires calculation of the energies of excited states. SE UV spectra are usually calculated with programs specifically parameterized for this purpose, such as INDO/S or ZINDO/S (section 6.2.4) [19], both of which are in, e.g. HyperChem [78]. ZINDO/S, which appears to have largely superseded INDO/S, is included in the primarily *ab initio* and DFT package Gaussian 98 [79]. Table 6.6 compares the UV spectrum of methylenecyclopropene calculated by ZINDO/S as implemented in Gaussian 98 for Windows (G98W) [80] with the *ab initio*-calculated (Table 5.20; the geometry for this was calculated by DFT – chapter 7) and the experimental spectra [81]. The ZINDO/S spectrum resembles the experimental spectrum

Table 6.6. Calculated and experimental [81] UV spectra of methylenecyclopropene

		Calculated			
ZINDO/S//AM1		RCIS/6-31 + G*//B3LYP/6-31G*		Experimental	
Wavelength (nm)	Relative intensity	Wavelength (nm)	Relative intensity	Wavelength (nm)	Relative intensity
288	12	224	15	309	13
224	0.2	209	6	242	0.6
213	100	196	0	206	100
204	1	194	8		
		193	100		

The SE calculations were done with ZINDO/S in G94W; the *ab initio* results are from Table 5.20.

Table 6.7. Some ionization energies (eV)

	ΔE			Koopmans'			
	AM1	PM3	*ab in.*	AM1	PM3	*ab in.*	Experiment
CH_3OH	10.5	10.7	10.6	11.1	11.1	12.1	10.9
CH_3SH	8.7	9.0	9.0	8.9	9.2	9.7	9.4
CH_3COCH_3	9.9	10.1	9.6	10.7	10.8	11.2	9.7

The ΔE values (cation energy minus neutral energy) correspond to adiabatic, and the Koopmans' theorem values to vertical IEs, but vertical IEs exceed adiabatic by only ca. 0.1 eV or less (Table 5.21). The *ab initio* energies are MP2(fc)/6-31G* (Table 5.21). Experimental values are adiabatic, from Refs. [83] (CH_3OH and CH_3COCH_3) and [84] (CH_3SH).

considerably better than does the *ab initio* one (the experimental 242 nm, and particularly the 309-nm band, are matched better than by the *ab initio* calculation). The times for the calculations, on a 200 MHZ PentiumPro running Windows NT, were 53 s (ZINDO) and 11 min (*ab initio*).

Ionization energies and electron affinities

The concepts of IE and EA were discussed in section 5.5.5. In Table 6.7, the results of some semiempirical calculations are compared with *ab initio* and experimental values, for the molecules of Table 5.21. This admittedly very small sample suggests that SE IEs calculated as energy differences might be comparable to *ab initio* values. Koopmans' theorem (the IE for an electron is approximately the negative of the energy of its molecular orbital; applying this to the HOMO gives the IE of the molecule) values are consistently bigger than those from energy differences using the same method (by 0.1–0.8 eV). No consistent advantage for any of the six methods is evident here, but a large sample would likely show the most accurate of these methods to be the energy difference using MP2(fc)/6-31G* (see Table 5.21 and accompanying discussion).

Calculations by Stewart on 256 molecules (of which 201 were organic), using Koopmans' theorem, gave mean absolute IE errors of 0.61 eV for AM1 and 0.57 eV for PM3; 60 of the AM1 errors (23%) and 88 of the PM3 (34%) were negative (smaller than the experimental values) [44]. Particularly large errors (2.0–2.9 eV) were reported for nine molecules: 1-pentene, 2-methyl-1-butene, acetylacetone, alanine (AM1), SO_3 (AM1), CF_3Cl (AM1), 1,2-dibromotetrafluoroethane, H_2SiF_2 (PM3), and PF_3 (AM1). For some these it may be the *experimental* results that are at fault; for example, there seems to be no reason why 2-methyl-1-butene and 2-methyl-2-butene should have such different IEs, and in the opposite order to those calculated: experimental, 7.4 and 8.7 eV; calculated, 9.7 and 9.3 (AM1), 9.85 and 9.4 eV (PM3), respectively. *Ab initio* (HF/3-21G) energy-difference calculations by the author give IEs in line with the AM1, rather than the claimed experimental, results: 2-methyl-1-butene, 9.4 eV; 2-methyl-2-butene, 9.1 eV. Calculations by the author on the first 50 of these 256 molecules (of these 50 all but H_2 and H_2O are organic) gave these mean absolute IE errors: AM1, 0.46 (12 negative); PM3, 0.58 (5 negative); *ab initio* HF/3-21G, 0.71 (11 negative). So for the set of 256 mostly organic molecules AM1 and PM3 gave essentially the same accuracy, and for the set of 50 molecules AM1 was slightly better than PM3 and the *ab initio* method was slightly worse than the semiempirical ones. The HF/3-21G level is the lowest *ab initio* one routinely used (or at least reported) nowadays; ionization energies and electron affinities comparable in accuracy to those from experiment can be obtained by high-accuracy *ab initio* calculations (sections 5.5.2.2b and 5.5.5), using the energy difference of the two species involved.

Dewar and Rzepa found that the MNDO (section 6.2.5.3) electron affinities of 26 molecules with delocalized HOMOs (mostly radicals and conjugated organic molecules) had an absolute mean error of 0.43 eV; for ten molecules with the HOMO localized on one atom, the error was 1.40 eV [82]. The errors from AM1 or PM3 should be less than for these MNDO calculations.

6.3.6 Visualization

Many molecular features that have been calculated semiempirically can be visualized, in a manner analogous to the case of *ab initio* calculations (section 5.5.6). Clearly, one wishes to be able to view the molecule, rotate it, and query it for geometric parameters. Semiempirically calculated vibrations, electrostatic potentials, and molecular orbitals also provide useful information when visualized, and little need be added beyond that already discussed for the visualization of *ab initio* results. AM1 and PM3 surfaces (van der Waals surfaces, electrostatic potentials, orbitals) are usually very similar in appearance to those calculated by *ab initio* methods, but exceptions occasionally occur. An example is the case of HCC^-, the conjugate base of ethyne (acetylene), Fig. 6.10. AM1 predicts that there is one HOMO and that it is of σ symmetry (symmetric about the molecular axis), but a HF/3-21G calculation predicts that there are two HOMOs of equal energy at right angles, each of π symmetry (having a nodal plane containing the molecular axis; one of these π-HOMOs is shown in Fig. 6.10). The 3-21G orbital pattern persists at the HF/6-31G* and MP2/6-31G* levels. Different orbital patterns at different calculational levels is not the rule, but is understandable since energetically close MOs may have their energetic priorities reversed on going to a different level.

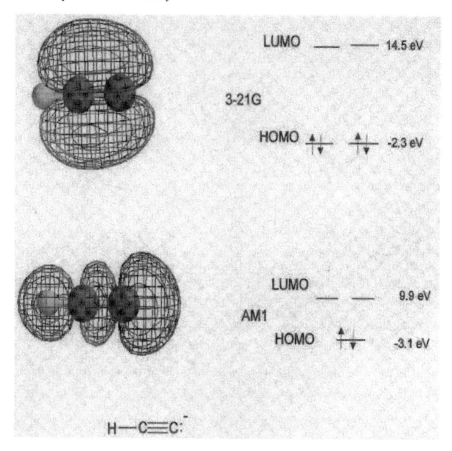

Figure 6.10. The HOMO of the ethyne conjugate base, calculated by AM1 and by HF/3-21G. AM1 predicts the HOMO to be unique and of σ symmetry (symmetrical about the molecular axis), but HF/3-21G predicts degenerate HOMO levels (the other is rotated 90° about the molecular axis) of π symmetry (with nodal planes containing the molecular axis). Only one of the degenerate 3-21G HOMOs is shown here. The orbitals were calculated and visualized with Spartan [54]. From your knowledge of the anion as a reagent in synthesis, which result do you think is more likely to be the correct one?

6.3.7   Some general remarks

Parameterized methods like ZINDO/S are probably the only way to calculate reasonably accurate UV spectra for large molecules. AM1 and PM3 have become extremely useful not only because they allow quantum mechanical calculations to be done on molecules which are still too big for *ab initio* or DFT (chapter 7) methods, but also as adjuncts to these latter methods, since they often allow a relatively rapid survey of a problem, such as an exploration of a potential energy surface: one can locate minima and transition states, then use the semiempirical structures (size permitting) as inputs for initial geometries, wavefunctions and hessians (section 2.4) in a higher-level geometry optimization. If geometry optimizations are not feasible, single-point calculations on AM1 or PM3 geometries (which are usually reasonably good) will likely give improved relative energies. The time is well past when SE calculations were regarded by many as "worthless" [85] , or, at best, as a poor substitute for *ab initio* calculations.

6.4   STRENGTHS AND WEAKNESSES OF SE METHODS

These remarks refer to AM1 and PM3 (and SAM1).

Strengths

Semiempirical calculations are very fast compared to *ab initio* and even to DFT (chapter 7), and this speed is often obtained with only an inconsequential loss of accuracy. Semiempirical geometries of normal molecules are entirely adequate for many purposes, and even transition state geometries are often adequate. Reaction and activation energies, although not accurate (except by chance cancellation of heat of formation errors), will probably expose any marked trends. Surprisingly, although they were parameterized using normal, stable molecules, AM1 and PM3 usually give fairly realistic geometries and relative energies for cations, radicals, anions, strained molecules, and even transition states.

Weaknesses

A major weakness of SE methods is that they must be assumed to be unreliable outside molecules of the kind used for their training set (the set of molecules used to parameterize them), until shown otherwise by comparison of their predictions with experiment or with high-level *ab initio* (or probably DFT) calculations. Although, as Dewar pointed out [86], the reliability of *ab initio* calculations, too, should be checked against experiment, the situation is somewhat different for these latter, at least at the higher levels; studies of exotic species, in particular, are certainly more trustworthy when done *ab initio* than semiempirically (see chapter 8). SE heats of formation are subject to errors of tens of $kJ\,mol^{-1}$, and thus heats (enthalpies) of reaction and activation could be in error by scores of $kJ\,mol^{-1}$. AM1 and PM3 underestimate steric repulsions, overestimate basicity and underestimate nucleophilicity, and can give unreasonable charges and structures; PM3 has been reported to tend to give more reliable structures, and AM1 better energies [76]. Neither AM1 nor PM3 are generally reliable in modelling hydrogen bonds [87,88], and SAM1 appears to be the semiempirical method of choice here [51].

In general, the accuracy of SE methods, particularly in energetics, falls short of that of current routine *ab initio* methods (this may not have been the case when AM1 was developed, in 1985 [86]). Parameters may not be available for the elements in the molecules one is interested in, and obtaining new parameters is something rarely done by people not actively engaged in developing new methods. SE errors are less systematic than *ab initio*, and thus harder to correct for.

6.5 SUMMARY OF CHAPTER 6

Semiempirical quantum mechanical calculations are based on the Schrödinger equation. This chapter deals with SCF SE methods, in which repeated diagonalization of the Fock matrix refines the wavefunction and the molecular energy (the SHM and EHM, in contrast, need only one matrix diagonalization because their matrix elements are not calculated using a wavefunction guess – see chapter 4). These calculations are much faster than *ab initio* ones, mainly because the number of integrals to be dealt with is greatly reduced by ignoring some, some integrals are approximated with the help of experimental quantities (hence "empirical"), and other integrals are calculated only approximately. In order of increasing sophistication, these SCF SE procedures have been developed: PPP, CNDO, INDO, and NDDO. The PPP method is limited to π electrons, while CNDO, INDO and NDDO use all the valence electrons. All four use the ZDO approximation, which sets the differential of the overlap integral equal to zero; this greatly reduces the number of integrals to be calculated. Traditionally, these methods were parameterized mostly using experimental quantities (usually ionization energies and electron affinities), but also (PPP and CNDO) making some use of minimal-basis-set (i.e. low-level) *ab initio* calculations. Of these original methods, only versions of INDO parameterized to reproduce UV spectra (INDO/S and its variant ZINDO/S) are much used nowadays. Today by far the most popular SCF SE methods are AM1 and PM3, which are NDDO-based, but carefully parameterized to reproduce experimental quantities (primarily heats of formation). AM1 and PM3 perform similarly and usually give quite good geometries, but less satisfactory heats of formation and relative energies. A modification of AM1 called SAM1, as yet relatively little-used, is said to be an improvement over AM1. AM1 and SAM1 represent work by the group of Dewar; PM3 is a version of AM1, by Stewart, differing mainly in a more automatic approach to parameterization.

REFERENCES

[1] (a) S. Weinberg, "Dreams of a Final Theory: the Search for the Fundamental Laws of Nature," Pantheon Books, New York, 1992. (b) Measuring the physical constants: A. Watson, Science, 2000, *287*, 1391.

[2] D. R. Hartree, Proc. Cambridge Phil. Soc., 1928, *24*, 89, 111, 426.

[3] The history of the development of computational chemistry (in the United States): J. D. Bolcer and R. B. Hermann, in Reviews in Computational Chemistry, vol. 8, K. B. Lipkowitz and D. B. Boyd, Eds., VCH, New York, 1994, chapter 1.

[4] Ref. [3, p. 12].

[5] M. J. S. Dewar, "The Molecular Orbital Theory of Organic Chemistry," McGraw-Hill, New York, 1969, p. 73.

[6] A single-point HF/6-31G* calculation on 1,3,5-triamino-2,4,6-trinitrobenzene, 282 basis functions; this was said to require 2.6 h on a 90 MHz pentium machine: Gaussian News, Summer 1995. This calculation on the C_1 conformation, using a 650 MHz Pentium III, took 13 and 46 min with two popular programs (February 2000).

[7] M. J. S. Dewar, "The Molecular Orbital Theory of Organic Chemistry," McGraw-Hill, New York, 1969, chapter 3.

[8] I. N. Levine, "Quantum Chemistry," 5th Ed., Prentice Hall, Upper Saddle River, New Jersey, 2000, chapter 16.

[9] W. Thiel in Advances in Chemical Physics, Volume XCIII, I. Prigogine and S. A. Rice, Eds., Wiley, New York, 1996.

[10] J. A. Pople and D. L. Beveridge, "Approximate Molecular Orbital Theory," McGraw-Hill, New York, 1970.

[11] R. Pariser and R. G. Parr, J. Chem. Phys., 1953, *21*, 466, 767.

[12] J. A. Pople, Trans. Faraday Soc., 1953, *49*, 1475.

[13] (a) Chemie in unserer Zeit, 1993, *12*, 21–31; (b) J. Griffiths, Chemistry in Britain, 1986, *22*, 997–1000.

[14] J. A. Pople and G A. Segal, J. Chem. Phys., 1966, *44*, 3289, and references therein.

[15] P. Coffey, Int. J. Quantum Chem., 1974, *8*, 263.

[16] Ref. [7, pp. 90–91].

[17] Ref. [10, p. 76].

[18] (a) J. A. Pople, D. L. Beveridge, and P. A. Dobosh, J. Chem. Phys., 1967, *47*, 2026; (b) R. N. Dixon, Mol. Phys., 1967, *12*, 83.

[19] INDO/S: M. Kotzian, N. Rösch, and M. C. Zerner, Theor. Chim. Acta, 1992, *81*, 201. (b) ZINDO/S is a version of INDO/S with some modifications, plus the ability to handle transition metals. The Z comes from the name of the late Professor Michael C. Zerner, whose group developed the suite of (mostly SE) programs called ZINDO, which includes ZINDO/S. ZINDO is available from, e.g. Molecular Simulations Inc., San Diego, CA., and CAChe Scientific, Beaverton, OR.

[20] J. A. Pople, D. P. Santry, and G. A. Segal, J. Chem. Phys., 1965, *43*, S129; J. A. Pople and G. A. Segal, J. Chem. Phys., 1965, *43*, S136; J. A. Pople and G. A. Segal, J. Chem. Phys., 1966, *44*, 3289.

[21] D. B. Boyd in Reviews in Computational Chemistry, vol. 6, K. B. Lipkowitz and D. B. Boyd, Eds., VCH, New York, 1995, chapter 5.

[22] (a) E. Wilson, Chemical & Engineering News, 1998, October 19, 12; (b) D. Malakoff, Science, 1998, *282*, 610; (c) Nobel lecture: Angew. Chem. Int. Ed., 1999, *38*, 1895.

[23] M. J. S. Dewar, "A Semiempirical Life," American Chemical Society, Washington, DC, 1992.

[24] Ref. [23, p. 131].

[25] M. J. S. Dewar, J. Am. Chem. Soc., 1975, *97*, 6591.

[26] M. J. S. Dewar, Science, 1975, *187*, 1037.

[27] T. A. Halgren, D. A. Kleier, and W, N. Lipscomb, Science, 1975, *190*, 591; response: M. J. S. Dewar, Science, 1975, *190*, 591.

[28] M. J. S. Dewar, J. Mol. Struct., 1983, *100*, 41.

[29] Ref. [23. p. 125].

[30] R. C. Bingham, M. J. S. Dewar, and D. H. Lo, J. Am. Chem. Soc., 1975, *97*, 1285.

[31] M. J. S. Dewar and G. Klopman, J. Am. Chem. Soc., 1967, *89*, 3089.

[32] N. C. Baird and M. J. S. Dewar, J. Chem. Phys., 1969, *50*, 1262.

[33] T. Clark, "A Handbook of Computational Chemistry," Wiley, New York, 1985, chapter 4.

[34] Ref. [8, p. 659].

[35] (a) First appearance of MNDO: M. J. S. Dewar and W. Thiel, J. Am. Chem. Soc., 1977, *99*, 4899. (b) Results of MNDO calculations on molecules with H, C, N, O: M. J. S. Dewar and W. Thiel, J. Am. Chem. Soc., 1977, *99*, 4907. (c) Results for molecules with B: M. J. S. Dewar and M. L. McKee, J. Am. Chem. Soc., 1977, 99, 5231.

[36] P. O'D. Offenhartz, "Atomic and Molecular Orbital theory," McGraw-Hill, New York, 1970, p. 325 (these matix elements are zero because the AO functions belong to different symmetry species, while the operator (kinetic plus potential energy) is spherically symmetric).

[37] M. J. S. Dewar and W. Thiel, Theor. Chim. Acta, 1977, *46*, 89.

[38] J. J. P. Stewart, J. Comp. Chem., 1989, *10*, 209.

[39] W. Thiel, Tetrahedron, 1988, *44*, 7393.

[40] For Dewar's very personal reminiscences of Austin see Ref. [23, pp. 111–120].

[41] M. J. S. Dewar, E. G. Zoebisch, E. F. Healy, and J. J. P. Stewart, J. Am. Chem. Soc., 1985, *107*, 3902.

[42] Note particularly Ref. [8, p. 662].

[43] Ref. [23, pp. 134, 135].

[44] J. J. P. Stewart, J. Comp. Chem., 1989, *10*, 221.

[45] J. J. P. Stewart, J. Comp. Chem., 1991, *12*, 320.

[46] M. J. S. Dewar, E. F. Healy, A. J. Holder, and Y.-C. Yuan, J. Comp. Chem., 1990, *11*, 541.

[47] J. J. P. Stewart, J. Comp. Chem., 1990, *11*, 543.

[48] Ref. [23, p. 185].

[49] A. J. Holder, R. D. Dennington, and C. Jie, Tetrahedron, 1994, *50*, 627.

[50] M. J. S. Dewar, C. Jie, and J. Yu, Tetrahedron, 1993, *49*, 5003.

[51] A. J. Holder and E. M. Evleth, in Chapter 7 in Modelling the Hydrogen Bond, D. A. Smith, Ed., American Chemical Society, Washington, DC, 1994.

[52] AMPAC 4.5 with Graphical User Interface: Semichem, P.O.Box 1649, Shawnee Misssion, KS 66216. www.Semichem.com.

[53] W. Thiel and A. A. Voityuk, J. Am. Chem. Soc., 1996, *100*, 616.

[54] Spartan: Wavefunction, Inc., 18401 Von Karman, Suite 370, Irvine CA. www.wavefun.com.

[55] R. Bosque and F. Maseras, J. Comp. Chem., 2000, *21*, 562.

[56] W. Thiel, J. Am. Chem. Soc., 1981, 103, 1413, 1421; A. Schweig and W. Thiel, J. Am. Chem. Soc., 1981, *103*, 1425.

[57] Ref. [8, chapters 16 and 17] and references therein.

[58] W. J. Hehre, "Practical Strategies for Electronic Structure Calculations," Wavefunction, Inc., Irvine, CA, 1995.

[59] J. J. P. Stewart, J. Mol. Struct. (Theochem), 1997, *410*, 195.

[60] MOZYME is a program in the suite of SE programs called MOPAC 2000, developed by J. J. P. Stewart: home.att.net/~mrmopac/. MOPAC has MINDO/3, MNDO, AM1 and PM3.

[61] J. J. P. Stewart, Int. J. Quantum Chem., 1996, *58*, 133.

[62] S. Schröder and W. Thiel, J. Am. Chem. Soc., 1985, *107*, 4422.

[63] W. J. Hehre, L. Radom, P. V. R. Schleyer, and J. A. Pople, "*Ab initio* Molecular Orbital Theory," Wiley, New York, 1986.

[64] W. Thiel, in "Computational Thermochemistry," K. K. Irikura and D. J. Frurip, Eds., American Chemical Society, Washington, DC, 1998, chapter 8.

[65] J. B. Pedley, "Thermochemical Data and Structures of Organic Compounds," Thermodynamics Research Center, College Station, Texas, 1994.

[66] (a) CO_2/N_2 copolymers: J. Bylykbashi and E. Lewars, J. Mol. Struct. (Theochem), 1999, *469*, 77. (b) Oxirenes: E. Lewars, Can. J. Chem., 2000, *78*, 297–306.

[67] $H_2C = CHOH$ *reaction* E. Lewars and I. Bonnycastle, J. Mol. Struct. (Theochem), 1997, *418*, 17 and references therein. *HNC reaction* V. S. Rao, A. Vijay, A. K. Chandra, Can. J. Chem., 1996, *74*, 1072. CH_3NC *reaction* The reported experimental activation energy is 161 kJ mol^{-1}: F. W. Schneider and B. S. Rabinovitch, J. Am. Chem. Soc., 1962, *84*, 4215; B. S. Rabinovitch and P. W. Gilderson, J. Am. Chem. Soc., 1965, *87*, 158. The energy of CH_3CN relative to CH_3NC by a high-level (G2) calculation is -98.3 kJ mol^{-1} (E. Lewars). An early *ab initio* study of the reaction: D. H. Liskow, C. F. Bender, H. F. Schaefer, J. Am. Chem. Soc., 1972, *95*, 5178. A comparison of CH_3CN, CH_3NC, other isomers and radicals, cations and anions: P. M. Mayer, M. S. Taylor, M. Wong, L. Radom, J. Phys. Chem. A, 1998, 102, 7074. *Cyclopropylidene reaction* H. F. Bettinger, P. R. Schreiner, P. v. R. Schleyer, H. F. Schaefer, J. Phys. Chem., 1996, *100*, 16147.

[68] Ref. [64, p. 157].

[69] Information supplied by Dr. R. Johnson of the National Institutes of Standards and Technology, USA (NIST): best fits to about 1100 vibrations of about 70 closed-shell molecules. An extensive collection of scaling factors is available on the NIST website (http://srdata.nist.gov/cccbdb/).

[70] A. P. Scott and L. Radom, J. Phys. Chem., 1996, *100*, 16502.

[71] A. J. Holder and R. D. Dennington II, J. Mol. Struct. (Theochem), 1997, *401*, 207.

[72] E. F. Healy and A. Holder, J. Mol. Struct. (Theochem), 1993, *281*, 141.

[73] M. B. Cooligde, J. E. Marlin and J. J. P. Stewart, J. Comp. Chem., 1991, *12*, 948.

[74] Ref. [58, pp. 74, 76–77, 80–82].

[75] A. C. Scheiner, J. Baker, and J. W. Andzelm, J. Comp. Chem., 1997, *18*, 775.

[76] N. T. Anh, G. Frisson, A. Solladié-Cavallo, and P. Metzner, Tetrahedron, 1998, *54*, 12841.

[77] G. Lendvay, J. Phys. Chem., 1994, *98*, 6098.

[78] Available from Hypercube Inc., Gainsville, FL.

[79] Available for several kinds of computers from Gaussian Inc., Pittsburgh, PA.

[80] Gaussian 98, revision A.6, M. J. Frisch, G. W. Trucks, H. B. Schlegel, G. E. Scuseria, M. A. Robb, J. R. Cheeseman, V. G. Zakrzewski, J. A. Montgomery, Jr., R. E. Stratmann, J. C. Burant, S. Dapprich, J. M. Millam, A. D. Daniels, K. N. Kudin, M. C. Strain, O. Farkas, J. Tomasi, V. Barone, M. Cossi, R. Cammi, B. Mennucci, C. Pomelli, C. Adamo, S. Clifford, J. Ochterski, G. A. Petersson, P. Y. Ayala, Q. Cui, K. Morokuma, D. K. Malick, A. D. Rabuck, K. Raghavachari, J. B. Foresman, J. Cioslowski, J. V. Ortiz, B. B. Stefanov, G. Liu, A. Liashenko, P. Piskorz, I. Komaromi, R. Gomperts, R. L. Martin, D. J. Fox, T. Keith, M. A. Al-Laham, C. Y. Peng, A. Nanayakkara, C. Gonzalez, M. Challacombe, P. M. W. Gill, B. Johnson, W. Chen, M. W. Wong, J. L. Andres, C. Gonzalez, M. Head-Gordon, E. Repogle, and J. A. Pople, Gaussian, Inc., Pittsburgh PA, 1998.

[81] J. B. Foresman and Æ. Frisch, "Exploring Chemistry with Electronic Structure Methods," Gaussian Inc., Pittsburgh, PA, 1996, p. 218.

[82] M. J. S. Dewar and H. S. Rzepa, J. Am. Chem. Soc., 1978, *100*, 784.

[83] R. D. Levin and S. G. Lias, "Ionization Potential and Appearance Potential Measurements, 1971–1981," National Bureau of Standards, Washington, DC, 1982.

[84] L. A. Curtiss, R. H. Nobes, J. A. Pople, and l. Radom, J. Chem Phys., 1992, *97*, 6766.

[85] Ref. [23, p. 180].

[86] For example, M. J. S. Dewar and D. M. Storch, J. Am. Chem. Soc., 1985, *107*, 3898.

[87] For a series of small, mostly nonbiological molecules AM1 seemed better than PM3, except for O–H/O hydrogen bonds: J. J. Dannenberg, J. Mol. Struct. (Theochem), 1997, *410*, 279.

[88] In model systems of biological relevance, mostly involving water, PM3 was superior to AM1: Y.- J. Zheng and K. M. Merz, J. Comp. Chem., 1992, *13*, 1151.

EASIER QUESTIONS

1. Outline the similarities and differences between the EHM on the one hand and methods like AM1 and PM3 on the other. What advantages does the EHM have over more accurate SE methods?
2. Outline the similarities and differences between MM, *ab initio*, and SE methods.
3. Both the simple Hückel and the PPP methods are π electron methods, but PPP is more complex. Itemize the added features of PPP.
4. What is the main advantage of an all-valence-electron method like, say, CNDO over a purely π electron method like PPP?
5. Explain the terms ZDO, CNDO, INDO, and NDDO, showing why the latter three represent a progressive conceptual improvement.
6. How does an AM1 or PM3 "total electron wavefunction" Ψ differ from the Ψ of an *ab initio* calculation?
7. *ab initio* energies are "total dissociation" energies (dissociation to electrons and atomic nuclei) and AM1 and PM3 energies are standard heats of formation. Is one of these kinds of energy more useful? Why or why not?
8. For certain kinds of molecules MM can give better geometries and relative energies than can even sophisticated SE methods. What kinds of properties can the latter calculate that MM cannot?
9. Why do transition metal compounds present special difficulties for AM1 and PM3?
10. Although both AM1 and PM3 normally give good molecular geometries, they are not too successful in dealing with geometries involving hydrogen bonds. Suggest reasons for this deficiency.

HARDER QUESTIONS

1. Why are even very carefully-parameterized SE methods like AM1 and PM3 not as accurate and reliable as high-level (e.g. MP2, CI, coupled-cluster) *ab initio* calculations?

2. Molecular mechanics is essentially empirical, while methods like PPP, CNDO, and AM1/PM3 are semiempirical. What are the analogies in PPP etc. to MM procedures of developing and parameterizing a forcefield? Why are PPP etc. only *semi*empirical?

3. What do you think are the advantages and disadvantages of parameterizing SE methods with data from *ab initio* calculations rather than from experiment? Could a SE method parameterized using *ab initio* calculations logically be called semi*empirical*?

4. There is a kind of contradiction in the Dewar-type methods (AM1, etc.) in that overlap integrals are calculated and used to help evaluate the Fock matrix elements, yet the overlap matrix is taken as a unit matrix as far as diagonalization of the Fock matrix goes. Discuss.

5. What would be the advantages and disadvantages of using the general MNDO/AM1 parameterization procedure, but employing a minimal basis set instead of a minimal valence basis set?

6. In SCF SE methods major approximations lie in the calculation of the H_{rs}^{core}, $(rs|tu)$, and $(ru|ts)$ integrals of the Fock matrix elements F_{rs} (Eq. (6.1)). Suggest an alternative approach to approximating one of these integrals.

7. Read the exchange between Dewar on the one hand and Halgren, Kleir and Lipscomb on the other [27]. Do you agree that SE methods, even when they give good results "inevitably obscure the physical bases for success (however striking) and failure alike, thereby limiting the prospects for learning why the results are as they are?" Explain your answer.

8. It has been said of SE methods: "They will never outlive their usefulness for correlating properties across a series of molecules . . . I really doubt their predictive value for a one-off calculation on a small molecule on the grounds that whatever one is seeking to predict has probably already been included in with the parameters." (A. Hinchliffe, "Ab Initio Determination of Molecular Properties," Adam Hilger, Bristol, 1987, p. x). Do you agree with this? Why or why not? Compare the above quotation with Ref. [23, pp. 133–136].

9. For common organic molecules Merck Molecular Force field geometries are nearly as good as MP(fc)/6-31G* geometries (section 3.4). For such molecules single-point MP(fc)/6-31G* calculations (section 5.4.2), which are quite fast, on the MMFF geometries, should give energy differences comparable to those from MP(fc)/6-31G*//MP(fc)/6-31G* calculations. Example: CH_2=CHOH/CH_3CHO, ΔE(MP2 opt, including ZPE) $= 71.6\,kJ\,mol^{-1}$, total time $\doteq 1064\,s$; ΔE(MP2 single point on MMFF geometries) $= 70.7\,kJ\,mol^{-1}$, total time $= 48\,s$ (G98 on a Pentium 3). What role does this leave for SE calculations?

10. Semiempirical methods are untrustworthy for "exotic" molecules of theoretical interest. Give an example of such a molecule and explain why it can be considered exotic. Why cannot SE methods be trusted for molecules like yours? For what other kinds of molecules might these methods fail to give good results?

Chapter 7

Density Functional Calculations

My other hope is that ... a basically new ab initio *treatment capable of giving chemically accurate results* a priori, *is achieved soon.*
                                    M. J. S. Dewar, *A Semiempirical Life, 1992.*

7.1 PERSPECTIVE

We have seen three broad techniques for calculating the geometries and energies of molecules: molecular mechanics (chapter 3), *ab initio* methods (chapter 5), and semiempirical methods (chapters 4 and 6). Molecular mechanics is based on a balls-and-springs model of molecules. *Ab initio* methods are based on the subtler model of the quantum mechanical molecule, which we treat mathematically starting with the Schrödinger equation. Semiempirical methods, from simpler ones like the Hückel and extended Hückel theories (chapter 4) to the more complex SCF semiempirical theories (chapter 6), are also based on the Schrödinger equation, and in fact their "empirical" aspect comes from the desire to avoid the mathematical problems that this equation imposes on *ab initio* methods. Both the *ab initio* and the semiempirical approaches calculate a molecular wavefunction (and molecular orbital energies), and thus represent *wavefunction methods*. However, a wavefunction is not a measurable feature of a molecule or atom – it is not what physicists call an "observable;" in fact there is no general agreement among physicists what, if anything, a wavefunction is [1].

Density functional theory (DFT) is based not on the wavefunction, but rather on the electron probability density function or electron density function, commonly called simply the electron density or charge density, designated by $\rho(x, y, z)$. This is a probability per unit volume; the probability of finding an electron in a volume element $dx\,dy\,dz$ centered on a point with coordinates x, y, z is $\rho(x, y, z)\,dx\,dy\,dz$. The units of ρ are logically volume^{-1} and since the units of $dx\,dy\,dz$ are volume, $\rho(x, y, z)\,dx\,dy\,dz$ is a pure number, a probability. However, if we regard the charge on the electron as our unit of charge then ρ has units of electronic charge volume^{-1} and $\rho(x, y, z)\,dx\,dy\,dz$ units of electronic charge. If we think of electronic charge as being smeared out in a fog around the molecule, then the variation of ρ from point to point (ρ is a function of x, y, z) corresponds to the varying density of the fog, and $\rho(x, y, z)\, dx\, dy\, dz$ centered on a point $P(x, y, z)$ corresponds to the amount of fog in the volume element $dx\, dy\, dz$. In a scatterplot of electron density in a molecule, the variation of ρ with position can be indicated by the density of the points. The electron density function is the basis not only of DFT, but of a whole suite of methods of regarding and studying atoms and molecules [2], and, unlike the wavefunction, is measurable, e.g. by X-ray diffraction or electron diffraction [3]. Apart from being an experimental observable and being readily grasped intuitively [4], the electron density has another property particularly suitable for any method with claims to being an improvement on, or at least a valuable alternative to, wavefunction methods: it is a function of position only, that is, of just *three* variables (x, y, z), while the wavefunction of an n-electron molecule is a function of $4n$ variables, three spatial coordinates and one spin coordinate, *for each electron*. No matter how big the molecule may be, the electron density remains a function of three variables, while the complexity of the wavefunction increases with the number of electrons. The term functional, which is akin to function, is explained in section 7.2.3.1. To the chemist, the main advantage of DFT is that in about the same time needed for an HF calculation one can often obtain results of about the same quality as from MP2 calculations (section 7.3). Chemical applications of DFT are but one aspect of an ambitious project to recast conventional quantum mechanics, i.e. wave mechanics, in a form in which "the electron density, and only the electron density, plays the key role" [5]. It is noteworthy that the 1998 Nobel Prize for chemistry was awarded to John Pople (section 5.3.3), largely for his role in developing practical wavefunction-based methods, and Walter Kohn[1], for the development of density functional methods [6].

A question sometimes asked is whether DFT should be regarded as a special kind of *ab initio* method. The case against this view is that the correct mathematical form of the DFT functional is not known, in contrast to conventional *ab initio* theory where the correct mathematical form of the fundamental equation, the Schrödinger equation, is (we think), known. In conventional *ab initio* theory, the wavefunction can be improved systematically by going to bigger basis sets and higher correlation levels, which takes us closer and closer to an exact solution of the Schrödinger equation, but in DFT there is so far no known way to systematically improve the functional (section 7.2.3.2); one must feel one's way forward with the aid of intuition and comparison of the results with experiment and of high-level conventional *ab initio* calculations. In this sense current DFT is semiempirical, but the limited use of empirical *parameters* (typically from zero to about 10), and the possibility of one day finding the exact functional makes it *ab initio* in spirit. Were the exact functional known, DFT might indeed give "chemically accurate results *a priori*."

---

[1] Walter Kohn, born in Vienna in 1923. B.A., B.Sc., University of Toronto, 1945, 1946. Ph.D. Harvard, 1948. Instructor in physics, Harvard, 1948–1950. Assistant, Associate, full Professor, Carnegie Mellon University, 1950–1960. Professor of physics, University of California at San Diego, 1960–1979; University of California at Santa Barbara 1979–present. Nobel Prize in chemistry 1998.

7.2 THE BASIC PRINCIPLES OF DENSITY FUNCTIONAL THEORY

7.2.1 Preliminaries

In the Born interpretation (section 4.2.6) the square of a one-electron wavefunction ψ at any point X is the probability density (with units of volume^{-1}) for the wavefunction at that point, and $|\psi|^2 dx\, dy\, dz$ is the probability (a pure number) at any moment of finding the electron in an infinitesimal volume $dx\, dy\, dz$ around the point (the probability of finding the electron *at* a mathematical point is zero). For a *multielectron* wavefunction Ψ the relationship between the wavefunction Ψ and the electron density ρ is more complicated (involving the summation over all spin states of all electrons of n-fold integrals of the square of the wavefunction), but it can be shown [7] that $\rho(x, y, z)$ is related to the component one-electron spatial wavefunctions ψ_i of a single-determinant wavefunction Ψ (recall from section 5.2.3.1 that the Hartree–Fock Ψ can be approximated as a Slater determinant of spin orbitals $\psi_i \alpha$ and $\psi_i \beta$) by

$$\rho = \sum_{i=1}^{n} n_i |\psi_i|^2 \qquad (7.1)$$

The sum is over n the occupied MOs ψ_i and for a closed-shell molecule each $n_i = 2$, for a total of $2n$ electrons. Equation (7.1) applies strictly only to a single-determinant wavefunction Ψ, but for multideterminant wavefunctions arising from configuration interaction treatments (section 5.4) there are similar equations [8]. A shorthand for $\rho(x, y, z)\, dx\, dy\, dz$ is $\rho(\mathbf{r})\, d\mathbf{r}$, where $\mathbf{r}$ is the position vector of the point with coordinates (x, y, z).

If the electron density ρ rather than the wavefunction could be used to calculate molecular geometries, energies, etc., this might be an improvement over the wavefunction approach because, as mentioned above, the electron density in an n-electron molecule is a function of only the three spatial coordinates x, y, z, but the wavefunction is a function of $4n$ coordinates. Density functional theory seeks to calculate all the properties of atoms and molecules from the electron density. The standard book on DFT is probably still that by Parr and Yang (1989) [9]; more recent developments are included in the book by Koch and Holthausen [10]. The textbook by Levine [11] gives a very good yet compact introduction to DFT, and among the many reviews are those by Friesner *et al.* [12], Kohn *et al.* [13], and Parr and Yang [14].

7.2.2 Forerunners to current DFT methods

The idea of calculating atomic and molecular properties from the electron density appears to have arisen from calculations made independently by Enrico Fermi and P. A. M. Dirac in the 1920s on an ideal electron gas, work now well-known as the Fermi–Dirac statistics [15]. In independent work by Fermi [16] and Thomas [17], atoms were modelled as systems with a positive potential (the nucleus) located in a uniform (homogeneous) electron gas. This obviously unrealistic idealization, the

Thomas–Fermi model [18], or with embellishments by Dirac, the Thomas–Fermi–Dirac model [18], gave surprisingly good results for atoms, but failed completely for molecules: it predicted all molecules to be unstable toward dissociation into their atoms (indeed, this is a theorem in Thomas–Fermi theory).

The Xα (X = exchange, α is a parameter in the Xα equation; see section 7.2.3.4a) method gives much better results [19, 20]. It can be regarded as a more accurate version of the Thomas–Fermi model, and is probably the first chemically useful DFT method. It was introduced in 1951 [21] by Slater, who regarded it [22] as a simplification of the Hartree–Fock (section 5.2.3) approach. The Xα method, which was developed mainly for atoms and solids, has also been used for molecules, but has been replaced by the more accurate Kohn–Sham type (section 7.2.3) DFT methods.

7.2.3 Current DFT methods: the Kohn–Sham approach

7.2.3.1 Functionals: The Hohenberg–Kohn theorems

Nowadays DFT calculations on molecules are based on the Kohn–Sham approach, the stage for which was set by two theorems published by Hohenberg and Kohn in 1964 (proved in Levine [23]). The first Hohenberg–Kohn [24] theorem says that all the properties of a molecule in a ground electronic state are determined by the ground state electron density function $\rho_0(x, y, z)$. In other words, given $\rho_0(x, y, z)$ we can in principle calculate any ground state property, e.g. the energy, E_0; we could represent this as

$$\rho_0(x, y, z) \rightarrow E_0 \tag{7.2}$$

The relationship (7.2) means that E_0 is a *functional* of $\rho_0(x, y, z)$. A *function* is a rule that transforms a number into another (or the same) number:

$$2 \xrightarrow{x^3} 8$$
$$1 \xrightarrow{x^3} 1$$

A *functional* is a rule that transforms a function into a number:

$$f(x) = x^3 \xrightarrow{\int_0^2 f(x)\,dx} \left.\frac{x^4}{4}\right|_0^2 = 4 \tag{7.3}$$

The functional $\int_0^2 f(x)\,dx$ transforms the function x^3 into the number 4. We designate the fact that the integral is a functional of $f(x)$ by writing

$$\int_0^2 f(x)\,dx = F[f(x)] \tag{7.4}$$

A functional is a function of a "definite" (cf. the definite integral above) function.

The first Hohenberg–Kohn theorem, then, says that any ground state property of a molecule is a functional of the ground state electron density function, e.g. for the energy

$$E_0 = F[\rho_0] = E[\rho_0] \tag{7.5}$$

The theorem is "merely" an *existence theorem*: it says that a functional F exists, but does not tell us how to find it; this omission is the main problem with DFT. The significance of this theorem is that it assures us that there *is* a way to calculate molecular properties from the electron density, and that we can infer that approximate functionals will give at least approximate answers.

The second Hohenberg–Kohn theorem [24] is the DFT analogue of the wavefunction variation theorem that we saw in connection with the *ab initio* method (section 5.2.3.3): it says that any trial electron density function will give an energy higher than (or equal to, if it were exactly the true electron density function) the true ground state energy. In DFT molecular calculations the electronic energy from a trial electron density is the energy of the electrons moving under the potential of the atomic nuclei. This nuclear potential is called the "external potential" and designated $v(\mathbf{r})$, and the electronic energy is denoted by $E_v = E_v[\rho_0]$ ("the E_v functional of the ground state electron density"). The second theorem can thus be stated

$$E_v[\rho_t] \geq E_0[\rho_0] \qquad (7.6)$$

where ρ_t is a trial electronic density and $E_0[\rho_0]$ is the true ground state energy, corresponding to the true electronic density ρ_0. The trial density must satisfy the conditions $\int \rho_t(\mathbf{r})d\mathbf{r} = n$, where n is the number of electrons in the molecule (this is analogous to the wavefunction normalization condition; here the number of electrons in all the infinitesimal volumes must sum to the total number in the molecule) and $\rho_t(\mathbf{r}) \geq 0$ for all $\mathbf{r}$ (the number of electrons per unit volume cannot be negative). This theorem assures us that any value of the molecular energy we calculate from the Kohn–Sham equations (below, a set of equations analogous to the Hartree–Fock equations, obtained by minimizing energy with respect to electron density) will be greater than or equal to the true energy (this is actually true only if the functional used were exact; see below). The Hohenberg–Kohn theorems were originally proved only for nondegenerate ground states, but have been shown to be valid for degenerate ground states too [25]. The functional of the inequality (7.6) is the correct, exact energy functional (the prescription for transforming the ground state electron density function into the ground state energy). The exact functional is unknown, so *actual* DFT calculations use approximate functionals, and are thus *not* variational: they can give an energy below the true energy. Being variational is a nice characteristic of a method, because it assures us that any energy we calculate is an upper bound to the true energy. However, this is not an essential feature of a method: Møller–Plesset and practical configuration interaction calculations (sections 5.4.2, 5.4.3) are not variational, but this is not regarded as a serious problem.

7.2.3.2 The Kohn–Sham energy and the KS equations

The first Kohn–Sham theorem tells us that it is worth looking for a way to calculate molecular properties from the electron density. The second theorem suggests that a variational approach might yield a way to calculate the energy and electron density (the electron density, in turn, could be used to calculate other properties). Recall that in wavefunction theory, the Hartree–Fock variational approach (section 5.2.3.4) led to the HF equations, which are used to calculate the energy and the wavefunction. An analogous variational approach led (1965) to the KS equations [26], the basis of current

molecular DFT calculations. The two basic ideas behind the KS approach to DFT are:
(1) To express the molecular energy as a sum of terms, only one of which, a relatively
small term, involves the unknown functional. Thus even moderately large errors in this
term will not introduce large errors into the total energy. (2) To use an initial guess of
the electron density ρ in the KS equations (analogous to the HF equations) to calculate
an initial guess of the KS orbitals (below); this initial guess is then used to refine these
orbitals, in a manner similar to that used in the HF SCF method. The final KS orbitals
are used to calculate an electron density that in turn is used to calculate the energy.

The Kohn–Sham energy

The strategy here is to regard the energy of our molecule as a deviation from an ideal
energy, which latter can be calculated exactly; the relatively small discrepancy contains
the unknown functional, whose approximation is our main problem. The ideal energy
is that of an ideal system, a fictitious *noninteracting reference system*, defined as one
in which the electrons do not interact and in which the ground state electron density ρ_r
is exactly the same as in our real ground state system: $\rho_r = \rho_0$. We are talking about
the *electronic* energy of the molecule; the total internal "frozen-nuclei" energy can be
found later by adding the internuclear repulsions, and the 0 K total internal energy by
further adding the zero-point energy, just as in HF calculations (section 5.2.3.6e).

The ground state electronic energy of our real molecule is the sum of the electron kinetic energies, the nucleus–electron attraction potential energies, and the
electron–electron repulsion potential energies (more precisely, the sum of the quantum-
mechanical average values or expectation values, each denoted (value)) and each is a
functional of the ground-state electron density:

$$E_0 = \langle T[\rho_0] \rangle + \langle V_{Ne}[\rho_0] \rangle + \langle V_{ee}[\rho_0] \rangle \tag{7.7}$$

Focussing on the middle term: the nucleus–electron potential energy is the sum over all
$2n$ electrons (as with our treatment of *ab initio* theory, we will work with a closed-
shell molecule which perforce has an even number of electrons) of the potential
corresponding to attraction of an electron for all the nuclei A:

$$\langle V_{Ne} \rangle = \sum_{i=1}^{2n} \sum_{nuclei A} -\frac{Z_A}{r_{iA}} = \sum_{i=1}^{2n} v(r_i) \tag{7.8}$$

where $v(r_i)$ is the external potential (explained in section 7.2.3.1, in connection with
Eq. (7.6)) for the attraction of electron i to the nuclei. The density function ρ can be
introduced into $\langle V_{Ne} \rangle$ by using the fact [27] that

$$\int \Psi \sum_{i=1}^{2n} f(r_i) \Psi \, dt = \int \rho(r) f(r) \, dr \tag{7.9}$$

where $f(r_i)$ is a function of the coordinates of the $2n$ electrons of a system and Ψ is
the total wavefunction (the integrations are over spatial and spin coordinates on the left
and spatial coordinates on the right). From Eqs (7.8) and (7.9), invoking the concept of

expectation value (chapter 5, section 5.2.3.3) $\langle V_{Ne} \rangle = \langle \Psi | \hat{V}_{Ne} | \Psi \rangle$, and since $\hat{V} = Vx$ we get

$$\langle V_{Ne} \rangle = \int \rho_0(\mathbf{r}) v(\mathbf{r}) \, d\mathbf{r} \tag{7.10}$$

So Eq. (7.7) can be written

$$E_0 = \int \rho_0(\mathbf{r}) v(\mathbf{r}) \, d\mathbf{r} + \langle T[\rho_0] \rangle + \langle V_{ee}[\rho_0] \rangle \tag{7.11}$$

Unfortunately, this equation for the energy cannot be used as it stands, since we do not know the functionals in $\langle T[\rho_0] \rangle$ and $\langle V_{ee}[\rho_0] \rangle$.

To utilize Eq. (7.11), Kohn and Sham introduced the idea of a reference system of noninteracting electrons. Let us define the quantity $\Delta\langle T[\rho_0] \rangle$ as the deviation of the real kinetic energy from that of the reference system:

$$\Delta\langle T[\rho_0] \rangle \equiv \langle T[\rho_0] \rangle - \langle T_r[\rho_0] \rangle \tag{7.12}$$

Let us next define $\Delta\langle V_{ee} \rangle$ as the deviation of the real electron–electron repulsion energy from a classical charge-cloud coulomb repulsion energy. This classical electrostatic repulsion energy is the summation of the repulsion energies for pairs of infinitesimal volume elements $\rho(\mathbf{r}_1) \, d\mathbf{r}_1$ and $\rho(\mathbf{r}_2) \, d\mathbf{r}_2$ (in a nonquantum cloud of negative charge) separated by a distance r_{12}, multiplied by one-half (so that we do not count the $\mathbf{r}_1/\mathbf{r}_2$ repulsion energy and again the $\mathbf{r}_2/\mathbf{r}_1$ energy). The sum of infinitesimals is an integral and so

$$\Delta\langle V_{ee}[\rho_0] \rangle = \langle V_{ee}[\rho_0] \rangle - \frac{1}{2} \iint \frac{\rho_0(\mathbf{r}_1)\rho_0(\mathbf{r}_2)}{r_{12}} d\mathbf{r}_1 \, d\mathbf{r}_2 \tag{7.13}$$

Actually, the classical charge-cloud repulsion is somewhat inappropriate for electrons in that smearing an electron (a particle) out into a cloud forces it to repel itself, as any two regions of the cloud interact repulsively. This physically incorrect *electron self-interaction* will be compensated for by a good exchange-correlation functional (below).

Using (7.12) and (7.13), Eq. (7.11) can be written as

$$E_0 = \int \rho_0(\mathbf{r}) v(\mathbf{r}) \, d\mathbf{r} + \langle T_r[\rho_0] \rangle + \frac{1}{2} \iint \frac{\rho_0(\mathbf{r}_1)\rho_0(\mathbf{r}_2)}{r_{12}} d\mathbf{r}_1 \, d\mathbf{r}_2$$
$$+ \Delta\langle T[\rho_0] \rangle + \Delta\langle V_{ee}[\rho_0] \rangle \tag{7.14}$$

The sum of the kinetic energy deviation from the reference system and the electron–electron repulsion energy deviation from the classical system is called the exchange-correlation energy functional or the exchange-correlation energy, E_{XC} (strictly speaking, the functional is the prescription for obtaining the energy from the electron density function):

$$E_{XC}[\rho_0] \equiv \Delta\langle T[\rho_0] \rangle + \Delta\langle V_{ee}[\rho_0] \rangle \tag{7.15}$$

The $\Delta\langle T \rangle$ term represents the kinetic correlation energy of the electrons and the $\langle \Delta V_{ee} \rangle$ term the potential correlation energy and the exchange energy, although exchange and correlation energy in DFT do not have exactly the same significance here as in HF theory [28]; E_{XC} is negative. So using Eq. (7.15), Eq. (7.14) becomes

$$E_0 = \int \rho_0(\mathbf{r})v(\mathbf{r})\, d\mathbf{r} + \langle T_r[\rho_0]\rangle + \frac{1}{2}\iint \frac{\rho_0(\mathbf{r}_1)\rho_0(\mathbf{r}_2)}{r_{12}}d\mathbf{r}_1\, d\mathbf{r}_2 + E_{XC}[\rho_0] \quad (7.16)$$

Let us look at the four terms in the expression for the molecular energy E_0 in Eq. (7.16).

1. The first term (the integral of the density times the external potential) is

$$\int \rho_0(\mathbf{r})v(\mathbf{r})d\mathbf{r} = \int \left[\rho_0(\mathbf{r}_1) \sum_{\text{nuclei A}} -\frac{Z_A}{r_{1A}}\right]d\mathbf{r}_1 = -\sum_{\text{nuclei A}} Z_A \int \frac{\rho_0(\mathbf{r}_1)}{r_{1A}}d\mathbf{r}_1$$

$$(7.17)$$

If we know ρ_0 the integrals to be summed are readily calculated.

2. The second term (the electronic kinetic energy of the noninteracting-electrons reference system) is the expectation value of the sum of the one-electron kinetic energy operators over the ground state multielectron wavefunction of the reference system (Parr and Yang explain this in detail [29]). Using the compact Dirac notation for integrals:

$$\langle T_r[\rho_0]\rangle = \left\langle \Psi_r \left| \sum_{i=1}^{2n} -\frac{1}{2}\nabla_i^2 \right| \Psi_r \right\rangle \quad (7.18)$$

Since these hypothetical electrons are noninteracting Ψ_r can be written exactly (for a closed-shell system) as a single Slater determinant of occupied spin molecular orbitals (section 5.2.3.1; for a *real* system, the electrons interact and using a single determinant causes errors due to neglect of electron correlation (section 5.4)). Thus, for a four-electron system,

$$\Psi_r = \frac{1}{\sqrt{4!}} \begin{vmatrix} \psi_1^{KS}(1)\alpha(1) & \psi_1^{KS}(1)\beta(1) & \psi_2^{KS}(1)\alpha(1) & \psi_2^{KS}(1)\beta(1) \\ \psi_1^{KS}(2)\alpha(2) & \psi_1^{KS}(2)\beta(2) & \psi_2^{KS}(2)\alpha(2) & \psi_2^{KS}(2)\beta(2) \\ \psi_1^{KS}(3)\alpha(3) & \psi_1^{KS}(3)\beta(3) & \psi_2^{KS}(3)\alpha(3) & \psi_2^{KS}(3)\beta(3) \\ \psi_1^{KS}(4)\alpha(4) & \psi_1^{KS}(4)\beta(4) & \psi_2^{KS}(4)\alpha(4) & \psi_2^{KS}(4)\beta(4) \end{vmatrix} \quad (7.19)$$

The 16 spin orbitals in this determinant are the KS *spin orbitals* of the reference system; each is the product of a KS spatial orbital ψ_i^{KS} and a spin function α or β. Equation (7.18) can be written in terms of the spatial KS orbitals by invoking a set of rules (the Slater–Condon rules [30]) for simplifying integrals involving Slater determinants:

$$\langle T_r[\rho_0]\rangle = -\frac{1}{2}\sum_{i=1}^{2n}\langle \psi_i^{KS}(1)|\nabla_1^2|\psi_i^{KS}(1)\rangle \quad (7.20)$$

The integrals to be summed are readily calculated. Note that DFT *per se* does not involve wavefunctions, and the KS approach to DFT uses orbitals only as a way to calculate the noninteracting-system kinetic energy and the electron density function; see below.

3. The third term in Eq. (7.16), the classical electrostatic repulsion energy term, is readily calculated if ρ_0 is known.

4. This leaves us with the exchange-correlation energy functional, $E_{XC}[\rho_0]$ (Eq. (7.15)) as the only term for which some method of calculation must be devised. Devising accurate exchange-correlation functionals for calculating this energy term from the electron density function is the main problem in DFT research. This is discussed in section 7.2.3.4.

Written out more fully, then, Eq. (7.16) is

$$E_0 = -\sum_{\text{nuclei A}} Z_A \int \frac{\rho_0(\mathbf{r}_1)}{r_{1A}} d\mathbf{r}_1 - \frac{1}{2}\sum_{i=1}^{2n}\langle\psi_i^{KS}(1)|\nabla_1^2|\psi_i^{KS}(1)\rangle$$
$$+ \frac{1}{2}\iint \frac{\rho_0(\mathbf{r}_1)\rho_0(\mathbf{r}_2)}{r_{12}}d\mathbf{r}_1\,d\mathbf{r}_2 + E_{XC}[\rho_0] \tag{7.21}$$

The term most subject to error is the relatively small $E_{xc}[\rho_0]$ term, which contains the (exactly) unknown functional.

The Kohn–Sham equations
The KS equations are obtained by utilizing the variation principle, which the second Hohenberg–Kohn theorem assures us applies to DFT. We use the fact that the electron density of the reference system, which is the same as that of our real system (see the definition at the beginning of the discussion of the KS energy), is given by [7]

$$\rho_0 = \rho_r = \sum_{i=1}^{2n}|\psi_i^{KS}(1)|^2 \tag{7.22}$$

where the ψ_i^{KS} are the KS spatial orbitals. Substituting the above expression for the orbitals into the energy expression of Eq. (7.21) and varying E_0 with respect to the ψ_i^{KS} subject to the constraint that these remain orthonormal (the spin orbitals of a Slater determinant are orthonormal) leads to *the KS* equations (the derivation is given by Parr and Yang [31]; the procedure is similar to that used in deriving the Hartree–Fock equations, section 5.2.3.4):

$$\left[-\frac{1}{2}\nabla_i^2 - \sum_{\text{nuclei A}}\frac{Z_A}{r_{1A}} + \int\frac{\rho(\mathbf{r}_2)}{r_{12}}d\mathbf{r}_2 + v_{XC}(1)\right]\psi_i^{KS}(1) = \epsilon_i^{KS}\psi_i^{KS}(1) \tag{7.23}$$

where ϵ_i^{KS} are the KS energy levels (the KS orbitals and energy levels are discussed later) and $v_{xc}(1)$ is the *exchange correlation potential*, arbitrarily designated here for electron number 1, since the KS equations are a set of one-electron equations (cf. the HF equations) with the subscript i running from 1 to n, over all the $2n$ electrons in the system. The exchange correlation potential is defined as the functional derivative of $E_{xc}[\rho(\mathbf{r})]$ with respect to $\rho(\mathbf{r})$:

$$v_{XC}(\mathbf{r}) = \frac{\delta E_{xc}[\rho(\mathbf{r})]}{\delta\rho(\mathbf{r})} \tag{7.24}$$

Functional derivatives, which are akin to ordinary derivatives, are discussed by Parr and Yang [32] and outlined by Levine [33]. We need the derivative v_{xc} for the KS equations (7.23), and the exchange-correlation functional itself for the energy equation (7.21).

The KS equations (7.23) can be written as

$$\hat{h}^{KS}(1)\psi_i^{KS}(1) = \epsilon_i^{KS}\psi_i^{KS}(1) \tag{7.25}$$

The KS operator $\hat{h}^{KS}$ is defined by Eq. 7.23; The significance of these orbitals and energy levels is considered later, but we note here that in practice they can be interpreted in a similar way to the corresponding HF and extended Hückel entities. Pure DFT theory has no orbitals or wavefunctions; these were introduced by Kohn and Sham only as a way to turn Eq. (7.11) into a useful computational tool, but if we can interpret the KS orbitals and energies in some physically useful way, so much the better.

The KS energy equation (7.21) is exact: if we knew the density function $\rho_0(\mathbf{r})$ and the exchange-correlation energy functional $E_{xc}[\rho_0]$, it would give the exact energy. The HF energy equation (Eq. (5.17), on the other hand, is an approximation that does not treat electron correlation properly. Similar considerations hold for the KS and HF equations, derived from the energy equations by minimizing the energy with respect to orbitals: even in the basis set limit, the HF equations would not give the correct energy, but the KS equations would, *if we knew the exact exchange-correlation energy functional*. In wavefunction theory we know how to improve on HF-level results: by using perturbational or configuration interaction treatments of electron correlation (section 5.4), but in DFT theory there is as yet no systematic way of improving the exchange-correlation energy functional. It has been said [34] that "while solutions to the [HF equations] may be viewed as exact solutions to an approximate description, the [KS equations] are approximations to an exact description!"

7.2.3.3 Solving the KS equations

First let us review the steps in carrying out a HF calculation (sections 5.2.3.6b–e). We start with a guess of the basis function coefficients c (cf. Eqs (7.26)), because the HF operator $\hat{F}$ involves the J and K integrals (section 5.2.3.6) which contain the wavefunction, and thus the c's (the wavefunction is composed of the c's and the basis functions). The operator is used with the basis functions to calculate the HF Fock matrix elements $F_{rs} = \langle \phi_r | \hat{F} | \phi_s \rangle$ which constitute the HF Fock matrix $\mathbf{F}$. An orthogonalizing matrix calculated from the overlap matrix $\mathbf{S}$ puts $\mathbf{F}$ into a form $\mathbf{F}'$ that satisfies $\mathbf{F}' = \mathbf{C}'\epsilon\mathbf{C}'^{-1}$ (section 5.2.3.6b). Diagonalization of $\mathbf{F}'$ gives a coefficients matrix $\mathbf{C}'$ and an energy levels matrix ϵ; transforming $\mathbf{C}'$ to $\mathbf{C}$ gives the matrix with the coefficients corresponding to the original basis set expansion of Eq. (7.26), and these are then used as a new guess to calculate a new $\mathbf{F}$; the process continues till it converges satisfactorily on the c's (i.e. the wavefunction) and the energy levels (which can be used to calculate the electronic energy); the procedure was shown in detail in section 5.2.3.6e.

The standard way of solving the KS eigenvalue equations, like that of solving the HF equations, which they resemble, is to expand the KS orbitals in terms of basis functions ϕ:

$$\psi_i^{KS} = \sum_{s=1}^{m} c_{si} \phi_s \quad i = 1, 2, 3, \ldots, m \tag{7.26}$$

This is exactly the same as was done with the HF orbitals in section 5.2.3.6a, and in fact the same basis functions are often used as in wavefunction theory, although as in all calculations designed to capture electron correlation, sets smaller than split-valence (section 5.3.3) should not be used; a popular basis in DFT calculations is the 6-31G*. Substituting the basis set expansion into the KS equations (7.23), (7.25) and multiplying by ϕ_1, ϕ_2, ..., ϕ_m leads, as in section 5.2.3.6a, to m sets of equations, each set with m equations, which can all be subsumed into a single matrix equation analogous to the HF equation $\mathbf{FC} = \mathbf{SC\epsilon}$. The key to solving the KS equations then becomes, as in the standard HF method, the calculation of Fock matrix elements and diagonalization of the matrix (section 5.2.3.6b). In a DFT calculation we start with a guess of the density $\rho(\mathbf{r})$, because this is what we need to obtain an explicit expression for the KS Fock operator $\hat{h}^{KS}$ (Eqs (7.23), (7.25), (7.24)). This guess is usually a noninteracting atoms guess, obtained by summing the electron densities of the individual atoms of the molecule, at the molecular geometry. The KS Fock matrix elements $h_{rs} = \langle \phi_r | \hat{h}^{KS} | \phi_s \rangle$ are calculated and the KS Fock matrix is orthogonalized and diagonalized, etc., to give initial guesses of the c's in the basis set expansion of Eq. (7.26) (and also initial values of the ϵ's). These c's are used in Eq. (7.23) to calculate a better density function (the orbitals ψ in Eq. (7.22) are composed of the c's and the chosen basis set – Eq. (7.26)). This new density function is used to calculate improved matrix elements h_{rs} which in turn give improved c's and then an improved density function, the iterative process being continued until the electron density etc. converge. The final density and KS orbitals are used to calculate the energy from Eq. (7.21).

The KS Fock matrix elements are integrals of the Fock operator over the basis functions. Because useful functionals are so complicated, these integrals, specifically the $\langle \phi_r | v_{xc} | \phi_s \rangle$ integrals, unlike the corresponding ones in HF theory, cannot be solved analytically. The usual procedure is to approximate the integral by summing the integrand in steps determined by a *grid*. For example, suppose we want to integrate e^{-x^2} from $-\infty$ to ∞. This could be done approximately, using a grid of width $\Delta x = 0.2$ and summing from -2 to 2 (limits at which the function is small):

$$\int_{-\infty}^{\infty} e^{-x^2} dx = \int_{-\infty}^{\infty} f(x)\, dx \simeq 0.2 f(-2+0.2) + 0.2 f(-2+0.4) + \cdots + 0.2 f(2)$$

$$= 0.2(9.80) = 1.96$$

The integral is actually $\pi^{1/2} = 1.77$. For a function $f(x, y)$ the grid would define the steps in x and y and actually look like a grid or net, approximating the integral as a sum of the volumes of parallelepipeds, and for the DFT function $f(x, y, z)$ the grid specifies the steps of x, y, and z. Clearly the finer the grid the more accurately the integrals are approximated, and reasonable accuracy in DFT calculations requires (but is not guaranteed by) a sufficiently fine grid.

7.2.3.4   The exchange-correlation energy functional

We have to consider the calculation of the fourth term in the KS operator of Eq. (7.23), the exchange-correlation potential $v_{xc}(\mathbf{r})$. This is defined as the functional derivative of the exchange-correlation energy functional, $E_{xc}[\rho(\mathbf{r})]$, with respect to the electron density functional (Eq. (7.23)). To help make sense of this, consider the simple one-dimensional function $x\,e^{-x^2}$, a function of the coordinate x. One functional of this function is the prescription "multiply the function by x and integrate from zero to infinity", giving

$$F[f(x)] = F[x\,e^{-x^2}] = \int_0^\infty x^2 e^{-x^2} dx = \frac{\pi^{1/2}}{4}$$

which is a number. The functional derivative of $F[x\,e^{-x^2}]$ with respect to the function $x\,e^{-x^2}$ involves various derivatives [32,33] of the function $x\,e^{-x^2}$ and of the integrand $x^2 e^{-x^2}$:

$$\frac{\delta F[f(x)]}{\delta f(x)} = \frac{\delta F[x\,e^{-x^2}]}{\delta(x\,e^{-x^2})} = \frac{d(x^2\,e^{-x^2})}{d(x\,e^{-x^2})} + \text{other terms}$$

i.e. the functional derivative is a function of x. Analogously, the exchange-correlation energy functional $E_{xc}[\rho(\mathbf{r})]$, a functional of $\rho(\mathbf{r})$, is a number which depends on the function $\rho(\mathbf{r})$ and on just what mathematical form the functional E_{xc} has, while the exchange-correlation potential $v_{xc}(\mathbf{r})$, the functional derivative of $E_{xc}[\rho(\mathbf{r})]$, is a function of the variable $\mathbf{r}$, i.e. of x, y, z. Clearly, $v_{xc}(\mathbf{r})$ depends on $\rho(\mathbf{r})$ and, like $\rho(\mathbf{r})$, varies from point to point in the molecule. Devising good functionals $E_{xc}[\rho(\mathbf{r})]$ (and thus their derivatives $v_{xc}(\mathbf{r})$) is the main problem in density functional theory.

7.2.3.4a   The local density approximation (LDA)

The simplest approximation to $E_{xc}[\rho(\mathbf{r})]$ is within the framework of the *local density approximation*, LDA; this applies to a uniform (homogeneous) electron gas (or one in which the electron density $\rho(\mathbf{r})$ varies only very slowly with position). The term *local* was perhaps used because for any point only the conditions (the electron density) *at that point* are considered, in contrast to so-called nonlocal methods (see below) in which for each point a gradient, which samples the region a bit beyond that point, is taken into account . For the LDA the exchange-correlation energy functional E_{xc}^{LDA} and its derivative v_{xc}^{LDA} can be accurately calculated [35–37]. The Xα method of Slater (section 7.2.2) [19–22] is a special case of the LDA, developed before the KS approach, in which the (relatively small) correlation part of the exchange-correlation functional is neglected and the exchange functional used is

$$E_{xc}^{X\alpha} = E_{x}^{X\alpha} = -\frac{9}{8}\left(\frac{3}{\pi}\right)\alpha \int [\rho(\mathbf{r})]^{4/3}\, d\mathbf{r} \tag{7.27}$$

The parameter α is empirical; values of 1 to $\frac{2}{3}$ give reasonable results for atoms. For the place of the $X\alpha$ method within the LDA and comparisons of atomic $X\alpha$ calculations with KS LDA and with HF calculations, see Parr and Yang [36].

7.2.3.4b *The local spin density approximation (LSDA)*

Better results than with the LDA are obtained by an elaboration of the LDA in which electrons of α and β spin in the uniform electron gas are assigned different spatial KS orbitals ψ_α^{KS} and ψ_β^{KS}, from which different electron density functions ρ^α and ρ^β follow. This "unrestricted" LDA method (cf. UHF, section 5.2.3.6e) is called the local spin density approximation, LSDA, and has the advantages that it can handle systems with one or more unpaired electrons, like radicals, and systems in which electrons are becoming unpaired, such as molecules far from their equilibrium geometries; even for ordinary molecules it appears to be more forgiving toward the use of (necessarily) inexact E_{xc} functionals [37]. For species in which all the electrons are securely paired, the LSDA is equivalent to the LDA. Like E_{xc}^{LDA} and its functional derivative v_{xc}^{LDA}, E_{xc}^{LSDA} and v_{xc}^{LSDA} can be accurately calculated [38,39]. LSDA geometries, frequencies and electron-distribution properties tend to be reasonably good, but (as with HF calculations) the dissociation energies are very poor, and uniform electron gas-type LSDA calculations appear to have been largely replaced by an approach that involves not just the electron density, but its gradient.

7.2.3.4c *Gradient-corrected functionals and hybrid functionals*

Gradient-corrected functionals
The electron density in an atom or molecule varies greatly from place to place, so it is not surprising that the uniform electron gas model has serious shortcomings. Most DFT calculations nowadays use exchange-correlation energy functionals E_{xc} that not only involve the LSDA, but also utilize both the electron density *and* its gradient or slope (first derivatives with respect to position). These functionals are called *gradient-corrected*, or said to use the *generalized-gradient approximation* (GGA). They are also called *nonlocal* functionals, in contrast to the older, local LDA and LSDA functionals. The term nonlocal may refer to the fact that calculating the gradient of $\rho(\mathbf{r})$ at a point amounts to sampling the value of ρ an infinitesimal distance beyond the "local" point of the coordinates $\mathbf{r}$, since the gradient is the change in ρ over an infinitesimal distance divided by that distance (i.e. in a Taylor series expansion of a function about a point the first, second etc. terms represent sampling the function increasingly nonlocally [40]). Nevertheless, it has been suggested [41] that the term nonlocal be avoided in referring to gradient-corrected functionals.

The exchange-correlation energy functional can be written as the sum of an exchange-energy functional and a correlation-energy functional, both negative, i.e. $E_{XC} = E_x + E_c$; $|E_x|$ is much bigger than $|E_c|$. For the argon atom E_x is -30.19 hartrees, while E_c is only -0.72 hartrees, calculated by the HF method [42]. Thus, it is not surprising that gradient corrections have proved more effective when applied to the exchange-energy functional, and a major advance in practical DFT calculations was the introduction of the Becke 88 functional [43], a "new and greatly improved functional for the exchange energy" [44]. Another example of a gradient-corrected exchange-energy functional is the Gill 1996 (G96) functional. Examples of gradient-corrected correlation-energy functionals are the Lee–Yang–Parr (LYP) and the Perdew 1986 (P86) functionals. All these functionals are commonly used with Gaussian-type (i.e. functions with $exp(-\mathbf{r}^2)$) basis functions for representing the KS orbitals (Eq. (7.26)). A calculation with B88 for E_x, LYP for E_c, and the 6-31G* basis set (section 5.3.3) would be designated as a B88LYP/6-31G* or B88LYP/6-31G* calculation (the Gaussian 98 program [45] performs this with the keyword BLYP/6-31G*). A calculation with B88 for E_x and P86 for E_c is called B88-P86 or B88P86 (BP86 in Gaussian 98). Other possible combinations of E_x, E_c and basis set are G96LYP/6-31 + G* and G96P86/6-311G**. References to the basis sets in Gaussian are given in the Gaussian manual and the online help. Sometimes rather than the analytical functions that constitute the standard Gaussian basis sets, numerical basis sets are used. A numerical basis function is essentially a table of the values that an atomic orbital wavefunction has at many points around the nucleus, with empirical functions fitted to pass through these points. The empirical functions are used in calculations instead of the analytical Gaussian-type functions. One such basis set is the DN* basis [46], available in some versions of the program Spartan [47]. A calculation designated by Spartan as BP/DN* uses the B88 and P86 (above) functionals with the DN* numerical basis. BP/DN* calculations are said [48] to give results similar to those from BP86/6-311G* (section 5.3.3) calculations, and results in section 7.3.2.2b support this. A numerical basis with polarization functions on hydrogen, designated DN**, is also available.

Hybrid functionals

Hybrid functionals augment the DFT exchange-correlation energy with a term calculated from HF theory. In HF theory, one expression for the electronic energy is Eq. (5.17)

$$E = 2\sum_{i=1}^{n} H_{ii} + \sum_{i=1}^{n}\sum_{j=1}^{n}(2J_{ij} - K_{ij}) \qquad (5.17 = 7.28)$$

where the sums are over n occupied spatial orbitals. If we remove the core energy (the first term, involving only electron kinetic energy and electron–nucleus attraction) and the coulomb potential energy (involving the coulomb integrals K) from this equation, we are left with the exchange energy, involving only the double sum of the exchange integrals J (section 5.2.3.2):

$$E_x = -\sum_{i=1}^{n}\sum_{j=1}^{n} K_{ij} \qquad (7.29)$$

Substituting into Eq. (7.29) the KS orbitals, which are quite similar to the HF orbitals (section 7.3.5), gives an expression, based on KS orbitals, for the HF exchange energy:

$$E_x^{HF} = -\sum_{i=1}^{n}\sum_{j=1}^{n} \left\langle \psi_i^{KS}(1)\psi_j^{KS}(2) \left| \frac{1}{r_{ij}} \right| \psi_i^{KS}(2)\psi_j^{KS}(1) \right\rangle \qquad (7.30)$$

Since the KS Slater determinant is an exact representation of the wavefunction of the noninteracting-electrons reference system, E_x^{HF} is the exact exchange energy for a system of noninteracting electrons with electron density equal to the real system (of course the basis set used imposes a limit on the accuracy of the molecular orbitals and thus on E_x^{HF}).

Including in an LSDA gradient-corrected DFT expression for E_{XC} ($E_{XC} = E_x + E_c$) a weighted contribution of the expression for E_x^{HF} gives a HF/DFT exchange-correlation functional, commonly called a hybrid DFT functional. The most popular hybrid functional at present (and in fact the most popular DFT functional) is based on an exchange-energy functional developed by Becke in 1993, and modified by Stevens *et al.* in 1994 by introduction of the LYP 1988 correlation-energy functional. This exchange-correlation functional, called the Becke3LYP or B3LYP functional [49] is

$$E_{xc}^{B3LYP} = (1-a_0-a_x)E_x^{LSDA} + a_0 E_x^{HF} + a_x E_x^{B88} + (1-a_c)E_c^{VWN} + a_c E_c^{LYP} \quad (7.31)$$

Here E_x^{LSDA} is the kind of accurate "pure DFT" LSDA non-gradient-corrected exchange functional alluded to in section 7.2.3.4b, E_x^{HF} is the KS-orbital-based HF exchange energy functional of Eq. (7.30), E_x^{B88} is the Becke 88 exchange functional mentioned above, E_c^{VWN} is the Vosko, Wilk, Nusair function (VWN, or Slater VWN, SVWN function), which forms part of the accurate functional for the homogeneous electron gas of the LDA and the LSDA (sections 7.2.3.4a and 7.2.3.4b), and E_c^{LYP} is the LYP correlation functional mentioned above; E_x and E_c of the last three terms are gradient-corrected. The parameters a_0, a_x and a_c are those that give the best fit of the calculated energy to molecular atomization energies. This is thus a gradient-corrected, hybrid functional. Of those functionals that have been around long enough to be well-tested, the B3LYP functional is, by and large, the most useful one (see the various applications of DFT, below). No doubt improved functionals will be discovered, and high hopes have been expressed for hybrid functionals by some of the chief pioneers in density functional theory: "A true marriage of density functional and Hartree-Fock ideas and technologies has emerged, and a potentially very beneficial cross-fertilization between DFT and traditional wave function methods has begun." [13].

DFT calculations with functionals incorporating gradient corrections, or gradient corrections and the HF exchange term (hybrid functionals), can be speeded up with only a little loss in accuracy by a so-called perturbation method [50]. Here the KS equations ((7.23), (7.25)) are solved using the derivative $v_{xc}(\mathbf{r})$ (Eq. (7.24)) corresponding to the LSDA functional, which is of course simpler than the gradient-corrected or hybrid functional. The energy is then calculated from Eq. (7.21), now using the gradient-corrected or hybrid functional. This is a perturbation approach because the "real" system with the better functional is regarded as a perturbation of the system with the approximate KS orbitals and corresponding electron density function. This perturbation method is available in some versions of Spartan [47] (see the discussion of gradient-corrected functionals, above) as a pBP/DN* or pBP/DN** calculation.

7.3 APPLICATIONS OF DENSITY FUNCTIONAL THEORY

Levine has compared geometries, energies, etc. from DFT with those from molecular mechanics, *ab initio*, and semiempirical methods [51]. Hehre [52] and Hehre and

Lou [48] have provided extensive, very useful compilations of *ab initio*, semiempirical, DFT, and some molecular mechanics results. Calculations in this chapter by the author were done with the programs Titan [53] (B3LYP), Spartan [47] (pBP and MP2), Gaussian 94 [54], and Gaussian 98 [45] (for BP86, high-accuracy energies, and IR, UV, and NMR spectra).

7.3.1   Geometries

Building on the general considerations about the calculation of molecular geometries in section 5.5.1, we can examine the results of DFT geometry optimizations of the same set of 20 molecules that were used in Fig. 5.23, and Fig. 6.2. The geometries in Fig. 7.1 are analyzed in Table 7.1, and Table 7.2 gives the dihedral angles for the same eight molecules as in Table 5.8, and Table 6.2 (Fig. 7.1 corresponds to Figs 5.23 and 6.2, Table 7.1 to Tables 5.7 and 6.1, and Table 7.2 to Tables 5.8 and 6.2). The DFT methods chosen were B3LYP/6-31G* and pBP/DN*, which were mentioned in section 7.2.3.4c. The gradient-corrected hybrid B3LYP functional with the 6-31G* basis set was chosen because this is at present the most popular DFT method, and is available in most commercial computational chemistry packages that include DFT. The perturbationally implemented BP86 (section 7.2.3.4c) functional with the DN* numerical basis represents a gradient-corrected, non-hybrid method. It is found in some versions of Spartan, and was chosen because it is fast (see below) and comparable in accuracy to B3LYP/6-31G*. As mentioned above, it should give similar results to B88P86/6-311G* (designated BP86/6-311G* in Gaussian 98) calculations; in section 7.3.2.2b we will see that pBP/DN*, B88P86/6-31G*, and B88P86/6-311G* actually do seem to give very similar results. Here, we compare geometries from B3LYP/6-31G*, pBP/DN*, MP2(FC)/6-31G* (the highest-level *ab initio* method in routine use), and experiment.

This survey suggests that B3LYP/6-31G* gives excellent geometries, quite similar to those from MP2/6-31G*, and pBP/DN* gives good geometries. Only for bonds from C to O, N, F, Cl, or S is B3LYP/6-31G* significantly worse than MP2, and about the same as pBP/DN*; apart from these, B3LYP/6-31G* bond lengths are mostly within 0.009 of experiment. pBP bond lengths are mostly within 0.026 of experiment. O–H bonds were always slightly too long by all three methods, and C–H bonds were always slightly too long for pBP/DN*. All three methods give bond angles usually within 1° of experiment (the worst error was in the HCN angle of MeNH$_2$, 2.2° for pBP/DN*). The dihedral angles by all three methods are all mostly within 1°–3° of experiment, but FCH$_2$CH$_2$F, FCH$_2$CH$_2$OH, and ClCH$_2$CH$_2$OH show somewhat larger errors (up to ca. 6°). The worst dihedral errors were ca. 8° for FCH2CH2F by the DFT methods, which is similar to the MP2/6-31G* error of 5.9° for the HOCC dihedral of ClCH$_2$CH$_2$OH.

The mean error in 39 ($13 + 8 + 9 + 9$) bond lengths is ca. 0.004–0.01 Å for B3LYP and MP2, and ca. 0.01–0.025 for pBP. The mean error in 18 bond angles is ca. 0.6° for the three methods. For the 10 dihedrals the mean error is ca. 2°.

Geometry errors for 108 molecules were reported by Scheiner *et al.* [55], comparing several *ab initio* and DFT methods. They found that Becke's original three-parameter function (which they denote ACM, for adiabatic connection method; B3LYP was developed as a modification of this [49]), with a 6-31G*-type and with the 6-31G**

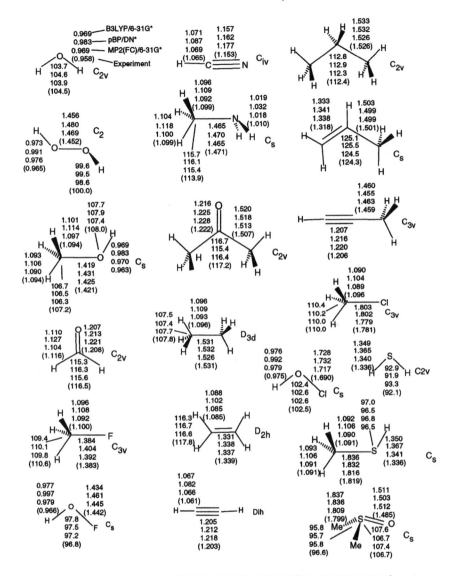

Figure 7.1. A comparison of some B3LYP/6-31G*, pBP/DN*, MP2(FC)/6-31G*, and experimental geometries. Calculations are by the author and experimental geometries are from [60]. Note that all CH bonds are ca. 1 Å, and all other bonds range from ca. 1.2–1.8 Å, and all bond angles (except for linear molecules) are ca. 90°–120°.

basis sets, gave average bond length errors of about 0.01 Å and bond angle errors of about 1.0°. They concluded that of the methods they examined ACM is the best choice for both geometries and reaction energies. St.-Amant *et al.* [56] also compared *ab initio* and DFT methods and found average dihedral angle errors of ca. 3° for 11 molecules using a perturbative gradient-corrected DFT method with an approximately

Table 7.1. Error in B3LYP/6-31G*, pBP/DN*, and MP2(FC)/6-31G* bond lengths and angles, from Fig. 7.1.

	Bond length errors, $r - r_{exp}$, Å			Bond angle errors, $a - a_{exp}$ angles
C–H	O–H, N–H, S–H	C–C	C–O, N, F, Cl, S	
Me₂OH -0.001/0.012/-0.004 0.007/0.020/ 0.003	H₂O 0.011/0.025/0.011	Me₂CO 0.013/0.011/0.006	MeOH -0.002/0.010/0.007	H₂O(HOH) -0.8/0.1/-0.6
HCHO -0.006/0.011/-0.012	H₂O₂ 0.008/0.026/0.011	CH₃CH₃ 0.000/0.001/-0.005	HCHO -0.001/0.005/0.013	H₂O₂(HOO) -0.4/-0.5/-1.4
MeF -0.004/0.008/-0.008	MeOH 0.006/0.020/0.007	CH₂CH₂ -0.008/-0.001/-0.002	MeF 0.001/0.021/0.009	MeOH (HCO) -0.5/-0.7/-0.9 (COH) -0.3/-0.1/-0.6
HCN 0.006/0.022/0.004	HOF 0.011/0.031/0.013	HCCH 0.002/0.009/0.015	HCN 0.004/0.009/0.024	HCHO (HCH) -1.2/-0.2/-0.9
MeNH₂ 0.005/0.019/0.001 -0.003/0.010/-0.007	MeNH₂ 0.009/0.022/0.008	CH₃CH₂CH₃ 0.007/0.006/0.000	MeNH₂ -0.006/-0.001/-0.006	MeF (HCH) -1.2/-0.5/-0.8
CH₃CH₃ 0.000/0.013/-0.003	HOCl 0.001/0.017/0.004	CH₂CHCH₃ 0.002/-0.002/-0.002 0.015/0.023/0.020	Me₂CO -0.006/0.003/0.006	HOF (HOF) 1.0/0.7/0.4
CH₂CH₂ 0.003/0.017/0.000	H₂S 0.013/0.029/0.004	HCCCH₃ 0.001/-0.004/0.004 0.001/0.010/0.014	MeCl 0.022/0.021/-0.002	MeNH₂(HCN) 1.8/2.2/1.5

CHCH	MeSH	MeSH	Me$_2$CO(CCC)
0.006/0.021/0.005	0.014/0.031/0.005	0.017/0.013/−0.003	−0.5/−1.8/−0.8
MeCl		Me$_2$SO	CH$_3$CH$_3$(HCH)
−0.006/0.008/−0.007		0.038/0.037/0.010	−0.3/−0.4/−0.1
MeSH			CH$_2$CH$_2$(HCH)
0.002/0.015/0.000			−1.5/−1.1/−1.2
0.001/0.015/−0.001			CH$_3$CH$_2$CH$_3$(CCC)
			0.4/0.5/−0.1
			CH$_2$CHCH$_3$(CCC)
			0.8/1.2/0.2
			MeCl (HCH)
			0.4/0.2/0.0
			H$_2$S(HSH)
			0.2/−0.2/1.2
			MeSH (CSH)
			0.5/0.0/0.3
8+, 4−, one 0	7+, 1−, one 0	5+, 4−, none 0	8+, 10−, none 0
13+, 0−, none 0	6+, 3−, none 0	6+, 10−, two 0	
4+, 7−, two 0	5+, 3−, one 0	6+, 11−, one 0	
mean of 13:	mean of 9:	mean of 9:	mean of 18:
0.004/0.015/0.004	0.005/0.007/0.008	0.011/0.013/0.009	0.8/0.6/0.7

Errors are given as B3LYP/pBP/MP2. In some cases (e.g. MeOH) errors for two bonds are given, on one line and the line below. A minus sign means that the calculated value is less than the experimental. The numbers of positive, negative, and zero deviations from experiment are summarized at the bottom of each column. The averages at the bottom of each column are arithmetic means of the absolute values of the errors

Table 7.2. B3LYP/6-31G*, pBP/DN*, MP2(FC)/6-31G* and experimental dihedral angles (degrees). In each case the starting structure was an AM1 geometry

Molecule	Dihedral Angles				Errors
	B3LYP	*pBP*	*MP2*	Exp.	
HOOH	119.3	116.4	121.3	119.1[a]	0.3/−2.7/2.2
FOOF	87.2	89.2	85.8	87.5[b]	−0.3/1.7/−1.7
FCH$_2$CH$_2$F	70.0	69.2	69.0	73[b]	−3.0/−4/−4
(FCCF)					
FCH$_2$CH$_2$OH					
(FCCO)	63.3	64.0	60.1	64.0°	−0.7/0.0/−3.9
(HOCC)	62.7	62.6	54.1	54.6°	8.1/8.0/−0.5
ClCH$_2$CH$_2$OH					
(ClCCO)	61.2	63.1	65.0	63.2[b]	−2.0/−0.1/1.8
(HOCC)	60.0	62.5	64.3	58.4[b]	1.6/4.1/5.9
ClCH$_2$CH$_2$F	66.7	69.6	65.9	68[b]	−1.3/1.6/−2.1
(ClCCF)					
HSSH	91.0	90.0	90.4	90.6[a]	0.4/−0.6/−0.2
FSSF	89.1	88.6	88.9	87.9[b]	1.2/0.7/1.0
					Deviations:
					5+, 5−/5+, 4−, one 0/4+, 6−
					mean of 10:
					1.9/2.4/2.3

[a]Hehre [60], pp. 151, 152. [b]M. D. Harmony, V. W. Laurie, R. L. Kuczkowski, R. H. Schwenderman, D. A. Ramsay, F. J. Lovas, W. H. Lafferty, and A. G. Makai, "Molecular Structures of Gas-Phase Polyatomic Molecules Determined by Spectroscopic Methods", J. Physical and Chemical Reference Data, **1979**, *8*, 619–721. [c]J. Huang and K. Hedberg, J. Am. Chem. Soc., **1989**, *111*, 6909.
Errors are given in the *Errors* column as B3LYP/pBP/MP. A minus sign means that the calculated value is less than the experimental. The numbers of positive and negative deviations from experiment and the average errors (arithmetic means of the absolute values of the errors) are summarized at the bottom of the *Errors* column. Calculations are by the author; references to experimental measurements are given for each measurement. Some molecules have calculated minima at other dihedrals in addition to those given here, e.g. FCH$_2$CH$_2$F at 180°

6-311G*-type basis set. These workers found average bond length errors of, e.g. 0.01 Å for C−H and 0.009 Å for C−C single bonds, and average bond angle errors of 0.5°. El-Azhary reported B3LYP with the 6-31G** and cc-pVDZ basis sets to give slightly better geometries than MP2, although MP2 avoids the occasional large errors given by B3LYP [57]. The effect of using different basis sets was minor. In a comparison of HF, MP2 and DFT (five functionals), Bauschlicher found B3LYP to be the best method overall [58].

B3LYP/6-31G*, pBP/DN*, and MP2(FC)/6-31G* geometries are further compared, for the species in four reaction profiles, in Fig. 7.2. These correspond to the *ab initio* comparisons of Fig. 5.21 and the semiempirical comparisons of Fig. 6.3. For the reactants and products, the DFT deviations from the MP2 geometries are not more than 0.026 Å (CH$_3$CN, 1.180 Å cf. 1.154 Å) and 0.8° (CH$_3$CN, 110.2° cf. 109.4°). For the transition states the maximum deviations from the MP results are −0.107 Å

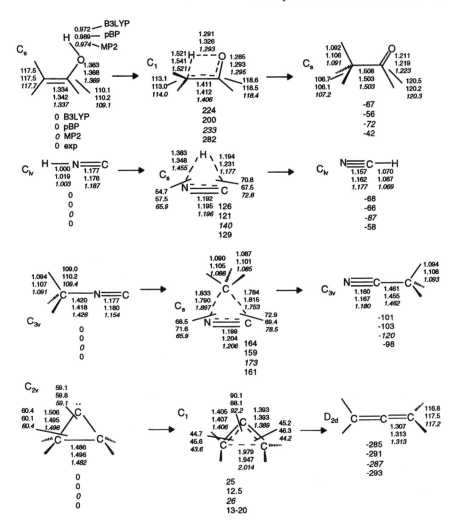

Figure 7.2. Geometries (Å and degrees) and relative energies (kJ mol⁻¹) for four reactions (most Hs are omitted for clarity). Geometries and energies are B3LYP/6-31G*, pBP/DN*, MP2(FC)/6-31G* and (for energies) experimental (the MP2 geometries and MP2 and experimental energies are as in Fig. 5.21). Calculated energies are corrected for ZPE (see text). The experimental energies are from [61]. Calculations are by the author.

(HNC reaction, 1.383 Å cf. 1.455 Å) and −11.2° (HNC reaction, 54.7° cf. 65.9°). We can assume that the deviations here from the true geometries are similar for the reactants and products to the cases of other normal molecules, as discussed above, i.e. that the DFT results are good to excellent (with B3LYP probably better than pBP). For transition states experimental geometries are not yet available, and we cannot be sure that the MP2 geometries are superior to the DFT ones. The B3LYP and pBP geometries are quite similar, with the largest discrepancies again being in the transition state for the HNC isomerization: 0.035 Å (1.383 Å cf. 1.348 Å) and 3.3° (70.8° cf. 67.5°). The consistency of the two DFT methods and their good agreement with MP2 (the "errors" for the HNC reaction transition state should not affect any qualitative conclusions one might draw) suggest that these DFT methods are quite comparable to MP2/6-31G* in calculating transition state geometries.

Hehre has compared bond lengths calculated by the DFT non-gradient-corrected SVWN method, B3LYP, and MP2, using the 6-31G (no polarization functions) and 6-31G* basis sets [59]. This work confirms the necessity of using polarization functions with the correlated (DFT and MP2) methods to obtain reasonable results, and also shows that for equilibrium structures (i.e. structures that are not transition states) there is little advantage to correlated over HF methods as far as geometry is concerned, a conclusion presented in section 5.5.1. Hehre and Lou [48] carried out extensive comparisons of HF, MP2, and DFT (SVWN, pBP, B3LYP) methods with 6-31G* and larger basis sets, and the numerical DN* and DN** bases. For a set of 16 hydrocarbons, MP2/6-311 + G(2d, p), B3LYP/6-311 + G(2d, p), pBP/DN**, and pBP/DN* calculations gave errors of 0.005, 0.006, 0.010, and 0.010 Å, respectively. HF/6-311 + G(2d, p) and SVWN calculations also gave errors of 0.010 Å. For 14 C–N, C–O and C=O bond lengths B3LYP and pBP (errors of 0.007 and 0.008 Å) were distinctly better than HF and SVWN (errors of 0.022 and 0.014 Å, respectively). The overall indication from the literature and the results in Figs 7.1 and 7.2 and Table 7.1 is that B3LYP/6-31G* calculations give excellent geometries and pBP/DN* calculations give good geometries. Larger basis sets may increase the accuracy, but the increase in time may not make this worthwhile. DFT calculations appear to be "saturated" more quickly by using bigger basis sets than are *ab initio* calculations: Merrill *et al.* noted that "Once the double split-valence level is reached, further improvement in basis set quality offers little in the way of structural or energetic improvement." [34]; Stephens *et al.* report that "Our results also show that B3LYP calculations converge rapidly with increasing basis set size and that the cost-to-benefit ratio is optimal at the 6-31G* basis set level. 6-31G* will be the basis set of choice in B3LYP calculations on much larger molecules [than $C_4H_6O_2$]." [49]. The results in section 7.3.2.2b, regarding Fig. 7.3, support the view that the 6-31G* basis nearly saturates gradient-corrected functionals.

7.3.2   Energies

7.3.2.1   Energies: preliminaries

Usually, we seek from a DFT calculation, as from an *ab initio* or semiempirical one, geometries (preceding section) and energies. Like an *ab initio* energy, a DFT energy is relative to the energy of the nuclei and electrons infinitely separated and at rest, i.e. it is the negative of the energy needed to dissociate the molecule into its nuclei and electrons. AM1 and PM3 semiempirical energies (section 6.3.2) are heats of formation, and by parameterization include zero-point energies. In contrast, an *ab initio* (section 5.2.3.6d) or DFT molecular energy, the energy printed out at the end of any calculation, is the energy of the molecule sitting motionless at a stationary point (section 2.2) on the potential energy surface; it is the purely electronic energy plus the internuclear

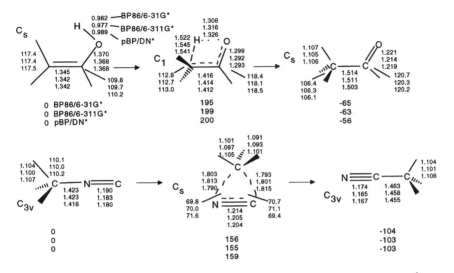

Figure 7.3. Comparison of geometries (Å and degrees) and relative energies (kJ mol⁻¹) from BP86/6-31G*, BP86/6-311G*, and pBP/DN* calculations (cf. Fig. 7.2). Calculations are by the author.

repulsion. In accurate work this "raw" energy should be corrected by adding the zero-point vibrational energy, to obtain the total internal energy at 0 K. Analogously to the HF equation in section 5.2.3.6d, Eq. (94) we have

$$E_{0K}^{total} = E_{DFT}^{total} + ZPE \tag{7.32}$$

(The Gaussian programs actually denote the DFT energy called here E_{DFT}^{total} as HF, e.g. (in hartrees or atomic units) HF = −308.86101). The main advantage of DFT over HF calculations is in being able to provide, in a comparable time, superior energy-difference results (reaction energies, activation energies, and electronic transition energies, i.e. UV spectra). We concentrate on energies from B3LYP/6-31G* and pBP/DN* calculations for the same reasons that these two methods were chosen for the geometry calculations in section 7.3.1.

7.3.2.2 Energies: calculating quantities relevant to thermodynamics and kinetics

7.3.2.2a Thermodynamics
Let us first see how DFT handles a case where HF fails badly: homolytic breaking of a covalent bond (section 5.4.1). Consider the reaction

$$H_3C-CH_3 + E_{diss} \rightarrow H_3C^{\bullet} \quad {}^{\bullet}CH_3$$

In principle, the dissociation energy can be found simply as the energy of two methyl radicals minus the energy of ethane. Table 7.3 (cf. Table 5.5) shows the results of HF/6-31G*, MP2(FC)/6-31G*, and two DFT (B3LYP/6-3G* and pBP) calculations.

Table 7.3. The C−C bond energy of ethane by HF (HF/6-31G*), MP2 (MP2(FC)/6-31G*), and DFT (B3LYP/6-31G* and pBP/DN*) calculations. For HF and MP2 the radical CH$_3^\bullet$ and the closed-shell molecule CH$_3$CH$_3$ were calculated by unrestricted and restricted methods, respectively; the LSDA of the DFT methods (which are also gradient-corrected) is analogous to the unrestricted method of *ab initio* calculations. The three numbers for each species are, respectively, the uncorrected energy, the ZPE, and the corrected energy. For the HF and MP2 calculations, the HF/6-31G* ZPE, corrected by a factor of 0.9136, was used, and for the DFT calculations the ZPE was calculated at that DFT level and is uncorrected (see text). Molecular energies are in hartrees and bond energies in hartrees and kJ mol^{-1} (2626× hartrees). The experimental C−C energy of ethane has been reported at from 368 to 377 kJ mol^{-1} [67]

Method/basis	Energy		
	CH$_3^\bullet$	CH$_3$CH$_3$	E(2CH$_3^\bullet$ − CH$_3$CH$_3$)
HF/6-31G*	−39.55899	−79.22876	0.09451
	0.02829	0.07285	248
	−39.53070	−79.15591	
MP2/6-31G*	−39.66875	−79.49474	0.14097
	0.02829	0.07285	370
	−39.64046	−79.42189	
B3LYP/6-31G*	−39.83829	−79.83042	0.13826
	0.02983	0.07524	363
	−39.80846	−79.75518	
pBP/DN*	−39.84527	−79.84241	0.13720
	0.02878	0.07223	360
	−39.81649	−79.77018	

The HF and MP2 ZPEs are from HF/6-31G* optimizations/frequency jobs, and were corrected by multiplying by 0.9136 [62]; the DFT ZPEs are from the B3LYP/6-3G* and pBP optimizations/frequencies and are uncorrected, since the correction factor should be close to unity [62]. As we saw in chapter 5, the HF method gives a very poor estimate of the bond energy, while the MP2 calculation provides a good value. Here we see that the B3LYP and pBP methods give a bond energy only slightly less than that from the MP2 calculation. Note, however, that in general, to get good (reliably within 10 or even 20 kJ mol^{-1}) atomization energies and reaction energies simply by subtraction requires "high-accuracy" methods, using higher-level electron correlation and bigger basis sets (section 5.5.2.2b). Martell *et al.* tested six functionals on 44 atomization energies and six reactions and concluded that the best atomization energies were obtained with hybrid functionals, but slightly better reaction enthalpies were obtained with non-hybrid ones [63]. St.-Amant *et al.* found that gradient-corrected functionals gave good geometries and energies for conformers; the dihedrals were on average within 4° of experiment and the relative energies were nearly as accurate as those from MP2 [56]. Scheiner *et al.* found that, as for geometries, Becke's original three-parameter function (usually called ACM, adiabatic connection method) gave the best reaction energies [55].

Table 7.4. Reaction energies and relative energies of isomers (B3LYP/6-31G* and pBP/DN*). The energies in hartrees include the ZPE. The calculations on O_2 are on triplet O_2. Calculations are by the author, experimental energies are from [52]

Reactants E, h	Products E, h	Reaction energy, or relative energy of isomers — Calculated, h/kJ mol^{-1}	Exp, kJ mol^{-1}
$H_2 + Cl_2$ $-1.16533 + (-920.34870)$ $= -921.51403$ $-1.16526 + (-920.50896)$ $= -921.67422$	2HCl $2(-460.78901) =$ $2(-460.86802) =$ -921.73604	$-921.57802 - (-921.51403)$ $-921.57802 = -0.06399/-168$ $-921.73604 - (-921.67422)$ $= -0.06182/-162$	-192
$2H_2 + O_2$ $2(-1.16533) + (-150.31626)$ $= -152.64692$ $2(-1.16526) + (-150.40153)$ $= -152.73205$	$2H_2O$ $2(-76.38779)$ $= -152.77558$ $2(-76.43797)$ -152.87594	$-152.77558 - (-152.64692)$ $= -0.12866/-338$ $-152.87594 - (-152.73205)$ $= -0.14389/-378$	-523
trans-2-butene -157.11841 -157.15607	*cis*-2-butene -157.11600 -157.15356	$-157.11600 - (-157.11841)$ $= 0.00241/6.3$ $-157.15356 - (-157.15607)$ $= 0.00251/6.6$	4.60
HCN -93.40621 -93.44340	HNC -93.38040 -93.41810	$-93.38040 - (-93.40621)$ $= 0.02581/67.8$ $-93.41810 - (-93.44340)$ $= 0.02530/66.4$	60.7

Table 7.5. Energy errors for hydrogenation reactions, isomerizations, bond separation reactions, and proton affinities, using four different methods; the basis set is 6-31G*. The errors, in $kJ\,mol^{-1}$, in each case the arithmetic mean of the absolute deviations from experiment of 10 reactions, were calculated from the data in Hehre [68]

Reaction	Method			
	HF	SVWN	MP2	B3LYP
Hydrogenation	15	20	17	23
Isomerization	15	19	16	17
Bond separation	11	5	4	10
Proton affinity	14	18	11	7

Some reaction energies and relative energies of isomers are compared in Table 7.4 (cf. Table 5.9, Table 6.3). The results for the HCl and H_2O formation reactions show no particular improvement over the HF calculations in Table 5.9. The 2-butene *cis/trans* energy difference error is about the same, but the HCN/HNC energy difference is closer to the experimental value. A large number of energy difference comparisons have been published for the pBP/DN* and pBP/DN** methods cf. HF, MP2 and experiment [48], and for B3LYP/6-31G* methods compared to HF, MP2 and experiment [52]. These comparisons involve homolytic dissociation, various reactions particularly hydrogenations, acid–base reactions, isomerizations, isodesmic reactions, and conformational energy differences. This wealth of data shows that while gradient-corrected DFT and MP2 calculations are vastly superior for homolytic dissociations, for "ordinary" reactions (involving only closed-shell species), their advantage is much less marked; for example, HF/3-21G, HF/6-31G*, SVWN/6-31G* (non-gradient-corrected DFT), all usually give energy differences similar to those from B3LYP/6-31G* and pBP/DN* and in fair agreement with experiment. Table 7.5 compares errors for hydrogenations, isomerizations, bond separation reactions (a kind of isodesmic reaction), and proton affinities; the methods are HF, SVWN, MP2, and B3LYP, all using the 6-31G* basis. In two of the four cases (hydrogenation and isomerization) the HF/6-31G* method gave the best results; in one case MP2 was best and in one case B3LYP. For the energy comparison of normal (not involving transition states) closed-shell organic species correlated methods like MP2 and DFT seem to offer little or no advantage, unless one needs accuracy within ca. $10–20\,kJ\,mol^{-1}$ of experiment, in which case high-accuracy methods should be used. The strength of gradient-corrected DFT methods appears to lie largely in their ability to give homolytic dissociation energies and activation energies with an accuracy comparable to that from MP2, but at a time cost comparable to that from HF calculations.

Bauschlicher *et al.* compared various methods and recommended B3LYP over HF and MP2, to a large extent on the basis of the performance of B3LYP with regard to atomization energies and transition metal compounds [58]. Wiberg and Ochterski compared HF, MP2, MP3, MP4, B3LYP, CBS-4 and CBS-Q with experiment in calculating energies of isodesmic reactions (hydrogenation and hydrogenolysis, hydrogen transfer, isomerization, and carbocation reactions) and found that while MP4/6-31G* and CBS-Q were the best, B3LYP/6-31G* was also generally satisfactory [64]. Rousseau and

Mathieu developed an economical way of calculating heats of formation by performing BP/DN* calculations on molecular mechanics geometries; rms deviations from experiment were about $16\,\mathrm{kJ\,mol^{-1}}$ for a variety of compounds [65]. Ventura *et al.* found DFT to be better than CCSD(T) (a high-level *ab initio* method) for studying the thermochemistry of compounds with the O–F bond [66].

7.3.2.2b Kinetics

Consider the reaction profiles in Fig. 7.2. In all four cases, the B3LYP/6-31G* and pBP/DN* activation energies differ by no more than $24\,\mathrm{kJ\,mol^{-1}}$ (224/200, 126/121, 164/159, 25/12.5 kJ mol^{-1}) and give a reasonable indication of the height of the barrier. The worst case is that of the ethenol (vinyl alcohol, H_2C=CHOH) isomerization, where the calculated barrier differs from the reported experimental one by from $58\,\mathrm{kJ\,mol^{-1}}$ (B3LYP) to $82\,\mathrm{kJ\,mol^{-1}}$ (pBP) (but the experimental barrier could be significantly in error [61a]).

Let us compare the activation energies for the reactions in Fig. 7.2 with high-accuracy (section 5.5.2.2d) energy difference values. We will compare the 0 K activation enthalpies, which is what these energy differences are (section 2.2), of the CBS-Q and G2 methods, two of the best high-accuracy methods, with the corresponding activation energies, in Fig. 7.2 (kJ mol^{-1}). For unimolecular reactions Arrhenius activation energies E_a exceed enthalpy differences $\Delta H^{\ddagger}$ by RT (Eq. (5.180)), but this is only $2.48\,\mathrm{kJ\,mol^{-1}}$ even at room temperature, 298 K.

H_2C=CHOH reaction

> CBS-Q, 237; G2, 237; B3LYP, 224; pBP, 200; exp, 282

HCN reaction

> CBS-Q, 128; G2, 124; B3LYP, 126; pBP, 121; exp 129

CH$_3$NC reaction

> CBS-Q, 164; G2, 161, B3LYP, 164; pBP, 159; exp, 161

Cyclopropylidene reaction

> CBS-Q, 22; G2, 22, B3LYP, 25; pBP, 12.5; exp 13 – 20

For the H_2C=CHOH isomerization the B3LYP/6-31G* activation energy is a little ($13\,\mathrm{kJ\,mol^{-1}}$) lower than the CBS-Q and G2 values (which are somewhat lower than the reported experimental value, which as stated above could be significantly in error). For the HCN isomerization the B3LYP/6-31G*, CBS-Q, G2 and experimental values are essentially the same. For the CH$_3$NC isomerization the B3LYP/6-31G*, CBS-Q, G2 and experimental values are again essentially the same, and for the cyclopropylidene isomerization the B3LYP/6-31G*, CBS-Q and G2 values are again essentially the same and equal to the higher end of the reported experimental values. In all cases the pBP/DN* activation energies are a little less than the CBS-Q, G2 and B3LYP values.

Figure 7.3 compares the results of BP86/6-31G*, BP86/6-311G*, and pBP/DN* calculations on the $H_2C=CHOH$ and CH_3NC reaction profiles. The BP86/6-31G* and BP86/6-311G* geometries are very similar and the relative energies are nearly identical, indicating that basis set saturation (section 7.3.1) has been essentially reached. The pBP/DN* geometries and energies resemble the BP86 ones quite closely, matching the BP86/6-311G* results a little more closely than the BP86/6-31G*. These results suggest that B3LYP/6-31G* activation energies are similar to those from the more time-demanding (below) CBS-Q and G2 methods, and are close to the experimental values; pBP/DN* and BP86/6-31G* (and BP86/6-311G*) activation energies may be a little (perhaps 5–25 kJ mol^{-1}) lower than those from B3LYP/6-31G*. Here are the times required for some DFT, CBS-Q, and G2 calculations (optimization + frequencies), in each case starting from an AM1 geometry; the jobs were run with Gaussian 94 [54] on a 600 MHz Pentium III computer:

Ethenol (vinyl alcohol, $H_2C=CHOH$)	
BP86/6-31G*	30.0 minutes, relative time 1
B3LYP/6-31G*	30.0 minutes, relative time 1.0
CBS-Q	53.4 minutes, relative time 1.8
G2	142.6 minutes, relative time 4.8

In a study of alkene epoxidation with peroxy acids, B3LYP/6-31G* gave an activation energy similar to that calculated with MP4/6-31G*//MP2/6-31G* but yielded kinetic isotope effects in much better agreement with experiment than did the *ab initio* calculation [69]. Even better activation energies than from B3LYP (which it is said tends to underestimate barriers [70,71]) have been reported for the BH&H-LYP functional [71–74]. In a study by Baker *et al.* [75] of 12 organic reactions using seven methods (semiempirical, *ab initio*, and DFT), B3PW91/6-31G* was best (average and maximum errors 15.5 and 54 kJ mol^{-1}) and B3LYP/6-31G* second best (average and maximum errors 25 and 92 kJ mol^{-1}). Jursic studied 28 reactions and recommended "B3LYP or B3PW91 with an appropriate basis set", but warned that highly exothermic reactions with a small barrier (ca. 10–20 kJ mol^{-1}) involving hydrogen radicals "are particularly difficult to reproduce." [76]. Barriers "above 10 kcal mol^{-1} [ca. 40 kJ mol^{-1}] should be reliable. Lower activation energies should be underestimated by 3–4 kcal mol^{-1} [ca. 13–17 kJ mol^{-1}]." [76]. As with thermodynamic energy differences (i.e. energy differences not involving a transition state), consistently obtaining activation energies accurate to 10–20 kJ mol^{-1} with some confidence requires one of the high-accuracy methods.

Density functional transition states and activation energies have their problems. Merrill *et al.* found that for the fluoride ion-induced elimination of HF from CH_3CH_2F none of the 11 functionals tested (including B3LYP) was satisfactory, by comparison with high-level *ab initio* calculations. Transition states were often looser and stabler than predicted by *ab initio*, and in several cases a transition state could not even be found. They concluded that hybrid functionals offer the most promise, and that "the ability of density functional methods to predict the nature of TSs demands a great deal more attention than it has received to date." [34]. Note that it is assumed here that high-level *ab initio* calculations are more trustworthy than DFT, presumably because current DFT functionals are really semiempirical.

7.3.3 Frequencies

The general remarks and theory about frequencies that were given in section 5.5.2.2d apply to DFT frequencies also. As with *ab initio* frequency calculations, but unlike semiempirical, one reason for calculating DFT frequencies is to get zero-point energies to correct the frozen-nuclei energies. The frequencies are also used to characterize the stationary point as a minimum, transition state, etc., and to predict the IR spectrum. As usual the wavenumbers ("frequencies") are the mass-weighted eigenvalues of the hessian, and the intensities are calculated from changes in dipole moment incurred by the vibrations.

Unlike *ab initio* and semiempirical frequencies, DFT frequencies are not always significantly lower than observed ones (indeed, calculated values slightly higher than experimental frequencies have been reported). Here are some correction factors that have been calculated for various functionals, as well as for some *ab initio* and semiempirical methods (slightly different factors were recommended for accurate calculations of ZPE) [62]; except for HF/3-21G the basis set for the *ab initio* and DFT methods is 6-31G*:

HF/3-21G	HF/6-31G*	MP2(FC)	AM1	BLYP	BP86	B3LYP	B3PW91
0.909	0.895	0.943	0.953	0.995	0.991	0.961	0.957

The BLYP/6-31G* and BP/86 (and probably the pBP/DN*) correction factors are very close to unity. For the frequencies of polycyclic aromatic hydrocarbons calculated by the B3LYP/6-31G* method, Bauschlicher multiplied frequencies below $1300\,cm^{-1}$ by 0.980 and frequencies above this by 0.967 [58]. In their paper introducing the modification of Becke's hybrid functional to give the B3LYP functional, Stephens *et al.* studied the IR and CD spectra of 4-methyl-2-oxetanone and recommended the B3LYP/6-31G* as an excellent and cost-effective way to calculate these spectra [49]. With six different functionals, Brown *et al.* obtained an agreement with experimental fundamentals of ca. 4–6%, except for BHLYP [77].

Let us examine the IR spectra of acetone, benzene, dichloromethane, and methanol, the same four compounds used in chapters 5 and 6 to illustrate calculated *ab initio* and semiempirical IR spectra. The DFT spectra in Figs 7.4–7.7 are based on the data in Table 7.6, and the MP2(FC)/6-31G* spectra are those used in Figs 5.33–5.36 and 6.5–6.8. The DFT methods chosen were B3LYP/6-31G* and BP86/6-31G*. The latter is used here as a substitute for pBP/DN*, which we often used for geometries and energies (for the reasons we concentrated on B3LYP and pBP, see the beginning of section 7.3.1), since as it is implemented in Spartan [47] intensities are not available from pBP/DN*. Recall (Fig. 7.3) that geometries and relative energies from B3LYP/6-31G*, B3LYP/6-311G*, and pBP/DN* were very similar; it is expected that IR spectra calculated by BP86/6-31G* and pBP/DN* will also be similar. From Figs 7.4–7.7 we see that the B3LYP/6-31G* and BP88/6-31G* spectra are very similar, and these DFT

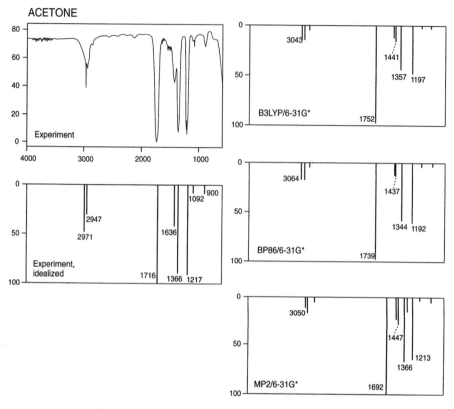

Figure 7.4. Experimental (gas phase) and DFT (B3LYP/6-31G* and BP86/6-31G*) and MP2(FC)/6-31G* calculated infrared spectra of acetone. The DFT spectra are based on the data in Table 7.6; the MP2 spectrum is that shown in Fig. 5.33.

spectra are quite similar to the MP2(FC)/6-31G* spectra. The MP2 spectra simulate the experimental spectra reasonably well, but as we saw in section 5.5.3, are not (for these "routine" molecules anyway) notably superior to the HF/3-21G and HF/6-31G* spectra.

7.3.4   Properties arising from electron distribution – dipole moments, charges, bond orders, atoms-in-molecules

The theory behind calculating dipole moments, charges, and bond orders, and using atoms-in-molecules analyses, was outlined in section 5.5.4; here the results of applying DFT calculations to these will be presented.

Dipole moments

Dipole moments seem to be more sensitive to geometry than most other properties, certainly more so than is energy (of course, for certain geometries and methods, errors may tend to cancel). Another point to note is that in the *perturbational* DFT method (7.2.3.4c) the gradient enters the calculation only to evaluate the energy (Eq. (7.21)), after the density has converged; properties like dipole moment that are calculated from

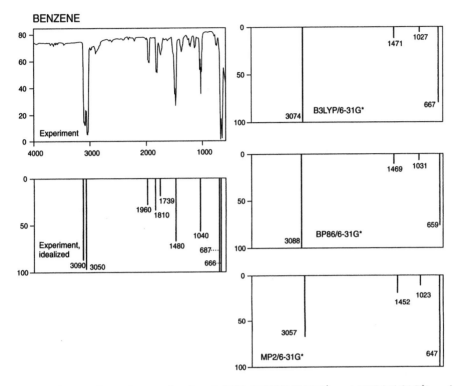

Figure 7.5. Experimental (gas phase) and DFT (B3LYP/6-31G* and BP86/6-31G*) and MP2(FC)/6-31G* calculated infrared spectra of benzene. The DFT spectra are based on the data in Table 7.6; the MP2 spectrum is that shown in Fig. 5.34.

the density will not be gradient-corrected. The gradient is applied only to energy and properties, like geometry, which are calculated from the energy. The dipole moment in such a calculation will reflect the gradient-correction only through the geometry optimization. In practice, the difference between perturbational and non-perturbational dipole moments is small [78]. Hehre [52] and Hehre and Lou [48] have provided quite extensive compilations of calculated dipole moments. These confirm that HF dipole moments tend to be larger than experimental, and show that electron correlation, through DFT or MP2, tends to lower the dipole moment, bringing it closer to the experimental value (e.g. for thiophene, from 0.80 D to 0.51 D for B3LYP; the MP2 value is 0.37 D and the experimental dipole moment is 0.55 D [48]).

Table 7.7 compares with experiment dipole moments, values calculated by B3LYP/6-31G*, pBP/DN*, by single-point B3LYP/6-31G* on AM1 geometries, and by MP2/FC/6-31G*, for 10 molecules (note that the pBP moments are not gradient-corrected, as explained above). The three DFT methods give quite similar average errors, ca. 0.1 D. Note that the smallest error, 0.09 D, is from the fast method of single-point B3LYP calculations on AM1 geometries, and the largest error, 0.34 D, is from the most time-consuming method, MP2 all the way. The DFT dipole moments are distinctly better than *ab initio* (errors ca. 0.3 D, Table 5.18) and semiempirical (errors ca.

DICHLOROMETHANE

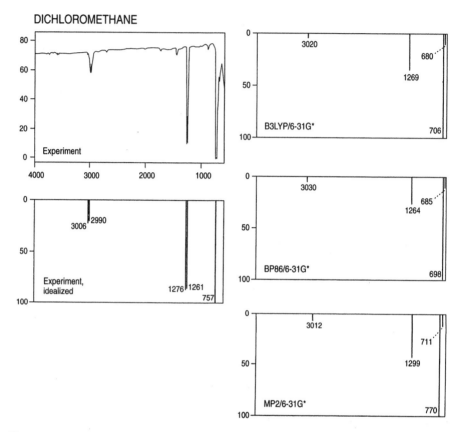

Figure 7.6. Experimental (gas phase) and DFT (B3LYP/6-31G* and BP86/6-31G*) and MP2(FC)/6-31G* calculated infrared spectra of dichloromethane. The DFT spectra are based on the data in Table 7.6; the MP2 spectrum is that shown in Fig. 5.35.

0.2–0.3 D, Table 6.5) moments. None of the methods consistently gives values accurate to within 0.1 D. Dipole moments of about the same accuracy as the DFT ones in Table 7.7 (possibly a bit less accurate than the B3LYP/6-31G*//AM1 ones) were obtained with single-point gradient-corrected calculations using a large basis set and geometries from perturbational gradient-corrected/6-311G**-type basis calculations. For 21 molecules the mean of the absolute deviations was 0.12 D [79]. Very accurate values (mean absolute deviation 0.06–0.07 D) can be obtained with gradient-corrected DFT and very large basis sets [55].

Charges and bond orders
The theory behind these was given in section 5.5.4. Recall that these parameters are not observables and so there are no experimental, "right" values to aim for; electrostatic potential charges and Löwdin bond orders are preferred to Mulliken charges and bond orders. The effect of various computational levels on atom charges has been examined [80].

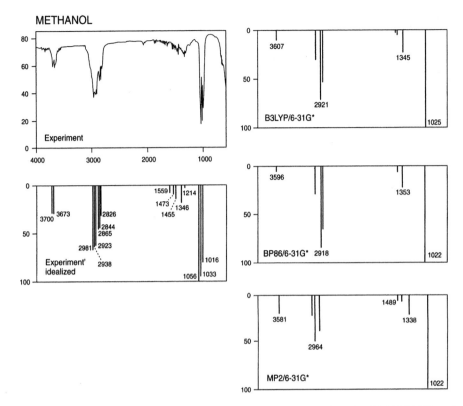

Figure 7.7. Experimental (gas phase) and DFT (B3LYP/6-31G* and BP86/6-31G*) and MP2(FC)/6-31G* calculated infrared spectra of methanol. The DFT spectra are based on the data in Table 7.6; the MP2 spectrum is that shown in Fig. 5.36.

Figure 7.8 shows charges and bond orders calculated for an enolate and a protonated enone system (the same as in Fig. 6.9), using B3LYP/6-31G* and HF/3-21G. The results are qualitatively similar regardless of whether one uses B3LYP or HF, or Mulliken vs. electrostatic potential/Löwdin. This is in contrast to the results in Fig. 6.9, where there were some large differences between the semiempirical and HF/3-21G values, and even between AM1 and PM3. For example, for the protonated species using the Mulliken method, AM1 and PM3 gave the oxygen a small negative charge, ca. -0.1, but the HF/3-21G method gave it a large negative charge, -0.63; even stranger, the terminal carbon had charges of 0.09, 0.23, and -0.25 by the AM1, PM3, and HF methods. In Fig. 7.8 the biggest differences among corresponding parameters is for the electrostatic potential charges in the protonated species, where the charges on the oxygen (-0.35 and -0.63) and on the carbonyl carbon (0.41 and 0.76) differ by a factor of about two. With both B3LYP and HF the terminal carbon of the enolate is counterintuitively assigned a bigger negative electrostatic potential charge than the oxygen, as was the case for AM1 and DFT. The calculated negative charge on the formally positive oxygen of the protonated molecule was commented on in section 6.3.4. As with the semiempirical values, bond orders are less variable here than are the charges, but even for this parameter

Table 7.6. Acetone, benzene, dichloromethane, and methanol, calculated IR spectra after discarding IR-inactive or very weak (less than 2 percent of the strongest band) frequencies, ignoring frequencies below $600\,cm^{-1}$, correcting frequencies, combining frequencies within $2\,cm^{-1}$ and their intensities, and normalizing intensities. Frequency correction factors [62]: B3LYP/6-31G*, 0.961; BP86/6-31G*, 0.991

	B3LYP/6-31G*		BP86/6-31G*	
	cm^{-1}	Intensity	cm^{-1}	Intensity
Acetone	860	4.3	848	5.6
	1087	2.8	1077	3.3
	1197	52	1192	58
	1357	45	1344	57
	1441	16	1437	18
	1457	13	1455	15
	1752	100	1739	100
	2932	4.9	2945	5.5
	2986	16	3006	16
	3043	14	3064	16
Benzene	667	75	659	75
	1027	6.2	1031	6.6
	1471	13	1469	11
	3074	100	3088	100
Dichloromethane	680	9.3	685	8.0
	706	100	698	100
	1269	34	1264	27
	3020	5.5	3030	5.4
Methanol	1025	100	1022	100
	1345	23	1353	21
	1452	6.2	1448	5.9
	1481	3.3	1481	3.5
	2880	53	2878	63
	2921	71	2918	80
	3010	29	3024	30
	3607	9.2	3596	5.9

there is one qualitative discrepancy: for the cation C/OH bond the Mulliken HF bond order is essentially single (1.18), while for the Löwdin B3LYP calculation the bond is essentially double (bond order 1.70). These results remind us that charges and bond orders are useful mainly for revealing *trends*, when a series of molecules, or stages along a reaction coordinate [81] are studied, all with the same methods (e.g. B3LYP/6-31G* and Löwdin bond orders).

Atoms-in-molecules

The atoms-in-molecules (AIM) analysis of electron density, using *ab initio* calculations, was considered in section 5.5.4. A comparison of AIM analysis by DFT with that by *ab initio* calculations by Boyd *et al.* showed that results from DFT and *ab initio*

Table 7.7. Some calculated dipole moments compared to experimental ones. Dipole moments are in Debyes. Calculations are by the author; experimental values are taken from [52] and [60]. For each method is given the number of positive and negative deviations from experiment and the arithmetic mean of the absolute values of the deviations. The basis set for the B3LYP and MP2 calculations is 6-31G*

	Computational method				
	B3LYP	pBP/DN*	B3LYP// AM1	MP2(fc)	Exp.
CH_3NH_2	1.47	1.54	1.31	1.60	1.31
H_2O	2.10	2.21	2.10	2.24	1.85
HCN	2.91	3.08	2.90	3.26	2.99
CH_3OH	1.70	1.84	1.68	1.95	1.70
Me_2O	1.28	1.37	1.25	1.60	1.30
H_2CO	2.19	2.34	2.23	2.84	2.34
CH_3F	1.72	2.01	1.65	2.11	1.85
CH_3Cl	2.09	2.01	1.91	2.21	1.87
Me_2SO	3.93	3.93	3.98	4.63	3.96
CH_3CCH	0.69	0.90	0.66	0.66	0.75
Deviations	3+, 6−, one 0	8+, 1−, one 0	4+, 6−, none 0	9+, 1−, none 0	
	mean 0.11	mean 0.14	mean 0.09	mean 0.34	

methods were similar, but gradient-corrected methods were somewhat better than the SVWN method, using QCISD *ab initio* calculations as a standard. DFT shifts the CN, CO, and CF bond critical points of HCN, CO, and CH_3F toward the carbon and increases the electron density in the bonding regions, compared to QCISD calculations [82].

7.3.5 Miscellaneous properties – UV and NMR spectra, ionization energies and electron affinities, electronegativity, hardness, softness and the Fukui function

UV spectra

In wavefunction theory, i.e. conventional quantum mechanics, UV spectra (electronic spectra) result from promotion of an electron from a molecular orbital to a higher-energy molecular orbital by absorption of energy from a photon: the molecule goes from the electronic ground state to an excited state. Since current DFT is said to be essentially a ground-state theory (e.g. references [11–14]), one might suppose that it could not be used to calculate UV spectra. However, there is an alternative approach to calculating the absorption of energy from light. One can use the time-dependent Schrödinger equation to calculate the effect on a molecule of a time-dependent electric field, i.e. the electric component of a light wave, which is an oscillating electromagnetic field, and can set the electron cloud of a molecule oscillating synchronously [83]. This is a semiclassical treatment in that it uses the Schrödinger equation but avoids equating the absorbed energy to *hv*, the energy of a photon. The calculation of UV spectra by

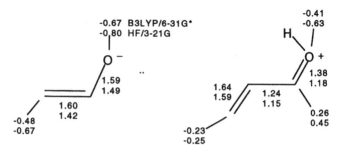

Mulliken charges and bond orders

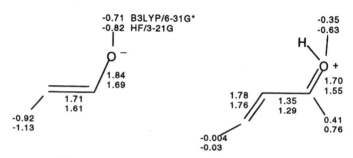

Electrostatic potential charges and Löwdin bond orders

Figure 7.8. Atom charges and bond orders calculated using B3LYP/6-31G* and HF/3-21G methods. Note that charges and bond orders involving hydrogens have been omitted.

DFT is based on the time-dependent KS equations, derived from the time-dependent Schrödinger equation [84]. The implementation of time-dependent DFT (TDDFT, or time-dependent density functional response theory, TD-DFRT) in Gaussian 98 [45] has been described by Stratman *et al.* [85]. Wiberg *et al.* used this implementation to study the effect of five functionals and five basis sets on the transition energies (the UV absorption wavelengths) of formaldehyde, acetaldehyde, and acetone [86]. Satisfactory results were obtained, and the energies were not strongly dependent on the functional, but B3P86 seemed to be the best and B3LYP the worst. The $6 - 311 + +G^{**}$ basis was recommended. Although these workers used MP2/6-311 + G** geometries, the results in Table 7.8 indicate that AM1 geometries, which can be calculated perhaps a thousand times faster, gives transition energies that are nearly as accurate (mean absolute errors of 0.12 and 0.18 eV, respectively). Table 7.9 compares with experiment the UV spectrum of methylenecyclopropene, calculated by *ab initio*, semiempirical, and DFT methods. The best of the three is the TDDFT calculation, which is the only one that reproduces the 308 nm band.

The HOMO-LUMO gap calculated with hybrid gradient-corrected functionals is approximately equal to the $\pi \rightarrow \pi^*$ UV transition of unsaturated molecules, and this could be useful in predicting UV spectra (see *ionization energies and electron affinities*, below).

Table 7.8. UV spectra (as transition energies in eV) of acetone, acetaldehyde, and formaldehyde, calculated by time-dependent DFT, using Gaussian 98 [45]. The results of using MP2/6-311 + G** [86] and (calculations by the author) AM1 geometries are compared; both sets of calculations are single-point B3P86/6-311 + +G**. For each molecule only six transitions, all singlets, are shown. The number of positive and negative deviations from experiment and the mean absolute errors are given

	MP2 geometry	AM1 geometry	Experiment
Acetone	4.41	4.26	4.43
	6.28	6.19	6.36
	7.26	7.17	7.41
	7.43	7.40	7.36
	7.67	7.59	7.49
	7.89	7.82	8.09
Acetaldehyde	4.29	4.14	4.28
	6.76	6.69	6.82
	7.29	7.26	7.46
	7.70	7.68	
	7.89	7.98	7.75
	8.35	8.16	8.43
Formaldehyde	3.95	3.83	4.1
	6.98	6.97	7.13
	7.93	7.95	8.14
	8.09	8.07	7.98
	8.81	8.84	
	9.23	8.87	
	5+, 10−	4+, 11−	
	mean of 15: 0.12	mean of 15: 0.18	

Table 7.9. Calculated (*ab initio*, semiempirical, DFT) and experimental [87] UV spectra of methylenecyclopropene, wavelength, nm (relative intensity). The recommended *ab initio* basis set [87] and DFT functional and basis set [86] are used. The *ab initio* results are from Table 5.20, and the semiempirical results are from Table 6.6

Calculated			
RCIS/6-31 + G*// B3LYP/6-31G*	ZINDO/S//AM1	TDDFT : B3P86/6 −311 + +G**//AM1	Experimental
224 (15)	228 (12)	309 (26)	308 (13)
209 (6)	224 (0.2)	226 (3)	242 (0.6)
196 (0)	213 (100)	210 (0)	206 (100)
194 (8)	204 (1)	208 (100)	
193 (100)		190 (0)	

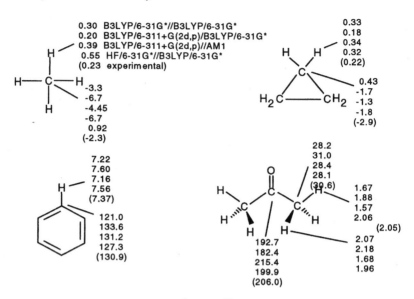

Figure 7.9. Calculated and experimental ^1H and ^{13}C NMR spectra: chemical shifts relative to TMS, H, and C, respectively. The calculations were with the default NMR method (GIAO) implemented in Gaussian 94W [54]. The experimental values are from [89] (a value of −3.5 has been given for the cyclopropane ^{13}C shift [90]).

NMR spectra

As with *ab initio* methods (section 5.5.5), NMR shielding constants can be calculated from the variation of the energy with magnetic field and nuclear magnetic moment. The chemical shift of a nucleus is its shielding value minus that of the TMS carbon or hydrogen nucleus (for ^{13}C or ^1H NMR spectra, respectively). The most accurate results are said to be obtained with MP2 calculations, but DFT is significantly better than HF [88]. Figure 7.9 compares with experiment ^{13}C and ^1H NMR spectra calculated at these levels: B3LYP/6-31G*//B3LYP/6-31G*, B3LYP/6-311 + G(2d, p)//B3LYP/6-31G*, B3LYP/6-311 + G(2d, p)//AM1, and HF/6-31G*//B3LYP/6-31G*. The mean absolute errors in the chemical shifts are:

B3LYP/6-31G*//B3LYP/6-31G*	^1H, 0.11; ^{13}C, 6.0
B3LYP/6-311 + G(2d, p)//B3LYP/6-31G*	^1H, 0.10; ^{13}C, 6.3
B3LYP/6-311 + G(2d, p)//AM1	^1H, 0.16; ^{13}C, 3.1
HF/6-31G*//B3LYP/6-31G*	^1H, 0.20; ^{13}C, 3.3

For such a very small sample (three ^1H and five ^{13}C atoms; the CH_3COCH_3 Hs were not included since the observed chemical shift is a weighted average) the results cannot reasonably be expected to do more than show up gross differences in accuracy, but they do suggest that for DFT NMR chemical shifts, if high accuracy is not needed there may be little point in using a bigger basis set than 6-31G*, and that AM1 geometries (for normal molecules, as usual) may be as good for these spectra as B3LYP/6-31G* geometries.

Ionization energies and electron affinities

These concepts were discussed in section 5.5.5. We saw that ionization energies (ionization potentials) and electron affinities can be calculated in a straightforward way as the energy difference between a molecule and the species derived from it by loss or gain, respectively, of an electron. Using the energy of the optimized geometry of the radical cation or radical anion (in the case where the species whose IE or EA we seek is a neutral closed-shell molecule) gives the adiabatic IE or EA, while using the energy of the ionized species at the geometry of the neutral gives the vertical IE or EA. Muchall *et al.* have reported adiabatic and vertical ionization energies and electron affinities of eight carbenes, calculated in this way by semiempirical, *ab initio*, and DFT methods [91]. They recommend B3LYP/6-31 + G*//B3LYP/3-21G(*) as the method of choice for predicting first ionization energies; the use of the small 3-21G(*) basis with B3LYP for the geometry optimization is unusual – see section 5.4.2 – usually the smallest basis used with a correlated method is 6-31G*. This combination is relatively undemanding and gives accurate (largest absolute error 0.14 eV) adiabatic and vertical ionization energies for the carbenes studied. Table 7.10 shows the results of applying this method to some other (non-carbene) molecules. The B3LYP/6-31 + G* energy differences are essentially the same with B3LYP/3-21G(*) and AM1 geometries; they are good estimates of the experimental IE, are somewhat better than the *ab initio* MP2 energy difference values, and are considerably better than the MP2 Koopmans' theorem (below) IEs. Of course, for unusual molecules (like the carbenes studied by Muchall *et al.* [91]) AM1 may not give good geometries, and for such species it would be safer to use B3LYP/3-21G(*) or B3LYP/6-31G* geometries for the single-point B3LYP/6-31 + G* calculations.

In wavefunction theory an alternative way to find IEs for removal of an electron from any molecular orbital is to invoke Koopmans' theorem: the IE for an orbital is the negative of the orbital energy; section 5.5.5. In chapters 5 and 6 both the energy difference and the Koopmans' theorem methods were used to calculate some IEs (Tables 5.21 and 6.7). The problem with applying Koopmans' theorem to DFT is that in DFT proper

Table 7.10. Some ionization energies (eV). The ΔE values (cation energy minus neutral energy) correspond to adiabatic and (in parentheses) vertical IEs; the Koopmans theorem values are vertical IEs. Experimental IEs are adiabatic (CH_3OH and CH_3COCH_3 [96], CH_3SH [97]). That the vertical IE is smaller than the adiabatic for the B3LYP/6-31 + G* calculation on CH_3SH is presumably due to a somewhat inaccurate geometry, probably for the cation (experimental vertical IEs are always larger than adiabatic since it takes energy to distort the relaxed-geometry cation to the geometry of the neutral)

	ΔE				
	B3LYP/6-31 + G* B3LYP/3-21G(*)	B3LYP/ 6-31 + G* AM1	MP2(FC)/ 6-31G*	Koopmans' (MP2(FC)/ 6-31G*)	Exp
CH_3OH	10.77 (10.92)	10.76 (10.85)	10.6	12.1	10.9
CH_3SH	9.40 (9.43)	9.53 (9.36)	9.0	9.2	9.4
CH_3COCH_3	9.60 (9.70)	9.67 (9.68)	9.6	11.2	9.7

there are no molecular orbitals, only electron density. The KS molecular orbitals (the orbitals ψ^{KS} that make up the Slater determinant of Eq. (7.19) were, as explained in section 7.2.3.2, introduced only to provide a practical way to calculate the energy (note Eqs (7.21), (7.22), and (7.26)). There was at one time a fair amount of argument over the physical meaning, if any, of the KS orbitals. Baerends and coworkers [92] and others have suggested that the KS orbitals, like the HF orbitals, provide a good basis for qualitative discussion. Stowasser and Hoffmann showed that the KS orbitals resemble those of conventional wavefunction (extended Hückel – chapter 4 – and HF *ab initio*) theory in shape, symmetry, and, usually, energy order [93]. They conclude that these orbitals can indeed be treated much like the more familiar orbitals of wavefunction theory. Furthermore, they showed that although the KS orbital energy values (the eigenvalues ϵ from diagonalization of the DFT Fock matrix – section 7.2.3.3) are not good approximations to the ionization energies of molecular orbitals (as revealed by photoelectron spectroscopy), there is a linear relation between $|\epsilon_i(KS) - \epsilon_i(HF)|$ and $\epsilon_i(HF)$. Salzner *et al.*, too, showed that in DFT, unlike *ab initio* theory calculations, negative HOMO energies are not good approximations to the IE [94], but, surprisingly, HOMO-LUMO gaps from hybrid functionals agreed well with the $\pi \rightarrow \pi^*$ UV transitions of unsaturated molecules.

Concerning electron affinities, in HF calculations the negative LUMO energy of a species M corresponds to the electron affinity not of M but rather of the anion M^- [95]. However, Salzner *et al.* reported that the negative LUMOs from LSDA functionals gave rough estimates of EA (ca. 0.3–1.4 eV too low; gradient-corrected functionals were much worse, ca. 6 eV too low) [94]. Brown *et al.* found that for eight medium-sized organic molecules the energy difference method using gradient-corrected functionals predicted electron affinities fairly well (average mean error less than 0.2 eV) [77].

Electronegativity, hardness, softness and the Fukui function
The idea of electronegativity was probably born about as soon as chemists suspected that the formation of chemical compounds involved electrical forces: metals and nonmetals were seen to possess opposite appetites for the "electrical fluid(s)" of eighteenth century physics. This "electrochemical dualism" is most strongly associated with Berzelius [98], and is clearly related to our qualitative notion of electronegativity as the tendency of a species to attract electrons. Parr and Yang have given a sketch of attempts to quantify the idea [99]. Electronegativity is a central notion in chemistry.

Hardness and softness as chemical concepts were apparently in the chemical literature as early as the 1950s [100], but did not become widely used till they were popularized by Pearson in 1963 [101]. In the simplest terms, the hardness of a species (atom, ion, or molecule) is a qualitative indication of how polarizable it is, i.e. how much its electron cloud is distorted in an electric field. In this sense the terms hard, and its opposite soft, were evidently suggested by D. H. Busch [101a] by analogy with the conventional use of these words to denote resistance to deformation by mechanical force. The hard/soft concept proved useful, particularly in rationalizing acid–base chemistry [102]. Thus, a proton, which cannot be distorted in an electric field (it has no electron cloud) is a very hard acid, and tends to react with hard bases. Bases in which sulfur electron pairs provide the basicity are soft, since sulfur is a big, fluffy atom, and they tend to react with soft acids. Perhaps because it was originally qualitative the hard–soft

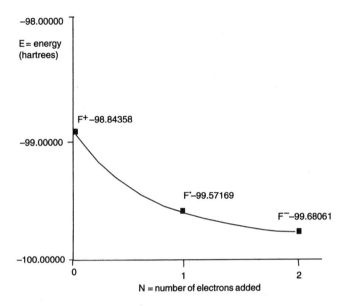

Figure 7.10. Change of energy (for F^+, $F^•$ and F^-) as electrons are added to a species. The energies were calculated at the QCISD(T)/6-311 + G* level. The slope of the curve at any point (first derivative) is the electronic chemical potential, and the negative of the slope the electronegativity of the species at that point. The curvature at any point (second derivative) is the hardness of the species). See also Table 7.11.

acid–base (HSAB) idea met with skepticism from at least one quarter: Dewar (of semiempirical fame) dismissed it as a "mystical distinction between different kinds of acids and bases" [103]. The hard/soft concept has been extensively used to interpret reactivity.

The Fukui function or frontier function was introduced by Parr and Yang in 1984 [104]. Its mathematical expression (below) defines it as the sensitivity of the electron density at various points in a species to a change in the number of electrons in the species. In a chemical reaction a change in electron number clearly involves removing electrons from or adding electrons to the HOMO or LUMO, respectively, i.e. the *frontier orbitals* whose importance was emphasized by Fukui[2]. This function thus measures changes in electron density that accompany chemical reactions. The Fukui function has been used to try to rationalize and predict the variation of reactivity from site to site in a molecule.

These concepts, which can be analyzed quantitatively using wavefunction theory, but are often treated in connection with DFT (perhaps because much of the underlying theory was formulated in this context [105]) will now be examined a bit more quantitatively. Consider the effect on the energy of a molecule, atom, or ion of adding electrons. Figure 7.10 shows how the energy of a fluorine cation F^+ changes as one

---

[2]Kenichi Fukui, born Nara, Japan, 1918. Ph.D. Kyoto Imperial University 1948, Professor Kyoto Imperial University 1951. Nobel Prize 1981. Died 1998.

and then two electrons are added, giving a radical F$^\bullet$ and an anion F$^-$. The number of electrons N we have added to F$^+$ (N is taken here as 0 for F$^+$, and is thus 1 for the atom and 2 for the anion) is integral (1, 2,. . .), but mathematically we can consider adding continuous electronic charge N; the line through the three points is then a continuous curve and we can examine $(\partial E/\partial N)_Z$, the derivative of E with respect to N at constant nuclear charge. Ca. 1875 Willard Gibbs studied theoretically the effect on the energy of a system of a change in its composition. The derivative $\mu = (\partial E/\partial n)_{T,p}$, is the change in energy caused by an infinitesimal change in the number of moles n, at constant temperature and pressure; it is called the chemical potential [106]. By analogy, $(\partial E/\partial N)_Z$, the change in energy with respect to number of added electrons at constant nuclear charge, is the electronic chemical potential (or in an understood context just the chemical potential) of an atom. For a molecule the differentiation is at constant nuclear framework, the charges and their positions being constant, i.e. constant external potential, v (section 7.2.3.1). So for an atom, ion, or molecule

$$\mu = \left(\frac{\partial E}{\partial N}\right)_v \qquad (7.33)$$

The electronic chemical potential of a molecular (including atomic or ionic) species, according to Eq. (7.33), is the infinitesimal change in energy when electronic charge is added to it. Figure 7.10 suggests that the energy will drop when charge is added to a species, at least as far as common charges (from about $+3$ to -1) go, and indeed, even for fluorine's electronegative antithesis, lithium, the energy drops along the sequence Li$^+$, Li$^\bullet$, Li$^-$ (QCISD(T)/6-311+G* gives energies of -7.23584, -7.43203, $-7.45448\ h$, respectively). Now, since one feels intuitively that the more electronegative a species, the more its energy should drop when it acquires electrons, we suspect that there should be a link between the chemical potential and electronegativity. If we choose for convenience to make (most) electronegativities positive, then since $(\partial E/\partial N)_v$ is negative we might define the electronegativity χ as the negative of the electronic chemical potential:

$$\chi = -\mu = -\left(\frac{\partial E}{\partial N}\right)_v \qquad (7.34)$$

From this viewpoint the electronegativity of a species is the drop in energy when an infinitesimal amount (infinitesimal so that it remains the same species) of electronic charge enters it. It is a measure of how hospitable an atom or ion, or a group or atoms in a molecule (section 5.5.4), is to the ingress of electronic charge, which fits in with our intuitive concept of electronegativity.

This definition of electronegativity was given in 1961 [107] and later (1978) discussed in the context of DFT [108]. Equation (7.34) could be used to calculate electronegativity by fitting an empirical curve to calculated energies for, e.g. M$^+$, M and M$^-$, and calculating the slope (gradient, first derivative) at the point of interest; however, the equation can be used to derive a simple approximate formula for electronegativity using a three-point approximation. For consecutive species M$^+$, M and M$^-$ (constant nuclear framework), let the energies be $E(M^+)$, $E(M)$, and $E(M^-)$. Then, by definition

$$E(M^+) - E(M) = I$$

the ionization energy of M, and

$$E(M) - E(M^-) = A$$

the electron affinity of M. Adding,

$$E(M^+) - E(M^-) = I + A$$

So, approximating the derivative at the point corresponding to M as the change in E when N goes from 0 to 2, divided by this change in electron number, we get

$$\left(\frac{\partial E}{\partial N}\right)_v = \frac{E(M^-) - E(M^+)}{2 - 0} = \frac{-(I + A)}{2}$$

i.e. using Eq. (7.34)

$$\chi = \frac{I + A}{2} \tag{7.35}$$

To use this formula one can employ experimental or calculated adiabatic (or vertical, if the species from removal or addition of an electron are not stationary points) values of I and A. This same formula (Eq. (7.35)) for χ was elegantly derived by Mulliken (1934) [109] using only the definitions of I and A. Consider the reactions

$$X + Y \to X^+ + Y^-$$

and

$$X + Y \to X^- + Y^+$$

If X and Y have the same electronegativity then the energy changes of the two reactions are equal, since X and Y have the same proclivities for gaining and for losing electrons, i.e.

$$I(X) - A(Y) = I(Y) - A(X)$$

i.e. $(I + A)$ for X $= (I + A)$ for Y.

So it makes sense to define electronegativity as $I + A$; the factor of $\frac{1}{2}$ (Eq. (7.35)) was said by Mulliken to be "probably better for some purposes" (possibly to make χ the arithmetic mean of I and A, an easily-grasped concept).

Electronegativity has also been expressed in terms of orbital energies, by taking I as the negative of the HOMO energy and A as the negative of the LUMO energy [110]. This gives

$$\chi = \frac{-(E_{HOMO} + E_{LUMO})}{2} \tag{7.36}$$

This expression has the advantage over Eq. (7.35) that one needs only the HOMO and LUMO energies of the species, which are provided by a one-pot calculation (i.e. by what is operationally a single calculation), but to use Eq. (7.35) one needs (cf. Fig. 7.10) the energy of the $(N - 1)$-, N- and $(N + 1)$-electron molecules (most simply at the geometry of the molecule whose electronegativity we are calculating; cf. the Fukui functions for SCN$^-$ later in this section). How good is Eq. (7.36)? $I = -E_{HOMO}$ is a fairly good approximation for the orbitals of wavefunction theory, but not for the KS orbitals, and $A = -E_{LUMO}$ is only a very rough approximation for the KS orbitals, and for wavefunction orbitals $-E_{LUMO}$ of M is said to correspond to the electron affinity of M^-, not of M (see *Ionization energies and electron affinities* in section 7.3.5). So how do the results of calculations using the formula of Eq. 7.36 compare with those using Eq. (7.35)? Table 7.11 gives values of χ calculated using QCISD(T)/6-311 + G^* (section 5.4.3) values of I and A, which should give good values of these latter two quantities, and compares these χ values with those from HOMO/LUMO energies calculated by *ab initio* (MP2(FC)/6-31G*) and by DFT (B3LYP/6-31G*). For the two cations the agreement between the three ways of calculating χ is good; for the other species it is erratic or bad, although the trends are the same for the three methods within a given family (hardness decreases from cation to radical to anion). There seems little doubt that Eq. (7.35) is the sounder way to calculate electronegativity. An exposition of the concept of electronegativity as the (negative) average of the HOMO and LUMO energies, and the chemical potential $(-\chi)$ as lying at the midpoint of the HOMO/LUMO gap, has been given by Pearson [110].

Chemical hardness and softness are much newer ideas than electronegativity, and they were quantified only fairly recently. Parr and Pearson (1983) proposed to identify the curvature (i.e. the second derivative) of the E vs. N graph (e.g. Fig. 7.10) with hardness, η [111]. This accords with the qualitative idea of hardness as resistance to deformation, which itself accommodates the concept of a hard molecule as resisting polarization – not being readily deformed in an electric field: if we choose to define hardness as the curvature of the E vs. N graph, then

$$\eta = \left(\frac{\partial^2 E}{\partial N^2}\right)_v = \left(\frac{\partial \mu}{\partial N}\right)_v = -\left(\frac{\partial \chi}{\partial N}\right)_v \tag{7.37}$$

where μ and χ are introduced from Eqs (7.33) and (7.34). The hardness of a species is then the amount by which its electronegativity – its ability to accept electrons – decreases when an infinitesimal amount of electronic charge is added to it. Intuitively, a hard molecule is like a rigid container that does not yield as electrons are forced in, so the pressure (analogous to the electron density) inside builds up, resisting the ingress of more electrons. A soft molecule may be likened to a balloon that can expand as it acquires electrons, so that the ability to accept still more electrons is not so seriously compromised. Softness is the reciprocal of hardness:

$$\sigma = \frac{1}{\eta} \tag{7.38}$$

and qualitatively, of course, it is the opposite in all ways.

To approximate hardness by I and A (cf. the approximation of electronegativity by Eq. (7.35)), we approximate the $E = f(N)$ curve (cf. Fig. 7.10) by a general quadratic (it *looks* like a quadratic):

$$E = aN^2 + bN + c$$
$$\frac{\partial^2 E}{\partial N^2} = 2a$$

Table 7.11. Electronegativity, χ, and hardness, η (cf. Fig. 7.10). For each species χ and η have been calculated in three ways: (1) From ionization energy (I) and electron affinity (A), using $\chi = 1/2(I + A)$ and $\eta = 1/2(I - A)$. I and A were calculated (QCISD(T)/6-311+G*) as the energy differences of the optimized-geometry species, i.e. adiabatic values. (2) From the MP2(FC)/6-31G* HOMO and LUMO, using $\chi = -1/2(E_{HOMO} + E_{LUMO})$ and $\eta = 1/2(E_{LUMO} - E_{HOMO})$. (3) From the B3LYP/6-31G* Kohn–Sham HOMO and LUMO, as for (2). All the numbers refer to units of eV

	I	A	HOMO, MP2 (HOMO, DFT)	LUMO, MP2 (LUMO, DFT)	$\chi: (I + A)/2$, HOMO/LUMO MP2, HOMO/LUMO DFT	$\eta: (I - A)/2$, HOMO/LUMO MP2, HOMO/LUMO DFT
F$^+$	36.0	19.8	−37.6(−30.0)	−17.7(−27.3)	27.9, 27.7, 28.7	8.1, 10.0, 1.4
F$^\bullet$	19.8	3.0	−19.5(−14.5)	−19.5(−14.5)	11.4, 19.5, 14.5	8.4, 0, 0
F$^-$	3.0	−14.0	−2.1(4.6)	42.1 (36.4)	−5.5, −20.0, −20.5	8.4, 22.1, 15.9
HS$^+$	20.2	11.3	−20.3(−16.8)	−10.7(−15.7)	15.8, 15.5, 16.3	4.5, 4.8, 0.6
HS$^\bullet$	11.3	1.7	−12.5(−8.7)	−12.5(−8.7)	6.5, 12.5, 8.7	4.8, 0, 0
HS$^-$	1.7	−6.4	−1.9(1.3)	12.3 (8.4)	−2.4, −5.2, −4.9	4.1, 7.1, 3.6

We will now let M denote any atom or molecule, and M^+ and M^- the species formed by removal or addition of an electron.

$E(M)$ corresponds to $N = 1$ and $E(M^-)$ corresponds to $N = 2$, so substituting into our quadratic equation

$$E(M) = a(1^2) + b(1) + c = a + b + c$$

and

$$E(M^-) = a(2^2) + b(2) + c = 4a + 2b + c$$

and so

$$2a = c + E(M^-) - 2E(M)$$

Since $E(0) = E(M^+) = a(0^2) + b(0) + c = c$,

$$2a = E(M^+) + E(M^-) - 2E(M) = [E(M^+) - E(M)] - [E(M) - E(M^-)] = I - A$$

i.e.

$$\eta = \left(\frac{\partial^2 E}{\partial N^2}\right)_v = I - A \tag{7.39}$$

Actually, the hardness is commonly *defined* as *half* the curvature of the E vs. N graph, giving

$$\eta = \frac{1}{2}\left(\frac{\partial^2 E}{\partial N^2}\right)_v = \frac{I - A}{2} \tag{7.40}$$

and from Eq. (7.37)

$$\eta = \frac{1}{2}\left(\frac{\partial^2 E}{\partial N^2}\right)_v = \frac{1}{2}\left(\frac{\partial \mu}{\partial N}\right)_v = -\frac{1}{2}\left(\frac{\partial \chi}{\partial N}\right)_v \tag{7.41}$$

The one-half factor is [110] to bring η into line with Eq. (7.35), where this factor arises naturally in applying the three-point approximation and the definitions of I and A to the rigorous Gibbs equation (Eq. (7.33)) for electronic chemical potential.

Electronegativity has also been expressed in terms of orbital energies, by taking I as the negative of the HOMO energy and A as the negative of the LUMO energy [110]. This gives

$$\eta = \frac{(E_{LUMO} - E_{HOMO})}{2} \tag{7.42}$$

Like the analogous expression for electronegativity (Eq. (7.36)), this requires only a "one-pot" calculation, of the HOMO and LUMO. Much of what was said about Eq. (7.36) applies to Eq. (7.42). Table 7.11 gives values of η calculated analogously to the χ values discussed above. The HOMO/LUMO hardness values are in even worse agreement with the I/A ones than are the HOMO/LUMO electronegativity values with the I/A values (the zero values for the HOMO/LUMO-calculated η of the radicals values arise from taking the half-occupied orbital (semioccupied MO, SOMO) as both HOMO and LUMO). The orbital view of hardness as the HOMO/LUMO gap is discussed by Pearson, who also reviews the principle of maximum hardness, according to which in a chemical reaction hardness and the HOMO/LUMO gap tend to increase, potential energy surface relative minima represent species of relative *maximum* hardness, and transition states are species of relative *minimum* hardness [110]. In recent papers these general ideas about hardness are expounded [112] and the reciprocal concept of softness is used (with the Fukui function) to rationalize some cycloaddition reactions [113].

The Fukui function (the frontier function) was defined by Parr and Yang [104] as

$$f(\mathbf{r}) = \left[\frac{\delta \mu}{\delta v(\mathbf{r})}\right]_N = \left[\frac{\partial \rho(\mathbf{r})}{\partial N}\right]_v \tag{7.43}$$

This says that $f(\mathbf{r})$ is the functional derivative (section 7.2.3.2, *The Kohn–Sham equations*) of the chemical potential with respect to the external potential (i.e. the potential caused by the nuclear framework), at constant electron number; and that it is also the derivative of the electron density with respect to electron number at constant external potential. The second equality shows $f(\mathbf{r})$ to be the sensitivity of $\rho(\mathbf{r})$ to a change in N, at constant geometry. A change in electron density should be primarily electron withdrawal from or addition to the HOMO or LUMO, the frontier orbitals of Fukui [114], hence the name bestowed on the function by Parr and Yang. Since $\rho(\mathbf{r})$ varies from point to point in a molecule, so does the Fukui function. Parr and Yang argue that a large value of $f(\mathbf{r})$ at a site favors reactivity at that site, but to apply the concept to specific reactions they define three Fukui functions ("condensed Fukui functions" [80]):

$$f^*(\mathbf{r}) = \left[\frac{\partial \rho(\mathbf{r})}{\partial N}\right]_v^* \qquad * = +, -, 0 \tag{7.44}$$

The three functions f^+, f_k^-, and f_k^0 refer to an electrophile, a nucleophile, and a radical. They are the sensitivity, to a small change in the number of electrons, of the electron density in the LUMO, the HOMO, and in a kind of average HOMO/LUMO half-occupied orbital. Practical implementations of these condensed Fukui functions are the "condensed-to-atom" forms of Yang and Mortier [115]:

$$f_k^+ = q_k(N+1) - q_k(N) \qquad \text{for atom k as an electrophile}$$

$$f_k^- = q_k(N) - q_k(N-1) \qquad \text{for atom k as a nucleophile} \tag{7.45}$$

$$f_k^0 = \tfrac{1}{2}[q_k(N+1) - q_k(N-1)] \qquad \text{for atom k as a radical}$$

Here $q_k(N)$ is the electron population (not the charge) on atom k, etc. (see below). Note that f_k^0 is just the average of f_k^+ and f_k^-. The condensed Fukui functions measure the sensitivity to a small change in the number of electrons of the electron density at atom k in the LUMO (f_k^+), in the HOMO (f_k^-), and in a kind of intermediate orbital (f_k^0); they provide an indication of the reactivity of atom k as an electrophile (reactivity toward nucleophiles), as a nucleophile (reactivity toward electrophiles), and as a free radical (reactivity toward radicals).

The easiest way to see how these formulas can be used is to give an example. Let us calculate f_k^- for the anion SCN$^-$. We shall calculate f_S^-, f_C^-, and f_N^-, to get an idea of the nucleophilic power of the S, C, and N atoms in this molecule. We need the electron population q on each atom or, what gives us the same information, the charge on each atom (for an atom in a molecule, electron population + charge = atomic number). We perform the calculations for the N-electron species (SCN⁻) and the $(N - 1)$-electron species (SCN•). If we were interested in the nucleophilic power of the atoms in a neutral molecule M, then to get f_k^- we would calculate the electron populations or charges on the atoms in M and in M⁺, and for the electrophilic power of the atoms in neutral M, to get f_k^+ we calculate the electron populations or charges on the atoms in M and M⁻. The calculations are performed for the two species at the same geometry. In introducing the condensed Fukui function, Yang and Mortier [115] used for each pair of species a single "standard" (presumably essentially average) geometry, with accepted, reasonable bond lengths and angles, and other workers do not specify whether they use for, say, M and M⁺, the neutral or the cation geometry. We will adopt the convention that for a calculation on M* (* = +, − or •), both geometries will be those of M*, the species of interest to us; this avoids the problem of trying to do a geometry optimization on a species that may not be a stationary point on the potential energy surface (assuming that M* is itself a stationary point – one will rarely be interested in something that is not), a situation that arises particularly for some anions.

Charges and electron populations from calculations on SCN⁻ and SCN• (and on CH₃CCH and CH₃CCH•⁺) are shown in Fig. 7.11. The anion SCN⁻ was optimized and then the AIM (section 5.5.4) electron population/charges were calculated (the AIM calculations were done with G98 using the keywords AIM = Charge). An AIM calculation was then done on the radical at the anion geometry. The optimization and both

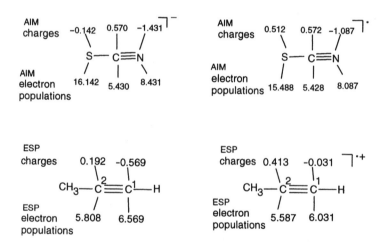

Figure 7.11. Charges on atoms and corresponding electron populations. For SCN⁻ and SCN• AIM (section 5.5.4) charges were used, and both species are at the optimized SCN⁻ geometry. For CH₃CCH and CH₃CCH•⁺ electrostatic potential charges (from Gaussian 98, keyword Pop=MK) were used, and both species are at the optimized CH₃CCH geometry. The method/basis for optimization and charge calculation is B3LYP/6-31+G* for SCN⁻ and SCN•, and B3LYP/6-311G** for CH₃CCH and CH₃CCH•⁺.

AIM calculations used the B3LYP/6-31 + G* method/basis (results for charges other than AIM and other methods/basis sets are shown shortly).

The condensed Fukui functions may now be calculated (see Fig. 7.11):

$$f^-(S) = q(S, \text{anion}) - q(S, \text{neutral}) = 16.142 - 15.488 = 0.654$$

$$f^-(C) = q(C, \text{anion}) - q(C, \text{neutral}) = 5.430 - 5.428 = 0.002$$

$$f^-(N) = q(N, \text{anion}) - q(N, \text{neutral}) = 8.431 - 8.087 = 0.344$$

This indicates that in SCN$^-$ the order of nucleophilicity is S > N ≫ C (which is what any chemist would expect). Sulfur is the softest atom here, and carbon the hardest. The results of such a calculation vary somewhat with the method/basis (e.g. HF/6-31G*, MP2/6-31G*, etc.), and especially with the way the charges/electron populations are calculated. Here are the f_k^- functions from the use of electrostatic potential charges (the G98 keyword Pop=MK was used) again using B3LYP/6-31 + G*:

$$f_k^-(S) = q(S, \text{anion}) - q(S, \text{neutral}) = 16.720 - 15.955 = 0.765$$

$$f_k^-(C) = q(C, \text{anion}) - q(C, \text{neutral}) = 5.542 - 5.707 = -0.165$$

$$f_k^-(N) = q(N, \text{anion}) - q(N, \text{neutral}) = 7.738 - 7.338 = 0.400$$

In this case the conclusions are unaffected.

In an extensive study, Geerlings *et al.* [80] showed that with AIM charges semiquantitatively similar results are obtained with a variety of correlation methods (HF, MP2, QCISD, and five DFT functionals), using bases similar to 6-31G*. The biggest deviation from QCISD (section 5.4.3; QCISD was taken as the most reliable of the methods used) was shown by MP2. For example, for CH$_2$CHO$^-$ all correlation methods except MP2 gave O a bigger f^- than C. If we disregarded the MP2 result as anomalous, this could be interpreted as indicating that the O is more nucleophilic than the C. Actually, in standard organic syntheses enolates usually react preferentially at the *carbon*, but the ratio of C:O nucleophilic attack can vary considerably with the particular enolate, the electrophile, and the solvent. To complicate things even more, the nucleophile is not always just the simple enolate: an ion pair or even aggregates of ion pairs may be involved [116]. Even for the case of an unencumbered enolate, the atom with the biggest f^- (the softest atom) cannot be assumed to be the strongest nucleophilic center, because, as Méndez and Gázquez point out in their study [117] of enolates using the Fukui function, one consequence of the HSAB principle is that an electrophile tends to react with a nucleophilic center of *similar* softness (soft acids prefer soft bases, etc.), not necessarily with the softest nucleophilic center. Thus, for the reaction of CH$_2$CHO$^-$ with the electrophile CH$_3$X, one might calculate, for CH$_2$CHO$^-$, $f^-(C)$ and $f^-(O)$, and for CH$_3$X, $f^+(C)$. The CH$_3$X C would be expected (in the absence of complications!) to bond to the atom (C or O) whose f^- value was closest to its $f^+(C)$ value. A study of the ethyl acetoacetate enolate using these and other concepts has been reported by Geerlings and coworkers [118]. This approach, which is applicable to any ambident species, is further illustrated below by the reaction of HNC with alkynes.

In a study of the reaction of alkynes with hydrogen isocyanide the condensed Fukui function was combined with the overall or global softness to try to rationalize the regioselectivity of attack on the triple bond [113]:

This reaction involves electrophilic attack by HNC on the alkyne, to give a zwitterion which reacts further. Can our concepts be used to predict which alkyne atom, C^1 or C^2 (using the designations of Nguyen *et al.* [113]) will be attacked – will the products be formed primarily through A or through B? Nguyen *et al.* approached this problem by first showing that the reaction is indeed electrophilic attack of HNC (i.e. acting as an electrophile) on the alkyne (acting as a nucleophile): the HOMO(alkyne)/LUMO(HNC) interaction has a smaller energy gap than the HOMO(HNC)/LUMO(alkyne) interaction. They then calculated the *local softness* or *condensed softness parameters* (not quite the same as the condensed-to-atoms parameters of Eqs (7.45) that we saw above; see below) of C^1 and C^2 of the alkyne and the C of HNC. For C^1 and C^2 of the alkyne the softness as a nucleophile, i.e. softness toward electrophiles, was calculated, with the aid of f_k^-, and for the HNC C softness as an electrophile, i.e. softness toward nucleophiles, was calculated, with the aid of f_k^+.

Illustrating how the calculations for CH_3CCH may be done:

(1) Optimize the structure of CH_3CCH and calculate its atom charges (and energy).

(2) Use the optimized geometry of CH_3CCH for a single-point (same geometry) calculation of the charges (and energy) for $CH_3CCH^{\bullet+}$. Steps (1) and (2) enable calculation of f_k^-.

(3) Use the optimized geometry of CH_3CCH for a single-point calculation of the energy of the anion $CH_3CCH^{\bullet-}$ (This is a radical anion).

Steps (1), (2), and (3) enable us to calculate the *global softness* (the softness of the molecule as a whole) of CH_3CCH. This is done by calculating the vertical ionization energy and electron affinity as energy differences, then calculating the global softness as the reciprocal of global hardness. From Eq. (7.38) this is $\sigma = 1/(I - E)$ or $\sigma = 2/(I - E)$, depending on whether we define hardness according to Eq. (7.39) or (7.40). Nguyen *et al.* use $\sigma = 1/(I - E)$, i.e they take hardness as $\eta = (I - E)$ rather than $\frac{1}{2}(I - E)$. The local softness of any atom of interest may now be calculated by multiplying f_k^- for that atom by σ. Let us look at actual numbers. The CH_3CCH B3LYP/6-311G** basis set and electrostatic potential charges (with the Gaussian keyword Pop=MK) were used. These gave the charges (and thus electron populations) shown in Fig. 7.11. From these populations,

$$f^-(C^1) = q(C^1, \text{neutral}) - q(C^1, \text{cation}) = 6.569 - 6.031 = 0.538$$

$$f^-(C^2) = q(C^2, \text{neutral}) - q(C^2, \text{cation}) = 5.808 - 5.587 = 0.221$$

The vertical ionization energy and vertical electron affinity are (here ZPEs have not been taken into account, as they should nearly cancel; in any case the significance of a calculated ZPE for the cation or anion at the geometry of the neutral is questionable, since the former two vertical species are not stationary points):

$$I = E(\text{cation}) - E(\text{neutral}) = -116.31237 - (-116.69077) = 0.37840 \, \text{h}$$

$$A = E(\text{neutral}) - E(\text{anion}) = -116.69077 - (-116.58078) = -0.10999 \, \text{h}$$

The softness is then $\sigma = 1/(I - A) = 1/[(0.37840 - (-0.10999)] = 2.048 \, \text{h}^{-1}$ So the local softness of the two carbons as nucleophiles (softness toward electrophiles) is

$$s^-(C^1) = 0.538(2.048) = 1.102$$

and

$$s^-(C^2) = 0.221(2.048) = 0.453$$

(Nguyen *et al.* report 1.096 and 0.460).

Since electron population is a pure number and global softness has the units of reciprocal energy, local softness logically has these units too, but the practice is to simply state that all these terms are in "atomic units".

Now consider analogous calculations on the HNC C, but for local softness as an *electrophile* (softness toward nucleophiles), using f_k^+. These calculations gave

$$s^+(\text{HNC C}) = 1.215$$

To predict which of the two alkyne carbons, C^1 or C^2, HNC will preferentially attack, one now invokes the "local HSAB principle" [119], which says that interaction is favored between electrophile/nucleophile (or radical/radical) of most nearly equal softness. The HNC carbon softness of 1.215 is closer to the softness of C^1 (1.102) than that of C^2 (0.453) of the alkyne, so this method predicts that in the reaction scheme above the HNC attacks C^1 in preference to C^2, i.e. that reaction should occur mainly by the zwitterion A. This prediction agreed with that from the more fundamental approach of calculating the activation energies as the difference of transition state and reactant energies. This kind of analysis worked for $-CH_3$ and $-NH_2$ substituents on the alkyne, but not for $-F$.

The concepts of hardness, softness, and of frontier orbitals, with which latter the Fukui function is closely connected, have been severely criticized [103,120]. It is also true that in some cases the results predicted using these methods can also be understood in terms of more traditional chemical concepts. Thus, in the alkyne–HNC reaction, resonance theory leads one to suspect that the zwitterion A, with the positive charge formally on the more substituted carbon, will be favored over B. Indeed, AM1, HF/6-31G* and pBP/DN* DFT calculations all show A to be of lower energy than B.

Also, the more rigorous method of locating the two transition states and comparing their energies was, in the author's hands, straightforward and not excessively demanding of time. Nevertheless, the large amount of work which has been done using these ideas suggests that they offer a useful approach to interpreting and predicting chemical reactivity. Even an apparently unrelated property, or rather complex of properties, namely aromaticity, has been subjected to analysis in terms of hardness [121]. As Parr and Yang say, "This is perhaps an oversimplified view of chemical reactivity, but it is useful" [122].

7.3.6   Visualization

The only cases for which one might anticipate differences between DFT and wavefunction theory as regards visualization (sections 5.5.6, 6.3.6) are those involving orbitals: as explained in section 7.2.3.2, *The Kohn-Sham equations*, the orbitals of currently popular DFT methods were introduced to make the calculation of the electron density tractable, and in a pure DFT theory orbitals do not exist. Thus, electron density, spin density, and electrostatic potential can be visualized in DFT calculations just as in *ab initio* or semiempirical work. However, visualization of orbitals, so important in wavefunction work (especially the HOMO and LUMO, which in frontier orbital theory [114] strongly influence reactivity) is not possible in a *pure* DFT approach. However, in currently popular DFT calculations one can visualize the KS orbitals, which are qualitatively much like wavefunction orbitals [93] (section 7.3.5, *Ionization energies and electron affinities*).

7.4   STRENGTHS AND WEAKNESSES OF DFT

Strengths
DFT includes electron correlation in its theoretical basis, in contrast to wavefunction methods, which must take correlation into account by add-ons (Møller–Plesset perturbation, configuration interaction, coupled-cluster) to *ab initio* HF theory, or by parameterization in semiempirical methods. Because it has correlation fundamentally built in, DFT can calculate geometries and relative energies with an accuracy comparable to MP2 calculations, in roughly the same time as needed for HF calculations. Aiding this, DFT calculations are basis-set-saturated more easily than are *ab initio*: limiting results are approached with smaller basis sets than for *ab initio* calculations. Calculations of post-HF accuracy can thus be done on bigger molecules than *ab initio* methods make possible. DFT appears to be the method of choice for geometry and energy calculations on transition metal compounds, for which conventional *ab initio* calculations often give poor results [58,123].

DFT works with electron density, which can be measured and is easily intuitively grasped [4], rather than a wavefunction, a mathematical entity whose physical meaning is still controversial.

Weaknesses
The exact exchange-correlation functional $E_{XC}[\rho_0]$, one of the terms in the DFT expression for the energy, is unknown, and no one knows how to systematically improve our approximations to it. In contrast, *ab initio* energies can be systematically lowered by

using bigger basis sets and by expanding the correlation method: MP2, MP3, ..., or more determinants in the CI approach. It is true that for a particular purpose 6-311G* may not be better than 6-31G*, and MP3 is certainly not necessarily better than MP2, but bigger basis sets and higher correlation levels will *eventually* approach an exact solution of the Schrödinger equation. The accuracy of DFT is being gradually improved by modifying functionals, not according to some grand theoretical prescription, but rather with the aid of experience and intuition, and checking the calculations against experiment. This makes DFT somewhat semiempirical. Some functionals contain parameters which must be fitted to experiment; these methods are even more heavily empirical. Since the functionals are not based purely on fundamental theory, one should be cautious about applying DFT to very novel molecules. Of course the semiempirical character of current DFT is not a fundamental feature of the basic method, but arises only from our ignorance of the exact exchange-correlation functional. Because our functionals are only approximate, DFT as used today is not variational (the calculated energy could be lower than the actual energy).

DFT is not as accurate as the highest-level *ab initio* methods, like QCISD(T) and CCSD(T) (but it can handle much bigger molecules than can these methods). Even gradient-corrected functionals apparently are unable to handle van der Waals interactions [124], although they do give good energies and structures for hydrogen-bonded species [125].

DFT today is mainly a ground-state theory, although ways of applying it to excited states are being developed.

7.5 SUMMARY OF CHAPTER 7

Density functional theory is based on the two Hohenberg–Kohn theorems, which state that the ground-state properties of an atom or molecule are determined by its electron density function, and that a trial electron density must give an energy greater than or equal to the true energy. Actually, the latter theorem is true only if the exact functional (see below) is used; with the approximate functionals in use today, DFT is not variational – it can give an energy below the true energy. In the Kohn–Sham approach the energy of a system is formulated as a deviation from the energy of an idealized system with noninteracting electrons. The energy of the idealized system can be calculated exactly since its wavefunction (in the Kohn–Sham approach wavefunctions and orbitals were introduced as a mathematical convenience to get at the electron density) can be represented exactly by a Slater determinant. The relatively small difference between the real energy and the energy of the idealized system contains the exchange-correlation functional, the only unknown term in the expression for the DFT energy; the approximation of this functional is the main problem in DFT. From the energy equation, by minimizing the energy with respect to the Kohn–Sham orbitals the Kohn–Sham equations can be derived, analogously to the HF equations. The molecular orbitals of the KS equations are expanded with basis functions and matrix methods are used to iteratively find the energy, and to get a set of molecular orbitals, the KS orbitals, which are qualitatively similar to the orbitals of wavefunction theory.

The simplest version of DFT, the LDA, which treats the electron density as constant or only very slightly varying from point to point in an atom or molecule, and also pairs two electrons of opposite spin in each KS orbital, is little used nowadays. It has been largely replaced by methods which use gradient-corrected ("nonlocal") functionals and which assign one set of spatial orbitals to α-spin electrons, and another set of orbitals to β-electrons; this latter "unrestricted" assignment of electrons constitutes the LSDA. The best results appear to come from so-called hybrid functionals, which include some contribution from HF type exchange, using KS orbitals. The most popular current DFT method is the LSDA gradient-corrected hybrid method which uses the B3LYP (Becke three-parameter Lee–Yang–Parr) functional.

Gradient-corrected and, especially, hybrid functionals, give good to excellent geometries. Gradient-corrected and hybrid functionals usually give fairly good reaction energies, but, especially for isodesmic-type reactions, the improvement over HF/3-21G or HF/6-31G* calculations does not seem to be dramatic (as far as the relative energies of normal, ground-state organic molecules goes; for energies and geometries of transition metal compounds, DFT is the method of choice). For homolytic dissociation, correlated methods (e.g. B3LYP, pBP/DN* and MP2) are vastly better than HF-level calculations; these methods also tend to give fairly good activation barriers.

DFT gives reasonable IR frequencies and intensities, comparable to those from MP2 calculations. Dipole moments from DFT appear to be more accurate than those from MP2, and B3LYP/6-31G* moments on AM1 geometries are good. Time-dependent DFT (TDDFT) is the best method (with the possible exception of semiempirical methods parameterized for the type of molecule of interest) for calculating UV spectra reasonably quickly. DFT is said to be better than HF (but not as good as MP2) for calculating NMR spectra. Good first ionization energies are obtained from B3LYP/6-31 + G*//B3LYP/3-21G(*) energy differences (using AM1 geometries makes little difference, at least with normal molecules). These values are somewhat better than the *ab initio* MP2 energy difference values, and are considerably better than MP2 Koopmans' theorem IEs. Rough estimates of electron affinities can be obtained from the negative LUMOs from LSDA functionals (gradient-corrected functionals give much worse estimates). For conjugated molecules, HOMO–LUMO gaps from hybrid functionals agreed well with the $\pi \to \pi^*$ UV transitions. The mutually related concepts of electronic chemical potential, electronegativity, hardness, softness, and the Fukui function are usually discussed within the context of DFT. They are readily calculated from ionization energy, electron affinity, and atom charges.

REFERENCES

[1] (a) A. Whitaker, "Einstein, Bohr and the Quantum Dilemma," Cambridge University Press, 1996. (b) P. Yam, Scientific American, June 1997, p. 124. (c) D. Z. Albert, Scientific America, May 1994, 58. (d) D. Z. Albert, "Quantum Mechanics and Experience," Harvard University Press, Cambridge, MA, 1992. (e) D. Bohm and H. B. Hiley, "The Undivided Universe," Routledge, New York, 1992. (f) J. Baggott, "The Meaning of Quantum Theory," Oxford University Press, New York, 1992. (g) M. Jammer, "The Philosophy of Quantum Mechanics," Wiley, New York, 1974.

[2] R. F. W. Bader, "Atoms in Molecules," Oxford, New York, 1990.

[3] Reference [2], pp. 7–8.

[4] G. P. Shusterman and A. J. Shusterman, J. Chem. Educ., 1997, *74*, 771.

[5] R. G. Parr and W. Yang, "Density-Functional Theory of Atoms and Molecules," Oxford, New York, 1989, p. 53.

[6] (a) E. Wilson, Chem. Eng. News, 1998, October 19, 12. (b) D. Malakoff, Science, 1998, *282*, 610.

[7] Cf. I. N. Levine, "Quantum Chemistry," 5th edn, Prentice Hall, Upper Saddle River, New Jersey, 2000, p. 422, equation (13.130); cf. p. 624, problem 15.67, for the KS orbitals. The multielectron wavefunction is treated on pp. 421–423.

[8] P.-O. Löwdin, Phys. Rev., 1955, *97*, 1474.

[9] R. G. Parr and W. Yang, "Density-Functional Theory of Atoms and Molecules," Oxford University Press, New York, 1989.

[10] W. Koch and M. Holthausen, "A Chemist's Guide to Density Functional Theory," Wiley-VCH, New York, 2000.

[11] I. N. Levine, "Quantum Chemistry," 5th edn, Prentice Hall, Upper Saddle River, New Jersey, 2000.

[12] R. A. Friesner, R. B. Murphy, M. D. Beachy, M. N. Ringnalda, W. T. Pollard, B. D. Dunietz, and Y. Cao, J. Phys. Chem. A, 1999, *103*, 1913.

[13] W. Kohn, A. D. Becke, and R. G. Parr, J. Phys. Chem. 1996, *100*, 12974.

[14] R. G. Parr and W. Yang, Annu. Rev. Phys. Chem., 1995, *46*, 701.

[15] E.g. D. J. Griffiths, "Introduction to Quantum Mechanics," Prentice-Hall, Engelwood Cliffs, NJ, 1995.

[16] Earlier work (1927) by Fermi was published in Italian and came to the attention of the physics community with a paper in German: E. Fermi, Z. Phys., 1928, *48*, 73. This appears in English translation in N. H. March, "Self-Consistent Fields in Atoms," Pergamon, New York, 1975.

[17] L. H. Thomas, Proc. Camb. Phil. Soc., 1927, *23*, 542; reprinted in N. H. March, "Self-Consistent Fields in Atoms," Oxford: Pergamon, New York, 1975.

[18] Reference [9], chapter 6.

[19] J. C. Slater, Int. J. Quantum Chem. Symp., 1975, *9*, 7.

[20] Reviews: (a) J. W. D. Connolly in "Semiempirical Methods of Electronic Structure Calculations Part A: Techniques," G. A. Segal, Ed., Plenum, New York, 1977. (b) K. H. Johnson, Adv. Quantum Chem., 1973, *7*, 143.

[21] J. C. Slater, Phys. Rev., 1951, *81*, 385.

[22] For a personal history of much of the development of quantum mechanics, with significant emphasis on the $X\alpha$ method, see: J. C. Slater, "Solid-State and Molecular Theory: A Scientific Biography," Wiley, New York, 1975.

[23] Reference [11], pp. 573–577.

[24] P. Hohenberg and W. Kohn, Phys. Rev. B, 1964, *136*, 864.

[25] Reference [9], section 3.4.

[26] W. Kohn and L. J. Sham, Phys. Rev. A, 1965, *140*, 1133.

[27] Reference [11], section 13.14.

[28] F. Jensen, "Introduction to Computational Chemistry," Wiley, New York, 1999, p. 180.

[29] Reference [9], sections 7.1–7.3.

[30] Reference [11], section 11.8.

[31] Reference [9], pp. 145–149, 151, 165.

[32] Reference [9], appendix A.

[33] Reference [11], p. 580.

[34] G. N. Merrill, S. Gronert, and S. R. Kass, J. Phys. Chem. A, 1997, *101*, 208.

[35] Reference [11], pp. 581–584.

[36] Reference [9], section 7.4 and appendix E.

[37] Reference [9], pp. 173–174.

[38] F. Jensen, "Introduction to Computational Chemistry," Wiley, New York, 1999, section 6.1.

[39] Reference [9], chapter 8.

[40] In mathematics the term "local" is used in much the same sense in which computational chemists speak of a local minimum: a minimum *in that region*, in contrast to a *global* minimum, the lowest point on the *whole* potential energy surface. A property of a mathematical set (which might have one member, a single function) at a point P which can be defined in terms of an arbitrarily small region around P is said to be a local property of the set. Thus, a derivative or gradient at a point is a local property of any function that is differentiable at the point (this is clear from the definition of a derivative as the limit of a ratio of finite increments, or as a ratio of infinitesimals). Mathematicians sometimes use "relative" and "absolute" instead of "local" and "global". See books on mathematical analysis, e.g. T. M. Apostol, "Mathematical Analysis," Addison-Wesley, Reading, MA, 1957, p. 73; R. Haggarty, "Fundamentals of Mathematical Analysis," Addison-Wesley, Reading, MA, 1989, p. 178. The word is also used in quantum physics to denote phenomena, or entities ("hidden variables"), that do not, or do not seem to, violate relativity. See e.g. A. Whitaker, "Einstein, Bohr and the Quantum Dilemma," Cambridge University Press, Cambridge, 1996; V. J. Stenger, "The Unconscious Quantum," Prometheus, Amherst, New York, 1995.

[41] A. St.-Amant, Chapter 5 in Reviews in Computational Chemistry, Volume 7, K. B. Lipkowitz and D. B. Boyd, Eds., VCH, New York, 1996, p. 223.

[42] Reference [11], p. 587.

[43] A. D. Becke, Phys. Rev. A, 1988, *38*, 3098.

[44] M. Head-Gordon, J. Phys. Chem., 1996, *100*, 13213.

[45] E.g., Gaussian 98, revision A.6, M. J. Frisch, G. W. Trucks, H. B. Schlegel, G. E. Scuseria, M. A. Robb, J. R. Cheeseman, V. G. Zakrzewski, J. A. Montgomery, Jr., R. E. Stratmann, J. C. Burant, S. Dapprich, J. M. Millam, A. D. Daniels, K. N. Kudin, M. C. Strain, O. Farkas, J. Tomasi, V. Barone, M. Cossi, R. Cammi, B. Mennucci, C. Pomelli, C. Adamo, S. Clifford, J. Ochterski, G. A. Petersson, P. Y. Ayala, Q. Cui, K. Morokuma, D. K. Malick, A. D. Rabuck, K. Raghavachari, J. B. Foresman, J. Cioslowski, J. V. Ortiz, B. B. Stefanov, G. Liu, A. Liashenko, P. Piskorz, I. Komaromi, R. Gomperts, R. L. Martin, D. J. Fox, T. Keith, M. A. Al-Laham, C. Y. Peng, A. Nanayakkara, C. Gonzalez, M. Challacombe, P. M. W. Gill, B. Johnson, W. Chen, M. W. Wong, J. L. Andres, C. Gonzalez, M. Head-Gordon, E. Repogle, and J. A. Pople, Gaussian, Inc., Pittsburgh PA, 1998.

[46] B. Delley and D. E. Ellis, J. Chem. Phys., 1982, *76*, 1949.

[47] Spartan is made by Wavefunction, Inc., 18401 Von Karman Avenue, Suite 370, Irvine CA 92612. The pBP calculations in this chapter were done with PC Spartan Pro, version 1.0, on a 600 MHz Pentium III running under Windows NT; MP2 optimizations were done with Spartan version 3.1.3 on an SGI R4000 Indigo workstation.

[48] W. J. Hehre and L. Lou, "A Guide to Density Functional Calculations in Spartan," Wavefunction Inc., Irvine CA, 1997.

[49] P. J. Stephens, F. J. Devlin, C. F. Chablowski, and M. J. Frisch, J. Phys. Chem., 1994, *98*, 11623, and references therein.

[50] L. Fan and T. Ziegler, J. Chem. Phys., 1991, *94*, 6057.

[51] Reference [11], chapter 17.

[52] W. J. Hehre, "Practical Strategies for Electronic Structure Calculations," Wavefunction, Inc., Irvine, CA, 1995.

[53] Titan Version 1 (run on a 600 MHz Pentium III under Windows NT) is a joint product of Wavefunction, Inc., 18401 Von Karman Avenue, Suite 370, Irvine CA 92612 and Schrödinger, Inc., 1500 SW First Avenue, Suite 1180, Portland, OR 97201.

[54] Gaussian 94 for Windows (G94W): Gaussian 94, Revision E.1, M. J. Frisch, G. W. Trucks, H. B. Schlegel, P. M. W. Gill, B. G. Johnson, M. A. Robb, J. R. Cheeseman, T. Keith, G. A. Petersson, J. A. Montgomery, K. Raghavachari, M. A. Al-Laham, V. G. Zakrzewski, J. V. Ortiz, J. B. Foresman, J. Cioslowski, B. B. Stefanov, A. Nanayakkara, M. Challacombe, C. Y. Peng, P. Y. Ayala, W. Chen, M. W. Wong, J. L. Andres, E. S. Replogle, R. Gomperts, R. L. Martin, D. J. Fox, J. S. Binkley, D. J. Defrees, J. Baker, J. P. Stewart, M. Head-Gordon, C. Gonzalez, and J. A. Pople, Gaussian, Inc., Pittsburgh PA, 1995.

[55] A. C. Scheiner, J. Baker, and J. W. Andzelm, J. Comput. Chem., 1997, *18*, 775.

[56] A. St.-Amant, W. D. Cornell, P. A. Kollman, and T. H. Halgren, J. Comput. Chem., 1997, *16*, 1483.

[57] A. A. El-Azhary, J. Phys. Chem., 1996, *100*, 15056.

[58] C. W. Bauschlicher, Jr., A. Ricca, H. Partridge, and S. R. Langhoff, in "Recent Advances in Density Functional Methods. Part II," D. P. Chong, Ed., World Scientific, Singapore, 1995.

[59] Reference [52], chapter 2.

[60] W. J. Hehre, L. Radom, P. v. R. Schleyer, J. A. Pople, "*Ab initio* Molecular Orbital Theory," Wiley, New York, 1986.

[61] (a) $H_2C-CHOH$ *reaction* E. Lewars and I. Bonnycastle, J. Mol. Struct. (Theochem), 1997, *418*, 17 and references therein. The experimental barrier could be significantly in error. (b) *HNC reaction* V. S. Rao, A. Vijay, A. K. Chandra, Can. J. Chem., 1996, *74*, 1072. (c) CH_3NC *reaction* The reported experimental activation energy is 161 kJ mol^{-1}: F. W. Schneider and B. S. Rabinovitch, J. Am. Chem. Soc., 1962, *84*, 4215; B. S. Rabinovitch and P. W. Gilderson, J. Am. Chem. Soc., 1965, *87*, 158. The energy of CH_3CN relative to CH_3NC by a high-level (G2) calculation is -98.3 kJ mol^{-1} (E. Lewars). An early *ab initio* study of the reaction: D. H. Liskow, C. F. Bender, and H. F. Schaefer, J. Am. Chem. Soc., 1972, *95*, 5178. A comparison of CH_3CN, CH_3NC, other isomers and radicals, cations and anions: P. M. Mayer, M. S. Taylor, M. Wong, and L. Radom, J. Phys. Chem. A, 1998, *102*, 7074. (d) *Cyclopropylidene reaction* H. F. Bettinger, P. R. Schreiner, P. v. R. Schleyer, H. F. Schaefer, J. Phys. Chem., 1996, *100*, 16147.

[62] A. P. Scott and L. Radom, J. Phys. Chem., 1996, *100*, 16502.

[63] J. M. Martell, J. D. Goddard, and L. Eriksson, J. Phys. Chem., 1997, *101*, 1927.

[64] K. B. Wiberg and J. W. Ochterski, J. Comp. Chem., 1997, *18*, 108.

[65] E. Rousseau and D. Mathieu, J. Comp. Chem., 2000, *21*, 367.

[66] O. N. Ventura, M. Kieninger, and R. E. Cachau, J. Phys. Chem. A, 1999, *103*, 147.

[67] 368 kJ mol^{-1}: R. T. Morrison and R. N. Boyd, "Organic Chemistry," 6th edn, Prentice Hall, Engelwood Cliffs, New Jersey, 1992, p. 21; 377 kJ mol^{-1}: K. P. C. Vollhardt and N. E. Schore, "Organic Chemistry," 2nd edn, Freeman, New York, 1987, p. 75.

[68] Reference [52]. The data are from pp. 54, 58, 64, and 70. In each case the first 10 examples were used.

[69] D. A. Singleton, S. R. Merrigan, J. Liu, and K. N. Houk, J. Am. Chem. Soc., 1997, *119*, 3385.

[70] M. N. Glukhovtsev, R. D. Bach, A. Pross, and L. Radom, Chem. Phys. Lett., 1996, *260*, 558.

[71] R. L. Bell, D. L. Tavaeras, T. N. Truong, and J. Simons, Int. J. Quantum Chem., 1997, *63*, 861.

[72] T. N. Truong, W. T. Duncan, and R. L. Bell, in "Chemical Applications of Density Functional Theory," B. B. Laird, R. B. Ross, and T. Ziegler, Eds., American Chemical Society, Washington, DC, 1996.

[73] Q. Zhang and R. L. Bell, J. Phys. Chem., 1995, *99*, 592.

[74] F. Eckert and G. Rauhut, J. Am. Chem. Soc., 1998, *120*, 13478.

[75] J. Baker, M. Muir, and J. Andzelm, J. Chem. Phys., 1995, *102*, 2063.

[76] B. S. Jursic, in "Recent Developments and Applications of Modern Density Functional Theory," J. M. Seminario, Ed., Elsevier, Amsterdam, 1996.

[77] S. W. Brown, J. C. Rienstra-Kiracofe, and H. F. Schaefer, J. Phys. Chem. A, 1999, *103*, 4065.

[78] Reference [48], pp. 25, 27.

[79] Calculated by the author from ref. 56, Table VII.

[80] P. Geerlings, F. De Profit, and J. M. L. Martin, in "Recent Developments and Applications of Modern Density Functional Theory," J. M. Seminario, Ed., Elsevier, Amsterdam, 1996.

[81] G. Lendvay, J. Phys. Chem., 1994, *98*, 6098.

[82] R. J. Boyd, J. Wang, and L. A. Eriksson, in "Recent Advances in Density Functional Methods. Part I," D. P. Chong, Ed., World Scientific, Singapore, 1995.

[83] Reference [11], sections 9.9, 9.10.

[84] M. E. Casida in "Recent Developments and Applications of Modern Density Functional Theory," J. M. Seminario, Ed., Elsevier, Amsterdam, 1996, and references therein.

[85] R. E. Stratman, G. E. Scuseria, and M. J. Frisch, J. Chem. Phys., 1998, *109*, 8218.

[86] K. B. Wiberg, R. E. Stratman, and M. J. Frisch, Chem. Phys. Lett., 1998, *297*, 60.

[87] J. B. Foresman and Æ. Frisch, "Exploring Chemistry with Electronic Structure Methods," Gaussian Inc., Pittsburgh, PA, 1996, p. 218.

[88] (a) M. J. Frisch, G. W. Trucks, and J. R. Cheeseman, in "Recent Developments and Applications of Modern Density Functional Theory," J. M. Seminario, Ed., Elsevier, Amsterdam, 1996. (b) Accurate ^{1}H NMR spectra have been reported for DFT calculations with empirical corrections: P. R. Rablen, S. A. Pearlman, and J. Finkbiner, J. Phys. Chem. A, 2000, *103*, 7357.

[89] R. M. Silverstein, G. C. Bassler, and T. C. Morrill, "Spectrometric Identification of Organic Compounds," 4th edn, Wiley, New York, 1981, pp. 226–270.

[90] G. C. Levy and G. L. Nelson, "Carbon-13 Nuclear Magnetic Resonance for Organic Chemists," Wiley, New York, 1972.

[91] H. M. Muchall, N. H. Werstiuk, and B. Choudhury, Can. J. Chem., 1998, *76*, 227.

[92] E. J. Baerends and O. V. Gritsenko, J. Phys. Chem., 1997, *101*, 5383.

[93] R. Stowasser and R. Hoffmann, J. Am. Chem. Soc., 1999, *121*, 3414.

[94] U. Salzner, J. B. Lagowski, P. G. Pickup, and R. A. Poirier, J. Comp. Chem., 1997, *18*, 1943.

[95] W. J. Hunt and W. A. Goddard, Chem. Phys. Lett., 1969, *3*, 414.

[96] R. D. Levin and S. G. Lias, "Ionization Potential and Appearance Potential Measurements, 1971–1981," National Bureau of Standards, Washington, DC.

[97] L. A. Curtis, R. H. Nobes, J. A. Pople, and I. Radom, J. Chem. Phys., 1992, *97*, 6766.

[98] J. J. Berzelius, "Essai sur la théorie des proportions chimiques et sur l'influence chimique de l'électricité," 1819; see M. J. Nye, "From Chemical Philosophy to Theoretical Chemistry," University of California Press, Berkeley, CA, 1993, p. 64.

[99] Reference [9]. pp. 92–95.

[100] R. S. Mulliken, J. Am. Chem. Soc., 1952, *64*, 811, 819.

[101] (a) R. G. Pearson, J. Am. Chem. Soc., 1963, *85*, 3533. (b) R. G. Pearson, Science, 1963, *151*, 172.

[102] (a) R. G. Pearson, "Hard and Soft Acids and Bases," Dowden, Hutchinson and Ross, Stroudenburg, PA, 1973. (b) T. L. Lo, "Hard and Soft Acids and Bases in Organic Chemistry," Academic Press, New York, 1977.

[103] M. J. S. Dewar, "A Semiempirical Life," American Chemical Society, Washington, DC, 1992; p. 160.

[104] R. G. Parr and W. Yang, J. Am. Chem. Soc., 1984, *106*, 4049.

[105] Reference [9], chapters 4 and 5 in particular.

[106] E.g. P. Atkins, "Physical Chemistry," 6th edn, Freeman, New York, 1998, p. 132.

[107] R. P. Iczkowski and J. L. Margrave, J. Am. Chem. Soc., 1961, *83*, 3547.

[108] R. G. Parr, R. A. Donnelly, M. Levy, and W. E. Palke, J. Chem. Phys., 1978, *68*, 3801.

[109] R. S. Mulliken, J. Chem. Phys., 1934, *2*, 782.

[110] R. G. Pearson, J. Chem. Educ., 1999, *76*, 267.

[111] R. G. Parr and R. G. Pearson, J. Am. Chem. Soc., 1983, *105*, 7512.

[112] A. Toro-Labbé, J. Phys. Chem. A, 1999, *103*, 4398.

[113] L. T. Nguyen, T. N. Le, F. De Proft, A. K. Chandra, W. Langenaeker, M. T. Nguyen, and P. Geerlings, J. Am. Chem. Soc., 1999, *121*, 5992.

[114] (a) K. Fukui, Science, 1987, *218*, 747. (b) I. Fleming, "Frontier Orbitals and Organic Chemical Reactions," Wiley, New York, 1976. (c) K. Fukui, Acc. Chem. Res., 1971, *57*, 4.

[115] W. Yang and W. J. Mortier, J. Am. Chem. Soc., 1986, *108*, 5708.

[116] F. A. Carey and R. L. Sundberg, "Advanced Organic Chemistry," 3rd edn, Plenum, New York, 1990, pp. 423–430.

[117] F. Méndez and J. L. Gázquez, J. Am. Chem. Soc., 1994, *116*, 9298.

[118] S. Damoun, G. Van de Woude, K. Choho, and P. Geerlings, J. Phys. Chem. A, 1999, *103*, 7861.

[119] J. L. Gázquez and F. Méndez, J. Phys. Chem., 1994, *98*, 4591.

[120] M. J. S. Dewar, "A Semiempirical Life," American Chemical Society, Washington, DC, 1992, p. 162.

[121] (a) Z. Zhou and R. G. Parr, J. Am. Chem. Soc., 1989, *111*, 7371. (b) Z. Zhou, R. G. Parr, and J. F. Garst, Tetrahedron Lett., 1988, *29*, 4843.

[122] Reference [9], p. 101.

[123] (a) Reference [10], part B, and references therein. (b) G. Frenking, J. Chem. Soc., Dalton Trans., 1997, 1653.

[124] (a) S. Kyistyan and P. Pulay, Chem. Phys. Lett., 1994, *229*, 175. (b) J. M. Perez-Jorda and A. D. Becke, Chem. Phys. Lett., 1995, *233*, 134.

[125] (a) M. Lozynski, D. Rusinska-Roszak, and H.-G. Mack, J. Phys. Chem. A, 1998, *102*, 2899. (b) C. Adamo and V. Barone in "Recent Advances in Density Functional Methods. Part II," D. P. Chong, Ed., World Scientific, Singapore, 1997. (c) F. Sim, A. St.-Amant, I. Papai, and D. R. Salahub, J. Am. Chem. Soc., 1992, *114*, 4391.

EASIER QUESTIONS

1. State the arguments for and against, regarding DFT as being more a semiempirical than an *ab initio*-like theory.

2. What is the essential difference between wavefunction theory and DFT? What is it that, in principle anyway, makes DFT simpler than wavefunction theory?

3. Why can't current DFT calculations be improved in a stepwise, systematic way, as can *ab initio* calculations?

4. Which of these prescriptions for dealing with a function are functionals: (1) square root of $f(x)$. (2) $\sin f(x)$. (3) $\sum_{x=1}^{3} f(x)$. (4) $\int f(x)\,dx$. (5) $\exp(f(x))$?

5. For which class(es) of functions is the nth derivative of $f(x)$ a functional?

6. Explain why a kind of molecular orbital is found in current DFT, although DFT is touted as an alternative to wavefunction theory.

7. What is fundamentally wrong with functionals that are not gradient-corrected?

8. The ionization energy of a molecule can be regarded as the energy required to remove an electron from its HOMO. How then would a pure density functional theory, with no orbitals, be able to calculate ionization energy?

9. Label these statements true or false: (1) For each molecular wavefunction there is an electron density function. (2) Since the electron density function has only x, y, z as its variables, DFT necessarily ignores spin. (3) DFT is good for transition metal compounds because it has been specifically parameterized to handle them. (4) In the limit of a sufficiently large basis set, a DFT calculation represents an exact solution of the Schrödinger equation. (5) The use of very large basis sets is essential with DFT. (6) A major problem in density functional theory is the prescription for going from the molecular electron density function to the energy.

10. Explain *in words* the meaning of the terms electronegativity, hardness, and the Fukui function.

HARDER QUESTIONS

1. It is sometimes said that electron density is physically more real than a wavefunction. Do you agree? Is something that is more easily grasped intuitively necessarily more real?

2. A functional is a function of a function. Explore the concept of a function of a functional.

3. Why is it that the HF Slater determinant is an inexact representation of the wavefunction, but the DFT determinant for a system of noninteracting electrons is exact for this particular wavefunction?

4. Why do we expect the "unknown" term in the energy equation ($E_{xc}[\rho_0]$, in Eq. (7.21)) to be small?

5. Merrill *et al.* have said that "while solutions to the [HF equations] may be viewed as exact solutions to an approximate description, the [KS equations] are approximations to an exact description!" Explain.

6. Electronegativity is the ability of an atom or molecule to attract electrons Why is it then (from one definition) the average of the ionization energy and the electron affinity (Eq. (7.14)), rather than simply the electron affinity?

7. Given the wavefunction of a molecule, it is possible to calculate the electron density function. Is it possible in principle to go in the other direction? Why or why not?

8. The multielectron wavefunction Ψ is a function of the spatial and spin coordinates of all the electrons. Physicists say that Ψ for any system tells us all that can be known about the system. Do you think the electron densiity function ρ tells us everything that can be known about a system? Why or why not?

9. If the electron density function concept is mathematically and conceptually simpler than the wavefunction concept, why did DFT come later than wavefunction theory?

10. For a spring or a covalent bond, the concepts of force and force constant can be expressed in terms of first and second derivatives of energy with respect to extension. If we let a "charge space" N replace the real space of extension of the spring or bond, what are the analogous concepts to force and force constant? Using the SI, derive the units of electronegativity and of hardness.

Chapter 8

Literature, Software, Books and Websites

The yeoman work in any science ... is done by the experimentalist, who must keep the theoretician honest.
Michio Kaku, Professor of Theoretical Physics, City University of New York.

8.1 FROM THE LITERATURE

A small smorgasbord of published papers will be discussed here, to show how some of the things that we have seen in previous chapters have appeared in the literature.

8.1.1 To be or not to be

8.1.1.1 Oxirene

Let us start with what looks like a simple problem: what can computational chemistry tell us about oxirene (oxacyclopropene, Fig. 8.1; the oxirene literature till 1983 has been reviewed [1]). Labeling one of the carbons of a diazo ketone $(R-C(N_2)-CO-R)$ can lead to a ketene with scrambled labeling. After excluding the possibility of scrambling in the diazo compound, this indicates that an oxirene species is formed. However, this does not tell us whether this species is an intermediate or merely a transition state (Fig. 8.2). A straightforward way to try to answer this question would seem to be to calculate the frequencies, at the level used to optimize the structure, and see if there are any imaginary frequencies – a relative minimum has none, while a transition state has one (section 2.5). In a preliminary investigation [2] Schaefer and coworkers found that oxirene was a minimum with the Hartree-Fock (HF) (SCF) method, and also when electron correlation was taken into account (section 5.4) with the CISD and CCSD methods, using double-zeta basis sets (section 5.3). However, in going from HF to CISD to CCSD, the ring-opening frequency fell from 445 to 338 to $262\,cm^{-1}$, which was said to be a much steeper drop than would be expected. A very comprehensive investigation with the above ("To be ...") title [3], in which the frequencies of oxirene were examined at 46 (!) different levels failed to definitively settle the matter: even using CCSD(T) calculations with large basis sets the results were somewhat quirky, and in fact of the six highest levels used, three gave an imaginary frequency and three

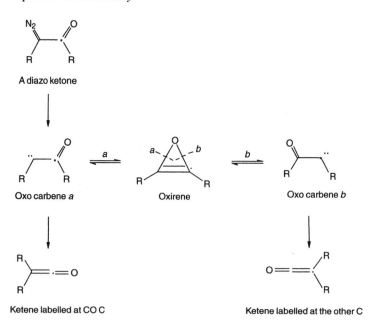

Figure 8.1. Generating an oxo carbene (a "ketocarbene") from a labelled diazo ketone sometimes leads to a ketene in which the label is scrambled. This indicates that a species with the symmetry of oxirene is formed.

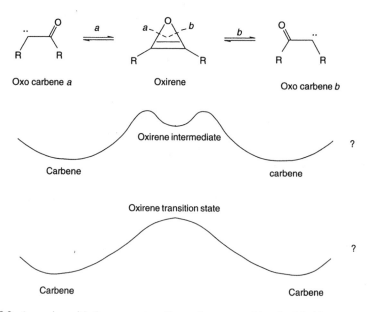

Figure 8.2. A species with the symmetry of an oxirene scrambles the label in an oxo carbene. But this does not tell us whether the oxirene is an intermediate or merely a transition state.

all real frequencies. At the two highest levels the ring-opening frequency was real, but uncomfortably low (139 and 163 cm^{-1}). Oxirene is the most notorious case of an unsolved computational "existence theorem".

8.1.1.2 Nitrogen pentafluoride

Nitrogen pentafluoride represents an interesting contrast to oxirene. Oxirene is, on paper, a reasonable molecule; there is no obvious reason why, however unstable it might be because of antiaromaticity [4] or strain, it should not be able to exist. On the other hand, NF_5 defies the hallowed octet rule; why should it be more reasonable than, say, CH_6? Yet a comprehensive computational study of this molecule left "little doubt" that it is a (relative) minimum on its potential energy surface [5]. The full armamentarium of post-HF methods, CASSDF, MRCI, CCSDT, CCSD(T), MP2 (section 5.4) and DFT (chapter 7) was employed here, and all agreed that D_{3h} (section 2.6) NF_5 is a minimum.

8.1.1.3 Pyramidane

If oxirene "should" exist and NF_5 "should" not, what are we to make of pyramidane (Fig. 8.3)? This molecule contradicts the traditional paradigm [6] of tetracoordinate carbon having its bonds tetrahedrally directed: the four bonds of the apical carbon point toward the base of a pyramid. Part of the calculated [7] potential energy surface of pyramidane is shown in Fig. 8.3. To improve the accuracy of the relative energies, the MP2 geometries were subjected to single-point calculations (section 5.5.2) using the QCI method (section 5.4.3), with the results shown (Fig. 8.3). At the QCISD(T)/6-31G*//MP(fc)/6-31G* level pyramidane is predicted to be a relative

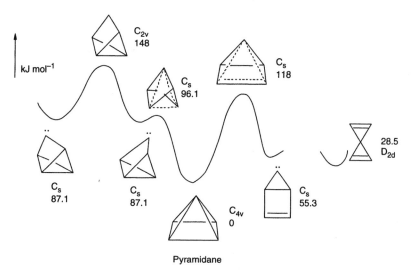

Pyramidane

Figure 8.3. (Part of) the pyramidane potential energy surface. QCICD (T)/6-31G*//MP2(fc)/6-31G* calculations.

1	2	3	4	5
Benzene	Dewar benzene	Benzvalene	Prismane	Bicyclopropenyl

Figure 8.4. Nitrogen analogs (CH → N) of these molecules have been investigated computationally.

minimum with a barrier of $96\,\text{kJ mol}^{-1}$ for its lowest-energy isomerization path, to the tricyclic carbene, which lies $87\,\text{kJ mol}^{-1}$ above it. This presents us with the astonishing possibility that the exotic hydrocarbon may be isolable at room temperature. The threshold barrier for isolability at room temperature is about $100\,\text{kJ mol}^{-1}$; for example, (*E*)-cycloheptene, with a barrier of $71\,\text{kJ mol}^{-1}$ to isomerization, has a room-temperature halflife of about 47 s [8], and the halflife rises steeply with the barrier. Other properties of pyramidane, including ionization energy and electron affinity (section 5.5.5), heat of formation (section 5.5.2.2c), and NMR spectra (section 5.5.5) were calculated [7b].

8.1.1.4   Beyond dinitrogen

There has in recent years been considerable interest in the possibility of making allotropes of nitrogen with more than two atoms per molecule. Curiously, almost all (the N_5 cation has been made [9, 10]) the work reported has been computational rather than experimental. These compounds are interesting because to any chemist with imagination the idea of a form of pure nitrogen that is not a gas at room temperature is fascinating, and because any such compound would be thermodynamically very unstable with respect to decomposition to dinitrogen.

Perhaps the first serious computational study of nitrogen oligomers was by Engelke, who studied the N_6 analogues of the benzene isomers in Fig. 8.4, first at the uncorrelated [11] then at the MP2 [12] level. The uncorrelated calculations suggested that 1–5 were "stable", i.e. kinetically stable, although thermodynamically much higher in energy than dinitrogen. However, on the MP2/6-31G* potential energy surface 1 is a hilltop (section 2.2) and 5 is a transition state (section 2.2). This illustrates the not-so-rare fact that optimistic predictions at low levels of theory may not be sustained at higher levels. Noncorrelated *ab initio*, and in particular, semiempirical (chapter 6) calculations, tend to be too permissive in granting reality to exotic molecules.

8.1.2   Mechanisms

We have seen, above, that computational chemistry can sometimes tell us with good reliability whether a molecule can exist. Another important application is to indicate how one molecule gets to be another: how chemical reactions occur. Indeed, the prime architect of one of the most useful computational tools, the AM1 method (chapter 6), questioned "whether the mechanism of *any* organic reaction was really known" [13] before the advent of computational chemistry! This skepticism was engendered by the difficulties and ambiguities in studying very transient intermediates, and the impossibility (at the time) of observing transition states.

8.1.2.1 The Diels–Alder reaction

This is one of the most important reactions in all of organic synthesis, as it unites two moieties in a predictable stereochemical relationship, with the concomitant formation of two carbon–carbon bonds (Fig. 8.5) [14]. The reaction has been used in the synthesis of complex natural products, for example, in an efficient synthesis of the antihypertensive drug reserpine [15]. Such a reaction seems to be well worth studying.

The Diels–Alder reaction and related pericyclic reactions, which can be treated qualitatively by the Woodward–Hoffmann rules (section 4.3.5), have been reviewed in the context of computational chemistry [16]. The reaction is clearly nonionic, and the main controversy was whether it proceeds in a concerted fashion (as indicated in Fig. 8.5) or through a diradical, in which one bond has formed and two unpaired electrons have yet to form the other bond. A subtler question was whether the reaction, if concerted, was synchronous or asynchronous: whether both new bonds were formed to the same extent as the reaction proceeded, or whether the formation of one ran ahead of the formation of the other. Using the CASSCF method (section 5.4.3) Li and Houk [17] concluded that the butadiene–ethene reaction is concerted and synchronous, and chided Dewar and Jie [18] for stubbornly adhering to the diradical (biradical) mechanism. A DFT (chapter 7) study also supported the concerted mechanism [19].

8.1.2.2 Abstraction of H from amino acids by the OH radical

This reaction seems more esoteric than the Diels–Alder, and although not "used" at all, may be very important. Proteins are combinations of amino acid residues, and

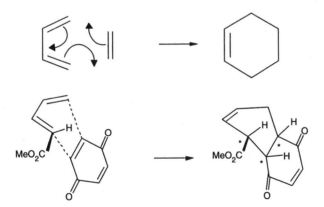

Figure 8.5. The prototypical Diels–Alder reaction is that between 1,3-butadiene and ethene, to form cyclohexene. The Diels–Alder reaction has been used in the synthesis of complex natural products; above, methyl 2,4-pentadienoate reacts with 1,4-benzoquinone to form an intermediate in the synthesis of the drug reserpine. In a one-pot reaction two carbon–carbon bonds are made and three chiral centers (*) are created with the correct relative orientations (i.e. essentially one diastereomer is formed).

Figure 8.6. Abstraction of a hydrogen atom from the α-C of an amino acid by hydroxyl radical has been investigated computationally.

oxidation of proteins by hydroxyl radicals may initiate Alzheimer's disease, cancer, and heart disease. The initial step in the destruction or modification of proteins by hydroxyl radical is likely to be abstraction of a hydrogen atom from the α-C (Fig. 8.6). In a very thorough study using MP2 (section 5.4) and DFT (chapter 7), Galano *et al.* (2001) calculated the geometries of the species (amino acid–OH complexes, transition states, and amino acid radicals) involved in the reactions of glycine and alanine (Fig. 8.6, R–H and CH_3, respectively) [20]. The rate constants were also calculated, using partition functions to calculate the preexponential factor (cf. section 5.5.2.2d). This paper provides a good account of how computational chemistry can be used to calculate absolute rate constants for reactions of molecules of moderate size.

8.1.3   Concepts

There are some very basic concepts in chemistry that have proved to be helpful in ratio-nalizing experimental facts, and which have been taught for perhaps the last fifty years, but which have nevertheless been questioned in the last decade or so; an example is the role of resonance in stabilizing species like carboxylate ions. Some newer concepts, intriguing but not as traditional, have also been scrutinized and questioned; an example is homoaromaticity.

8.1.3.1   Resonance vs. inductive effects

Beginning organic chemistry students learn that carboxylic acids are stronger acids than alcohols because of resonance stabilization of the conjugate base (which is more important than the charge-separation resonance in the acid), while resonance does not figure in either an alcohol or its conjugate base. This traditional wisdom was apparently first questioned by Thomas and Siggel, on the basis of photoelectron spectroscopy [21]. They concluded that the relatively high acidity of carboxylic acids is largely inherent in the acid itself, as a consequence of the polarization of the COOH group caused by the electronegative carbonyl group pulling electrons from the hydrogen atom. This idea was taken up by Streitwieser, and applied to other acids, e.g. nitric and nitrous acids, dimethyl sulfoxide and dimethyl sulfone [22]. The results for carbonyl compounds were interpreted in accord with another iconoclastic idea, namely that the carbonyl group is better regarded as $>C+ {}^-O-$ than as $>C{=}O$ [23]. This polarization interpretation was arrived at largely with the aid of atoms-in-molecules (AIM) analysis of the electron populations on the atoms involved (section 5.5.4), and a simpler variation of AIM (the projection function difference plot) developed by Streitwieser and coworkers [24]. Work by others also supports the view that it is "initial-state electrostatic polarization" that is largely responsible for the acidity of several kinds of compounds, including carboxylic acids [25]. Streitwieser cautions that "merely reproducing experiment does not teach us much unless the results are analyzed to provide understanding of the important contributions to the numbers in terms of reference systems and simpler models." [26]. Note that such analysis is presented to a large extent visually (section 5.5.6).

8.1.3.2 Homoaromaticity

Aromaticity [27] is associated with the delocalization of (in the simplest version) π electrons (the role of these π electrons in imposing symmetry on the prototypical aromatic species, benzene, is being questioned, but that is another story [28]). A Hückel number of cyclically delocalized electrons confers aromaticity on a molecule (section 4.3.5). The idea behind homoaromaticity (homologous aromaticity) is that if a system is aromatic, then if we interpose one or more atoms between adjacent p orbitals of the π system, provided overlap is not lost, the aromaticity may persist (Fig. 8.7). While there is little doubt about the reality of homoaromaticity in ions, neutral homoaromaticity has been elusive [29].

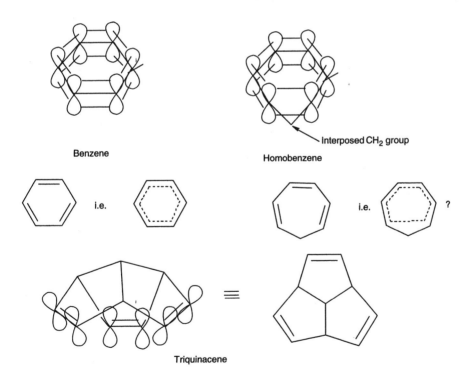

Figure 8.7. Homoaromaticity. Interposing a CH_2 group between one pair of formal double bonds of benzene gives monohomobenzene. Is this delocalized like benzene, or is it just cycloheptatriene? Is triquinacene, with a CH group interposed between each pair of formal double bonds, a trishomobenzene?

One molecule that might be expected to be homoaromatic, if the phenomenon can exist in neutral species, is triquinacene (Fig. 8.7): the three double bonds are held rigidly in an orientation which appears favorable for continuous overlap with concomitant cyclic delocalization of six π electrons. Indeed, its potential aromaticity was one of the reasons cited for the synthesis of this compound [30]. A measurement of the heat of hydrogenation of triquinacene found a value 18.8 kJ mol^{-1} lower than that for each of the next two steps (leading to hexahydrotriquinacene) [31]. This was taken as proof of homoaromaticity in the triene, i.e. that the compound was 18.8 kJ mol^{-1} stabler than expected for an unstabilized species (note that this is a small stabilization energy compared to the resonance energy of benzene, a recent computational estimate of which is 105.9 kJ mol^{-1} [32]). However, a recent experimental and computational study of this question led to the conclusion that triquinacene is not homoaromatic [33]. This was shown by (*a*) redetermination of its heat of formation (which had been calculated from the heat of hydrogenation in the earlier work [31]) using the measured heat of combustion, (*b*) by calculation of the heat of hydrogenation of a double bond in triquinacene and in its di- and tetrahydro derivatives (**1, 2, 3**, Fig. 8.8), and by (*c*) calculation of magnetic properties of the triene and related molecules. The heat of formation was about 17 kJ mol^{-1} higher than the previously reported [31] value, removing the supposed stabilization energy. The heats of hydrogenation of the double bonds were calculated with the aid of homodesmotic reactions, a kind of isodesmic reaction (section 5.5.2.2a) which preserves the number of each kind of bond, and so in which correlation errors should cancel well; for **1, 2**, and **3** the calculated hydrogenation energies of a double bond are all essentially the same, showing that a double bond of **1** is an ordinary cyclopentene double bond. Note that using cyclopentane (Fig. 8.8) rather than, say, ethane – which would also preserve bond types – to (conceptually) hydrogenate **1, 2**,

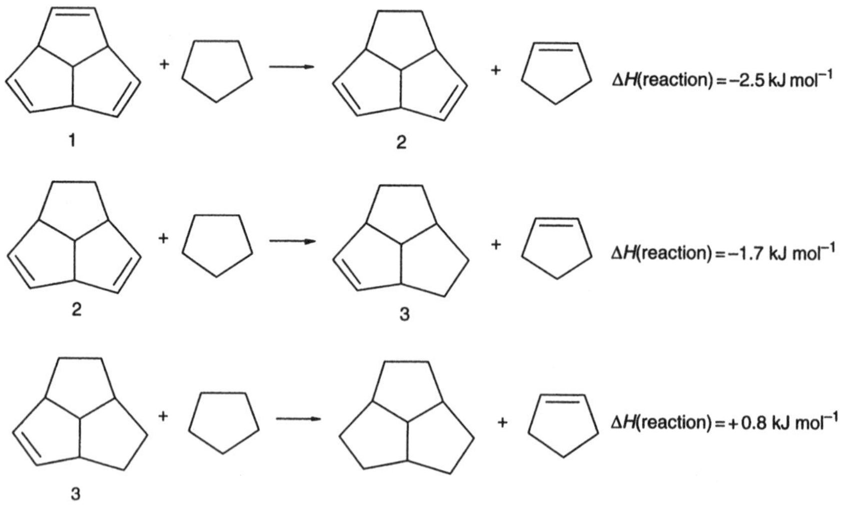

Figure 8.8. The heat of hydrogenation of a double bond in triquinacene is essentially the same as that of a double bond in dihydrotriquinacene and in tetrahydrotriquinacene, and is about the same as in cyclopentene, indicating that triquinacene is not homoaromatic.

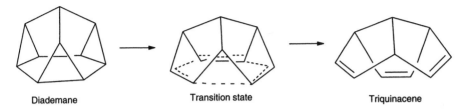

Diademane   Transition state   Triquinacene

Figure 8.9. The isomerization of diademane to triquinacene proceeds through an aromatic transition state, as shown by the magnetic susceptibility and NICS values for the three species.

and 3 should largely cancel out energy differences due to ring strain. The two magnetic properties calculated [33] arise from the presence of a diatropic ring current [27], which tends to push an aromatic molecule out of a magnetic field (calculated property: magnetic susceptibility, χ), and which exerts NMR shielding on a proton at or above the ring center (calculated property: nucleus-independent chemical shift, NICS). NICS values are obtained from the calculated NMR shielding (section 5.5.5) of a "ghost nucleus" [34] with no charge, electrons or basis functions, placed at or above [35] the ring center. Calculation of the changes in χ and in NICS along the reaction coordinate for the known reaction, the isomerization of diademane to triquinacene (Fig. 8.9), showed that the transition state, but neither the reactant nor the product, was aromatic. Homoaromaticity in a few neutral, ground-state molecules has been claimed [36].

The absence of homoaromaticity in triquinacene is probably due to the three pairs of nonbonded carbons being too far apart, 2.533 Å, from X-ray diffraction; in the transition state (Fig. 8.9), in contrast, the nonbonded CC distance has been reduced to 1.867 Å according to a B3LYP/6-311+G** calculation (section 7.2.3.4c).

Significantly, the measured C=C length, 1.319 Å, is close to the normal C=C length (calculated and measured parameters of triquinacene are cited in [33]).

8.2   TO THE LITERATURE

A feast of information on computational chemistry is available, a small selection of which is given below.

8.2.1   Books

Listed in chronological order; readers should use their judgement to decide in which order to read them.

Essentials of Computational Chemistry. Theories and Models, C. J. Cramer, Wiley, New York, 2002.
Covers a wide range of topics. The level is sometimes quite advanced. Critical discussions of the literature.

Computational Chemistry: A Practical Guide for Applying Techniques to Real World Problems, D. Young, Wiley, New York, 2001.

A "meta-book" in that it lists several books on computational chemistry; it also lists many websites concerned with computational chemistry, and many computational chemistry programs. The material of this book is available at http://server.ccl.net/cca/documents/dyoung/

A Chemist's Guide to Density Functional Theory, W. Koch and M. C. Holthausen, Wiley-VCH, New York, 2000.
Detailed introduction to the theory and applications of DFT.

Quantum Chemistry, 5th edn, I. N. Levine, Prentice Hall, Upper Saddle River, NJ, 2000.
Chapters 15, 16, and 17 give many references to the original literature, to books, to programs, and to websites. Enormously useful book on quantum chemistry in general.

Series of books from Wavefunction, Inc, makers of the Spartan computational chemistry program 2000 and earlier. For available books contact Wavefunction, http://www.wavefun.com/
These books, oriented toward Wavefunction's Spartan program, are very useful introductions to the practice of computational chemistry.

Introduction to Computational Chemistry, F. Jensen, Wiley, New York, 1999.
Good general introduction. Goes fairly deeply into theory.

Computational Thermochemistry, K. K. Irikura and D. J. Frurip, Eds., American Chemical Society, Washington, DC, 1998.
Useful source of information on the calculation of energy quantities: heats of formation, reaction energies, bond energies, activation energies, high-accuracy methods (G2, CBS, etc.), energies of solvation.

The Encyclopedia of Computational Chemistry, 5 volumes, P. von R. Schleyer, Ed., Wiley, New York, 1998.
A convenient source of information, but pricey (ca. $3000).

Molecular Mechanics Across Chemistry, A. K. Rappé and C. J. Casewit, University Science Books, Sausalito, CA, 1997.
Detailed presentation of the applications of MM, particularly in biochemistry and drug design.

Exploring Chemistry with Electronic Structure Methods, 2nd edn, J. Foresman and Æ. Frisch, Gaussian, Inc., Pittsburgh, PA, 1996.
Very useful hands-on guide; oriented toward Gaussian 94, but very useful for Gaussian 98 too.

Molecular Modelling. Principles and Applications, A. R. Leach, Longman, Essex, England 1996.
Good general introduction. Goes reasonably deeply into theory.

Modern Quantum Chemistry. Introduction to Advanced Electronic Structure Theory, A. Szabo and N. S. Ostlund, 1st edn, revised, McGraw-Hill, New York, 1989.
A detailed, very advanced introduction to basic Hartree–Fock, CI, and MP theory.

Ab Initio *Molecular Orbital Theory*, W. J. Hehre, L. Radom, P. von R. Schleyer, and J. A. Pople, Wiley, New York, 1986.
Still a good introduction to *ab initio* calculations, although one should realize that there have been considerable advances since 1986. Basic theory, advice, and extensive collections of calculated and experimental geometries, energies, and frequencies.

A Handbook of Computational Chemistry, T. Clark, Wiley, New York, 1985.
Still useful, although dated. A revised edition will be welcome.

Book series:
Reviews in Computational Chemistry, K. B. Lipkowitz and D. B. Boyd, Eds., Wiley-VCH, New York; volume 1 appeared in 1990, volume 17 is currently (July 2001) in preparation.
A volume in this series typically has from four to eleven chapters, each a kind of tutorial on the theory and application of some computational method. For tables of contents and other information see http://chem.iupui.edu/~boyd/rcc.html.

8.2.2   The WorldWide Web

Information on even specialized scientific topics can often be obtained from ordinary search engines. For example, a popular search engine gave information (ten hits for each) on these five topics, using the keywords shown: Hartree Fock, potential energy surface, molecular mechanics, Huckel, Extended Huckel. In several cases the hypertext leads one to tutorials, and to free programs.

Many websites are given in the books by Young and by Levine, above; some useful ones are:

http://ccl.osc.edu/ccl/cca.html
CCL, the computational chemistry list. A truly extraordinarily helpful forum for exchanging ideas, asking questions and getting help. If you join the network you can expect typically 5–10 messages a day.

http://www.chem.swin.esu.au/chem_ref.html
Gives links to sites for general chemistry, chemistry education, computational chemistry, etc.

qcldb.ims.ac.jp/index.html
A database of the literature of *ab initio* and DFT calculations.

www.ccdc.cam.ac.uk/
The Cambridge Crystallographic data Centre; contains the Cambridge Structural Database, which has X-ray or neutron diffraction structures of more than 230 000 compounds.

8.3   SOFTWARE AND HARDWARE

Many programs are described in the books by Young and by Levine, above; I mention here only a few that may be of particular interest to people getting into computational chemistry.

8.3.1 Software

SPARTAN

Wavefunction, http://www.wavefun.com/

This is a suite of programs with MM (SYBYL and MMFF), *ab initio*, semiempirical (MNDO, AM1, PM3), and DFT, with its own superb graphical user interface (GUI) for building molecules for calculations, and for viewing the resulting geometries, vibrational frequencies, orbitals, electrostatic potential distributions, etc. SPARTAN is a complete package in the sense that one does not need to buy add-on programs like, say, a GUI. The program is very easy to use and its algorithms are robust – they usually accomplish their task, e.g. the sometimes tricky job of finding a transition state usually works with SPARTAN. Versions of the program are available for PCs running under Windows NT and LINUX, for Macs, and for UNIX workstations. It has some high-level correlated *ab initio* methods and is nevertheless extremely useful for research, not to mention teaching.

GAUSSIAN

www.gaussian.com/

The most widely used computational chemistry program. Actually a suite of programs with MM (AMBER, DREIDING, UFF), *ab initio*, semiempirical (CNDO, INDO, MINDO/3, MNDO, AM1, PM3) and DFT, and all the usual high-level correlated *ab initio* methods. The common high-accuracy methods are available simply by keywords. There is a large number of basis sets and functionals. Electronically excited states can be calculated. GAUSSIAN has appeared in improved versions every few years from 1970 (...G92, G94, G98). It is now available in versions for PCs running under Windows NT and LINUX, and for UNIX workstations. GAUSSIAN does not have an integrated GUI, but there are several graphics programs for creating input files and for viewing the results of calculations. GaussView, expressly designed for GAUSSIAN 98, is highly recommended as the solution to all GAUSSIAN graphics problems.

GAMESS (General Atomic and Molecular Electronic Structure System)

www.msg.ameslab.gov/GAMESS/GAMESS.html

Not as many options as GAUSSIAN but free. Versions are available for PCs, Macs, UNIX workstations and supercomputers.

HyperChem

http://www.hyper.com

Has MM (MM+, AMBER, BIO+, OPLS), semiempirical (extended Hückel, CNDO, INDO, MINDO/3, MNDO, ZINDO/1, ZINDO/S, AM1, PM3), Hartree–Fock, and single-point MP2. Available for PCs with Windows 95, 98, NT and 2000, and UNIX workstations.

Q-Chem

www.q-chem.com/

"The first commercially available quantum chemistry program capable of analyzing large structures in practical amounts of time." For *ab initio* (including high-level correlated methods) and DFT. Q-Chem is available for PCs running under LINUX, for UNIX workstations, and for supercomputers.

JAGUAR

www.psgvb.com

Made by Schrödinger, Inc., JAGUAR is an *ab initio* (Hartree Fock and MP2) and DFT package that uses sophisticated algorithms to speed up *ab initio* calculations. It is said to be particularly good at handling transition metals, solvation, and conformational searching. The Jaguar algorithms combined with the SPARTAN GUI are available as TITAN from the makers of JAGUAR and SPARTAN.

ACES II

www.qtp.ufl.edu/Aces2/

Particularly recommended for MP2 calculations and for CCSD(T) optimizations + frequencies, which latter are perhaps the most reliable calculations that can currently be done on molecules of reasonable size (up to about 10 heavy atoms). CCSD(T) optimizations and frequencies tend to be considerably slower with some other programs. Available for UNIX workstations and supercomputers.

MOLPRO

www.tc.bham.ac.uk/molpro/

Intended for high-level correlated *ab initio* calculations (multiconfiguration SCF, multi-reference CI, and CC). "The emphasis is on highly accurate computations . . . accurate *ab initio* calculations can be performed for much larger molecules than with most other programs." MOLPRO has been run on a variety of machines with UNIX-type operating systems.

8.3.2 Hardware

Someone beginning computational chemistry, who intends to use it extensively enough to warrant having one's own machine (strongly recommended), might wish to get a high-end PC running under Windows NT or LINUX: such a machine is fairly cheap and it will do even sophisticated correlated *ab initio* calculations. A 1.5 MHz Pentium with 1 GB of memory and 40 GB or more of disk space is now not unusual (soon it may be substandard). While this is a reasonable choice for general computational chemistry, certain jobs will run faster on other configurations of machine and operating system. Using standard Gaussian 94 test jobs and various operating systems, and varying software and hardware parameters, Nicklaus *et al.* comprehensively compared a wide range of "commodity computers" [37]. These are personal computers like those of the Pentium series, and machines in a similar price range (the costliest was about U.S. $5000 and most were less than $3000, ca. 1998). They concluded that "commodity-type computers have . . . surpassed in power the more powerful workstations and even supercomputers . . . .Their price/performance ratios will make them extremely attractive for many chemists who do not have an unlimited budget, . . ."

8.3.3   Postscript

A few years ago the president of a leading computational chemistry software firm told the author that "In a few years you will be able to have a Cray [a leading supercomputer brand] on your desk for $5000." Supercomputer performance is a moving target, but the day has indeed come when one can have on one's desk for a few thousand dollars computational power that was not long ago available only to an institution, and for a good deal more than $5000. A corollary of this is that computational chemistry has become an important, indeed sometimes essential, auxiliary to experimental work. More than that, calculations have become so reliable that not only can parameters like geometries and heats of formation often be calculated with an accuracy rivalling or exceeding that of experiment, but where high-level calculations contradict experiment, the experimentalists might be well advised to repeat their measurements. The implications for the future of chemistry of the happy conjunction of affordable supercomputer power and highly sophisticated software hardly needs to be stressed.

REFERENCES

[1] E. Lewars, Chem. Rev., 1983, *83*, 519.

[2] G. Vacek, B. T. Colegrove, and H. F. Schaefer, Chem. Phys. Lett., 1991, *177*, 468.

[3] G. Vacek, J. M. Galbraith, Y. Yamaguchi, H. F. Schaefer, R. H. Nobes, A. P. Scott, and L. Radom, J. Phys. Chem., 1994, *98*, 8660.

[4] V. I. Minkin, M. N. Glukhovtsev, and B. Ya. Simkin, "Aromaticity and Antiaromaticity," Wiley, New York, 1994; N. L. Bauld, T. L. Welsher, J. Cassac, and R. L. Holloway, J. Am. Chem. Soc., 1978, *100*, 6920.

[5] H. F. Bettinger, P. von R. Schleyer, and H. F. Schaefer, J. Am. Chem. Soc., 1998, *120*, 11439.

[6] J. H. van't Hoff, Bull. Soc. Chim. Fr., 1875, II *23*, 295; J. A. LeBel, Bull. Soc. Chim. Fr., 1874, II *22*, 337.

[7] (a) E. Lewars, J. Mol. Struct. (Theochem), 1998, *423*, 173. (b) E. Lewars, J. Mol. Struct. (Theochem), 2000, *507*, 165. Coupled-cluster calculations gave very similar results to Fig. 8.3 for the relative energies of pyramidane, the transition states, and the carbenes: J. P. Kenny, K. M. Krueger, J. C. Rienstra-Kiracofe, and H. F. Schaefer, J. Phys. Chem. A, 2001, *105*, 7745.

[8] Y. Inoue, T. Ueoka, T. Kuroda, and T. Hakushi, J. Chem. Soc., Perkin II, 1983, 983.

[9] R. Dagani, Chemical and Engineering News, 2000, 14 August, 41.

[10] R. Rawls, Chemical and Engineering News, 1999, 25 January, 7.

[11] R. Engelke, J. Phys. Chem., 1989, *93*, 5722.

[12] R. Engelke, J. Phys. Chem., 1992, *96*, 10789.

[13] M. J. S. Dewar, "A Semiempirical Life," American Chemical Society, Washington, DC, 1992, p. 125.

[14] M. B. Smith and J. March, "Advanced Organic Chemistry," 5th edn, Wiley, New York, 2001, pp. 1062–1075.

[15] R. B. Woodward, F. E. Bader, H. Bickel, A. J. Frey, and R. W. Kierstead, Tetrahedron, 1958, *2*, 1.

[16] K. N. Houk, Y. Li, and J. D. Evanseck, Ang. Chem. Int. Ed. Engl., 1992, *31*, 682.

[17] Y. Li and K. N. Houk, J. Am. Chem. Soc., 1993, 115, 7478.

[18] M. J. S. Dewar and C. Jie, Acc. Chem. Res., 1992, *25*, 537.

[19] E. Goldstein, B. Beno, and K. N. Houk, J. Am. Chem. Soc., 1996, *118*, 6036.

[20] A. Galano, J. R. Alvarez-Idaboy, L. A. Montero, and A. Vivier-Bunge, J. Comp. Chem., 2001, *22*, 1138.

[21] M. R. F. Siggel and T. D. Thomas, J. Am. Chem. Soc., 1986, *108*, 4360.

[22] A. Streitwieser, "A Lifetime of Synergy with Theory and Experiment," American Chemical Society, Washington, DC, 1996, pp. 166–170, and references therein.

[23] K. B. Wiberg, Acc. Chem. Res., 1999, *32*, 922.

[24] A. Streitwieser, "A Lifetime of Synergy with Theory and Experiment," American Chemical Society, Washington, DC, 1996, pp. 157–170, and references therein.

[25] F. Bökman, J. Am. Chem. Soc., 1999, *121*, 11217.

[26] A. Streitwieser, "A Lifetime of Synergy with Theory and Experiment," American Chemical Society, Washington, DC, 1996, p. 170.

[27] V. I. Minkin, M. N. Glukhovtsev, and B. Ya. Simkin, "Aromaticity and Antiaromaticity," Wiley, New York, 1994; M. Glukhovtsev. Chem. Educ., 1997, *74*, 132; P. von R. Schleyer and H. Jiao. Pure and Appl. Chem., 1996, *68*, 209; D. Lloyd. J. Chem. Inf. Comput. Sci., 1996, *36*, 442.

[28] J. J. Mulder, J. Chem. Ed., 1998, *75*, 594; A. Shurki and S. Shaik, Ang. Chem. Int. Ed. Engl., 1997, *36*, 2205; P. C. Hiberty, D. Danovich, A. Shurki, and S. Shaik, J. Am. Chem. Soc., 1995, *117*, 7760. For skepticism about this demotion of the role of the π electrons in imposing D_{6h} symmetry on benzene: H. Ichikawa and H. Kagawa, J. Phys. Chem., 1995, *99*, 2307; E. D. Glendening, R. Faust, A. Streitwieser, K. P. C. Vollhardt, and F. Weinhold, J. Am. Chem. Soc., 1993, *115*, 10952.

[29] M. B. Smith and J. March, "Advanced Organic Chemistry," 5th edn, Wiley, New York, 2001, p. 70 and references therein.

[30] R. B. Woodward, T. Fukunaga, and R. C. Kelly, J. Am. Chem. Soc., 1964, *86*, 3162.

[31] J. F. Liebman, L. A. Paquette, J. R. Peterson, and D. W. Rogers, J. Am. Chem. Soc., 1986, *108*, 8267.

[32] M. N. Glukhovtsev, R. D. Bach, and S. Laiter, J. Mol. Struct. (Theochem), 1997, *417*, 123.

[33] S. P. Verevkin, H.-D. Beckhaus, C. Rüchardt, R. Haag, S. I. Kozhushkov, T. Zywietz, A. De Meijere, H. Jiao, and P. Von R. Schleyer, J. Am. Chem. Soc., 1998, *120*, 11130.

[34] Æ. Frisch and M. J. Frisch, "Gaussian 98 User's reference," 2nd edn, Gaussian, Inc., Pittsburgh, PA, 1998, p. 245.

[35] (a) The original NICS paper, which also compares NICS with χ: P. von R. Schleyer, C. Maerker, A. Dransfeld, H. Jiao, and N. J. R. van Eikema Hommes, J. Am. Chem, Soc., 1996, *118*, 6317. (b) Subsequent modifications and applications of NICS: R. West, J. Buffy, M. Haaf, T. Müller, B. Gehrus, M. F. Lappert, and Y. Apeloig, J. Am. Chem, Soc., 1998, *120*, 1639; T. Veszprémi, M. Takahashi, B. Hajgató, J. Ogasarawa, K. Sakamoto, and M. Kira,

J. Phys. Chem. A., 1998, *102*, 10530; I. Morao and F. P. Cossio, J. Org. Chem., 1999, *64*, 1868.

[36] H. Jiao, R. Nagelkerke, H. A. Kurtz, R. V. Williams, W. T. Borden, and P. von R. Schleyer, J. Am. Chem, Soc., 1997, *119*, 5921.

[37] M. C. Nicklaus, R. W. Williams, B. Bienfait, E. S. Billings, and M. Hodošček, J. Chem. Inf. Comput. Sci., 1998, *38*, 893.

INDEX